Contents

Preface

Message from SADCC

SADCC is the acronym for the Southern Africa Development Coordination Conference. Presently, there are nine participating countries: Angola, Botswana, Lesotho, Malawi, Mozambique, Swaziland, Tanzania, Zambia, and Zimbabwe, (Namibia joined SADCC in 1990).The objective of SADCC is to strengthen the well-being of their countries and decrease economic dependence on other countries. There are several important programs in SADCC, agriculture being one. Within agriculture there is a food security component based in Zimbabwe, and agricultural research is coordinated by Botswana.

Several international centers act as executing agents for programs approved by SADCC. ICRISAT has one regional program for the improvement of sorghum and millets in Bulawayo and another for the improvement of groundnuts based at the Chitedze Research Station near Lilongwe, Malawi.

Though the dominant cereal of the SADCC region is maize, sorghum and millets are traditional and important, particularly in the dry areas. Research to improve these crops is recent compared to that for maize, wheat, soybeans, and tobacco.

The disease situation is of concern; downy mildew is found widely in the region and leaf diseases are severe in our nurseries in parts of the area. Grain molds and *Striga* are also of concern. We now have a much better idea of which diseases are of priority concern than when we began in 1984, and are developing strategies to work on them. These will be stated in greater detail in this book.

L.R. House
Executive Director,
SADCC/ICRISAT Regional Sorghum
and Millets Improvement Program,
PO Box 776, Bulawayo, Zimbabwe.

Message from INTSORMIL

Sorghum and millet are vital life-sustaining crops in many parts of the world. The southern African region is one of the major areas of focus for our collaborative research activities, and many INTSOR-MIL scientists are involved in projects across many disciplines and countries in the region.

INTSORMIL cooperates with ICRISAT in the long-term degree training program for SADCC scientists, and many of us enjoy the opportunity of serving as major professors for SADCC graduate students at our universities in USA. This training program is well underway with 25 SADCC graduate students currently undertaking MS and PhD degree training in USA. We believe that the long-term investment in development of the scientific talent in the region will pay handsome dividends in the future.

I would like to concentrate on two major topics here.

The first concerns fundamentals of the Collaborative Research Support Programs (CRSPs), the model for collaborative research. The CRSP model is one of several possible means of bringing the agricultural research capacity of the more developed nations, particularly USA, to bear effectively and efficiently on the key problems of agriculture in developing nations. It is not or cannot be a short-term panacea or a long-term solution, but it does hold a great deal of promise as a cost-effective, productive means of making significant contributions to agricultural development in the developing countries as a mid-term, interim measure.

Despite all the complexities of the CRSPs, bureaucratic and otherwise, the basics are simple and straightforward. The CRSP model is based on three fundamental assumptions.

1. There are problems that are common in important ways to agriculture in USA and in the developing nations.
2. The assumptions are that collaborative research among agricultural scientists in USA and their counterparts in the developing nations will be highly complementary. This is to say that the results of such efforts will be useful in accelerating the agricultural development process in the developing countries through contributing to the upward shifting of farm-level production functions and the relaxation of other constraints. It is also to say that it will be complementary in the sense of providing knowledge, information, and technology useful to the solution of productivity and adjustment constraints faced by U.S. agriculture.
3. The model assumes that, from the fixed stock of research resources available to agriculture worldwide, the scientific output will be greater as a result of such collaborative work. A corollary assumption of the above is that the expanded output will be mutually beneficial to developing and developed nations.

My second topic is INTSORMIL: what it is and how it functions. INTSORMIL is the CRSP program for improvement of sorghum and millet production and utilization in countries where these cereals are important staple foods. The program is funded jointly by the U.S. Agency for International Development and by five U.S. universities—Purdue, Texas A&M, Nebraska, Mississippi State, and Kansas State. Research scientists at these universities develop collaborative research projects with their counterparts in national agricultural research programs. We believe that strong national research programs [in less developed countries (LDCs) and the USA alike] are the key to long-term improvements in the productivity, marketing, and utilization of these important staple cereals. We also believe that national research programs (in the LDCs and the USA) can function effectively only with strong multidisciplinary teams of well-trained and highly motivated research scientists. Our approach to helping achieve these objectives is through development of mutually beneficial collaborative research projects between LDC scientists and U.S. scientists. Collaboration involves exchanges of scientists germplasm, research technology, and information. Enhanced professional development of all collaborators (from the LDCs or the USA) in the INTSORMIL program will lead to increased efficiency and productivity of all of our research programs. The collaborative research projects that result will enhance the lives and economic well-being of all people who depend on sorghum and millet for sustenance.

In summary, INTSORMIL is a scientist-to-scientist collaborative research program with the goals to increase the productivity and output of LDC research programs, as well as enhance our research efforts at U.S. institutions. Its success depends to a very large extent on the development of close collegial relationships between scientists and we plan to continue developing these relationships.

J.D. Axtell
Professor of Agronomy,
Purdue University, West Lafayette,
IN 47906, USA.

Message from ICRISAT

This book is important for three reasons:

1. The widening food gap in the third world, especially in sub-Saharan Africa, is of great concern to humanity. The only hope of narrowing this gap is by the use of improved production technology. Disease-resisting high-yielding genotypes are the kingpins of this technology.
2. This book reflects 10 years of intense activity in sorghum research and we should pause and take stock of the world situation and plan new strategies for research and development in the war against diseases and pests.
3. ICRISAT wishes to benefit from the matters discussed in this book to evolve its strategic plan.

Increasing severity of diseases of sorghum and millet, particularly of pearl millet in Africa, is a matter of great concern. Moreover, a second generation of problems accompanied the introduction of new technologies based on shorter duration and higher-yielding genotypes; these problems call for new strategies for removing the stress of diseases and pests. The new problems call for more intensive and interdisciplinary efforts of research for achieving a breakthrough in the control of diseases for increasing productivity.

We critically review, in this book, the advances in knowledge and the progress which we have made in understanding, managing, and reducing the incidence of the diseases and developing genotypes resistant to them.

In 1978, grain mold and stalk rots were identified as major disease problems of improved sorghum. Ten years later, these diseases continue to be at the top of our list. ICRISAT scientists claim that they have developed sound techniques for identification of resistant material and selected stable sources of resistance, but this information is only of academic interest unless sorghum breeders make effective use of this material, incorporating its resistance into agronomically superior sorghums.

Five years ago, an international workshop on stalk rot of sorghum critically reviewed the stalk rot situation and made specific recommendations. I understand suitable techniques for screening for disease resistance have been developed, but we have to go a long way to transfer this resistance to sorghums with agronomically desirable backgrounds.

Sorghum leaf diseases are thought to be more serious in Africa. Some are endemic to certain regions; in some cases, we do not know if they reduce yields significantly. Without understanding the mechanisms of disease resistance or the processes and benefits of national and international cooperation, success in building up disease resistance is difficult.

Certain diseases, such as ergot and viruses, are affecting the hybrids more than the varieties, and steps need to be taken to develop resistance against them.

Some pathologists claim that downy mildew of sorghum is no problem. Sources of resistance are available, as is the knowledge for controlling this disease in sorghum. Judging from the incidence of this disease in sorghum fields at the Bulawayo facility, I believe we have a long way to go to achieve success. In pearl millet, downy mildew still remains the severest enemy of the crop. We need to know more about the mechanism of resistance of this disease. We cannot ignore the fact that there are more virulent pathotypes in Africa than in India, and that resistance to disease is short-lived and breaks down quickly. While the breeders are developing newer and newer genotypes resistant to the disease, the pathogen is evolving more-virulent types. Surprisingly, the wild relatives of pearl millet show no evidence of resistance to the new pathotypes. There is perhaps a silver lining in the dark cloud, however, ICRISAT pathologists have found that, even in the susceptible hybrids and cultivars of pearl millet, an odd plant will show recovery resistance to the pathogen. In the initial stages, these plants may develop disease under severe disease pressure but later on will outgrow the disease and produce a normal yield. Some varieties show greater percentages of recovery than others. S.B. King and J. Craig explain it more authoritatively. This is an important mechanism of survival in cultivars, and needs to be exploited. I suggest that a team of geneticists and pathologists study this phenomenon critically, as it seems to hold the key to success in dealing with this disease.

Resistance to ergot in pearl millet, especially in the hybrids, remains elusive. Because of health hazards associated with ergot, in addition to reductions in yield, adoption of hybrids is restrained. Innovative strategic and basic research is needed to tackle this disease. Biotechnology may offer an avenue; the possibility needs examination.

Smut and rust of millet are not so intractable as ergot and downy mildew, but even so, international cooperation in making use of sources of resistance and evaluating the final products of breeders' fields is in order. Of course, this is not possible without joint research by pathologists and breeders, working side by side.

It seems that nematodes are affecting millet in sandy soils of western Africa. I do not know if this pest is receiving the attention of pathologists, or of the entomologists, or if it is being passed over by both.

Some disease problems are amplified by nutritional and drought stress. Stalk rot of sorghum and *Striga* in sorghum and millet are typical cases. Pathologists, breeders, and plant nutritionists/agronomists working together could be very effective in developing varieties or techniques for managing *Striga* and stalk rot in sorghum and millet.

In our research in diseases, we have not made adequate use of biochemistry. I understand ergosterol content in the grain is a good indicator of colonization of grain by grain mold fungi; also an elevated content of 2-4 flavanol in the grain and even in the leaves and stems is a measure of the ability of a genotype to resist molds. Could such techniques be used routinely in screening for resistance?

I know that plant pathologists will name large numbers of sorghum and millet diseases in each ecological zone, but we must develop a practical approach to identify and prioritize the most serious diseases as targets of research in the next decade. Our major efforts should be the transfer of multiple-disease resistance to agronomically superior genotypes, or development of a system of disease management which is practicable and economical to use, even by resource-poor farmers.

The ideas and findings promulgated in this book should lead to development of collaborative projects between national programs, ICRISAT, INTSORMIL, and institutes in the developed countries for applied, strategic, and basic research on priority diseases. Active participation and input into the Cereal Cooperative Research Networks (CCRNs) by pathologists of the national research organization is essential, as it is these scientists who are in the best position to record and report the incidence of a disease and the identities of genotypes resistant to it. Screening for disease resistance at hot spots and use of standard techniques in evaluation is essential, so as to get useful data. ICRISAT has a gene bank with more than 30 000 accessions in sorghum and 19 000 accessions in pearl millet. It is a rich source, which needs to be exploited for its sources of resistance to diseases and the other desirable traits that enable crop improvement.

Our training facilities at ICRISAT Center is another resource for strengthening crop improvement in many nations of the world, including capability in disease research. I would suggest that the training constraints in disease management be identified; this information will be useful to ICRISAT. Special training programs can be organized to meet urgent needs.

Finally, let us not forget that sorghum and millet in the semi-arid tropics are staple foods of the peasant farmer of limited means, and he needs multiple-disease resistant and high-yielding seed to grow. He is not interested in pathologists' findings or breeders' claims, but he needs the product of their joint efforts so that he can produce crops with higher and more consistent yields year after year.

J.S. Kanwar
Deputy Director General Emeritus,
International Crops Research Institute
for the Semi-Arid Tropics (ICRISAT).
Patancheru, A.P. 502 324, India.

Part 1

Regional and Country Reports
(Sorghum)

Sorghum and Pearl Millet Pathology in Zimbabwe

E. Mtisi[1]

Abstract

The important sorghum diseases in Zimbabwe are leaf blight (Exserohilum turcicum), ergot (Sphacelia sorghi), and downy mildew (Peronosclerospora sorghi). In pearl millet the major problem diseases are ergot (Claviceps sp) and smut (Tolyposporium penicillariae). Methods of screening for resistance to these major diseases are outlined and plans for future work on sorghum and pearl millet pathology in Zimbabwe presented.

Introduction

Maize (*Zea mays*), sorghum (*Sorghum bicolor*), and pearl millet (*Pennisetum americanum*) are the principal sources of carbohydrates for a majority of the Zimbabwean population. Sorghum and pearl millet are particularly important in natural regions III, IV, and V (Fig. 1). These areas are marginal in rainfall and soil fertility; maize sowings here often fail. Most of the land occupied by communal farmers lies in these regions. Some commercial farmers grow sorghum in the high-rainfall areas, where the crop out-yields maize in late sowings (Sanderson 1967, pp. 17–23). Land-race pearl millet cultivars in the communal areas yield around 0.5 t ha⁻¹ (Muza et al. 1986). Sorghum produces from 0.5 to 0.9 t ha⁻¹ (Kyomo and Keswani 1987; Mushonga, personal communication). Low yields are also attributed to additional constraints, including insect pests, nematodes, diseases, and bird damage.

Zimbabwe's progress in improvement of varieties and cultural practices has increased yields to about 2 t ha⁻¹ for pearl millet (Muza et al. 1986) and 2.5 to 3 t ha⁻¹ for sorghum in areas with rainfall averaging 500 mm per annum (Brown 1973). Where rainfall is adequate, sorghum yields of 7 t ha⁻¹ have been achieved (Brown 1973; Mushonga 1984). The present need is to sustain and improve this yield, perhaps by improving disease resistance in plants.

In the past, research in sorghum and pearl millet pathology was on a small scale (Mushonga and Appa Rao 1986), and most of the attention focused on diseases of commercial crops. However, our plant breeders at Zimbabwe's Department of Research and Specialist Services have tried to incorporate disease resistance into our national sorghum and pearl millet improvement programs. Most screening for disease resistance in these crops has been carried out under natural epiphytotics (Mushonga 1984; Muza et al. 1986). Screening for ergot (*Claviceps fusiformis*) resistance in pearl millet is an exception (Mushonga 1983).

Since Zimbabwe's independence in 1980 and with the establishment at Matopos of the SADCC/ICRISAT Sorghum and Millets Improvement Program in 1984, sorghum and pearl millet pathology are receiving more attention.

Sorghum and Pearl Millet Disease Situations

National surveys of disease-pest incidence and severity during 1985, 1986, and 1987 revealed that sorghum is subject to attack by leaf blight

---

1. Plant Pathologist, Plant Protection Research Institute, Research and Specialist Services, Box 8100, Causeway, Harare, Zimbabwe.

Mtisi, E. 1992. Sorghum and pearl millet pathology in Zimbabwe. Pages 3–7 *in* Sorghum and millets diseases: a second world review. (de Milliano, W.A.J., Frederiksen, R.A., and Bengston, G.D., eds). Patancheru, A.P. 502 324, India: International Crops Research Institute for the Semi-Arid Tropics.

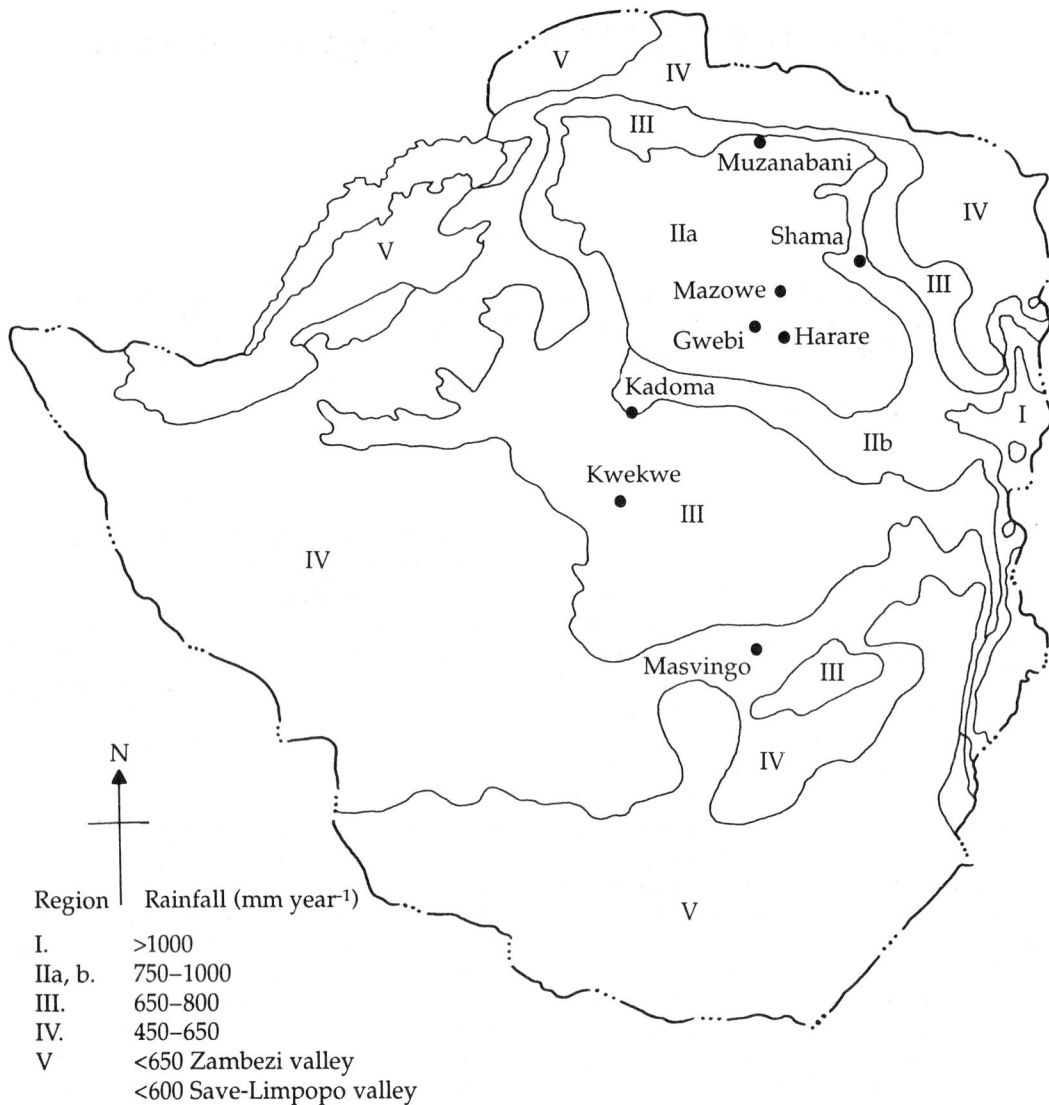

Figure 1. Rainfall zones of Zimbabwe.

Region	Rainfall (mm year^{-1})
I.	>1000
IIa, b.	750–1000
III.	650–800
IV.	450–650
V	<650 Zambezi valley
	<600 Save-Limpopo valley

(*Exserohilum turcicum*), downy mildew (*Peronosclerospora sorghi*), ergot (*Sphacelia sorghi*), rust (*Puccinia purpurea*), and smuts—in particular covered kernel smut (*Sporisorium sorghi*) and to a lesser extent loose kernel smut (*Sphacelotheca cruenta*) and head smut (*Sporisorium reilianum*) (Page et al. 1985). Earlier observations at Matopos identified some of these diseases as potential problems (Sanderson 1967, pp. 17-23). Visits to farmers' fields in the high-rainfall areas of region II indicated that leaf blight, ergot, and downy mildew are major problems in susceptible varieties, confirming Graham's observations (1972). Ergot sclerotia were observed in Zimbabwe by de Milliano and Mtisi while surveying farmers' fields in 1987. The role of sclerotia in the survival and spread of the pathogen in Zimbabwe should be investigated.

Diseases recorded on sorghum but thought to be of minor importance in Zimbabwe include gray leaf spot (*Cercospora sorghi*), ladder spot (*C. fusimaculans*), anthracnose (*Colletotrichum gram-*

inicola), head mold (*Curvularia lunata*), ear blight (*Fusarium heterosporum*), leaf spots caused by *Phyllosticta sorghiphila*, *Mycosphaerella holci*, and *Gloeocercospora sorghi*), sooty stripe (*Ramulispora sorghi*), stalk rots caused by *Gibberella fujikuroi*, *Macrophomina phaseolina*, and *Nigrospora oryzae*, and bacterial stripe (*Pseudomonas andropogonis*) (Rothwell 1983).

In 1985, the Plant Pathology Advisory Service at Zimbabwe's Plant Protection Research Institute reported the first record of long smut (*Tolyposporium enhrenbergii*) on sorghum. Virus-like symptoms, most likely due to maize dwarf mosaic virus, have been noticed on a number of lines in the SADCC/ICRISAT Sorghum Disease Observation Nursery grown at the Henderson Research Station, at Mazoe since 1985 [as reported by Mughogho (Obilana 1986)]. Local sorghum varieties—Red Swazi, Serena, DC 75, SV 1, and SV 2—grown at research stations have exhibited similar symptoms. However, these symptoms have not yet been reported in farmers' fields in the communal areas.

In pearl millet, ergot (*Claviceps fusiformis*) and smut (*Tolyposporium penicillariae*) are the major diseases. Downy mildew (Sclerospora graminicola), zonate leaf spot (*Gloeocercospora* sp), leaf spots caused by various fungi (*Cochlibolus bicolor*, *Colletotrichum* sp), rust (*Puccinia penicillariae*), and false mildew (*Beniowskia sphaeroidea*) are considered of minor importance. False mildew is still classified as a minor disease, because false mildew has not yet been recorded in farmers' fields. But serious epiphytotics of false mildew have been reported in research stations in regions II and III (Muza et al. 1986; Mushonga and de Milliano, personal communication). Screening work carried out during the 1986/87 season revealed that most of the local germplasm and our recently released variety, PMV 1, are susceptible to false mildew and ergot. Fortunately some sources of resistance to these diseases are available in the global germplasm, and it may be that these can be incorporated in our breeding lines.

During the past 2 years, a new leaf spot of pearl millet caused by *Bipolaris urochloae* was recorded in Zimbabwe. Observations on a susceptible line, 7042, in a disease-monitoring nursery at Henderson Research Station have clearly indicated that this leaf spot is a potentially important disease. Line 7042 succumbed to the disease before heading.

Yield-loss studies have not yet been carried out in Zimbabwe, but it has been noticed that losses due to these diseases can be substantial, especially in the higher-rainfall areas (Sanderson 1967, pp. 17–23).

Zimbabwe's Plant Pathology Research Programs

Our research is mainly aimed at producing for communal farmers of limited resources, cultivars with stable resistance to the major diseases. Communal farmers are Zimbabwe's main growers of these crops.

The following objectives are sought:

1. Development of reliable field-screening techniques for major sorghum and pearl millet diseases.
2. Identification of sources of resistance and incorporation of this resistance by breeders into genotypes with other desirable characters, such as high yield and drought tolerance.
3. Determination of stability of resistance through multilocational trials within Zimbabwe and neighboring countries.

Developing reliable screening

With the help of SADCC/ICRISAT, methods to establish a 'sick plot' for resistance screening against sorghum downy mildew were tested and found to be effective. In the 1st year, spreader rows were established from artificially infected seedlings. In the 2nd year, the ratoon of the spreader rows was used to infect newly-sown spreader rows and the test sorghums. Now under test (1987/88) for downy mildew resistance at SADCC/ICRISAT's Matopos Research Station are 72 advanced lines and 9 released varieties. Of the nine released varieties, SV 1, SV 2, PNR 8311, and PNR 8544 were found to be resistant with severity scores of 0 to 5%. We are in the process of establishing our own 'sick plot' at Henderson Research Station, about 30 km northeast of Harare.

Leaf blight in sorghum. Epiphytotics of leaf blight have been successfully created by inoculating sorghum plants with infected ground leaves prior to heading. Overhead irrigation was provided during dry spells. Spreader rows of susceptible varieties were grown at the borders and inside the nursery. Considerable success has been achieved in increasing the inoculum on sorghum seed, and methods of artificial inoculation are now being investigated. Locally bred advanced lines are also being tested for leaf-blight resistance.

Ergot in pearl millet. With the help of SADCC/ ICRISAT, a procedure for inoculating pearl millet with the ergot pathogen was established. Heads of test lines were bagged at the booting stage, inoculated with a spore suspension at time of maximum stigma emergence, and re-bagged to maximize humidity and ensure infection.

Work remaining

1. Develop reliable screening techniques for rusts and smuts in these cereals.
2. Conduct screening programs in the laboratory or greenhouse in order to control environmental variables.
3. Conduct epidemiological studies to gain knowledge useful in development of disease-management packages for communal farmers.
4. Seek cultural practices that may reduce disease incidence.
5. Determine yield losses caused by the major diseases.
6. Study seedborne pathogens of sorghum and millet.

Discussion

Since the beginning of Zimbabwe's sorghum-improvement program, leaf blight, ergot, downy mildew, and covered kernel smut have been identified as the major and most widespread sorghum diseases in the country (Sanderson 1967, pp. 17–23; Rattray 1968; Brown 1973; Graham 1972). The widespread occurrence of these diseases may be in part because some (such as downy mildew) also occur in maize. Close rotation of these cereals with maize enhances the chances of survival and spread of the pathogens. In communal areas, especially in region III, maize is also grown and because land is

limited, rotation with maize is likely. In regions IV and V, rainfall is more limited, and the close rotation with maize is less common. This practice tends to create natural 'sick plots' for endemic pathogens like *Sporisorium reilianum*. Also contributing may be the seedborne nature of these pathogens (Noble et al. 1968). Most communal farmers keep their seed and do not treat it, permitting the pathogen to remain endemic. Ergot, widespread in the Mashonaland Central Province, is serious in cytoplasmic male-sterile lines. Consequently, hybrid seed production in this area can be impaired severely. The hazards of ergot in livestock feed cannot be overemphasized. A practical method for control of ergot must be found.

Seed-health testing (plant quarantine) has been neglected in Zimbabwe. It is possible that new diseases may be introduced by pathogens arriving on or in seed. A number of the pathogens of pearl millet and sorghum have been reported to be seedborne (Noble and Richardson 1968). There is insufficient evidence, however, to confirm seedborne dissemination of a number of pathogens of these cereals.

Recent pest and disease surveys in communal areas by Plant Protection Research Institute (1987) personnel have revealed a close association of nematodes and stunting in both sorghum and pearl millet. *Pratylenchus zea, P. branchyurus, Rotylenchus parvus,* and *Paralongidorus* sp were identified during these surveys. *Meloidogyne javanica* caused serious stunting and chlorosis to finger millet (Page et al. 1985). However, the general economic importance of nematodes on sorghum and pearl millet is an open area for investigation in Zimbabwe.

Conclusion

The phytopathological research on sorghum and millets diseases in our country has been slow and insufficient. This is mainly due to shortage of trained manpower, especially in the areas of phytobacteriology and plant virology. Shortages of transport and equipment are logistical constraints. The Government of Zimbabwe is aware of these needs and is making an effort to provide resources, and we feel the situation will improve considerably in the near future.

Acknowledgment. I wish to express gratitude

to W. A. J. de Milliano and S. D. Singh of the SADCC/ICRISAT installation at Bulawayo for assistance in developing sorghum and pearl millet screening methodologies, and to C.L. Keswani for his critical review and valuable suggestions. I am grateful for the assistance and cooperation of J.N. Mushonga, team leader of Zimbabwe's Sorghum and Millet Improvement Program, Department of Research and Specialist Services, University of Zimbabwe, for valuable comments and observations on sorghum and millet diseases.

References

Brown, J. 1973. Grain sorghum planting for the 1973/1974 season. Department of Conservation and Extension (CONEX) Leaflet 22/75. Salisbury, Rhodesia: Rhodesia Ministry of Agriculture. 2 pp.

Graham, R.N. 1972. Grain Sorghum. Agricultural handbook leaflet number 113.01. Salisbury, Rhodesia: Rhodesia Ministry of Agriculture. 2 pp.

Kyomo, M.L., and Keswani, C.L. 1987. Structure and land distribution and corresponding yields of field crops in eastern and southern Africa. Paper presented at FAO/SIDA Seminar on Food Production Through Low-Cost Food-Crop Technology, 2–17 Mar 1987. Harare, Zimbabwe.

Mushonga, J.N. 1983. Screening of ergot in pearl millet. Five inbreds and one hybrid for a breeding program. Zimbabwe Agricultural Journal 80:239–241.

Mushonga, J.N. 1984. Sorghum and pearl millet research in Zimbabwe. Pages 49–54 in Proceedings, First Regional Workshop for Sorghum and Millet in Southern Africa, 23–26 Oct 1984, Harare, Zimbabwe. Bulawayo, Zimbabwe: SADCC/ICRISAT Sorghum and Millet Improvement Program.

Mushonga, J.N., and Appa Rao, S. 1986. Progress in sorghum improvement. Pages 103–118 in Proceedings, Second Regional Workshop on Sorghum and Millets for Southern Africa, 23–27 Sep 1985. Gaberone, Botswana. Bulawayo, Zimbabwe: SADCC/ICRISAT SMIP.

Muza, F.R., Mushonga, J.N., and Appa Rao, S. 1986. Pearl millet improvement in Zimbabwe. Pages 119–126 in Proceedings, Second Regional Workshop on Sorghum and Millets for Southern Africa, 23–27 Sep 1985, Gaberone, Botswana. Bulawayo, Zimbabwe: SADCC/ICRISAT SMIP.

Noble, M., and Richardson, M.J. 1968. An annotated list of seed-borne diseases. 2nd edn. Phytopathological Papers, No. 8. Kew, England: Commonwealth Mycological Institute. 191 pp.

Obilana, A.T. 1986. Improving sorghum for the southern African region; report of 1984/85. Pages 175–185 in Proceedings, Second Regional Workshop on Sorghum and Millets for Southern Africa, 23–27 Sep 1985. Gaberone, Botswana. Bulawayo, Zimbabwe: SADCC/ICRISAT SMIP.

Page, S.L.J., Mguni, C.M., and Sithole, S.Z. 1985. Pests and diseases of crops in communal areas of Zimbabwe. Overseas Development Administration Technical report. 203 pp.

Plant Protection Research Institute. 1987. Advisory Plant Pathology. Page 115 in 1984/85 Annual Report, Department of Research and Specialist Services. Harare, Zimbabwe: Zimbabwe Ministry of Agriculture.

Rattray, A.G.H. 1968. Sorghums in the Salisbury Research Station. Rhodesia Agricultural Journal 65:58.

Rothwell, A. 1983. A revised list of plant diseases occurring in Zimbabwe. Kirkia 12:233–351.

Sanderson, K.W. 1967. Principles and practice of sorghum production. Modern Farming (Feb).

Sorghum Diseases in Southern Africa

W.A.J. de Milliano[1]

Abstract

In the last 15 years, 10 of the 12 countries of southern Africa have produced about 12 to 15% of the continent's sorghum grain harvest; this is about 2 to 3% of the world's production. Tanzania, Republic of South Africa, Mozambique, and Zimbabwe had the largest areas under production. National grain yields were low and only occasionally exceeded 1 t ha^{-1}. There is no indication that fluctuation in hectarages devoted to sorghum, grain yields, or total production are due to high disease incidence. During this period, varieties and hybrids have begun to replace landraces.

There is still a need to identify the diseases of sorghum in southern Africa, and study their occurrence and incidence. Data showing losses caused by diseases are not available. The southern African nations have many ecological zones and some countries have as many as 20 diseases. Witchweeds, covered smut, downy mildew, leaf blight, sooty stripe, grain molds, anthracnose, rust, charcoal rot, and ergot, all are considered to be of importance. Incidence of ergot and virus diseases appear to be on the increase. Disease research in relation to sorghum improvement was stimulated by the initiation of the SADCC/ICRISAT Sorghum and Millets Improvement Program in 1984. The present status of information on diseases and research results is reported and areas for further research are proposed.

Introduction

In the last 15 years, sorghum has been grown in the 12 countries of southern Africa for its grain, sweet stems, and total biomass. Mozambique, South Africa, Tanzania and Zimbabwe have the largest areas under production (Table 1). Annual production by 10 of these nations, from 1969 to 1986, was approximately 12% to 14% of all sorghum grain produced in Africa. This represents approximately 2 to 3% of the total world production. Production data from Angola and Namibia are not available.

The area under production appeared to be stable in Lesotho and Malawi, and showed increases in the Republic of South Africa and in Zaire (Table 1). Tanzania more than doubled its sorghum area. In Zimbabwe, the area sown to sorghum in 1979–81 was only about one-half that sown during the 1969–71 seasons, but sowings during the 1984–86 seasons had increased to more than two and approximately tripled between 1979 and 1983 harvest years to the 1974 area (Central Statistical Office 1987, p. 147). The causes for these changes may include:

1. Farmers' response to successive droughts (according to Tanzanian scientists).
2. Economic factors, e.g., relative price of sorghum compared to maize (Central Statistical Office 1987, p. 147).
3. The availability of improved varieties (such as Serena, Dwarf Lulu, and Tegemeo in Tanzania and hybrids such as DC 99 and DC 75 in Zimbabwe), that encouraged farmers to grow sorghum.
4. Successful extension and promotion via mass communication.

1. Principal Cereals Pathologist, SADCC/ICRISAT Sorghum and Millets Improvement Program, P.O. Box 776, Bulawayo, Zimbabwe. Present address: Section Leader, Phytopathology, Zoaduniz Westende 62, PO Box 26, 1600 AA Enkhuizen, The Netherlands.

de Milliano, W.A.J. 1992. Sorghum diseases in southern Africa. Pages 9–19 *in* Sorghum and millets diseases: a second world review. (de Milliano, W.A.J., Frederiksen, R.A., and Bengston, G.D., eds). Patancheru, A.P. 502 324, India: International Crops Research Institute for the Semi-Arid Tropics. (CP 731)

Table 1. Area ('000 ha) sown to sorghum and proportion (%) of the total area sown to sorghum of 10 southern African countries and Africa, 1969–71, 1979–81, and 1984–86.[1]

	1969–71		1979–81		1984–86	
Country	Area ('000 ha)	Proportion (%)	Area ('000 ha)	Proportion (%)	Area ('000 ha)	Proportion (%)
Botswana	133.1	8	98.4	5	55.7	3
Lesotho	75.0	5	57.5	3	60.7	3
Malawi	106.7	6	126.7	7	135.3	7
Mozambique	260.9	16	254.7	14	200.0	10
Swaziland	6.8	1	2.1	1	2.0	1
South Africa	362.8	22	359.8	20	467.0	25
Tanzania	310.0	19	713.3	39	700.0	37
Zaire	28.7	2	31.1	2	38.0	2
Zambia	75.8	5	38.8	2	34.2	2
Zimbabwe	290.9	17	139.9	8	200.0	11
Total area sown to sorghum						
Ten southern African countries	1,650.7		1,822.3		1,892.9	
Africa	13,803.9		12,535.1		15,686.0	

1. Source: FAO tapes, 1986 (1988) for the Republic of South Africa, for the other countries, FAO 1972, 1981, pp. 106–109, and 1986, pp. 120–121.

Distribution of Sorghums

Red- and white-grained sorghums could be found in adjacent fields in Botswana, Lesotho, and Zimbabwe. In most areas of Malawi and in Mozambique, sorghum grains are predominantly white. In Tanzania and Zambia, whites and reds tend to be grown in different regions of the country (SADCC/ICRISAT Sorghum and Millets Improvement Program 1985).

Plant Pathology Research

In most countries in southern Africa, there are very few phytopathologists. One of the countries had a phytopathologist working solely on a single crop (sorghum). National pathologists were seldom able to concentrate on a specific crop species. There were a few papers, usually of a general nature, published in regional magazines. Rarely does a paper on sorghum in southern Africa appear in international journals, as did Molefe's (1975). Botswana. Lesotho, and Swaziland have not published updated lists of sorghum pathogens in a scientific journal (Johnston and Booth 1983, pp. 208–224). However,

recent lists have been published locally (Qhobela et al. 1986, pp. 9–12).

In a small country like the Netherlands (300 by 200 km), with some 14 million people, there are more than 300 phytopathologists with MSc degrees from the State Agricultural University, and annually 5 to 10 are awarded degrees. In contrast Zambia, with an area 23 times that of the Netherlands, has some 10 national phytopathologists with MSc degrees; one or two graduates with BSc degrees each year. To earn the MSc degree, Zambian students must study abroad. Only Zimbabwe and South Africa have employed a number of phytopathologists and both countries have universities offering MSc degree programs in plant pathology. Various foreign-aid programs are assisting scientists from the region with advanced study in cereals pathology. INTSORMIL has trained two to date; an additional five are in training.

The main activities of phytopathologists in the southern African nations include identification of sorghum diseases and scoring of disease incidences and severities. Identification of sorghums for further testing was performed mainly by breeders. An exception is ZSV 1, identified by a plant pathologist.

In southern Africa, the sorghum-growing area lies from 2°N latitude to almost 32°S latitude and from 12 to 41°E longitude at altitudes from sea level to 2600 m. Mean annual rainfall in these areas will range from less than 300 mm to as much as 1500 mm. Within and among the countries the topography and rainfall may vary considerably; consequently, several distinct ecological areas for sorghum production exist, and many sorghum diseases (Table 2). For example, Beck (Williams et. al. 1980) reported as many as 20 pathogens affecting sorghum in Malawi.

A beginning has been made to identify diseases, and record their occurrence. Since 1984,

Table 2. Sorghum diseases/pathogens in 10 countries of southern Africa.[1]

Disease/pathogen	ANG[2]	BOT[2]	LES[2]	MAL[2]	MOZ[2]	RSA[2]	SWA[2]	TAN[2]	ZAM[2]	ZIM[2]
Acremonium wilt	N	N	N	N	N	N	O	O	N	O
Anthracnose	O	R	O	R	R	R	O	R	R	R
Bacterial leaf stripe	N	N	O	N	N	N	N	O	N	R
Bacterial leaf streak	O	N	O	O	N	R	N	O	N	N
Banded leaf (sheath) blight	O	N	N	N	R	N	O	N	N	N
Crazy top	N	N	N	N	R	N	N	N	N	N
Covered kernel smut	R	R	R	R	R	R	O	R	R	R
Charcoal rot	O	R	O	O	N	R	N	R	O	R
Downy mildew	O	R	N	R	R	R	O	R	R	R
Ear blight	N	N	O	N	N	R	N	N	N	R
Ergot	N	R	O	O	R	R	O	R	R	R
Grain molds	N	O	O	R	R	R	O	R	R	R
Gray leaf spot	O	N	N	R	R	N	O	R	R	R
Giberella fujikuroi	N	N	N	N	N	R	N	R	N	R
Head smut	N	R	O	R	N	R	O	R	R	R
Ladder leaf spot	N	N	N	O	N	N	N	O	O	O
Leaf blights	N	R	O	O	N	R	O	R	R	R
Leaf sheath spot	N	N	N	R	N	N	N	N	N	N
Long smut	N	R	N	R	N	R	N	R	O	R
Loose smut	N	R	N	R	R	N	O	R	R	R
Milo disease	N	N	N	N	N	R	N	N	N	N
Mycosphaerella sp	N	N	N	N	N	N	N	N	R	R
Oval leaf spot	O	R	O	R	N	N	N	R	O	N
Phoma leaf spots	N	O	N	N	N	R	N	O	R	N
Phyllosticta blight	N	N	N	N	R	N	N	N	R	R
Rust	R	O	O	R	R	R	O	R	R	R
Rough leaf spot	N	N	N	O	N	N	N	O	R	N
Septoria leaf stripe	N	N	N	N	R	N	N	N	N	R
Smut	N	N	N	N	R	N	N	N	N	N
Sooty stripe	N	R	N	R	R	N	O	R	R	R
Virus	N	R	O	R	N	R	O	R	O	O
Zonate leaf spot	N	O	O	R	R	N	N	R	R	R
Striga asiatica	R	R	N	R	R	R	R	R	R	R
Striga forbesii	N	R	N	R	N	N	O	R	O	R
Striga hermonthica	R	A	A	A	O	A	A	R	A	A

1. O = reported, but identity not confirmed; R = reported and identity confirmed; A = absent; N = no information.
2. ANG = Angola, BOT = Botswana, LES = Lesotho, MAL = Malawi, MOZ = Mozambique, RSA = South Africa, SWA = Swaziland, TAN = Tanzania, ZAM = Zambia, ZIM = Zimbabwe.

annual regional meetings have been held so that local scientists could report to others the disease situations in their countries. In many cases, the reporting scientists did not have time to identify the pathogen with certainty. Confirmation of a number of the reported diseases is required. In many cases, there is no information as to the presence of a disease (Table 2). Listing the sorghum diseases present in southern Africa should be completed as soon as possible. Mapping disease distribution in relation to the agroecological regions should be undertaken promptly in all areas, as is now being done in Zimbabwe. Beginning with the 1984/85 season, a team of Zimbabwe plant-protection scientists has surveyed and recorded pest and disease incidence in each of its five agroecological regions, with the view of obtaining a historic picture of crop-protection problems. The incidence of each disease on sorghum is assessed in each agroecological region. This is an exemplary effort, and the experience in conducting this survey and learning to use its data could be of great use to other countries. However, with a single sampling period, the strong variations in disease incidences between years, the limited knowledge of the relation between disease scores and yield losses, and possible disease interactive effects (related perhaps to changing situational variables), it may be difficult for me to draw, after only 3 years, conclusions useful for extrapolation to the future.

Importance of Plant Diseases

Plant diseases reduce grain and dry matter yields. Certain diseases, such as grain molds and ergot, reduce grain quality and cause health hazards to people and cattle as they produce toxic fungal structures and toxic exudates. In some areas of southern Africa, it is difficult to collect statistics on yield, but it is even more difficult to collect loss data, and this has not been done in the last 10 years. In some cases, therefore, I must present only "qualified guesses." It is believed that:

1. Not one of the 12 southern African countries had to take areas out of production because of a specific disease (Table 1).
2. The national grain yields did not show drastic changes as a result of a specific disease.

3. Loss of dry matter is of some importance, because common sorghums are not absolutely resistant to the common leaf diseases.
4. Quality loss was limited because of the photoperiod sensitivity; in photoperiod-sensitive sorghums, the heads ripen during a drier part of the season.
5. There is little data to assess the importance of plant-disease hazards to man and animals.

King (1975) and Frederiksen (1986) estimated sorghum yield losses in Africa due to diseases to be from 10 to 15%. These estimates may be applicable to sorghum-yield losses in southern Africa. As long as agriculture is taking place in a closed system, it is to be expected that diseases and their hosts have a natural balance, because of elimination of high virulence and high host susceptibility. But the agricultural systems have opened up during the last 10 years, and the new situation may serve to increase the importance of diseases in the region and cause yield losses exceeding 10 to 15%. Characteristics of the new system include:

1. Increased communication and exchange of germplasm between countries and within countries. Large numbers of germplasm lines are now available. Most national programs have only limited facilities and few staff members to test this germplasm. New material tends to differ from the local germplasm, and often gives distinctively higher yields. It was observed, e.g., by Doggett (in Williams et al. 1980) that new introductions may become severely affected by diseases, such as charcoal rot, that are unimportant on local germplasm. ZSV 1 became affected by a sorghum virus previously not noticed in Zambia.
   Because of the superior yields of introduced germplasm, expansion programs were accelerated and the time until release became less than 5 years. National facilities in many of the countries were inadequate to test the new material in all the country's ecological zones before the material reached the farmers. At the same time, farmers were finding it easier to get "improved material." There was improved communication throughout much of the land, and drought relief and other assistance programs were making it a point to "get better seed to the farmer."

2. In the past, farming was in the wild ecosystem and all fields, even in large-scale farming, were surrounded by bush. Now human populations have increased rapidly and new governments are pushing expansion of agriculture. In some countries (such as Lesotho, Malawi, and the Republic of South Africa) large areas are covered by agricultural crops. The sorghum fields are close to each other, and spread of diseases is a natural consequence.

Southern Africa also saw changes that should be helpful in avoiding crop losses from disease.

Through the establishment of the Southern Africa Development Coordination Conference (SADCC), regional coordination and cooperation increased and an enthusiastic effort in regional testing and resistance screening of national and international sorghums began. Promising sorghums are intensively tested at hot spots and in disease sick plots throughout the region. In the 1985/86 rainy season, a regional cereals pathology program was initiated with four major objectives:

1. Identification of the major diseases and the major-disease hot spots.
2. Adaptation and development of screening techniques effective for major sorghum diseases in southern Africa.
3. Identification of sources of resistance to these diseases.
4. Disease monitoring and surveillance.

Most national programs (Botswana, Lesotho, Malawi, Swaziland, Tanzania, Zambia, and Zimbabwe) began to develop expertise to test disease resistances, and a beginning was made to identify regional crop vulnerability through the SADCC/ICRISAT Sorghum and Millets Improvement Program.

Sorghum Diseases in Southern Africa

My comments on the importance of disease groups in relation to sorghum production in southern Africa will include (1) seed rots and seedling blight, (2) foliar diseases, (3) stalk and root rots, (4) inflorescence diseases, and (5) witchweeds. Nonparasitic diseases, such as aluminum toxicity, iron chlorosis, salt damage, and zinc deficiency, do occur in the region. Agrono-

mists and breeders have considered these in crop improvement programs, but the discussion here is limited to pathogen-related diseases and *Striga*.

Seed rots and seedling blight

Fungi damaging the endosperm and embryo of seed and causing seed rots include: *Aspergillus* spp, *Fusarium* spp, *Penicillium* spp, and *Rhizopus* spp. Seedling blights were caused by *Exserohilum turcicum*, *Fusarium* spp, *Macrophomina phaseolina*, *Mycosphaerella holci*, *Phyllosticta sorghiphila*, *Peronosclerospora sorghi*, *Pythium* spp, *Rhizopus* spp and *Rhizoctonia* spp (McLaren 1983; Rothwell 1983). Nematodes that caused seedling blight were *Pratylenchus* spp (Page et al. 1985) and *Meloidogyne* spp. Other fungi and nematodes may be encountered once attention can be given to detailed study of these diseases. The importance of seed rots and seedling blights is still largely unknown. Poor stand establishment occurred in cool and wet soils in Lesotho (Ndimande, personal communication 1982) and the Republic of South Africa (McLaren, personal communication 1988). Other causes for poor stands, however, still need to be determined. In downy mildew (DM) sick plots in Zambia and Zimbabwe, seedling blights reduced stands by more than 50%.

Foliar Diseases

Foliar diseases in southern Africa are caused by viruses, bacteria, fungi, and nematodes.

A. Viral diseases

Advanced research on detection and identification of viruses has been done in South Africa. In 1985, Zambia and SADCC/ICRISAT Sorghum and Millets Improvement Program initiated a cooperative effort with Texas A&M University, USA to identify viruses; in 1987, the project was expanded to include ICRISAT Center in India, Kansas State University (INTSORMIL) in USA, and ICRISAT's EARSAM network as cooperators. Since 1986, the testing of differential nurseries and identification of sources of resistance is ongoing in a virus sick plot, making use of early sowing of the ZSV 1 spreader rows and later sowing (by about 3 weeks) of the test entries. QL 11 was symptomless for 3 years in the sick plot; QL 3 was symptomless in 1986 and 1988, but showed a mild mosaic in 1987. Local germplasm, locally developed varieties, hybrids, and foreign introductions, showed chlorosis at altitudes from 100 to 1670 m in several countries. In Zambia, Johnsongrass (*Sorghum halepense*) showed chlorosis after natural infection in 1986. In South Africa, both maize dwarf mosaic and sugarcane mosaic virus have been reported, and the latter appears to be increasing in importance (McLaren, personal communication 1988). Virus has not been officially reported in sorghum in Lesotho, Swaziland, Zambia, or Zimbabwe. Virus-like symptoms were observed in Zimbabwe from 1986 to 1988 and in Swaziland in 1987. It was established that the virus in Zambia differs serologically from the virus on sorghum in India (Reddy 1987).

Of the 106 grass species tested for resistance to virus by Rosenkranz (1987), 29 are present in Zimbabwe (National Herbarium and Natural Botanic Garden, personal communication 1987). Of these, six were susceptible and seven moderately susceptible to the viruses in Mississippi (Rosenkranz 1987). Several of the grasses, such as *Echinochloa colona*, *Digitaria ciliaris*, and *Eragrostis ciliaris*, are common in the region. It needs to be determined if these grasses can serve as reservoirs for viruses infective to sorghum.

B. Bacterial diseases

Bacterial leaf streak (*Xanthomonas campestris* pv *holcicola*). It is considered important in Angola (Marcelino, personal communication 1988) and in the Republic of South Africa (McLaren, personal communication 1988). Bacterial leaf streak and possibly leaf stripe symptoms were common on sorghum varieties and sudangrass in Lesotho (Qhobela et al. 1986, pp. 9–12).

Bacterial stripe (*Pseudomonas* sp). It occurred in several other countries (Rothwell 1983; Shao 1984). Plant quarantine regulations of Malawi and the Republic of South Africa require that imported seed be free from *Pseudomonas andropogonis* (Williams et al. 1980; McLaren, personal communication 1988). In 1986, cooperative efforts by SADCC/ICRISAT and Kansas State University, USA got underway; its purpose is to learn more about pathogenic bacteria affecting sorghum. In Zambia, bacterial symptoms were observed but the pathogen or pathogens were not identified (Angus 1965, 1966). The loss in biomass caused by bacteria on sorghum in the region requires quantification.

C. Fungal diseases

Eight fungal foliar pathogens: *Cercospora sorghi*, *Colletotrichum graminicola*, *Exserohilum turcicum*, *Gloeocercospora sorghi*, *Peronosclerospora sorghi*, *Ramulispora sorghi*, *R. sorghicola* and *Puccinia purpurea* commonly occur in the region (Angus 1965, 1966; Williams et al. 1980; Rothwell 1983; Plumb-Dhindsa and Mondjane 1984; Shao 1984; Qhobela et al. 1986; Botswana: Ministry of Agriculture 1987; and McLaren, personal communication 1988). Their economic importance is not known. In 1986, cooperative efforts with Texas A&M University and ICRISAT got underway, working with downy mildew and the isolation, culturing, and screening of leaf spot diseases.

Anthracnose. It is recorded in seven countries of southern Africa (Table 2). In 1987 at Mansa, Zambia, a good number of entries from the Zambian program and ICRISAT were susceptible to anthracnose. In the 1987/88 season, with the assistance of an ICRISAT Center scientist, SADCC/ICRISAT began screening for resistance to anthracnose in Zambia. In 1988, it was confirmed that certain sorghums, including several lines resistant to anthracnose in India during the last 10 years, were susceptible at several research stations in Zimbabwe.

Leaf blight. Since the 1986/87 season, the SADCC/ICRISAT program is assisting scientists with the Zimbabwe Sorghum Improvement Program; the purpose is to develop expertise in screening for resistance to leaf blight. In Zimbabwe, Gwebi (17°41'S latitude, 30°52'E longitude, 1448 m altitude, 826 mm rainfall) and Henderson (17°35'S latitude, 30°58'E longitude, 1292 m altitude, 869 mm rainfall) regularly had leaf blight epiphytotics. Some white-grained sorghums showing resistance to leaf blight at Gwebi during the last 2 years were Siloe Best, Nyakosaba Best, and Tenant White (Lesotho),

Segaolane, Marupantse and Kanye (Botswana), SAR 1, QL 3, (148 × Framida) -sel, KU 302; KU 417, (SC 108-3 × CS 3541) -sel and (IS 12611 × SC 108-3) -sel. Red-grained sorghums with resistance to leaf blight were the Botswana varieties 65 D and Town. The variety Framida and the hybrids PNR 8544 and PNR 8311 were susceptible to leaf blight.

Downy mildew. Epiphytotics occurred regularly in Botswana (Botswana: Ministry of Agriculture, 1987), Zimbabwe, and Zambia at locations situated at altitudes from sea level to 1448 m and rainfall of more than 400 mm. During a 29-day drought period in January and February 1987, severities increased uniformly, likely because leaf wetness caused by dew persisted long enough for successful infection. In Zimbabwe, research was initiated to study yield loss, using chemicals to control the disease in the control plots. With the assistance of ICRISAT scientists, moderate-scale sorghum downy mildew sick plots of 2.5 hectares each, were developed by adapting techniques for use at the SADCC/ICRISAT Sorghum and Millets Improvement Program at Bulawayo, Zimbabwe (20°24'S latitude, 28°28'E longitude, 1342 m altitude, 594 mm rainfall) in the 1986/87 season and at Golden Valley, Zambia, in the 1987/88 season. At Bulawayo it was established that the ratoon and the downy mildew pathogen can survive the cold dry season, and that spreader rows can be well established with inoculum from the surviving ratoon. By late September 1987, there were about 10 stems m^{-2}; about 50% of the stems were showing symptoms of systemic infection.

Sandala released in Tanzania, ZSV 1 (= ICSV 2 = SPV 386) released in Zambia in 1983, and SV 1 (= ICSV 112 = SPV 475) released in 1985 in Zimbabwe, were all resistant to downy mildew following inoculation. Framida, Serena, Red Nyoni, SV 2 (Zimbabwe), and Tegemeo (Tanzania) appeared moderately susceptible. Marupantse, Kanye, 65 D, Town and Segaolane (least susceptible) from Botswana, Siloe Best, Nyakosa Best and Tenant White from Lesotho, and Red Swazi (from Zimbabwe) were susceptible. The commercial hybrids PNR 8311 and PNR 8544 appeared resistant in 1987, but only moderately resistant in 1988. DC 75 was moderately susceptible. Resistance was found in local collections from many southern African countries, although not in Botswana and Swaziland. Resistance was

found, however, in germplasm that had been newly introduced to these countries. Many entries which were identified as resistant at ICRISAT Center, such as QL 3 and entries from the International ICRISAT Sorghum Downy Mildew Nursery, appeared resistant in Zimbabwe. Considering that a good number of advanced entries appeared to be resistant (38–75% of the entries in the Zimbabwean national nurseries at Matopos in 1987), it does not appear to be essential to develop a separate program to transfer genes from highly resistant material to locally adapted material.

Two forms of incomplete resistance were observed which will receive further attention to find practical uses: In one, symptoms were present on all leaves, but head exertion and fertility were normal. In the other, symptoms appeared immediately after infection in the seedling stage, but disappeared during crop development, and seed production appeared normal.

It was furthermore observed at Matopos, Zimbabwe, that symptoms of systemic infection disappeared during the cold dry season, but reappeared when it was warmer in late September.

Considering the crop's vulnerability, as seen in Botswana and Lesotho, and the common occurrence of downy mildew in countries such as Zambia and Zimbabwe, it is important for farmers of the southern African region that Zambia and Zimbabwe continue the regional screening activities initiated with the assistance of SADCC/ICRISAT Sorghum and Millets Improvement Program. Chemical seed treatment to control downy mildew, initiated in Botswana and Zimbabwe, needs attention relative to use in other areas.

Sooty stripe. Severe epiphytotics have occurred at several locations in Tanzania, Zambia, and Zimbabwe. A Sooty Stripe and Downy Mildew Resistance Nursery was established in Zambia in 1986 for further testing in the region. ZSV 1 (a Zambian release) and Tegemeo (Tanzania) have shown outstanding resistance. Germplasm has been collected from western and eastern Africa and from Texas A&M University (INTSORMIL) for confirmation of resistance at this nursery.

Rust. Severe epiphytotics occur regularly at several locations in Malawi, Swaziland, Tanzania,

Zambia, and Zimbabwe. Several entries with outstanding resistance have been identified and will be tested to confirm their resistance.

Other leaf spots. Epiphytotics of gray leaf spot, oval leaf spot, and zonate leaf spot, occurred regularly at several locations in Malawi, Zambia, and Zimbabwe. Selections for leaf health were made where epiphytotics occurred. Sources of resistance identified elsewhere (India, western and eastern Africa, and Texas A&M University (INTSORMIL) are being tested.

Stalk and root rots. Fungi affecting the stalk included *Macrophomina phaseolina*, *Fusarium* spp, and *Colletotrichum graminicola* (Angus 1965, 1966, pp. 65–68; Rothwell 1983; Shao 1984; Botswana: Ministry of Agriculture 1987; McLaren, personal communication 1988). *Fusarium* spp caused root rots (Shao 1984, McLaren, personal communication, 1988). Charcoal rot occurred in areas where the crop was exposed, especially following anthesis, to drought stress and high temperatures such as occurred in Botswana, Lesotho, Tanzania, and Zimbabwe. In several countries charcoal rot is not officially reported on sorghum, (e.g., in Angola, Lesotho, Malawi, Swaziland, or Zambia). Considering the wide host range of the pathogen, it may be present there, and this requires investigation.

Certain high-yielding varieties [e.g., SV 2 (A 6460), SAR 1 (ICSV 145) -1-1-1, SPV 475, and ZSV 1)], and hybrids (e.g., PNR 8544), appeared to be susceptible to charcoal rot in Zimbabwe. Lines identified as resistant elsewhere by ICRISAT and INTSORMIL are being tested regionally.

Identification of the pathogens causing stalk and root rots in these areas requires additional study.

Inflorescence Diseases

Inflorescence diseases are caused by viruses and fungi.

Viral. The maize dwarf mosaic and sugarcane mosaic viruses cause stunting, fewer seeds and reduced seed mass.

Fungal. Important fungal diseases that attack sorghum inflorescence include: ergot, grain molds, and smuts.

A. Ergot.

From Tanzania (Doggett in Williams et al. 1980) to South Africa (McLaren, personal communication 1988), this disease is of particular importance in male-sterile lines. It is a particular problem in late-sown sorghum and in the Republic of South Africa, where sorghum is grown under center-pivot irrigation (McLaren, personal communication 1988). In 1987, ergot epiphytotics occurred for the first time in Lesotho and Malawi. Cooperative work with the Imperial College in London, and the University of Zimbabwe, was initiated to learn more about the ergot life cycle, the toxicity of sclerotia, and the development of artificial screening techniques. In cooperation with ICRISAT's EARSAM network, attempts to identify germplasm with resistance to ergot will begin.

In Zimbabwe, year-old conidia were able to cause infection. Time to production of honeydew was similar (7–15 days) to that recorded in the United Kingdom. Infection occurred even during extended drought in 1987.

Sclerotia were formed at most locations with ergot infections in Botswana, Malawi, Swaziland, Tanzania, Zambia, and Zimbabwe. In darkened growth chambers, mature sclerotia harvested at the end of the rainy season (March), germinated over a wide range of temperatures (4–37°C). Germination percentages were low (less than 20% of sclerotia).

The importance of ergot is related to the use of hybrids. It is expected that hybrids will become more popular and that therefore this disease may become more important because of its interference with seed production.

B. Grain molds

Several newly released and prereleased varieties, such as SV 1 (Zimbabwe) and certain hybrids (e.g., PNR 8544), were susceptible to grain molds. The hybrid DC 75 was relatively resistant. It is expected that with the increase of popularity of the nonphotosensitive varieties and hybrids, grain molds will increase in importance.

In the last 10 years, ICRISAT grain mold resistance nurseries have been operating, but the material lacked agronomic eliteness and was not used by breeders. Agronomically elite material with grain mold resistance was recently introduced from ICRISAT for testing in the southern African region. Screening methods need to be adapted to the region and local pathogens and sources of resistance need to be identified.

C. Smuts

Covered smut (*Sporisorium sorghi*), head smut (*S. reilianum*), long smut (*Tolyposporium ehrenbergii*) and loose smut (*Sphacelotheca cruenta*) were all reported in the region.

Covered smut. This sorghum smut, commonly found in most of the 12 southern African countries on local and introduced germplasm alike, is of importance to small-scale farmers in several countries (Ndimande, personal communication 1982, Botswana: Ministry of Agriculture 1987). Chemical control of covered smut has been recommended in Tanzania since the 1950s (Doggett in Williams et al. 1980), and appears to be of importance in all 12 southern African countries. Lesotho has actively tested control methods since 1985, making use of artificial inoculation at sowing. This disease can be controlled effectively with seed-dressing fungicides; low-cost seed dressings have been identified and many others are available for evaluation.

Head smut. Head smut, confirmed in six nations (Table 2), can be expected in all southern African countries (Sundaram in Williams et al. 1980). In the last 10 years, incidence levels have been very low. Sources of resistance (Texas A&M University, INTSORMIL) were introduced to test for local adaptation and performance. Considering the complicated development of resistance due to high chance of the development of new races, continuous monitoring for this disease is required.

Long smut. This smut is commonly found in areas affected by drought. In recent droughts it was observed in Botswana, Lesotho, Malawi, Tanzania, Zambia (first record, 1987) and Zimbabwe (first record, 1985). In 1987, at Hombolo in Tanzania, up to 40% of the head area in susceptible genotypes was affected. In susceptible genotypes, in other countries, less than 10% of the head area was affected. Verification of the resistance of entries tested in Kenya is important in particular for use in Tanzania.

Loose smut. Loose smut affects local and introduced germplasm in many countries, before and after ratooning, but has not reached high incidence.

Witchweeds. Three witchweeds were of importance in the region: *Striga asiatica, S. hermonthica,* and *S. forbesii. S. asiatica* were reported in most countries (Riches et al. in SADCC/ICRISAT Sorghum and Millets Improvement Program 1987), in particular on poorly drained soils with a low nutrient status. *S. hermonthica* occurred in specific areas mainly in Tanzania, and *S. forbesii* in specific areas in Tanzania and Zimbabwe. In the 1985/86 season, SADCC/ICRISAT began researching the *Striga* spp. *S. asiatica* was studied intensively in Botswana by the Department of Agricultural Research (Botswana: Ministry of Agriculture 1987). ICRISAT provided sources of resistance to *S. asiatica* from India, and the New Dominion University, USA, provided scientific cooperation. A field identification guide for the region was published and distributed (Obilana et al. 1987).

Conclusion

Doggett (in Williams et al. 1980) hypothesized that the importance of diseases is positively related to yield levels. Sorghum yields surpassing 3 t ha^{-1} have been achieved under experimental conditions. In Malawi, Swaziland, South Africa, Tanzania, Zambia, and Zimbabwe, yields in trial plots were 3 to 12 times higher than the national sorghum grain yields. In developed countries, chemical control was used to obtain high yields in wheat, but in sorghum, disease-resistant lines appear to be the most promising method to increase grain yields. With Doggett's hypothesis in mind, it appears that this is an appropriate time to put more efforts into developing disease resistance, with increased yields.

Acknowledgment. D. Rohrbach assisted with the preparation of the Introduction and the material on Grain Yield.

References

Angus, A. 1965. Annotated list of plant pests and diseases in Zambia Parts 1–2. Mount Makulu, Zambia: Plant Pathology Library, Mount Makulu Research Station. (Limited distribution.) 384 pp.

Angus, A. 1966. Annotated list of plant pests and diseases in Zambia (supplement). Mount Makulu, Zambia: Plant Pathology Library, Mount Makulu Research Station. (Limited distribution.)

Botswana: Ministry of Agriculture. 1987. Annual Report for the Division of Arable Crops Research, 1985–1986. Gaborone, Botswana: Botswana Ministry of Agriculture. (Limited distribution.) 274 pp.

Central Statistical Office (Zimbabwe) 1987. Statistical Yearbook 1987. Harare, Zimbabwe: Central Statistical Office (Limited distribution).

FAO. 1972. Report to the Government of Botswana on plant diseases on the work of G.I. Nilsson. UNDP Report No. TA 3057. Rome, Italy: Food and Agricultural Organization of the United Nations. 34 pp.

FAO. 1981. FAO Production yearbook 1980. Rome, Italy: Food and Agricultural Organization of the United Nations.

FAO. 1986. FAO Production yearbook 1985. Rome, Italy: Food and Agricultural Organization of the United Nations.

FAO Tapes, 1986. 1988. Rome, Italy: Food and Agricultural Organization of the United Nations.

Frederiksen, R.A. 1986. Sorghum Diseases. The potential for control. Pages 324–337 *in* Proceedings of the 4th Regional Workshop on Sorghum and Millet Improvement in Eastern Africa, 22–26 Jul 1985. Gebrekidan, B. (ed.). Nairobi, Kenya: OAU/SAFGRAD. (Limited distribution.)

Johnston, A., and **Booth, C.** (eds.). 1983. Regional and country lists of plant diseases. Plant Pathologist's Pocket Book. Kew, England: Commonwealth Mycological Institute.

King, S.B. 1975. Sorghum diseases and their control. Pages 411–434 *in* Proceedings of the Sorghum in the Seventies Conference, 1972. (Rao, N.G.P., and House, L.R., eds.) New Delhi, India.

McLaren, N.W. 1983. The effect of herbicides, cultivars, and fungicides on pre- and post-emergence damping-off and seedling blight of sorghum (*Sorghum bicolor* (L) Moench.). Crop Production 12:101–103.

Molefe, T.L. 1975. Occurrence of ergot on sorghum in Botswana. Plant Disease Reporter 59(9):751–753.

Obilana, A.B., Riches, C.R., de Milliano, W.A.J., Knepper, D., and **House, L.R.** 1987. Witchweeds (*Striga* species) of southern Africa. A field identification guide. Bulawayo, Zimbabwe: SADCC/ICRISAT Sorghum and Millets Improvement program, in cooperation with the Department of Agricultural Research, Botswana. 3 pp.

Page, S.L., Mguni, C.M., and **Sithole, S.Z.** 1985. Pests and diseases of crops in communal areas of Zimbabwe, 1984–85 growing season. Overseas Development Administration Technical Report. Harare, Zimbabwe: Plant Protection Research Institute, Department of Research and Specialist Services (Limited distribution.) 203 pp.

Plumb-Dhindsa, P., and **Mondjane, A.M.** 1984. Index of plant diseases and associated organisms of Mozambique. Tropical Pest Management 30(4):407–429.

Qhobela, M., Ramasike, M., and **Lepheana, T.** 1986. A guide to the common pests and diseases of Lesotho's food crops. Agricultural Research Technical Information Handbook RD-H-4. Maseru, Lesotho: Ministry of Agriculture 9–12. (Limited distribution.)

Reddy, D.V.R. 1987. Report of a visit to Malawi and Zambia, March 5–22, 1987. Patancheru, P.O., Andhra Pradesh 502 324, India: International Crops Research Institute for the Semi-Arid Tropics (Limited distribution.) 36 pp.

Rosenkranz, E. 1987. New hosts and taxonomic analysis of the Mississippi native species tested for reaction to maize dwarf mosaic and sugarcane mosaic viruses. Phytopathology 77:598–607.

Rothwell, A. 1983. A revised list of plant diseases occurring in Zimbabwe. Kirkia 12(II):288–289.

SADCC/ICRISAT Sorghum and Millets Improvement Program.1985. Proceedings, First Regional Workshop. Oct 23–26, 1984. Harare, Zimbabwe. Bulawayo, Zimbabwe: SADCC/ICRISAT Sorghum and Millets Improvement Program (Limited distribution.) 89 pp.

SADCC/ICRISAT Sorghum and Millets Improvement Program. 1987. Proceedings, Third Regional Workshop. Lusaka, Zambia, Oct 6–10, 1986. Bulawayo, Zimbabwe: SADCC/ICRISAT Sorghum and Millet Improvement Program (Limited distribution.) 476 pp.

Shao, F.M. 1984. Diseases of sorghum in Tanzania. Pages 43–54 *in* Proceedings of the Third Regional Workshop on Sorghum and Millet Improvement in Eastern Africa. 5–8 Jun 1984. Morogoro, Tanzania. (Gebrekidan, B., ed.) Nairobi, Kenya: SAFGRAD/ICRISAT Program. (Limited distribution.)

Williams, R.J., Frederiksen, R.A., and Mughogho, L.K. (eds.) 1980. Sorghum diseases, a world review: proceedings of the International Workshop on Sorghum Diseases, 11–15 Dec 1978, ICRISAT, Hyderabad, India. Patancheru, Andhra Pradesh 502 324, India: International Crops Research Institute for the Semi-Arid Tropics. 469 pp.

Sorghum Diseases in Eastern Africa

Mengistu Hulluka[1] and J.P.E. Esele[2]

Abstract

Sorghum diseases in eastern Africa, last reviewed by Doggett in 1978 continue to be important constraints to production of this cereal. Of the head blights and grain diseases, covered smuts, grain mold, and ergot appear to be economically important. The major leaf diseases are anthracnose, leaf blight, gray leaf spot, and downy mildew. Charcoal rot is the important stalk rot disease in the region. Surveys of sorghum diseases have been undertaken in most countries of eastern Africa. Priorities have been determined. In most instances, the current condition dictates that the Striga problem receives more attention, as present research consists mainly of screening for disease resistance. In some countries more detailed work is underway, including breeding to incorporate resistance factors into elite local lines by using exotic materials as donors. More attention needs to be given on specific research on sorghum disease management in the future.

Introduction

Sorghum improvement has been significant in eastern Africa during the past decade. Some countries, like Sudan, are producing hybrid sorghum on a large scale, and have others in the pipeline. The programs of pathological studies receive better attention than ever before. These developments were made possible because of the increasing demand to improve sorghum production in eastern Africa.

Eastern African countries included in this report are Burundi, Ethiopia, Kenya, Rwanda, Somalia, Uganda, Sudan, and part of Tanzania. These countries collectively produce about 4 million tonnes of grain sorghum per year on just over 6 million hectares of land (Guiragossian 1986). Almost 90% of the area sown to sorghum in eastern Africa is located in Sudan, Ethiopia, Tanzania, and Somalia (Table 1).

The Current Situation

During the 1978 Workshop, Doggett (1980) and Hulluka and Gebrekidan (1980), outlined a list of sorghum diseases commonly occurring in this region. Not much has changed in regard to types of diseases, but some have gained in importance and demand closer attention.

According to Guiragossian (1986), disease problems in the region are not so urgent as to warrant priority over breeding, agronomic, and entomological problems. Landrace sorghum populations constitute about 90% of the production, and disease problems are minimal with landraces. However, when improved varieties are considered, the case is different. Some diseases need attention, and could become priority problems. Doggett (1980) mentioned this 10 years ago when he said that diseases such as leaf blight, anthracnose, charcoal rot, ergot, and grain mold are troublesome on improved lines.

1. Associate Professor, Debre Zeit Agricultural Research Center (AUA), PO Box 32, Debre Zeit, Ethiopia.
2. Research Officer, (UAFRO), Sorghum and Millet Unit, Serere Research Station, PO Saoroti, Uganda.

Hulluka, M., and Esele, J.P.E. 1992. Sorghum diseases in eastern Africa. Pages 21–24 *in* Sorghum and millets diseases: a second world review. (de Milliano, W.A.J., Frederiksen, R.A., and Bengston, G.D., eds). Patancheru, A.P. 502 324, India: International Crops Research Institute for the Semi-Arid Tropics.

Table 1. Average annual sorghum production of eight eastern African countries, 1980-84.[1]

Country	Area sown ('000 ha)	Sorghum production ('000 t)	Grain yield (t ha⁻¹)
Burundi	160	200	1.2
Ethiopia	900	1320	1.5
Kenya	167	170	1.0
Rwanda	169	195	1.2
Somalia	500	235	0.5
Uganda	233	400	1.7
Sudan	3500	1450	0.4
Tanzania	650	450	0.5
Total	6279	4420	
Mean			992.5

1. Source: Guiragossian (1986)

Important Sorghum Diseases

Since 1982, with the establishment of the yearly Regional Workshop on Sorghum Improvement in Eastern Africa, it is now possible to get a better picture of disease patterns (Table 2) in each country (Esele 1985; Elzein 1986). The information on sorghum diseases has been compiled from reports of the ICRISAT/SAFGRAD team on their surveys in 1986.

These data indicate that sorghum in all countries of the region seem to be suffering from diseases of similar nature. Variation has been observed in some countries however, mainly due to different ecological zones and to the level of improvement in sorghum lines.

In considering only the major diseases of the region, the following are characteristically important. Among the panicle and head diseases, ergot, grain mold, covered smut, and loose smut are important. Of the foliar diseases, anthracnose, leaf blight, and gray leaf spot are rampant over all the humid regions when high rainfall occurs. Rust and the sorghum downy mildew are next in importance. The prominent stalk rot disease, especially on improved lines under drought stress conditions, is charcoal rot.

Table 2. Sorghum disease problems identified by the ICRISAT/SAFGRAD surveys in seven eastern African countries.[1]

Disease	Burundi	Ethiopia	Kenya	Rwanda	Somalia	Sudan	Uganda
Grain mold	x[2]	xxx	xxx	x	x	x	xxx
Downy mildew	x	xx[3]	xx	x	x	x	xx
Charcoal rot	xx	xx	xxx[4]	x	x	xxx	xxx
Leaf blight	x	xx	xxx	xxx	x	xxx	xx
Rust	xx	xx	xx	x	x	x	xx
Gray leaf spot	x	xx	xxx	x	x	xx	xx
Anthracnose	xx	xxx	xxx	xx	x	xx	xxx
Sooty stripe	xx	xx	xx	xx	x	xx	xx
Zonate leaf spot	x	x	x	x	x	x	x
Bacterial stripe	x	xx	xx	xx	x	xx	xx
Ergot	xxx	xx	xxx	xxx	x	x	xxx
Head smut	x	x	xxx	x	xx	x	xx
Long smut	x	x	xxx	x	x	xxx	xx
Covered smut	x	xx	x	xx	xx	x	xx
Loose smut	x	xx	x	xxx	xx	x	xx
Nematodes	-[5]	-	xx	-	x	x	x
Striga hermonthica	xxx	xxx	xxx	xxx	x	xxx	xx

1. Source: Guiragossian (1986).
2. x = low priority; 3. xx = intermediate priority; 4. xxx = high priority;
5. (-) = no information.

The *Striga* spp are invariably a major problem in almost all countries in eastern Africa, and heavy crop losses from severe infestation by these parasitic weeds are common. In Uganda, losses due to *Striga* can be as much as 30% (Esele 1986); in western Tanzania, the parasite caused up to 40% loss in yield (Shao 1985). In Sudan and Ethiopia, the widespread *Striga* spp have forced farmers to abandon their land (Elizen 1986).

At country levels within the region, the following five diseases are ranked in the order listed (Guiragossian 1986).

Burundi:	Ergot, anthracnose, rust, charcoal rot, and *Striga*.
Ethiopia:	*Striga*, anthracnose, grain mold, covered smut, and ergot.
Kenya:	*Striga*, grain mold, leaf blight, covered kernel smut, and head smut.
Rwanda:	Leaf blight, ergot, covered smut, downy mildew, and *Striga*.
Somalia:	Covered smut, loose smut, head smut, charcoal rot, and anthracnose.
Sudan:	*Striga*, charcoal rot, long smut, leaf blight, and anthracnose.
Uganda:	Grain molds, smuts, anthracnose, leaf blight, and *Striga*.

Research Activity

Research attention to sorghum diseases is much more evident than it was 10 years ago. Surveys to identify diseases have been carried out in most countries, and the major diseases have been categorized according to importance. Specific studies of sorghum diseases have received less emphasis.

Breeders were highly involved in selecting breeding materials, using disease resistance as one criterion in selection. Special research projects set up to develop resistant lines exist in Ethiopia, Uganda, and Kenya. However, within the past 10 years, pathologists have been included in the Sorghum Improvement Program, and some progress has been made in searching out and incorporating resistance into the breed-

ing lines. Cases in point include screening for resistance to sorghum anthracnose, grain molds, and downy mildews. International Disease and Insect Nursery programs are in operation.

The assistance of ICRISAT in this regard is invaluable. Much collaborative work is underway.

A backcrossing program has even been initiated in Ethiopia in an attempt to incorporate resistance to anthracnose from exotic sources to indigenous elite lines. Some success has been achieved in restoring genes for resistance into selected Ethiopian lines.

Independent studies likewise have been carried out to evaluate indigenous as well as exotic lines for resistance to different isolates of *Colletotrichum graminicola* and to other leaf diseases.

However, except in a few cases, factors conditioning resistance have not been worked out, and in most cases the stability and durability of the lines is not clearly understood, perhaps due to the changing virulence of the pathogen. To reduce the influence of variability in virulence of the isolates, some workers are testing advanced lines in a wide range of ecological zones.

This closer attention to pathological problems reflects the growing importance given to sorghum in eastern Africa, and efforts to utilize its tolerance to drought-stress and its adaptability to wider ecological zones to meet the growing demand for staple food.

Sorghum Disease Control

Among the many alternative measures listed by Frederiksen (1985), selecting for disease resistance has been the most important approach so far taken. Evaluating on the basis of reaction of elite lines to the prevailing important diseases before release for large-scale production is now a common practice.

Seed-dressing with chemicals against smuts is practiced in many instances, though it is not widely accepted by peasant farmers.

In Somalia, where damage by covered and loose kernel smuts has at times reached the 40% level (Alahaydoian and Ali Nur Duale 1985), evaluation of seed-dressing chemicals was initiated.

Some countries, such as Kenya, have a strong quarantine procedure to exclude diseases considered to be dangerous. However, with more frequent and wider germplasm exchanges between countries, the volume of materials arriving sometimes necessitates relaxation of the quarantine measures. Diseases like downy mildew, and some viruses have found their way into many research stations.

To our knowledge, the use of other control measures is not so significant as to warrant presentation here.

Future Prospects

We envision more and more people getting involved in sorghum pathology work. Great interest has been shown by national program scientists to increase sorghum production through selection and breeding. With crop improvement we expect less diversity of genotypes and thus increased susceptibility to various diseases, in turn calling for more resistant factors to be incorporated. Pathology research programs are expected to be strengthened as well, perhaps providing valuable sources for resistance that will be of use to our breeders.

To protect these efforts, stricter quarantine measures may have to be applied to the exchange of materials, especially to limit new and virulent biotypes of the existing pathogens and to avoid potential importation of new diseases. More research work on the other control measures for sorghum diseases may be advantageous.

References

Alahaydoian, E.K., and Ali Nur Duale. 1985. The Somalia sorghum improvement project. Pages 128–135 in Proceedings, Third Regional Workshop on Sorghum and Millet Improvement in Eastern Africa, 5–8 Jun 1984, Soroti, Uganda.

Doggett, H. 1980. Sorghum diseases in East Africa. Pages 33–35 in Sorghum diseases, a world review: proceedings of the International Workshop on Sorghum Diseases, 11–15 Dec 1978, Hyderabad, India. Patancheru, Andhra Pradesh 502 324, India: International Crops Research Institute for the Semi-Arid Tropics. 469 pp.

Elzein, Ibrahim N. 1986. An overview of sorghum improvement in the Sudan. Pages 320–326 in Proceedings, Fifth Regional Workshop on Sorghum and Millet Improvement in Eastern Africa, 5–12 Jul 1986, Bujumbura, Burundi.

Esele, J.P.E. 1985. An overview of sorghum diseases in Uganda. Pages 91–99 in Proceedings, Fourth Regional Workshop on Sorghum and Millet Improvement in Eastern Africa, 22–26 Jul 1985, Soroti, Uganda.

Esele, J.P.E. 1986. Crop protection aspects of sorghum in Uganda. Pages 244–258 in Proceedings, Fifth Regional Workshop on Sorghum and Millet Improvement in Eastern Africa, 5–12 Jul 1986, Bujumbura, Burundi.

Frederiksen, R.A. 1985. Sorghum diseases: their potential for control. Pages 324–327 in Proceedings, Fourth Regional Workshop on Sorghum and Millet Improvement in Eastern Africa, 22–26 Jul 1985, Soroti, Uganda.

Guiragossian, V. 1986. Sorghum production constraints and research needs in eastern Africa. Pages 28–46 in Proceedings, Fifth Regional Workshop on Sorghum and Millet Improvement in Eastern Africa, 5–12 Jul 1986, Bujumbura, Burundi.

Hulluka, M., and Gebrekidan, B. 1980. Diseases of sorghum in Ethiopia. Pages 36–39 in Sorghum diseases, a world review: proceedings of the International Workshop on Sorghum Diseases, 11–15 Dec 1978, Hyderabad, India. Patancheru, Andhra Pradesh, 502 324, India:International Crops Research Institute for the Semi-Arid Tropics. 469 pp.

Shao, F.M. 1985. Disease of sorghum in Tanzania. Pages 43–54 in Proceedings, Third Regional Workshop on Sorghum and Millet Improvement in Eastern Africa, 5–8 Jun 1984, Soroti, Uganda.

Sorghum Diseases in Western Africa

M.D. Thomas[1]

Abstract

In western Africa sorghum (Sorghum bicolor) is attacked by several fungi, bacteria, viruses, and by the parasitic weed Striga hermonthica. Sorghum is grown mainly in the Sudanian Zone (500–1000 mm rainfall) and in the northern Guinean Zone (1000–1200 mm rainfall). Some sorghum is also grown in the Sahelian Zone (400–500 mm rainfall). The importance of any of the sorghum diseases depends to a large extent on the zone in which the crop is grown. In local and introduced genotypes alike, diseases such as gray leaf spot and leaf anthracnose are more prevalent in the northern Guinean and Sudanian zones; sooty stripe and long smut occur more frequently in the Sudanian Zone. Grain mold is found mostly in introduced genotypes which mature during periods of relatively high rainfall. Oval leaf spot, zonate leaf spot, leaf blight, head smut, and covered smut occur at low levels in both the northern Guinean and Sudanian Zones. Head smut is a particular problem in farmers' fields. Striga hermonthica occurs abundantly in all zones and attacks all genotypes. Preliminary data indicate that symptoms of gray leaf spot first appear between flowering and 10 days after in Burkina Faso. The progress of the disease is slow and exceeds 75% leaf area infected only at maturity. In contrast, symptoms of leaf anthracnose appear early, the disease progresses rapidly, and kills top leaves during grain formation. Soil shading with groundnut suppressed Striga in farmers' fields.

Introduction

Sorghum [*Sorghum bicolor (L) Moench*] is grown widely in many countries in semi-arid western Africa. In several of these countries, sorghum can be regarded either as the staple food or the second most important source of energy for the population. Most of the sorghum is grown in the Sudanian Zone (500–1000 mm rainfall) and the northern Guinean Zone (1000–1200 mm rainfall). Some sorghum is also grown in the southern Sahelian Zone (400–500 mm rainfall).

In western Africa sorghum is attacked by several fungi, bacteria, viruses, and the parasitic weed *Striga hermonthica* (Del.) Benth (Table 1). The situation with nematodes is not quite clear and there appears to be no report of nematode infection of sorghum in western Africa. There is variation in incidence and severity of the dis-eases according to the agroecological zones mentioned above. Thus, the importance of any one of the sorghum diseases in western Africa depends to a large extent on the zone where the crop is growing. Importance is also related to whether a disease occurs on local landraces or on introduced genotypes. There appears to be "pockets" within western Africa for the occurrence of some diseases. For example, symptoms of downy mildew and virus-like symptoms have been reported in some countries (Tyagi 1980, Zummo 1984), whereas in many countries these two categories of symptoms are absent, or at least have not been reported. In general the importance of a disease is related to several other factors, such as the progress of the disease during the different growth stages of the host and the effect of the disease on grain yield.

---

1. Principal Sorghum Pathologist, ICRISAT. B.P. 910, Bobo-Dioulasso, Burkina Faso.

Thomas, M.D. 1992. Sorghum diseases in western Africa. Pages 25–29 *in* Sorghum and millets diseases: a second world review. (de Milliano, W.A.J., Frederiksen, R.A., and Bengston, G.D., eds). Patancheru, A.P. 502 324, India: International Crops Research Institute for the Semi-Arid Tropics. (CP 733).

Table 1. Diseases of sorghum in western Africa.[1]

Disease	Causal agent	Disease	Causal agent
Fungal foliar		Long smut	*Tolyposporium ehrenbergii*
Gray leaf spot	*Cercospora sorghi*	Loose smut	*Sphacelotheca cruenta*
Leaf anthracnose	*Colletotrichum graminicola*	**Root and stalk**	
Leaf blight	*Exserohilum turcicum*	Acremomium wilt	*Acremonium strictum*
Leaf spot	*Phoma insidiosa*	Charcoal rot	*Macrophomina phaseolina*
	Phyllosticta sorghiphila	Fusarium stalk rot	*Fusarium moniliforme*
Oval leaf spot	*Ramulispora sorghicola*	Milo disease	*Periconia circinata*
Rough leaf spot	*Ascochyta sorghina*	Pokkah boeng	*Fusarium subglutinans*
Rust	*Puccinia purpurea*	Root rot	*Fusarium* spp
Sooty stripe	*Ramulispora sorghi*		*Macrophomina phaseolina,*
Zonate leaf spot	*Gloeocercospora sorghi*		*Rhizoctonia sorghi,*
Leaf sheath			*Phoma sorghina*
Sheath blight	*Rhizoctonia solani*	Stem anthracnose	*Colletotrichum graminicola*
Panicle		**Seedling**	
Ergot	*Sphacelia sorghi*		*Fusarium moniliforme*
Covered smut	*Sporisorium sorghi*		*Phoma sorghina*
False smut	*Cerebella sorghi-vulgaris*	Downy mildew	*Peronosclerospora sorghi*
Grain mold	*Phoma sorghina*	**Bacterial[2]**	
	Curvularia lunata	Bacterial leaf	*Pseudomonas syringae*
	Colletotrichum graminicola	spot	
	Fusarium culmorum	Bacterial leaf	*Xanthomonas holcicola*
	F. moniliforme	streak	
	F. oxysporum	Bacterial leaf	*Pseudomonas andropogoni*
	F. semitectum	stripe	
	Helminthosporium halodes	Bacterial soft rot	*Erwinia* sp
	Aspergillus niger	Red stripe and	*Xanthomonas rubrilineans*
	Chaetomium sp	top rot	
	Penicillium funiculosum	Yellow leaf blotch	*Pseudomonas* sp
	Cladosporium spp	**Viral**	
	Rhizopus stolonifer	Mosaic	Sugarcane mosaic virus
	Aspergillus flavus		
	Nigrospora sphaerica	**Parasitic plant**	
	Alternaria alternata	Striga	*Striga hermonthica*
	Sclerotium rolfsii		
Head blight	*Fusarium moniliforme*		
Head smut	*Sporisorium reilianum*		

1. Source: Zummo (1984).
2. Leaf-infecting agents, except *Erwinia* spp, a root-infector, and top rot (*X. rubrilineans*) a disease of the stalk.

The occurrence rather than importance of sorghum diseases in western Africa is emphasized. Prevalence is used in relative terms to mean a higher incidence of a disease. Based on observations from six countries (Burkina Faso, Cameroon, Côte d'Ivoire, Ghana, Mali, and Niger), some features of the more prevalent diseases in local and in introduced genotypes are presented. A summary of research efforts for the past 3 years will conclude this review.

Diseases Prevalent in Local Sorghums

Leaf diseases

Gray leaf spot and leaf anthracnose are the more common leaf diseases in farmers' fields, and are more serious in the northern Guinean Zone. Most often gray leaf spot is severe in the top four or six leaves only towards maturity. Abundant sporulation occurs in very wet areas.

It would appear that leaf anthracnose causes more damage than gray leaf spot. All leaves can be heavily infected during grain formation. In many instances, complete drying of the top leaves occurs before the grain matures. However, there are many local genotypes in which leaf anthracnose appears only as midrib infections; abundant tiny dark brown necrotic spots on the leaf lamina resemble a hypersensitive-type reaction. In this type of host reaction, the top four or six leaves are rarely killed whereas the lower leaves are completely dry. Formation of acervuli is rare in either type of infection.

High incidence of oval leaf spot, zonate leaf spot, and leaf blight occurs but not at severe levels, in many farmers' fields in the Northern Guinean Zone and the Sudanian Zone. Bacterial stripe has been observed in many local genotypes and is mostly restricted to the lower leaves.

Panicle diseases

Grain molds are rare in farmers' fields. Most local sorghum varieties in normal years develop the grain after the rains cease. This enables the plant to produce grain virtually free from seed molds (Zummo 1984). Head smut, long smut, and covered smut occur at low levels. Head smut is, however, more common.

Parasitic weeds

Striga hermonthica occurs abundantly in farmers' fields in all the agroecological regions of western Africa. Some genotypes do not show typical *Striga* symptoms, even under heavy infection.

Diseases Prevalent in Introduced Sorghums

Leaf diseases

Gray leaf spot and leaf anthracnose are the more prevalent leaf diseases in introduced genotypes in the northern Guinean Zone and in very wet areas of the Sudanian Zone. In the absence of conclusive data from western Africa, it is diffi-

cult to comment on the effect of these two leaf diseases on grain yield. Experiments show that in highly susceptible genotypes (eg., IS 18696, ICSV 20-1 BF, and ICSV 16-3 BF) grown under favorable conditions for disease development, first symptoms of gray leaf spot can appear anytime within 10 days after infection. The disease progresses slowly, at about 20 to 30 days after flowering. Infection usually affects less than 50% of the leaf area in the top four leaves. The progress of the disease increases slightly after this; at physiological maturity the leaf area infected (LAI) usually exceeds 75% in any of the top four leaves. In many other susceptible genotypes, symptoms appear even later and disease progress is even slower (unpublished data 1987). This pattern of disease progress implies that gray leaf spot may not cause serious loss in yield. Odvody (1986) suggested that the economic impact of gray leaf spot is hard to assess, because epidemic conditions only develop near crop maturity.

With leaf anthracnose, the situation is completely different and perhaps less ambiguous. In susceptible genotypes—IS 18442, IS 2139, IS 4585, or IS 1552—leaf anthracnose symptoms appear early and the disease progresses rapidly and kills top leaves during grain formation. The same is true for IS 18696 growing under conditions favorable for leaf anthracnose. For example, LAI in the top four leaves for IS 2139 and IS 1552 was between 26% and more than 75% at flowering. Within 3 weeks after flowering, the four top leaves were dead. The same general pattern of infection and disease progress was evident in IS 18696 and IS 18442 (unpublished data 1987). This supports the view that anthracnose is a major threat to sorghum production in some areas of western Africa, because of the rapid rate at which the anthracnose pathogen can damage sorghum plants approaching maturity (Frederiksen 1984).

When conditions are favorable, damage from sooty stripe can be serious. More than 50% LAI in the top four leaves may occur between flowering and maturity. Dry conditions in the Sudanian Zone appear to favor development of sooty stripe. However, preliminary observations at Farako-Bâ and Kamboinse in Burkina Faso and at Cinzana in Mali suggest that sustained high humidity may be necessary for successful infection and subsequent initiation of symptoms during early stages of the disease. This might be related to the abundant formation of conidia when conditions are warm and moist. (Bandyopadhyay 1986).

Panicle diseases

Grain molds and weathering effects are major problems with introduced genotypes maturing during relatively high rainfall. Long smut is more severe in the Sudanian and southern Sahelian zones.

Parasitic weeds

As with some locals, many introduced sorghums are susceptible to *Striga*. Identifying unequivocally stable lines resistant to *Striga* is a complex process.

Research efforts

Study of the development of gray leaf spot, sooty stripe, and leaf anthracnose in various introduced genotypes and the effect on grain yield is now underway. The work on sooty stripe and leaf anthracnose is in collaboration with the National Program of Burkina Faso. More basic work on the biology of the fungi and the epidemiology of these diseases is planned. The major objective is to understand these diseases in order to develop effective screening techniques. Effort is being made to identify sources of resistance to these diseases through a regional nursery. Preliminary inheritance studies suggest that susceptible reaction to gray leaf spot and resistance to leaf anthracnose are dominant (Murty and Thomas, unpublished data 1987).

ICRISAT's sprinkler irrigation method for grain mold screening has been successfully tested for 2 years at Farako-Bâ, Burkina Faso. This technique will now be used to routinely screen genotypes for grain mold in western Africa.

Soil shading with a nonerect spreading groundnut line (59–426) has proved successful in reducing *Striga* in several of our experiments conducted in farmers' fields in 1986 and 1987. Differences in soil temperature and moisture between shaded and unshaded plots were negligible (Thomas, M.D., unpublished data 1986,

1987). Work will continue on the mechanism of *Striga* suppression, using crops in addition to groundnut.

In a limited survey of five localities, the National Program of Burkina Faso has determined that the most common species of fungus found in sorghum seeds is *Colletotrichum graminicola*, the anthracnose fungus. Many *Curvularia* species were detected in all parts of the grains, including the embryo (Kabore and Couture 1983). Using the toothpick method, Neya, and Kabore (1987) observed between 7, 9, and 46% loss in grain yield of Gnofing, a Burkina Faso local variety. They showed a positive correlation between leaf and stem anthracnose.

In Senegal, Louvel and Arnaud (unpublished data) suggested that grain mold resulted in poor germination of seeds and poor seedling vigor. *Fusarium* spp and *Curvularia* spp were the dominant fungi in the molded grains they studied.

Acknowledgment. I wish to thank Y. Paco and A. Neya of the National Program in Burkina Faso for their help in our studies on sooty stripe and leaf anthracnose, respectively.

References

Bandyopadhyay, R. 1986. Sooty stripe. Pages 13–14 *in* Compendium of sorghum diseases (Frederiksen, R.A., ed). St. Paul, MN 55121, USA: The American Phytopathological Society.

Frederiksen, R.A. 1984. Disease problems in sorghum. Pages 263–271 *in* Sorghum in the Eighties: proceedings of the International Symposium on Sorghum, 2–7 Nov 1981, ICRISAT Center, India. Patancheru, Andhra Pradash 502 324, India: International Crops Research Institute for the Semi-Arid Tropics.

Kabore, K.B., and **Couture, L.** 1983. Mycoflora des semences du sorgho cultivé en Haute-Volta. (In Fr. Summary in Eng). Le Naturaliste Canadien 110: 453–457.

Neya, A., and **Kabore, K.B.** 1987. Mesure de l'incidence de l'anthracnose et de la pourriture rouge des tiges causés par le *Colletotrichum graminicola* chez le sorgho. (In Fr. Summary in Eng.). Phytoprotection 68: 121–126.

Odvody, G.N. 1986. Gray leaf spot. Pages 11–12 *in* Compendium of sorghum diseases (Frederiksen, R.A. ed). St. Paul, MN 55121, USA: The American Phytopathological Society.

Tyagi, P.D. 1980. Sorghum diseases in Nigeria. Pages 45–52 *in* Sorghum diseases, a world review: proceedings of the International Workshop on Sorghum Diseases, 11–15 Dec. 1978, Hyderabad, India. Patancheru, Andhra Pradesh 502 324, India: International Crops Research Institute for the Semi-Arid Tropics. 469 pp.

Zummo, N. 1984. Sorghum diseases in western Africa. Washington DC 20250, USA: USDA, US-AID. 32 pp.

Sorghum and Sorghum Diseases in Japan

Takashi Kimigafukuro[1]

Abstract

Sorghum has been a crop in Japan since the 14th century, when the cereal was introduced from China. It became a food crop in high districts not suited for paddy rice. Following the world war II, sowings dwindled; in 1965 the government dropped sorghum production from its statistical records. With the development of Japan's livestock industry, sorghum became a valued animal feed and production records were reinitiated in 1973, when 1 334 000 t of forage were produced on 19 300 ha. By 1985 hectarage had nearly doubled, with 35 000 ha producing 2 385 000 t.

Sugarcane mosaic and bacterial leaf stripe (Pseudomonas andropogonis) have been recorded on sorghum in Japan. Most numerous are the fungal diseases, including two that may be unknown in other sorghum-growing areas. Hayake-byou (Kabatiella sorghi) occurs in most parts of Japan; yellow spot (Cercospora koepkei var sorghi) was first reported in Japan in 1940. A major thrust of the sorghum improvement program in Japan is varietal resistance to banded leaf and sheath blight, (Rhizoctonia solani), as this disease is common in Japan and is of the same R. solani group responsible for important diseases of rice and corn in the country.

Introduction

Sorghum was probably introduced into Japan from China in the 14th century. Since then, the cereal has been cultivated as grain sorghum for food in mountain districts where paddy rice could not be cultivated. However, the cultivation of sorghum decreased considerably after World War II, a consequence of the stabilization of Japan's food supply. In 1965, sorghum statistics were no longer listed in government statistical records.

With the development of a livestock industry in Japan around 1960, forage sorghum has attracted attention as an animal feed, because of its wide range of adaptability to various environmental conditions, high yields, suitability for mechanization, and lodging resistance. Hectarage sown to sorghum reappeared in the statistical records relating to feed crops in 1973. The hectarage sown to sorghum increased rapidly until 1981, but has remained at the same level since 1982 (Table 1).

In Japan, sorghum competes with maize as a summer feed crop and production is increasing in the southwestern districts of Japan. In this region sorghum is one of the best options for the farmer, as it is more resistant to drought, heat, and lodging than is maize. The cultivation of sorghum has not become a practice in all regions of Japan, as sorghum does not grow as well under low temperature as does maize. Furthermore the digestibility and palatability of silage made of sorghum are inferior to those of maize silage. Sorghum, with its high yielding ability, is the better forage crop when conditions are suitable. To expand the area sown to sorghum, it is necessary to develop hybrid varieties which are more adapted to the Japanese climate and to improve methods for preparing sorghum silage.

1. Plant Pathologist, National Grassland Research Institute, Japan.

Kimigafukuro, T. 1992. Sorghum and sorghum diseases in Japan. Pages 31–34 *in* Sorghum and millets diseases: a second world review. (de Milliano, W.A.J., Frederiksen, R.A., and Bengston, G.D., eds). Patancheru, A.P. 502 324, India: International Crops Research Institute for the Semi-Arid Tropics.

Table 1. Forage sorghum production in Japan from 1976 to 1985.[1]

Year	Area ('000 ha)	Production ('000 t)
1976	19.3	1 334
1977	21.3	1 530
1978	27.2	1 847
1979	30.0	2 144
1981	36.9	2 534
1982	37.5	2 496
1983	37.3	2 547
1984	36.2	2 492
1985	35.5	2 385

1. Source: Japan: Ministry of Agriculture, Forestry and Fisheries, (1986).

The sorghum breeding program in Japan was initiated at the Chugoku National Agricultural Experiment Station from 1962 to 1971 and at two government-funded stations set up in Prefectural Agricultural Experiment Stations. At these stations, major emphasis is placed on improving yield, palatability and digestibility, tolerance to low temperatures, and resistance to lodging, bird injury, pests, and diseases.

Diseases of Sorghum in Japan

As in the case of many other crops of which the cultivation has been intensified, sorghum in Japan has recently developed disease problems. The major diseases (Table 2) are leaf blight, anthracnose, zonate leaf spot, target leaf spot, banded leaf and sheath blight, bacterial leaf stripe.

Two of these, yellow spot (Cercospora koepkei) and hayake-byou (Kabatiella sorghi), do not appear in the Compendium of Sorghum Diseases published by the American Pathological Society in 1986. Symptoms and agents of these diseases are outlined as follows:

Hayake-byou

This disease, caused by *Kabatiella sorghi* (Nishihara and Yokoyama 1971), occurs in many areas of Japan.

Symptoms. The disease occurs mainly on leaf blades and rarely on leaf sheath, stems, and heads. The symptoms consist of large numbers of small circular, elliptical, or spindle-shaped spots 5 mm across. Lesions are water-soaked at first, and then become yellowish-brown or brownish-purple with light yellow haloes. When the spots are numerous, the whole leaf becomes brownish-purple and dies prematurely.

Causal organism. The acervuli are indistinct without setae. Hyaline clavate conidiophores, 10–20 by 3–5 μm in size, emerge from the stomata or through the epidermis. The conidia are borne in a cluster at the tip of the conidiophore. The conidia are hyaline, falcate, nonseptate, and from 10–21 by 1.8–3.0 μm in size. The fungus is similar to the corn eyespot fungus [*Kabatiella zeae* Narita and Hiratsuka (1959)], but conidia and conidiophore size and type of parasitism are different.

Yellow spot

This disease of sorghum was first observed in China prefecture near Tokyo in 1940.

Table 2. Sorghum diseases recorded in Japan.

Common name	Causal agent
Viral disease	
Sugarcane mosaic	Sugarcane mosaic virus
Bacterial disease	
Bacterial leaf stripe	*Pseudomonas andropogonis*
Fungal disease	
Anthracnose	*Colletotrichum graminicola*
Banded leaf and sheath blight	*Rhizoctonia solani*
Ergot	*Claviceps* sp
Gray leaf spot	*Cercospora sorghi*
Leaf blight	*Exserohilum turcicum*
Rust	*Puccinia purpurea*
Covered kernel smut	*Sporisorium reilianum*
Loose kernel smut	*Sphacelotheca cruenta*
Head smut	*Sporisorium sorghi*
Sooty stripe	*Ramulispora sorghi*
Southern sclerotial rot	*Sclerotium rolfsii*
Target leaf spot	*Bipolaris cookei*
Zonate leaf spot	*Gloeocercospora sorghi*
Hayake-byou	*Kabatiella sorghi*
Yellow spot	*Cercospora koepkei* var *sorghi*

Symptoms. The disease is characterized by the presence of chlorotic and rectangular spots with a white powdery appearance (conidial formation), especially on the undersurface of the leaf blades. Later, the spots change to light brownish necrotic lesions without conspicuous border coloration.

Causal organism. Conidiophores extrude through stomata singly or in tufts of 2 to about 12. They are geniculate then denticulate towards the apex, 22–80 by 3.6–6.9 μm in size, and with or without septa. Septa may be up to five in number. Conidia are obclavate with an obconic then truncate bottom and attenuated then blunt apex, from 21–55 by 3–6 μm in size, with one to six (mostly three) septa.

The morphology of the fungus is similar to that of the yellow spot fungus (*Cercospora koepkei* Kruger) of sugarcane, in many cases without parasitism. Therefore, the sorghum fungus was designated as *C. koepkei* Kruger var. *sorghi* (Goto et al. 1962).

Banded leaf and sheath blight

Banded leaf and sheath blight, caused by *Rhizoctonia solani*, is a minor disease of sorghum in other countries. However, it is very common in Japan, due to the warm and humid weather that is conducive to its development and spread in summer. It is anticipated that the disease will become more important, as the causal organism belongs to the same strains (anastomosis group 1) of *R. solani* Kühn that cause banded leaf and sheath blight, important diseases of rice and corn. The cultivation of sorghum is expected to expand into areas currently planted to rice and corn.

Resistance to banded leaf and sheath blight is an important objective in the breeding of new sorghum varieties in Japan. At the Nagano Prefecture Chushin Agricultural Experimental Station, studies based on the percentage of plant length infected to total plant length were initiated to estimate the degree of resistance to banded leaf and sheath blight. Estimation of disease incidence by this procedure revealed a highly positive correlation to the yield decrease caused by the disease (Table 3).

The percentage of plant length infected to total plant length in 72 varieties and of sorghum infected artificially with *R. solani* are presented in Table 4.

Table 3. Correlation between the percentage of plant length infected with *Rhizoctonia solani* to total plant length and yield decrease.[1]

Sampling date	Coefficients		
	Foliage	Ear	Total
22 Jul	0.866**	0.881**	0.923**
12 Aug	0.418	0.824**	0.771**
2 Sep	0.533**	0.873**	0.816**

1. Source: Kasuga and Takizawa (1984).

Table 4. Grouping of varieties or strains of sorghum based on the percentage of plant length infected artificially by *Rhizoctonia solani* to total plant length.[1]

Plant length infected (%)	Varieties or strains
<10	(932251)A, Kokkaku 2, M 91034, P 956
10–20	(932233)A, Daikoukaku, F 6-3A-1, F 6-3A-5, F 6-3A-9, Fuji A, IS 10420A, M 36001, M 90306, M 90322, M 90386, M 90394, M 90975, Rancher, Red amber, Rozoku, SPV 354, Zairaisyu
20–80	406 A, (954149)A, BR 48, Darset, DN E7, GS 401, Earlyhalo 926, Gazera 6020, IS 517, IS 2830A, IS 8722, IS 10288, IS 10378, JN 1-1-2, JN 1-1-3, JN 8, JN 9, Kyoukou 382, M 39335, M 90325, Resistant W.A., Rox orange, Sekisyokuzairaisyu, Senkinshiro, Setokou 3, Setokou 5, Suzuho, Tousankou 7, Tousankou 8, Tx 2727, Waikai (1), Westland, Wheatland A
80–90	BJ 12, JN 1-1-1, JN 2-2-3, JN 3, JN 6, IS 3048 A
>90	JN 2-2-1, JN 2-2-2, JN 4, JN 5, JN 7, KS-57A, SD 102

1. Source: Kasuga and Takizawa (1984).

References

Japan: Ministry of Agriculture Forestry and Fisheries. 1986. The 61st Statistical Yearbook of Ministry of Agriculture Forestry and Fisheries of Japan, 1984-85. Statistics and Information Department. 627 pp.

Nishihara, N., and **Yokoyama, T.** 1971. Studies on *Kabatiella sorghi* Nishihara et Yokoyama, the causal fungus of a new leaf spot disease of sorghum. (In Jap.) Annals Phytopathological Society Japan 40:170 (Abstract).

Narita, T., and **Hiratsuka, Y.** 1959. Studies on *Kabatiella zeae* n. sp., the causal fungus of new leaf spot disease of corn. (In Jap., Summaries in En.) Annals Phytopathological Society Japan 24:147-153.

Goto, K., Hirano, K., and **Fukatsu, R.** 1962. Yellow spot of sorghum, a new disease. (In Jap., summaries in En.) Annals Phytopathological Society Japan 27:49–52.

Hashiba, T.,Uchiyamada, H., and **Kimura, K.** 1981. A method to estimate the disease incidence based on the height of the infected part in rice sheath blight disease. (In Jap. summaries in En.) Annals Phytopathological Society Japan 47:194–198.

Kasuga, K., and **Takizawa, Y.** 1984. Studies on the field resistance to sheath blight in sorghum. 1. Varietal differences of effects on growth and yield. (In Jap.) Bulletine Nagano Prefecture Chushin Agricultural Experimental Station 3:64–69.

Diseases of Sorghum in the Philippines

N.G. Tangonan[1] and S.C. Dalmacio[2]

Abstract

Use of sorghum for human food is not yet widespread in the Philippines—rice and corn are the preferred cereals. Efforts to promote sorghum in various food products are beginning, but the major part of the crop goes into livestock feed. Production is not expanding, even though the crop is well suited to Philippine conditions. in general, there has been little change in the disease situation in the past 10 years. Research efforts have intensified; the government has provided legislation for PHILCOSORI (Philippine Corn and Sorghum Research Institute) but not yet funded it. The International Sorghum Multiple Disease Resistance Nursery (120 genotypes) was screened at the University of Southern Mindanao Agricultural Research Center in 1987. Grain mold was the most severe disease noted. The hybrid Pioneer 8258 with high resistance to target and gray leaf spots, and some resistance to Rhizoctonia *disease was released.*

Introduction

In the Philippines, sorghum was introduced almost 20 years ago and is grown and used mainly as grain to supplement the feed requirements of the growing poultry and livestock industries. The crop is well adapted to the agroclimatic conditions of the country. As a source of carbohydrates and protein, sorghum compares with maize in nutritive value.

Rice is the main staple food of Filipinos; many just now are developing a taste for sorghum. Delicacies made from sorghum and/or with sorghum as a main ingredient are taking the form of fortified foods, rice-like products, breakfast, and convenience foods. Notwithstanding, Filipinos prefer rice and then maize before sorghum as a cereal food.

As a feed-grain supplement, sorghum is used more as an alternate or substitute for maize. Lopez noted that commercial feeds contain only 2.8% sorghum (Oliva 1986).

Sorghum Production Trends and Current Situation

Since the introduction of sorghum into the Philippines almost two decades ago, neither hectarage devoted to its production nor yield have increased substantially. Total sorghum production peaked from cropping season 1977/78 to 1978/79 but has since declined (Fig. 1). Ferrer and Almeda (1986) reported that the farmer's average hectarage in sorghum is 1.44; with corresponding yield of 1.94 t ha[-1]. The general annual cropping patterns followed are rice-maize-sorghum or rice-sorghum (Oliva 1986).

Although sorghum has been included in the feed grains and maize production programs of the Philippine government in past years, specific program components and guidelines for its promotion or support were absent (Oliva 1986). The low status of sorghum could be attributed to various problems, such as nonavailability of seeds, very low support prices, and other mar-

1. Professor and Chairman, Department of Plant Pathology, College of Agriculture, University of Southern Mindanao, Kabacan, Cotabato, the Philippines 9407.
2. Plant Pathologist, Pioneer Overseas Corporation, the Philippines, and Visiting Professor, Department of Plant Pathology, University of the Philippines at Los Baños, College, Laguna, the Philippines 4031.

Tongonan, N.G., and Dalmacio, S.C. 1992. Diseases of sorghum in the Philippines. Pages 35–40 *in* Sorghum and millets diseases: a second world review. (de Milliano, W.A.J., Frederiksen, R.A., and Bengston, G.D., eds). Patancheru, A.P. 502 324, India: International Crops Research Institute for the Semi-Arid Tropics.

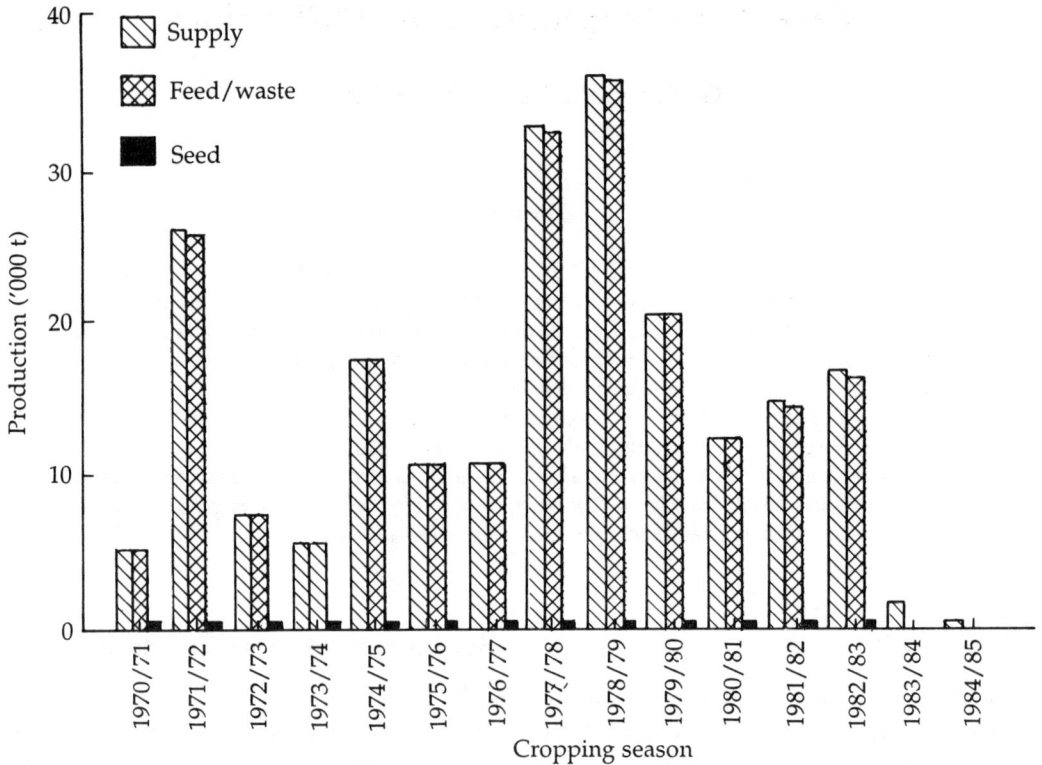

Figure 1. Sorghum supply and use in the Philippines, 1970/71 to 1984/85 cropping seasons. Source: Ferrer and Almeda (1986).

keting problems, absence of financing schemes, damage from pests and diseases, and poor post-harvest technologies. Farmers have expressed willingness, however, to produce sorghum if these problems are remedied (Clarete 1984, Oliva 1986). Growing sorghum as a feed grain will definitely offer great benefits to farmers considering the recent necessity for imports of maize by the Philippine government (Alunan 1988). Sorghum can augment feedmillers' supplies.

PHILCOSORI

On 18 Nov 1987, House Bill No. 3752–also known as the Corn and Sorghum Act of 1987–was filed in Congress (Lower House of representatives) by Congressman R.B. Gutang of the 1st District of Catabato.

The bill provides for the establishment of the Philippine Corn and Sorghum Research Institute (PHILCOSORI), proposed for the University of Southern Mindanao at Kabacan, Cotabato. If and when the bill is approved, PHILCOSORI will ensure the funding for maize and sorghum research and development.

Sorghum Pathology

During the First International Workshop on Sorghum Diseases, at Hyderabad, India, in December 1978, Dalmacio (1980) reported 13 diseases affecting sorghum in the Philippines. In terms of incidence or of new diseases recorded, changes in the sorghum pathology situation in the country are minimal.

The diseases considered as limiting to sorghum production are, as they were in 1978, *Phy-*

Table 1. Diseases of sorghum in the Philippines, season of likely occurrence, relative importance, and effect on yield.[1]

Disease	Season of occurrence[2]	Relative importance[3]	Estimated yield loss[4] (%)
Sugarcane mosaic	both	minor	55
Sugarcane fiji	both	minor	–
Bacterial leaf stripe	both	minor	–
Bacterial stalk rot	both	minor	–
Rust	dry	major	30
Tar spot	both	major	50
Gray leaf spot	wet	major	–
Zonate leaf spot	both	minor	–
Anthracnose/red rot	wet	minor	–
Target leaf spot	wet	major	50
Rhizoctonia	both	major	75
Charcoal rot	dry	minor	–
Head blight	wet	minor	–
Head molds	both	major	–
Seedling blight	wet	minor	–

1. Source: Dalmacio (1982).
2. Maybe applicable in areas where thee are distinct wet and dry seasons. Wet season sowing is April to June. Dry season sowing is October to December.
3. Based on frequency of observation, severity of disease, and potential for rapid spread.
4. From available experimental data.

llachora tar spot, gray leaf spot, head/grain molds, rust, and *Rhizoctonia* banded leaf and sheath blight (Table 1). Recent trends, however, show the emerging importance and potential threats of zonate and target leaf spots as well as stalk rots. The incidence of these diseases have been increasing in alarming proportions on some commercial hybrids and varieties in Southern and Central Mindanao (Exconde 1982).

Other diseases affecting sorghum in the Philippines include sugarcane mosaic (red leaf disease), charcoal and bacterial stalk rots, bacterial stripe, anthracnose, sugarcane fiji disease, and seedling/head blights (Table 1).

Experiments have demonstrated that sugarcane mosaic can reduce yield by as much as 55%. Rust on sorghum in the Philippines may decrease yields by 30%, target leaf spot by 22%, *Rhizoctonia* disease by 75% (Karganilla and Elazegui 1973, Benigno and Vergara 1977), and tar spot by 50%. During the 1978 wet season, severe incidence of target leaf spot in the disease nursery at Los Baños Central Experiment Station reduced yield by 50%; in a contemporary trial, under shade, tar spot prevented measurable yields (Dalmacio 1980).

Screening for resistance to sorghum diseases has been carried out by Dalmacio and his co-workers since 1978. Sources of resistance to rust and target leaf spot (201 and 198 lines, respectively) seem to be more common (Table 2), as compared to those of gray leaf spot (27 resistant

Table 2. Summary of results on the screening for resistance to six sorghum diseases in the Philippines.[1]

Disease	Line/variety Tested (no.)	Line/variety Resistant (no.)
Gray leaf spot	2525	27
Rhizoctonia sheath blight	88	0
Rust	555	201
Tar spot	2268	51[2]
Target leaf spot	2484	198
Head molds	50	0

1. Source: Dalmacio et al. (1981).
2. Includes two resistant lines (with a rating of 1) and 49 moderately resistant lines (with a rating of 2).

lines) and tar spot (51 resistant lines). Not one of the lines or varieties was completely free of gray leaf spot or tar spot, but expression of symptoms is delayed in resistant plants and fewer lesions appear. However, when 88 sorghum lines were inoculated with *Rhizoctonia solani* all of the lines were susceptible. Similarly, 50 lines tested for resistance to head molds showed infection (Dalmacio et al. 1981, Dalmacio 1982). Sources of resistance to most of the important sorghum diseases have been identified.

Furthermore, Paderes and Dalmacio (1984) examined resistance to tar spot by evaluating components of rate-reducing resistance to *Phyllachora sorghi* (Syn *P. sacchari*). The findings reveal that resistance prolonged the latent period (LP) by 12%, shortened the infectious period (IP) by 69%, reduced stomata size (SS) by 85%, and reduced sporulation capacity (SC) by 93%. Among these components, IP, SS, SC, and infection frequency (IF) may be considered more important than LP. Correlation analysis showed that LP, SS, and SC were more strongly associated with each other than with IP.

Screening work of 120 sorghum genotypes of the International Sorghum Multiple Disease Resistance Nursery (ISMDRN) established at the Southern Mindanao Agricultural Research Center (SMARC) showed mean severity ratings ranging from 1.3 for zonate leaf spot and *Rhizoctonia* disease to 4.5 for grain molds. (Table 3).

Table 3. Disease reactions of 120 ISMDRN'87 sorghum entries screened at SMARC-USM, Tangonan, Kabacan, Cotabato, the Philippines, 1987.

Disease	Mean severity ratings[1]
Zonate leaf spot	1.3
Gray leaf spot	1.9
Target leaf spot	1.7
Rhizoctonia banded leaf and sheath blight	1.3
Anthracnose	1.5
Tar spot	3.5
Grain mold	4.5
Root/stalk rots	1.8

1. Scored on a 1-5 scale, where 1 = no symptoms and 5 = severe infection.

Phyllachora tar spot and head/grain mold were severe and sorghum hybrids and cultivars tested were susceptible to these two diseases. Other foliar diseases had ratings ranging from 1.3 to 1.9, and the sorghum genotypes were resistant to most foliar diseases except tar spot. Head/grain mold infection had a mean severity of 4.5 (Table 3).

The incidence of a stalk rot complex in sorghum fields at SMARC-USM was also observed to be infecting most hybrids and cultivars tested. It was believed to have been caused by the combined effect of *Colletotrichum graminicola, Fusarium moniliforme,* and *Rhizoctonia solani.*

An interesting development was the release of a new sorghum hybrid by Pioneer Overseas Corporation, Philippines. This hybrid, Pioneer 8258, was noted to have excellent resistance to target and gray leaf spots and some degree of resistance to *Rhizoctonia* disease, although *Rhizoctonia* resistance seems to be highly influenced by plant height.

Research Needs

During the First National Symposium and Workshop on Corn and Sorghum Crop Protection, held at the University of Southern Mindanao, 17–20 Mar 1982 [sponsored by the Philippine Council for Agriculture and Resources Research and Development (PCARRD) and the Ministry of Agriculture and Food (now the Department of Agriculture)], Dalmacio and Exconde listed the Philippine research needs in sorghum pathology, as follows:
1. Monitor disease occurrence on sorghum sowings during the dry season and determine the relative importance.
2. Search for sources of resistance to various diseases and develop resistant varieties.
3. Establish disease nurseries in various sorghum-growing areas.

While yield loss has been demonstrated in field experiments, it is necessary to determine the precise yield losses in cultivars and hybrids, at different disease intensities, during each cropping or sowing season (Exconde 1982).

Table 4 summarizes the research areas and current status of sorghum disease management for 13 sorghum diseases in the Philippines. Much remains to be studied in the areas of yield-loss assessment, epidemiology, host-pathogen

Table 4. Areas with research potentials and current status of sorghum disease management in the Philippines.[1]

Research area	Root and stalk rot — Bacterial	Root and stalk rot — Fungal	ANT	BLS	GLS	GM	HB	RUS	RZD	SB	SFD	SM	TLS	TS	ZLS
Etiology	/	/	/	/	/	/	/	/	/	/	/	/	/	/	/
Assessment of yield loss															
Farmers' fields	x	x	x	x	x	x	x	x	x	x	x	x	x	x	
Experimental	x	x	x	x	x	x	x	/	/	x	x	/	/	/	x
Epidemiology	x	0	x	x	x	x	x	x	x	x	x	x	x	x	x
Host-pathogen interaction	x	0	x	x	x	x	x	/	x	x	x	x	x	/	x
Strains/races	x	x	x	x	x	x	x	x	x	x	x	x	x	x	x
Host range	x	x	x	x	x	x	x	x	x	x	x	x	x	x	x
Control															
Host resistance	x	x	0	x	0	0	x	0	0	x	x	x	0	0	0
Chemical	x	0	x	x	x	x	x	x	x	x	x	x	x	/	x
Cultural	x	/	x	x	x	x	x	x	x	x	x	x	x	x	x
Biological	x	x	x	x	x	x	x	x	x	x	x	x	x	x	x

1. x = Studies needed research gaps; / = studied; 0 = studies ongoing. ANT = anthracnose/red rot; BLS = bacterial leaf stripe; GLS = gray leaf spot; HB = head blight; GM = grain mold; RUS = rust; RZD = *Rhizoctonia* disease; SB = seedling blight; SFD = sugarcane fiji disease; SM = sugarcane mosaic; TLS = target leaf spot; TS = tar spot; ZLS = zonate leaf spot.

interactions, strains or physiological races, host range, and disease management and control.

Acknowledgment. We are deeply indebted to Pioneer Overseas Corporation, Philippines, for much-needed materials and assistance.

References

Alunan, M.M. 1988. NFA (National Food Authority) okays another importation of corn. Business Bulletin, 14 Jan 1988. p. 21.

Benigno, D.R.A., and Vergara, D.C. 1977. Red stripe disease of sorghum in the Philippines. Philippine Agriculturist 61:157–165.

Clarete, C.L. 1984. Status of sorghum production technology and marketing in Mindanao (Terminal report) Kabacan, Cotabato, Philippines: Southern Mindanao Agricultural Research Center, University of Southern Mindanao. 100 pp.

Dalmacio, S.C. 1980. Sorghum diseases in the Philippines. Pages 70–71 *in* Sorghum diseases, a world review: proceedings of the International Workshop on Sorghum Diseases, 11–15 Dec 1978, Hyderabad, India. Patancheru, Andhra Pradesh 502 324, India: International Crops Research Institute for the Semi-Arid Tropics. 469 pp.

Dalmacio, S.C. 1982. Breeding for disease and insect resistance in sorghum. SMARC Monitor 3(1):21–26.

Dalmacio, S.C., Dayan, M.P., and Pascual, C.B. 1981. Identification of sources of resistance to some major diseases of sorghum in the Philippines. Philippine Phytopathology 17:38–46.

Exconde, O.R. 1982. Disease control in corn and sorghum. SMARC Monitor 3(1):29–32.

Ferrer, R.M., and Almeda, J.P. 1986. Economic analysis of sorghum production, 1984. Economic Research Report No. 28 (Dec1986). Quezon City, Philippines: Bureau of Agricultural Economics. 30 pp.

Karganilla, A.D., and Elazegui, F.A. 1973. Sorghum diseases. Upland Crops Annual Reports 1971–73, Philippines. Los Baños, Laguna, Philippines: University of the Philippines at Los Baños.

Oliva, L.P. 1986. On-farm research on sorghum in Southern Mindanao (Terminal report). Kabacan, Cotabato, Philippines: Southern Mindanao Agricultural Research Center, University of Southern Mindanao. 75 pp.

Paderes, E.P., and Dalmacio, S.C. 1984. Components of rate-reducing resistance in some sorghum lines to *Phyllachora sorghi*, the cause of sorghum tar spot. Philippine Agriculturist 67(3): 279–287.

Sorghum Diseases in Thailand

T. Boon-Long[1]

Abstract

Sorghum is cultivated in Thailand as a second crop or mixed with other crops. In 1987, it occupied about 193 000 ha with a production of 211 258 tonnes. Diseases include grain mold, ergot, gray leaf spot, and other leaf diseases. Leaf disease is less important than grain disease, because diseases usually appear late, after the grain has filled. The downy mildew of maize in Thailand does not infect sorghum.

Grain mold disease resistance has been tested at Suwan Farm and Supanbusi research center. The prevalent organisms are Curvularia lunata, Colletotrichum graminicola, C. gloeosporoides, Fusarium spp, and Phoma sp. Ergot in Thailand was first identified in 1983, and spread on male-sterile sorghums used for hybrid production in 1984. The hosts are Sorghum vulgare, S. almum, S. sudanensis, S. halepense, Dichanthium annulatum, Brachiatia mutica, and Panicum maximum.

Introduction

Sorghum is third of the important cereal crops in Thailand; in 1987, about 193 000 ha produced about 211 million kg (Table 1) (Thailand: Ministry of Agriculture and Cooperatives 1987). Sorghum is grown in scattered locations throughout the central plain and northeastern Thailand. The important sorghum-growing provinces are Nakornsawan, Lopburi, Saraburi, and Nakornrajseema. It is cultivated mostly after corn as a second crop, or mixed with other crops. In 1987, the average yield was about 1.15 t ha^{-1} in Thailand. Yields vary from year to year; the best yield (2.5 t ha^{-1}) was harvested in 1965 (Pupipat 1980) and the annual yield over the last decade is just a bit more than 1.08 t ha^{-1}. Eighty percent of the production is exported; the remainder is used for animal feed in the country. Rainfall in most of the sorghum-growing area ranges from 900–1600 mm, with good distribution from May through October. Mean, maximum, and minimum temperatures are 28 to 29, 40, and 19°C,

Table 1. Sorghum production in Thailand, 1978–87.[1]

Year	Area sown ('000 ha)	Area harvested ('000 has)	Production ('000 t)	Yield (t ha^{-1})
1978	169.9	150.6	126	0.84
1979	175.7	150.7	216	1.40
1980	189.1	160.6	199	1.20
1981	247.4	234.1	237	1.00
1982	279.8	266.1	274	1.00
1983	245.4	236.2	236	1.00
1984	265.1	250.9	327	1.30
1985	294.1	283.7	374	1.30
1986	309.6	291.7	404	1.40
1987	193.9	183.7	211	1.20

1. Source: Thailand: Ministry of Agriculture and Cooperatives (1987).

1. Sorghum Pathologist, Plant Pathology and Microbiology Division, Thailand Department of Agriculture, Bangkok 10900, Thailand.

Boon-Long, T. 1992. Sorghum diseases in Thailand. Pages 41–43 *in* Sorghum and millets diseases: a second world review. (de Milliano, W.A.J., Frederiksen, R.A., and Bengston, G.D., eds). Patancheru, A.P. 502 324, India: International Crops Research Institute for the Semi-Arid Tropics.

respectively. Elevation above mean sea level is between 100 and 150 m.

Three institutions—Kasetsart University located in Bangkok, Khon, Kaen University located in the Northeast, and Thailand's Department of Agriculture (with field-crop stations all over the country)—are now conducting research in sorghum improvement.

Farmers are using the cultivars Early and Late Hegari, KU 439, U-thong I (or DA 80), and the recently released Department of Agriculture's Supanburi 60, an open-pollinated release with red pigment. Hybrid varieties are sold by private sectors, mostly to the higher income farmers or big farms. The demands for hybrids will continue as long as the higher price of sorghum grain prevails.

Disease has the potential of being one of the factors limiting sorghum production in Thailand (Table 2). Grain diseases, such as grain mold and ergot, can effect quality and yield. Leaf diseases are considered less important than grain disease in Thailand mostly because they occur after seed-set and the grain has already filled.

Many sorghum diseases have been recorded in Thailand (Pupipat 1980). These diseases are still occurring, and spread occasionally if the weather favors pathogen development. However, severe yield losses seldom occur, perhaps due to the cropping system; farmers grow sorghum following maize, and maize downy mildew does not infect sorghum in Thailand. Sorghum is also sown as a mixed crop with cotton, beans, and sesame; these species are not alternate hosts to sorghum pathogens

Research on Sorghum Diseases

Grain mold

Grain mold on sorghum has been studied by Kasetsart University and the Plant Pathology and Microbiology Division in Thailand's Department of Agriculture. Disease resistance was tested at Suwan Farm. Diseased grains were isolated and the causal fungi found to be from 13 genera. The most prevalent are *Curvularia lunata*, *Fusarium* spp, *Colletotrichum graminicola*, *C. gloeosporoides*, and *Phoma* sp. Of these, only *Curvularia lunata* and *Fusarium* spp appeared in Pupipat's listing (1980), which reported *Aspergillus* spp as well.

Aspergillus is a strange mold. Mutation was induced (using gamma radiation at 35 and 40 Kr), but disease resistance has not yet been found. The project continues.

Rust

The screening technique for selecting resistance to rust has been to inoculate at 20–22°C and 95% RH, including a trace amount of polysorbate 20 in the spore suspension. Infection on sorghum seedlings takes place uniformly within 10 days. Urediospores were viable 30 days after collection from the field.

Ergot

Sorghum ergot, first found in Thailand in 1983 on Sudax (Boon-Long 1983), was identified as *Sphacelia sorghi* (Sudax is a commercial pasture crop developed from crosses between selected male-sterile sorghum and sudan grasses). In

Table 2. Sorghum diseases found in Thailand.

Disease	Pathogen	Disease prevalence[1]
Grain mold	*Curvularia lunata*	1
	Fusarium spp	2
	Colletotrichum graminicola	1
	C. gloeosporoides	2
	Phoma sp	2
Ergot	*Sphacelia sorghi*	1
Rust	*Puccinia purpurea*	2
Gray leaf spot	*Cercospora sorghi*	1
Anthracnose	*C. graminicola*	2
	C. gloeosporoides	2
Zonate leaf spot	*Gloeocercospora sorghi*	3
Charcoal rot	*Macrophomina phaseolina*	3
Leaf blight	*Exerohilum turcicum*	2
Tar spot	*Phyllachora sorghi*	3

1. Prevalence: 3 >2 >1

1984, the disease spread to male-sterile sorghum lines used to develop hybrid seeds in Saraburi Province (the central plain area) (Boon-Long et al. 1984). Male-sterile sorghum was quickly infected, and the infection developed rapidly. Currently ergot is the most damaging disease of male-sterile sorghum. Open-pollinated lines do not become infected, however. Chemical control was tested and lines were screened for resistance, but without success. The ergot hosts were tested by doing Koch's postulates and cross inoculations. These hosts are *Sorghum vulgare*, *S. almum*, *S. sudanensis*, *S. halepense*, *Dichanthium annulatum*, *Brachiaria mutica*, and *Panicum maximum*.

Tar spot

Tar spot, endemic in October and November of 1984, caused no yield loss. Dissected infected leaf tissue (parafin method), stained with safranin and fast green, showed that *Phyllachora sacchari* P. Henn infected only the palisade cells of leaves and did not infect cells of the vascular bundle.

Future Plans and Needs for Collaboration

Thailand's research in sorghum diseases will concentrate on grain mold and ergot. Resistance nurseries for grain mold will be conducted at Supanburi Agricultural Research Station and disease-resistant materials (from ICRISAT and other parts of the world) will be needed for these trials. Research on resistance to ergot is also of high priority. Thailand is a tropical country, and its climate favors anthracnose development, so we shall have to be careful about this problem, even though severe damage by anthracnose is not yet occurring.

References

Boon-Long, T. 1983. Ergot of Sudan in Thailand. Journal of the Thai Phytopathological Society 4:219–221.

Boon-Long, T., Glinsukorn Thirayudh, and Chandrasrikul, Anong 1984. Ergot - A new disease of sorghum in Thailand. Agricultural Research Journal 2:146–150.

Pupipat, U. 1980. Sorghum disease in Thailand. Pages 72–73 *in* Sorghum diseases, a world review: proceedings of the International Workshop on Sorghum Diseases, 11–15 Dec 1978, Hyderabad, India. Patancheru, Andhra Pradesh 502 324, India: International Crops Research Institute for the Semi-Arid Tropics.

Thailand: Ministry of Agriculture and Cooperatives. 1987. Agricultural Statistics of Thailand, Crop Year 1986/87. 60–65 pp.

Sorghum Diseases in India:
Knowledge and Research Needs

K.H. Anahosur[1]

Abstract

India has the world's largest area devoted to sorghum cultivation. FAO reports world average yield to be some 1.5 t ha⁻¹ while the average yield in India is reported to be only about two-thirds t ha⁻¹. Most of India's sorghum is found in eight states, where it is cultivated in rainy, postrainy, and summer seasons. Of these, the rainy season is the most productive.

Sorghum-disease research in India has been underway, under the sponsorship of the All-India Coordinated Sorghum Improvement Project (AICSIP), since 1970. Regional distribution of diseases has been studied and reported, and responsibility for appropriate research assigned to AICSIP research installations in seven states. Dharwad Centre in Karnataka is the major AICSIP installation. Screening for disease-resistant material and development of resistant lines is a common activity for all diseases. Crop-loss estimation, chemical control, and control by crop management are studied.

Comparisons of chemical treatments for controlling grain molds, sorghum downy mildew, ergot, and rust are reported. Multiple-disease and insect resistance is now an AICSIP goal, and several lines resistant to several diseases have been identified. Of these, no line is more than moderately resistant to ergot.

Introduction

Sorghum is an important grain and fodder crop, ranking fourth after paddy, wheat, and maize in the world. The grain is used as staple food by humans in several forms—baked, cooked, or fried. Fodder is used as cattle feed. The cereal is grown in tropical and subtropical climates. Worldwide, it is cultivated on 50 191 million hectares, producing 77 452 million t—an average yield of 1.54 t ha⁻¹. India, with the world's largest sorghum area, sows 15.3 million ha and harvests 10.3 million t, an average yield of 0.673 t ha⁻¹. In India, sorghum is grown where rainfall ranges from 500 to 1000 mm and temperatures from 26 to 32°C. In some states, it is grown in rainy (*kharif*), postrainy (*rabi*), and summer seasons. It is cultivated in the states of Maharashtra, Karnataka, Madhya Pradesh, Andhra Pradesh, Rajasthan, Gujarat, Tamil Nadu, and Uttar Pradesh. More than 90% of India's sorghum production is in these states.

Disease Distribution

Sorghum is subject to several destructive diseases that reduce grain and stover yields. Some diseases are major and others are minor; some are location-specific. Sundaram (1977) and Ravindranath (1980) have reviewed sorghum pathology research in India. The distribution in India of sorghum diseases and their causal organism is presented in Table 1. Grain molds, charcoal rot, fusarium root and stalk rot, ergot, downy mildew, and rust are most destructive and research work on the control of these diseases continues under the sponsorship of the All

1. Professor and Head, Department of Plant Pathology, University of Agricultural Sciences, Hubli, Dharwad, Karnataka 580 005, India, and Principal Investigator (Sorghum Pathology), All India Coordinated Sorghum Improvement Project.

Anahosur, K.H. 1992. Sorghum diseases in India: knowledge and research needs. Pages 45–56 *in* Sorghum and millets diseases: a second world review. (de Milliano, W.A.J., Frederiksen, R.A., and Bengston, G.D., eds). Patancheru, A.P. 502 324, India: International Crops Research Institute for the Semi-Arid Tropics.

Table 1. Regional distribution of sorghum diseases in India.

Disease	Causal organism	Occurrence
Grain molds	*Fusarium, Curvularia* spp	Parts of Andhra Pradesh, northern Karnataka (Belgaum, Dharwad, and Shimoga).
	Phoma, Alternaria,	Marathawada in Maharashtra.
	Exserohilum, Gonatobotrytis	Coimbatore, North Arcot, Salem districts of Tamil Nadu.
	Trichothecium, and others	Indore of Madhya Pradesh; Gujarat.
Charcoal rot	*Macrophomina phaseolina*	Northern parts of Karnataka; Marathawada districts of Maharashtra, parts of Gujarat and Andhra Pradesh.
Fusarium root and stalk rot	*Fusarium moniliforme*	Northern parts of Karnataka; parts of Marathawada in Maharashtra and Andhra Pradesh
Ergot	*Claviceps sorghi*	Maharashtra, northern Karnataka, and Mysore; parts of Andhra Pradesh, Tamil Nadu, Gujarat, and Madhya Pradesh.
Downy mildew	*Peronosclerospora sorghi*	Parts of Karnataka, Tamil Nadu, Maharashtra, and Andhra Pradesh, sporadic in Madhya Pradesh and Uttar Pradesh.
Rust	*Puccinia purpurea*	All sorghum-growing regions of India.
Zonate spot	*Gloeocercospora sorghi*	All sorghum-growing regions of India.
Anthracnose	*Colletotrichum graminicola*	Parts of Karnataka, Madhya Pradesh, Uttar Pradesh, Andhra Pradesh, Tamil Nadu, and Rajasthan.
Sooty stripe	*Ramulispora sorghi*	Parts of Maharashtra, Andhra Pradesh, Rajasthan, and Karnataka.
Rough leaf spot	*Ascochyta sorghina*	Parts of Andhra Pradesh and Madhya Pradesh.
Gray leaf spot	*Cercospora sorghi*	Parts of Karnataka, Madhya Pradesh, Rajasthan, Uttar Pradesh, and Andhra Pradesh
Sorghum red stripe	Virus (SCMV)	In parts of Maharashtra, Karnataka, and Andhra Pradesh.
Grain smut	*Sporisorium sorghi*	Sporadic in different regions.
Loose smut	*Sphacelotheca cruenta*	Parts of Maharashtra and Karnataka.
Head smut	*Sporisorium reilianum*	Parts of Maharashtra, Karnataka, Andhra Pradesh, and Tamil Nadu.
Long smut	*Tolyposporium ehrenbergii*	Occasionally sporadic in Rajasthan, Tamil Nadu, Karnataka, Andhra Pradesh, and Indore.
Tar spot	*Phyllachora sacchari*	Parts of Karnataka.
Wilt	*Acremonium strictum*	Parts of northern India.
Leaf blight	*Setosphaeria turcica*	Parts of Madhya Pradesh, Rajasthan, Tamil Nadu, Karnataka, and Andhra Pradesh.
Crazy top	*Sclerophthora macrospora*	One report in Maharashtra.

India Coordinated Sorghum Improvement Project (AICSIP) at 10 centers. This project began in 1970 as part of Indian Council of Agricultural Research (ICAR). It was upgraded to "National Research Centre", guided by a project director. All Indian states participate through their agricultural universities. It is a multidisciplinary-multilocational program in which pathologists actively contribute (ICAR 1987). The Centers pursue work on particular groups of diseases in line with the diseases' importance (Table 2).

Present Knowledge

Grain molds

The principal grain mold fungi in India are *Fusarium moniliforme, Curvularia lunata, Phoma sorghina, Alternaria alternata, Exserohilum speciferum, Gonatobotrytis* sp, and *Aspergillus* spp. To identify the resistance to these fungi, three screening methods were tested:

Table 2. Research problems studied at the All India Coordinated Sorghum Improvement Project (AICSIP) centers, India.

AICSIP Center	Major emphasis
Dharwad (Karnataka)	Sorghum downy mildew, grain molds, charcoal rot, fusarium stalk rot, ergot, rust, testing and breeding for resistance, inoculation techniques, crop-loss estimates, control by chemicals and agronomic practices, genetics and mechanism of resistance, multiple-disease resistance.
Parbhani (Maharashtra)	Charcoal rot, leaf spots, red stripe, grain molds, screening and breeding for resistance, red stripe control.
Indore (Madhya Pradesh)	Leaf spot, grain molds, screening for individual leaf spots and identifying multiple leaf spot resistant lines, breeding for resistance to leaf spots, testing and breeding for resistance to grain molds
Udaipur (Rajasthan)	Screening and breeding for resistance to leaf spot diseases, studies on *Phoma sorghina* and long smut, estimation of loss due to individual diseases.
Coimbatore (Tamil Nadu)	Screening and breeding for resistance to SDM, control of SDM, grain molds, screening for resistance to ergot, control of ergot.
Palyam (Andhra Pradesh)	Screening for resistance to grain molds, control of grain molds.
Akola (Maharashtra)	Screening and breeding for resistance to ergot, screening and control of grain molds.
Navsari (Gujarat)	Screening for leaf-spot resistance, screening and breeding for resistance to head molds, screening for resistance and control of charcoal rot, and ergot.
Bijapur (Karnataka)	Screening and breeding for resistance to charcoal rot.
Rahuri (Maharashtra)	Screening and breeding for resistance to charcoal rot, control of charcoal rot.

1. Spray-inoculation of earheads with spore suspension of *Fusarium*, *Curvularia*, and *Phoma*; cover with paper bag.
2. Spray earheads with water; cover with paper bag.
3. Leave earheads exposed to natural infection during the rainy season.

Analysis of data showed these three methods to be on par with each other.

Extensive testing of breeder's material and germplasm, identified the following as resistant on the basis of less reduction in field grade, less reduction in threshed grain grade, maximum germination (90–95%), and no loss in 100-mass. The lines are SPV-126, 312, 346, 351, 386, 472, and 2219 B; DMS 1 B; MR 750 and 849; IS-3443, 3547, 14322, 10892, 14375, 14380, 24995, and 24996. Some of these (IS-14375, 14380, 24995, 24996, and 10892) are brown sorghums and were completely free from molds. The lines IS-3443, 24995, and 24996 have been utilized in developing white-grained high-yielding sorghums, and the developing generations were tested at Dharwad.

Efforts were made to identify effective fungicides for grain mold control. It was found that a combination of captan (0.2%) + aureofungin (200 ppm) was best (Table 3), followed by mancozeb (0.27) + captan (0.2%), and then thiram (0.2%) + carbendazim (0.5%).

Charcoal rot

Extensive isolations from roots and stalks of diseased plants collected during the rainy and post-rainy seasons yielded *Macrophomina phaseolina*, *Fusarium moniliforme*, and other *Fusarium* spp, separately and in mixtures. Most of the high-yielding varieties and hybrids presently cultivated in India are highly susceptible. Grain-yield losses have been reported to range from 23 to 64% (Mughogho and Pande 1983, pp. 11–24) and 15 to 55% (Anahosur and Patil 1983a). Plants receiving 80 kg ha[-1] or more nitrogen fertilization showed more disease than those receiving from 40 to 80 kg ha[-1] or those receiving less than 40 kg ha[-1] (Anahosur et al. 1977). Further, Avadhani et al. (1979), and Mote and Ramshe (1980) reported similar results, i.e., excess nitrogen predisposes the plants to susceptibility. Effect of sowing date on charcoal rot incidence in the rainy and postrainy seasons revealed that rainy season sowings made in July suffered

Table 3. Experiments with six chemical control formulations for control of grain molds of CV 296 B (Average of four locations).

Treatment	Field grade (1-5)[2]	Threshed-grain grade (1-5)[2]	Grain yield (t ha[-1])	Rank[1]
Captan 0.2% + aureofungin (200 ppm)	1.5	1.4	3.2	1
Mancozeb 0.2% + captan (0.2%)	1.6	1.5	3.1	2
Thiram (0.2%) + carbendazim (0.05%)	1.7	1.6	2.0	3
Captan (0.2%)	1.7	1.5	2.8	4
Carbendazim (0.1%) (Bavistin®)	1.8	1.6	2.8	5
Captafol (0.2%) (Difolatan®)	1.8	1.6	2.6	6
Control	4.2	3.7	2.5	7

1. Based on effectiveness in preventing grain yield losses.
2. Scored on a 1-5 scale: where 1 = least disease and 5 = most disease.

more from disease than did those made in June, and sowings made during September and the first fortnight of October (postrainy season) suffered more disease (Anahosur and Patil 1981a). Patil et al. (1982) and Mughogho and Pande (1983, pp. 11–24) reported that increased plant densities are associated with increased infection.

Effective chemical control techniques for this disease are not yet available; use of resistant varieties and hybrids is the most promising approach. Screening to identify resistant genotypes presently follows three procedures: the toothpick method (Hsi 1961), stem tape inoculation method (Mayee and Garud 1978), and the sick plot method. The toothpick method, with modifications, has been followed at different Centers (Mughogho and Pande 1983, pp. 11–24; Anahosur 1983). The screening procedure is as follows: in the sick plots, during the second fortnight of September, two rows of the test entry are sown between two rows of the susceptible control (CHS 6) in three replications. Twenty plants in one of the rows are inoculated in the second internode with *M. phaseolina*-coated tooth picks 15 days after anthesis. Plants in the other row are exposed to natural infection. Lodging percentage, number of nodes crossed, length of spread in the stalk, and differences in grain mass are recorded and compared. Although infection success is not always consistent, this method seems to be the most suitable available.

Among several genotypes tested, E 36–1 has shown stable resistance. Tolerant genotypes are A 1, Afzalpur local, SPV 249, SPV 428, and SPV 488 (Anahosur and Patil 1983a). Breeding charcoal rot resistant varieties using CSV 5 as the resistant source produced resistant high-yielding lines such as SPV 248 and SPV 488 (Gowda et al. 1981). Testing of lines developed from crosses SPV 86 × E 36–1 and E 36–1 × SPV 86 is in progress at Dharwad Centre. Inheritance of resistance studies revealed that resistance was partially dominant and controlled by polygenes and that heritability was poor (Rana et al. 1982a). Venkatrao and Shinde (1985) reported that susceptibility was dominant and resistance was controlled by polygenes and by nonallelic interaction. Additive × dominance and dominance × dominance effects predominated in resistance. The quantity of root and stalk sugars, phenols, and amino acids in a resistant sorghum (E 36–1) was more than in a susceptible (CSH 6). Further,

the quantity of sugars and phenols in diseased root and stalk was considerably less than in healthy plants of the susceptible genotypes (Anahosur et al. 1982; Anahosur and Naik 1985).

Fusarium root and stalk rot caused by *Fusarium moniliforme* and *Fusarium moniliforme* var *subglutinanse* was destructive in postrainy season in Karnataka (Anahosur et al. 1980a). During an epiphytotic of *F. moniliforme* stalk rot on 16 ha in Bagalkot the loss in grain yield was 15–40% in M 35–1 and Muguthi.

Screening of several varieties by tooth-pick method in the postrainy season at Dharwad indicated that CSV 5, SPV–386, 391, and 392 possessed some resistance.

Downy mildew

AICSIP's sorghum downy mildew work is concentrated at Dharwad, Mysore, and Coimbatore. This disease has very high potential for economic losses. Crop-loss estimates in genotypes of varying susceptibility in 1985 and 1986 (Table 4) indicated that maximum grain loss (78.6%) occured in the highly susceptible DMS 652, whereas the smallest grain loss (2.7%) was seen in CSV 4. Maximum disease pressure in 1985 was 73.8%. In 1986, the disease pressure was greater (94.2%) and thus grain loss (93.5%) maximum in highly susceptible DMS 652 and minimum (9.6%) in CSV 4. Other genotypes showed greater losses in yield in 1986 than in 1985.

Table 4. Average loss (%) in grain yield because of sorghum downy mildew (SDM) in 1985 and 1986.

Genotype	SDM incidence (%)	Grain yield loss (%)
CSV 4	5.19	6.16
CSV 10	11.42	12.98
SPV 475	37.35	46.47
CSH 5	11.40	18.33
CSH 9	12.78	25.82
SPH 196	18.70	29.73
296 B	50.95	52.83
DMS 652	83.88	86.06

Effective screening techniques for mass screening of sorghum genotypes have been developed (Anahosur 1980, 1987; Anahosur and Hegde 1979). Genotypes are exposed to oospores and to conidia. Some 800 varieties, 400 hybrids, 50 B-lines, and about 25 R-lines were tested by 1987. The resistant genotypes (showing less than 5% systemic infection) are SPV-35, 81, 104, 126, 166, 291, 312, 346, 351, 462, 736, 775; CSH-2, 3, 4, 5, 6; SPH-10, 11, 12, 30, 39, 59, 68, 159, 176, and 185; DMS 1 B, 2219 B, 3660 B, 3691 B, 1202 B, 1807 B, M 12 B; CSV 4, PVK 3, PVK 10, and MR 750.

A selection from Nandyal, in Andhra Pradesh, DMRS 1 (IS 27042) was developed and found to possess stable resistance to downy mildew (Anahosur et al. 1984a). From International Sorghum Downy Mildew Nursery (ISDMN), the stable resistant genotypes are IS-3443, 27042, 22230, 22231, and 18757; these were completely free from infection.

Breeding resistant varieties using IS 3443 and IS 27042 is in progress at Dharwad. The cross SB 905(S) × DMRS 1 (R) has yielded several high-yielding progenies coupled with high levels of resistance; 10 of these are in an advance stage of evaluation. In India, hybrids developed on 2296 A (17.2–63.6% diseased) and 199 A (34–41.6% diseased) were susceptible, whereas those produced on 2219–A and 2077–A and DMS 1 B have shown resistance (4.5–8.9%, 3.4–11.4%, and 1.7–4.6% disease incidence, respectively).

Rajasab et al. (1979, 1980a, 1980b) reported the dispersal of conidia as far as 80 m for as long as 5 hours after midnight. Ramlingam and Rajasab (1981) reported least infection from oospores and maximum infection from conidia; thus conidia are known to play major roles in epiphytotics.

Prabhu et al. (1983, 1984) reported the seed-borne nature of downy mildew in sorghum. These workers detected mycelium in 40% of the seeds and in 5% of the endosperm. Mycelium was not found in the embryo. Kaveriappa and Safeeulla (1978) expressed doubt as to the survival of mycelium when seeds are sundried.

Shetty and Safeeulla (1981) reported oospore germination by branched aseptate germ tubes from weathered oospores. A study of the inheritance of resistance to downy mildew (Rana et al. 1982a) reported that resistance is dominant and controlled by three genes. Bhat et al. (1982), using crosses 296(S) × QL 3 (R) and 296 (S) ×

Uchv 2(R), reported that resistance is dominant and controlled by four genes. We have studied this aspect, using crosses DMRS 1(R) × QL 3(R), 269 B(S) × IS 3547(R), 296 B(S) × QL 3(R), and IS 3443(R) × QL 3 (R) and found that resistance is dominant and controlled by three genes 57 (R) : 7(S) or by four genes 243(R) : 13(S).

Effective control of downy mildew by different formulations of metalaxyl was achieved as seed dress (Venugopal and Safeeulla 1978), and as seed dress + spray (Anahosur and Patil 1980, 1983b). Sorghum downy mildew cure was also reported (Anahosur and Patil 1980, 1981b) and subsequently a spray schedule for cure was developed. Metalaxyl applied to seed at the rate of 1 or 2 g a.i. kg^{-1} seed was not phytotoxic, but when applied at 4, 6, or 8 g a.i. kg^{-1} of seed it was phytotoxic. Seeds treated at 1 or 2 g a.i. kg^{-1} could be preserved at room conditions for 9 months without affecting either the efficacy of fungitoxicant or seed germination (Anahosur 1986a).

Ergot

Systematically estimated grain-yield losses from ergot are not available, but heavy grain losses are realized in commercial and seed-production plots. All A-lines are highly susceptible. Extensive screening of varieties, hybrids, A- and B-lines, R- lines, and some germplasms by artificial inoculation show the less-susceptible (30% infection) genotypes to be CS 3541, IS 3443, IS 14332, IS 3547, SPH 232, and DMS 1 B. The screening procedure involves four evening spray inoculations of a rich conidial suspension on newly emerged earheads on alternate days and then covering the heads with paper bags. This helped to induce high disease pressure. Evaluation was on a scale of 1 to 5, where 1= no disease, and 5 = at least 51% spikelets infected.

Early sowings in Maharashtra, Karnataka, and Gujarat avoided the disease (Sangitrao et al. 1979; Anahosur and Patil 1982a). The meteorological conditions favoring honeydew development were minimum temperatures of 18–31°C, relative humidities of 65–85%, and drizzling rains. Dry weather following the honeydew stage favored sclerotia development (Sangitrao and Bhade 1979; Anahosur et al. 1984b). In the Karnataka sowings, growth of Cerebella sorghi suppressed sclerotia production.

Sundaram (1977) reported control of ergot with three sprays of ziram (0.2%); Anahosur (1979) reported mancozeb (0.2%) gave better control than ziram (0.2%). In subsequent years, systemic fungicides like carboxin (Vitavax®) and thiophanate (Topsin®) have shown some promise at some locations.

Study of pollen management and effective pollination using a spray of cytokinins, sucrose, urea, and growth hormones is now underway.

Rust

Rust is destructive in the rainy and postrainy seasons. Extensive screening of sorghum genotypes by adopting the spreader row technique (Sharma and Jain 1975; Sharma 1980) and four spray inoculations with uredospores on alternate days beginning on the 45th day of sowing helped to create high disease pressure and identify resistant genotypes. Disease scoring (1 = no disease and 5 = more than 51% leaf area damaged) showed SPV-247, 1622 B, 2947 B, SB 401 B, IS 3443, IS 3547, and IS 18758 were immune to rust. Resistant genotypes were SPV-81, 126, 245, 241, 312, 346, 351, 386, 459, 462, 472, 736, 775, SPH-159, 176, 185, 187, 361, 364, MSH-51, 52, DMS 1 B, M 9 B, M 12 B, MR 750, R 12 8, R 259, RSR 9, IS 13442, E 36-1, IS 8185, IS 8607, and IS 7254. Most of these are of the tan plant type. The level of resistance in the newly released genotypes, compared to earlier genotypes, is very high. Anahosur et al. (1981) reported on several of the germplasm lines tested, and the majority of the resistant lines were of the tan plant color.

Sorghum sown before the 3rd week of June escaped from rust (Anahosur et al. 1982), whereas in later sowings, made through October, the disease was very high. Sorghum sown in November and December showed less disease.

Gangadharan et al. (1976) reported effective control of rust and leaf blight with zineb spray (0.2%). Agarwal and Kotasthane (1973) reported control of rust by seed dressing with oxycarboxin (Plantvax®) followed by two sprays at 0.1%. Sharma (1980) reported ziram (0.2%), captan (0.2%), and thiram (0.2%) as foliar sprays gave effective control. Anahosur (1981) reported that four sprays of mancozeb at 0.2% or 0.4% at 10-day intervals from the 35th day of sowing gave significantly better control than ziram, captan, and zineb.

With regard to inheritance of resistance studies, Rana et al. (1976) reported that susceptibility is dominant and resistance is governed by three major genes. Indira et al. (1983) showed that tan × tan crosses showed more resistance than tan × purple or purple × purple crosses. Tan parents transmitted resistance. Incomplete dominance was seen in some tan × tan crosses in which one parent was highly resistant. Our study involved crosses 296 B(S) × IS 3547 (R), IS 3443(R) × QL (S), and 296 B(S) × QL 3 (S) and indicated that resistance is dominant and governed by one gene 3(R) : 1(S) or two genes 15(R) : 1(S).

Leaf spots

Resistance evaluation against zonate leaf spot, rough spot, leaf blight, gray leaf spot, and sooty stripe showed the following genotypes to be resistant at all AICMIP centres: SPV-81, 126, 312, 346, 351, 386, 462, 472, 544, 736, 775, APH-159, 176, 185, 187, 264, 361, 362, 411, M 6 B, M 10 B, M 12 B, DMS 1 B, SB 101 B, AKMS 3 B, ICS 4 B, CS 3541, SB 1085, R 250, R 390, R 1036, IS 3443, IS 8283, IS 8607, IS 3547, and IS 18758.

Anahosur (1986b) reported control of rust, anthracnose, and zonate leaf spot by mancozeb (0.2%) spray, whereas Sharma (1980) reported control of leaf spot diseases by zineb (0.2%) or ziram (0.2%).

Anahosur et al. (1980b) reported single-gene resistance for gray leaf spot; susceptibility is dominant.

Inheritance of resistance to zonate leaf spot in crosses 296 B(S) × IS 3547(R), IS 3443(R) × QL 3(S), and 296 B(S) × QL 3(S) indicated that resistance is a dominant trait and governed by three genes 57(R) : 7(S) or four genes 243(R) : 13(S)

Red stripe

Usually sporadic in Maharashtra and Karnataka, this virus disease appeared in a severe form in Karnataka, Andhra Pradesh, Maharashtra, and elsewhere in postrainy season, 1987.

Mali and Garud (1978) identified this virus as a Johnsongrass strain of sugarcane mosaic virus (SCMV). Several genotypes were tested; Indian sorghums 73, 844, 952, 1040, 1518, 2301, 2522, 2601, 5169, 7499, 7998, 9227, and Gobwani were immune. The vectors of this virus are *Rhopalosiphum maidis* and *Melanaphis sacchari*.

Bandyopadhyay et al. (1987) reported wilt of sorghum from north India caused by *Acremonium strictum* and its transmission through seed.

Multiple Disease and Insect Resistance

A number of germplasm lines were tested at several locations for multiple-disease resistance (Table 5). Plants were exposed to 10 diseases; IS 3443, 3547, and 8283 showed resistance to all but ergot. All lines in the trial, however, showed only moderate resistance to ergot. IS 14332 was only moderately resistant to charcoal rot.

All breeders' materials in this multilocational trial were resistant to anthracnose, downy mildew, leaf blight, rough leaf spot, rust, sooty stripe, and zonate leaf spot. All were moderately resistant to grain mold as well as to ergot. SPV 346 and SB 1085 were resistant to charcoal rot; the other breeders' lines were either susceptible or only moderately resistant to this disease.

In a study of selecting for inheritance of multiple-disease resistance, Rana et al. (1982b) discussed the possibility of selecting progenies possessing resistance to downy mildew, rust, and charcoal rot. Our studies involving crosses IS 3443(R) × QL 3(S), 296 B(S) × IS 3547(R), and 296 B(S) × QL 3(R) indicated that it is possible to incorporate resistance for sorhum downy mildew (SDM), rust, and zonate leaf spots in the new progenies.

Table 5. Multiple-disease resistance in sorghum in All India Coordinated Sorghum Improvement Project (AICSIP) trials, India.[1]

Germplasm	GM	SDM	CR	Ergot	Rust	Zn	Anth.	LB	SS	RLS
Indian sorghum										
IS 3443	R	R	R	MR	R	R	R	R	R	R
IS 3547	R	R	R	MR	R	R	R	R	R	R
IS 6365	R	R	R	MR	S	R	R	S	R	R
IS 8383	R	R	R	MR	R	R	R	R	R	R
IS 18758	R	S	MR	MR	R	R	R	R	R	R
IS 14332	R	R	MR	MR	R	MR	R	R	R	R
IS 2328	R	MR	MR	MR	R	R	R	R	R	R
Breeders' material										
CSF 4	MR	R	S	MR	R	R	R	R	R	R
HPV 126	MR	R	S	MR	R	R	R	R	R	R
SPV 81	MR	R	S	MR	R	R	R	R	R	R
SPV 346	MR	R	R	MR	R	R	R	R	R	R
SPV 351	MR	R	S	MR	R	R	R	R	R	R
SPV 386	MR	R	MR	MR	R	R	R	R	R	R
SPV 462	MR	R	MR	MR	R	R	R	R	R	R
SPV 736	MR	R	MR	MR	R	R	R	R	R	R
SPV 775	MR	R	MR	MR	R	R	R	R	R	R
DMS 1 B	MR	R	S	MR	R	R	R	R	R	R
MR 750	MR	R	MR	MR	R	R	R	R	R	R
SB 1085	MR	R	R	MR	R	R	R	R	R	R

1. R = Resistant, MR = Moderately resistant, S = Susceptible; GM = Grain molds, SDM = Sorghum downy mildew, CR = Charcoal rot, Zn = Zonate leaf spots, Anth = Anthracnose, LB = Leaf blight, SS = Sooty stripe, and RLS = Rough leaf spot.

Multiple disease resistance insect nursery

Multiple-disease resistance combined with insect resistance is needed and additional effort at national and international levels is desirable. Nurseries such as ISDMN and International Sorghum Multiple Disease Resistance Nursery (IS-MDRN) have been beneficial and need to be strengthened. New research nurseries, such as an international multiple-disease-CVM-insect resistance nursery may be started. Development of integrated management programs for major diseases could be of great value.

Research Needs

Grain molds

1. Incorporation of resistance into high-yielding commercial genotypes.
2. Study of the mechanisms and genetics of resistance.
3. Identify effective systemic fungicides.

Charcoal rot, fusarium root, and stalk rot

1. Development of effective screening techniques.
2. Study of the stalk-rot complex, its development, and the role and interactions of different fungi.
3. Study of the genetics of resistance and development of resistance sources.
4. Incorporation of resistance in commercial sorghums.
5. Determination of yield losses and conditions favoring disease development.
6. Biological control methods and integrated control programs.

Downy mildew

1. Incorporation of resistance into B-lines and development of new resistant A-lines.
2. Development of resistant hybrids involving resistant parents.
3. Study of germination and role of oospores.
4. Study of the effect of oospore and/or conidia concentrations on resistance.

Ergot

1. Standardization of screening techniques, identification of resistant sources.
2. Incorporation of resistance sources into commercial sorghum.
3. Clarification of the role and value of pollen management in dealing with ergot disease.
4. Identify effective systemic fungicides.

Leaf diseases

1. Incorporation of multiple-disease resistance into commercial sorghums.
2. Determination of yield losses due to the major leaf diseases.
3. Identification of systemic fungicides to control the major foliar diseases.
4. Clarification of the mechanisms of inheritance of resistance to leaf spot diseases.

References

Agarwal, S.C., and **Kotasthane, S.R.** 1973. Efficacy of systemic fungicides and antibiotics in checking the rust of *Sorghum vulgare* L. Science and Culture 39:235–236.

Anahosur, K.H. 1979. Chemical control of ergot of sorghum. Indian Phytopathology 32:487–489.

Anahosur, K.H. 1980. Current sorghum downy mildew research in the All India Sorghum Project. Pages 200–206 *in* Sorghum diseases, a world review: proceedings of the International Workshop on Sorghum Diseases, 11–15 Dec 1978, Hyberabad, India. Patancheru, Andhra Pradesh 502 324, India: International Crops Research Institute for the Semi-Arid Tropics.

Anahosur, K.H. 1981. Chemical control of sorghum rust. Indian Journal of Agricultural Sciences 51:111–112.

Anahosur, K.H. 1983. Comparison of the tooth pick and sick plot method on the incidence of charcoal rot in sorghum. Sorghum Newsletter 26:108–109.

Anahosur, K.H. 1986a. Effect of metalaxyl seed treatment on seedling emergence and downy mildew incidence in sorghum. Indian Phytopathology 39:75–77.

Anahosur, K.H. 1986b. Chemical control of foliar diseases of sorghum. Indian Phytopathology 39:526–528.

Anahosur, K.H. 1987. Evaluation of sorghums to downy mildew resistance. Paper, Seventeenth AICSIP workshop, Maharashtra Agricultural University, Parbhani, Maharashtra, India, 25–27 May 1987. 13 pp.

Anahosur, K.H., and Hedge, R.K. 1979. Assessment of the techniques used in screening sorghum genotypes to downy mildew. Mysore Journal of Agricultural Sciences 13:449–451.

Anahosur, K.H., and Naik, S.T. 1985. Relationship of sugars and phenols of root and stalk of sorghum in charcoal rot. Indian Phytopathology 38:131–134.

Anahosur, K.H., and Patil, S.H. 1980. Chemical control of sorghum downy mildew in India. Plant Disease 64:1004–10

Anahosur, K.H., and Patil, S.H. 1981a. Date of planting and incidence of charcoal rot of sorghum. Sorghum Newsletter 24:118–119.

Anahosur, K.H., and Patil, S.H. 1981b. Sorghum downy mildew cure by metalaxyl (25 WP). Sorghum Newsletter International. Working group on graminacious downy mildews 3:5.

Anahosur, K.H., and Patil, S.H. 1982a. Effect of date of sowing on the incidence of ergot of sorghum. Indian Phytopathology 35:507–509.

Anahosur, K.H., and Patil, S.H. 1983a. Evaluation of sorghum genotypes for charcoal rot resistance. Indian Phytopathology 36:407–408.

Anahosur, K.H., and Patil, S.H. 1983b. Effective spray schedule of metalaxyl (25 WP) for sorghum downy mildew therapy. Indian Phytopathology 36:407–408.

Anahosur, K.H., Rajsekhar, B.G., and Goudreddy, B.S. 1977. Effect of nitrogenous fertilizer on the incidence of charcoal rot of sorghum. Sorghum Newsletter 20:23–24.

Anahosur, K.H., Kulkarni, K.A., Gowda, B.T.S., and Parameswarappa, R. 1980a. Fusarium stalk rot of sorghum. Sorghum Newsletter 23:42.

Anahosur, K.H., Gowd, B.T.S., and Patil, S.H. 1980b. Inheritance of resistance to Cercospora sorghi causing gray leaf spot on sorghum. Current Science 49:637.

Anahosur, K.H., Parameswarappa, R., and Gowda, B.T.S. 1981. Reaction of sorghum genotypes to rust under Dharwad conditions. Mysore Journal of Agricultural Sciences 15:429–430.

Anahosur, K.H., Patil, S.H., and Naik, S.T. 1982. Effect of date on sowing on the incidence of sorghum rust. Indian Phytopathology 35:247–250.

Anahosur, K.H., Patil, S.H., and Naik, S.T. 1984a. A new sorghum genotype as a source of stable resistance to downy mildew. Current Science 53:873.

Anahosur, K.H., Lakshman, M., and Naik, S.T. 1984b. Relationship of weather conditions with the development of ergot disease of sorghum. Sorghum Newsletter 27:116.

Anahosur, K.H., Patil, S.H., and Hedge, R.K. 1982. Relationship of stalk sugars and phenols with charcoal rot of sorghum. Indian Phytopathology 35:335–337.

Avadhani, K.K., Patil, S.S., Mallanagouda, B., and Parwatikar, S.R. 1979. Nitrogen fertilization and its influence on charcoal rot. Sorghum Newsletter 22:119–220.

Bandyopadhyay, R., Mughogho, L.K., and Satyanarayana, M.V. 1987. Systemic infection of sorghum by Acremonium strictum and its transmission through seed. Plant Disease 77:647–650.

Bhat, M.G., Gowda, B.T.S., Anahosur, K.H., and Goud, J.V. 1982. Inheritance of plant pigmentation and downy mildew resistance in sorghum. SABRAO Journal 14(53):59.

Gangadharan, K., Subramanian, N., Mohanraj, D., Kandaswamy, T.K., and Sundaram, N.V.

1976. Control of foliar diseases of sorghum. Madras Agricultural Journal 63:413–414.

Gowda, B.T.S., Anahosur, K.H., and Parameswarappa, R. 1981. Breeding charcoal rot resistant sorghum genotypes. Mysore Journal of Agricultural Sciences 15:503–506.

Hsi, D.C.H. 1961. An effective technique for screening sorghum for resistance to charcoal rot. Phytopathology 5:340–341.

ICAR (Indian Council of Agricultural Research). 1987. Progress report, 1975–1987, of the All India Coordinated Sorghum Improvement Project. New Delhi, India: ICAR.

Indira, S., Rana, B.S., Rao, V.J.M., Jayamohan Rao, V.J., and Rao, N.G.P. 1983. Host plant resistance to rust in sorghum. Indian Journal of Genetics 43:193–199.

Kaveriappa, K.M., and Safeeulla, K.M. 1978. Seedborne nature of Sclerospora sorghi in sorghum. Proceedings, Indian Academy of Sciences 87:303–308.

Mali, V.R., and Garud, T.B. 1978. Sorghum red stripe — a Johnsongrass strain of sugarcane mosaic virus in India. FAO Plant Protection Bulletin 26:28–29.

Mayee, C.D., and Garud, T.B. 1978. An assured method for evaluating sorghum to charcoal rot. Indian Phytopathology 31:121.

Mote, U.N., and Ramshe, D.G. 1980. Nitrogen application increases the incidence of charcoal rot in rabi sorghum cultivars. Sorghum Newsletter 23:129.

Mughogho, L.K., and Pande, S. 1983. Sorghum root and stalk rots, A critical review. Patancheru, A.P. 502 324, India. International Crops Research Institute for the Semi-Arid Tropics. (Limited distribution.)

Patil, R.C., Deshmane, N.B., and Pandhare, T.M. 1982. Effect of plant density and row spacing on charcoal rot incidence in four cultivars of sorghum. Sorghum Newsletter 25:110.

Prabhu, M.S.C., Safeeulla, K.M., and Shetty, H.S. 1983. Penetration and establishment of downy mildew mycelium in sorghum seeds and its transmission. Proceedings, Indian Academy of Sciences 49:459–465.

Prabhu, M.S.C., Safeeulla, K.M., Venkatsubbiah, P., and Shetty, H.S. 1984. Detection of seedborne inoculum of Peronosclerospora sorghi in sorghum. Phytopathologische Zeitschrift 111:174–178.

Rajasab, A.H., Shenio, M.H., and Ramlingam, A. 1979. Epidemiology of sorghum downy mildew III, dispersal and deposition of inoculum. Kavaka 7:63–68.

Rajasab, A.H., Shenoi, M.H., and Ramlingam, A. 1980a. Epidemiology of sorghum downy mildew IV. Incidence of local lesion infection. Proceedings, Indian National Sciences Academy 46:207–214.

Rajasab, A.H., Shenoi, M.H., and Ramlingam, A. 1980b. Epidemiology of sorghum downy mildew V. Incidence of systemic infection. Proceedings, Indian National Sciences Academy 46:552–561.

Ramlingam, A., and Rajasab, A.H. 1981. Epidemiology of sorghum downy mildew VI. Relative importance of oospores and conidia in epidemics of systemic infection. Proceedings, Indian National Sciences Academy 47:725–730.

Rana, B.S., Anahosur, K.H., Rao, M.J.V., Rao, V.J.H., Parameswarappa, R., and Rao, N.G.P. 1982a. Inheritance of field resistance to sorghum downy mildew. Indian Journal of Genetics 42:70–74.

Rana, B.S., Anahosur, K.H., Rao, V.J.H., Parameswarappa, R., and Rao, N.G.P. 1982b. Inheritance of field resistance to sorghum charcoal rot and selection for multiple disease resistance. Indian Journal of Genetics 42:302–310.

Rana, B.S., Tripathi, D.P., and Rao, N.G.P. 1976. Genetic analysis of some exotic × Indian crosses in sorghum XV. Inheritance of resistance to sorghum rust. Indian Journal of Genetics 36:244–249.

Ravindranath, V. 1980. Sorghum diseases in India. Pages 57–66 *in* Sorghum diseases, a world review: proceedings of the International Workshop on Sorghum Diseases, 11–15 Dec 1978, Hyderabad, India. Patancheru, Andhra Pradesh 502 324, India: International Crops Research Institute for the Semi-Arid Tropics.

Sangitaro, C.S., and Bhade, G.H. 1979. Meteorological factors associated with honey dew development and sclerotial stage in sorghum ergot. Sorghum Newsletter 22:107–108.

Sangitrao, C.S., Taley, Y.M., and Moghe, P.G. 1979. Effect of planting date on the appearance of sugary disease. Sorghum Newsletter 22:111.

Sharma, H.D. 1980. Screening of sorghum for leaf-disease resistance in Sorghum Diseases. Pages 249–263 *in* Sordhum diseases, a world review: proceedings of the International Workshop on Sorghum Diseases, 11–15 Dec 1978, Hyderabad, India. Patancheru, Andhra Pradesh 502 324, India: International Crops Research Institute for the Semi-Arid Tropics.

Sharma, H.C., and Jain, N.K. 1975. Effect of leaf diseases on grain yields of some varieties of sorghum. Proceedings, Indian Academy of Sciences 81:(5)223–227.

Shetty, H.A., and Safeeulla, K.M. 1981. Studies on the oospore germination of *Peronosclerospora sorghi*. Kavaka 8:47–51.

Sundaram, N.V. 1977. Pathological research in India. Paper, International Sorghum Workshop, 6–13 Mar 1977, ICRISAT, Hyderabad, India, Patancheru, Andhra Pradesh 502 324, India. 19 pp.

Venkatrao, D.N., and Shinde, V.K. 1985. Inheritance of charcoal rot resistance in sorghum. Journal of the Maharashtra Agricultural University 10:54–56.

Venugopal, M.N., and Safeeulla, K.M. 1978. Chemical control of downy mildew of pearl millet, sorghum, and maize. Indian Journal of Agricultural Sciences 48:537–539.

Sorghum Diseases in Brazil

C.R. Casela[1], A.S. Ferreira[2], and R.E. Schaffert[3]

Abstract

Sorghum in Brazil is subject to damage by several diseases. Anthracnose (Colletotrichum gram-inicola) is the most damaging disease occurring in all areas of the country. Rust (Puccinia purpurea) is also becoming an important sorghum disease in Brazil, and charcoal rot (Macro-phomina phaseolina) has been observed in the semi-arid northeastern region, as well as in late (February) sowings in central Brazil. Downy mildew (Peronosclerospora sorghi) persists as an important disease in the southern parts of the country. Zonate leaf spot (Gloeocercospora sorghi), ladder spot (Cercospora fusimaculans), leaf blight (Exserohilum turcicum), and stalk rot (Colletotrichum graminicola), have been causing only moderate damage.

Introduction

Sorghum, a relatively new crop in Brazil, is increasing both in production and area sown. Production increased 39% and the area sown increased by 38% from 1982 to 1984, and average yield was 1.8 t ha^{-1} (Veiga 1986) (Table 1).

The states of Rio Grande do Sul, São Paulo, and Parana, produce more than 75% of the sorghum grain in Brazil, with an average yield of 2 t ha^{-1}.

In Rio Grande do Sul, sorghum is becoming a traditional crop, sown mostly in a crop-rotation system with soybeans (Veiga 1986). In the semi-arid northeast, sorghum production is concentrated in the states of Pernambuco, Bahia, Ceara, and Rio Grande do Norte. Sorghum, due to its drought tolerance and higher yields, is one of the alternatives to maize for farmers in this region.

In central Brazil, sorghum is now becoming an important option for farmers; it can be sown in February after harvest of the principal crop, usually soybeans. This is becoming a traditional system in the states of São Paulo and Parana in southern Brazil (Vianna et al. 1986).

Most of the grain sorghum cultivated in Brazil are with hybrids, except in the northeastern areas, where varieties are used.

Grain sorghum in Brazil is used principally in swine and poultry rations, but the National Maize and Sorghum Research Center is developing sorghum cultivars for human consumption. Several studies in Brazil have shown that sorghum can be mixed with wheat flour in proportions varying from 15 to 50% for baking in breads, cakes, cookies, etc., without significant change in the quality of the products.

In central Brazil, sorghum for this purpose should be sown in February, just after harvest of the soybean. In the semi-arid northeastern region, the drier conditions permit good grain quality (Schaffert 1986). Forage sorghums are used principally for silage production for dairy cattle; however the silage can also be fed to beef cattle.

Experiments in Brazil indicate that sweet sorghum can be used to produce alcohol for fuel, to be used pure or mixed with gasoline. The National Maize and Sorghum Research Center has recently released new varieties of sweet sorghum.

1. Sorghum Breeder for Disease Resistance, National Maize and Sorghum Research Center-EMBRAPA (NMSRC), Sete Lagoas, Minas Gerais, Brazil.
2. Sorghum Pathologist at the above address.
3. Sorghum Breeder at the above address.

Casela, C.R., Ferreira, A.S., and Schaffert, R.E. 1992. Sorghum diseases in Brazil. Pages 57–62 *in* Sorghum and millets diseases: a second world review. (de Milliano, W.A.J., Frederiksen, R.A., and Bengston, G.D., eds). Patancheru, A.P. 502 324, India: International Crops Research Institute for the Semi-Arid Tropics.

Table 1. Sorghum production and productivity in Brazil, 1982-84.

State	Sorghum area (ha)			Production (t ha⁻¹)			Yield (t ha⁻¹)		
	1982	1983	1984	1982	1983	1984	1982	1983	1984
Rio Grande do Sul	50 423	51 638	65 964	105 634	105 687	136 695	2.10	2.05	2.07
São Paulo	34 970	31 273	30 000	69 940	62 546	60 000	2.00	2.00	2.00
Paraná	5 904	12 320	15 054	16 285	33 092	39 574	2.79	2.69	2.63
Pernambuco	6 284	4 233	9 916	4 713	1 516	14 775	0.75	0.36	1.50
Bahia	7 511	24 172	19 194	12 764	13 804	13 354	1.64	0.57	0.70
Ceará	5 400	2 692	6 028	6 750	1 618	9 464	1.25	1.60	1.60
Mato Grasso do Sul	3 123	3 371	4 803	3 684	6 874	7 760	1.20	2.04	1.61
Rio Grande do Norte	7 424	3 555	9 884	3 851	497	12 355	0.52	0.14	1.25
Góias	1 115	2 272	3 290	1 964	5 231	8 160	1.61	2.30	2.48
Minas Gerais	280	471	4 419	436	692	8 046	1.56	1.47	1.54

1. Source: Veiga (1986).

Average alcohol production ranged from 2500 to 3500 L ha⁻¹.

Sorghum Diseases in Brazil

Annual surveys in Brazilian sorghum fields, show that anthracnose (*Colletotrichum graminicola*), rust (*Puccinia purpurea*), downy mildew (*Peronosclerospora sorghi*), charcoal rot (*Macrophomina phaseolina*), and grain weathering, are the major sorghum diseases in Brazil. Several other sorghum diseases are found in Brazil (Table 2); the principal diseases are discussed below.

Anthracnose

Anthracnose is the most important disease of sorghum in Brazil. It is found all over the country and can cause severe losses in yield and in quality. All forms of infection are found in Brazil, but foliar infections predominate over stalk and panicle infections.

Currently, Brazilian research on *C. graminicola* has focused on characterizing the pathogen variability and host resistance. Several races have been reported as important in Brazil (Casela et al., this volume).

The durability of the hypersensitivity resistance to sorghum anthracnose is questionable, considering the plasticity of *C. graminicola*. At-

Table 2. Sorghum diseases in Brazil.

Disease	Pathogen
Foliar	
Anthracnose	*Colletotrichum graminicola*
Rust	*Puccinia Purpurea*
Sorghum downy mildew	*Peronosclerospora sorghi*
Leaf blight	*Exerohilum turcicum*
Ladder spot	*Cercospora fusimaculans*
Zonate leaf spot	*Gloeocercospora sorghi*
Sooty stripe	*Ramulispora sorghi*
Root and stalk	
Charcoal rot	*Macrophomina phaseolina*
Red stalk rot	*Colletotrichum graminicola*
Head	
Grain weathering	Various fungi
Covered smut	*Spahacelotheca sorghi*
Head smut	*Sporisorium reilianum*
Virus	
Mosaic	Sugarcane mosaic virus

tention is therefore being given to a type of resistance expressed by a lower rate of disease development over a certain period of time. The evaluation of sorghum genotypes for this resistance is made by creating artificial epidemics in microfield test plots by inoculating a spreader row of a susceptible cultivar with a spore suspension of the pathogen. The inoculum multiplies and disseminates toward the screening plots. The interplot interference in the screening plots is reduced by using a border row of a resistant cultivar (Fig. 1). The disease intensity at three different points in the screening plot (0.5, 3.0, and 5.5 m from the source of inoculum) is measured on regular schedule as percentage of leaf area affected by the pathogen, in accordance to a standard disease scale (Sharma 1983). The data were transformed into AUDPC (Area Under Disease Progress Curve) (Vanderplank 1963). AUDPC values of some experimental hybrids are listed in Table 3.

Values in the same column not followed by the same letter are significantly different (P=0.05), according to Duncan's test.

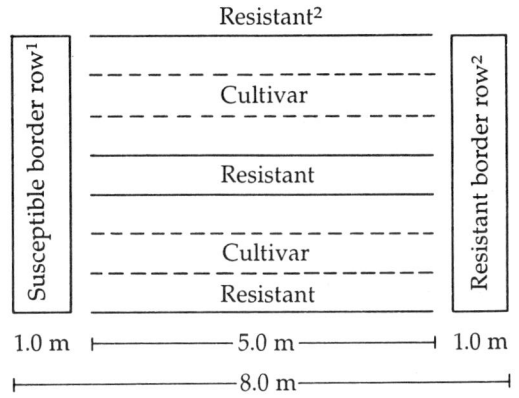

1. BR 001 (Tx 623) - Inoculated 40-50 days after sowing with a spore suspension of 10^6 conidia mL^{-1}

2. SC 283

Figure 1. Diagram of field plots used to evaluate resistance to anthracnose in field conditions under a disease gradient.

Table 3. Area under anthracnose progress curve of 12 sorghum cultivars, evaluated under decreasing levels of inoculum, at three distances (0.5 m, 3.0 m, and 5.5 m) from a spreader row.[1]

	AUDP curve		
	Distance from susceptible border row		
Cultivar	(0.5 m)	(3.0 m)	(5.5 m)
Tx 623	2501.0a	1890.0a	1885.3a
CMSXS 359	2080.0b	882.0b	858.7b
CMSXS 357	1660.2c	623.0c	591.5c
CMSXS 354	1473.5d	417.7d	327.8ef
CMSXS 366	1415.2de	306.8e	257.8g
CMSXS 352	1321.8c	417.7d	366.3e
CMSXS 356	1181.8f	563.5c	450.3d
CMSXS 358	995.2g	311.5e	291.7fg
CMSXS 357	976.5g	168.0f	112.0h
BR 302	962.5g	399.0d	364.0e
BR 007 B	862.2gh	430.5d	260.2g
BR 300	621.8i	179.7d	110.8h

1. Values in the same column not followed by the same letter are significantly different (P = 0.05) according to Duncan's Multiple Range test.

59

Rust (*Puccinia purpurea*) is also becoming an important sorghum disease in Brazil. Preliminary tests in greenhouse conditions and field observations from the National Sorghum Uniform Disease Nursery suggest the existence of races of *P. purpurea* in Brazil. Changes in reaction of some cultivars, such as Brandes (BR 501), with time indicate this possibility. A high degree of "slow rusting" has been detected on some sorghum genotypes in Brazil.

A natural infection method is used in the field to determine degrees of resistance to sorghum rust. The cultivar Theis is used in the spreader row. The interplot interference between screening plots is reduced by using border rows of a resistant cultivar, such as SC 326–6 (Fig. 2). The disease intensity in each cultivar is scored at three points in the screening plot at 7– or 15–day intervals according to a standard disease scale. The data of each genotype, similar to the leaf anthracnose screening procedure, are transformed to AUDPC. The AUDPC values for some experimental hybrids and varieties in Brazil are presented in Table 4.

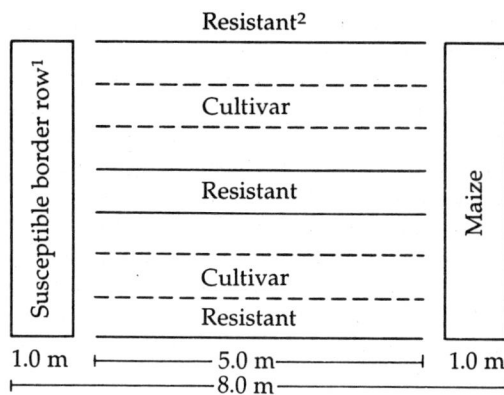

1. Theis (BR 503)

2. SC 326-6 (BR 005)

Figure 2. Diagram of field plots used to evaluate resistance to rust in field conditions under a disease gradient.

Charcoal rot

Sorghum charcoal rot (*Macrophomina phaseolina*) has been observed in the semi-arid regions of Brazil (northeastern Brazil), where the water

Table 4. Area under rust progress curves of 14 sorghum cultivars, evaluated under decreasing levels of inoculum at 0.5 m, 3 m, and 5.5 m from a spreader row.[1]

| | | AUDP curve | | |
| | | Distances, from susceptible border row | | |
Cultivar	Pedigree	(0.5 m)	(3.0 m)	(5.5 m)
CMSXS 351	BR 007A × CMSXS 178	911.5a	698.0a	658.3a
BR 001	.	778.2b	558.5b	638.3a
BR 302	BR 001A × BR 011R	757.7bc	593.0b	524.0b
BR 008		695.0c	331.0c	238.3c
SC 112-14		603.3d	279.0d	192.0d
CMSXS 180	BRP 3 R × SC 326-6	524.0e	69.5f	42.8f
CMSXS 354	BR 008 A × CMSXS 179	382.7f	122.0e	58.2c
CMSXS 179	BRP 3 R × SC 326-6	262.0g	42.0fg	31.5efg
CMSXS 359	CMSXS 156 × CMSXS 180	104.0h	31.0fg	10.5g
CMSXS 181	BRP 3 R × SC 326-6	15.7hi	31.5fg	17.5fg
CMSXS 156		60.7hi	10.5g	10.5g
CMSXS 182	BRP 3 R × SC 326-6	33.7i	10.5g	10.5g
CMSXS 178	SC 326-6 × SC 748-5	32.3i	10.5g	10.5g
CMSXS 157		21.0i	10.5g	10.5g

1. Values in the same column followed by different letters are significantly different (*P* = 0.05) according to Duncan's Multiple Range test.

deficit and high temperatures are conducive to development of this disease. Late sowings (February) in double-cropping systems in central Brazil are also favorable to charcoal rot development. High incidence of the disease has been observed in some of these areas.

Because of the importance of sorghum in the northeastern and central regions, the National Maize and Sorghum Research Center (NMSRC) is beginning a program to identify genotypes resistant to this disease, hoping to incorporate this resistance into elite cultivars. This approach is being used at the NMSRC-Experimental Station in northern Minas Gerais State, a semi-arid area.

Sorghum downy mildew

The sorghum downy mildew (*Peronosclerospora sorghi*) is commonly found in areas of the states of Rio Grande do Sul, São Paulo, Santa Catarina, Parana, and Minas Gerais.

Results from downy mildew trials (Table 5) have been obtained from the states of São Paulo and Rio Grande do Sul. The reactions of some sorghum genotypes in Jaboticabal-Brandes show some to be susceptible to pathotype 4 of *Peronosclerospora sorghi*, a new race of the fungus from Palotina, PR. (Nakamura and Fernandes 1984).

Other Diseases

Other diseases occurring in Brazilian sorghum fields which vary from year to year and from region to region are included in Table 2. Grain "weathering" is especially important in white-endosperm genotypes developed for human food.

Selection for resistance to leaf blight (*Exserohilum turcicum*), zonate leaf spot (*Gloeocercospora sorghi*), ladder spot (*Cercospora fusimaculans*), and other diseases is conducted in the National Sorghum Uniform Disease Nursery, breeding trials, and the anthracnose and rust resistant nurseries.

Studies of F_1, F_2 and F_3 generations of crosses between QL 3 (resistant) and susceptible genotypes indicate that one dominant gene determines resistance to Sugarcane mosaic virus (Pinto 1984).

Sorghum Disease Research in Brazil

Disease resistance studies are undoubtedly essential for continued efficient sorghum production in Brazil. Development of resistant genotypes is pursued in Brazil's sorghum breeding program. Good hybrids recently developed by the NMSRC, as well as those by private seed companies, will be released soon. However, accurate basic information about sorghum dis-

Table 5. Downy mildew incidence in 10 systematically infected sorghum genotypes in the 1983/84 National Forage Trial at Jaboticabal, São Paulo, Brazil.

Cultivar	Hybrid or variety	Source	Incidence (%)
AG 2001	H[1]	AGROCERES	0.0
Sordan NK 77	H	BRAZISUL	8.4
Contisilo	H	CONTIBRASIL	0.0
Brandes	V[2]	EMBRAPA	0.0
CMSXS 615	V	EMBRAPA	0.0
BR 601	H	EMBRAPA	0.0
BR 601	H	EMBRAPA	0.0
BR 603	H	EMBRAPA	0.0
Sart	V	AGROCERES	4.7
IPA 7301158	V	IPA	26.2

1. H = hybrid, 2. V = variety.

eases, their pathogens and host pathogen relationships, is needed and will be researched.

The existence of races of *Colletotrichum graminicola* clearly shows the importance of continuous monitoring of that pathogen population in the principal production areas of Brazil, especially in view of the highly virulent race 31H detected in Jatai, Goias. A better understanding of the *Sorghum bicolor–Colletotrichum graminicola* relationship is essential, considering that it fits in a "gene-for-gene" disease category.

Studies of the inheritance of vertical resistance to *Colletotrichum graminicola* have been initiated at the NMSRC to evaluate segregating generations of crosses made between the differential cultivars and inoculations with appropriate races. This information, associated with continuing race surveys of *Colletotrichum graminicola*, will permit more accurate determination of genetic associations in the host. This should be of value in developing a more stable vertical resistance or, in other words, the major virulence dissociation in the pathogen (Vanderplank 1982).

Stable resistance to *Colletotrichum graminicola* also implies the identification of dilatory resistance, characterized by a reduced rate of epidemic development, in spite of a susceptible reaction class. Field evaluations conducted at the NMSRC have indicated the existence of variability for this form of resistance to *Colletotrichum graminicola* in sorghum genotypes. Some observations have shown small, but significant, interactions between races of the pathogen and sorghum cultivars, but not to the extent that extreme changes in the ranking of cultivars are suggested.

The increasing incidence of rust in Brazil indicates that it will become a serious disease. The appearance of *Puccinia purpurea* on some previously resistant cultivars, such as Brandes, indicates that detailed study of the physiological specialization of the pathogen is necessary. The "slow rusting" phenomenon in many sorghum hybrids indicates that success in developing stable resistance to rust is likely.

A continuous screening for resistance to sorghum downy mildew in southern regions of Brazil is necessary, due to the area's importance as a grain producer.

The increasing importance of sorghum in double-cropping systems, mostly following soybeans, demonstrates the importance of charcoal rot (*Macrophomina phaseolina*). This pathogen can attack soybeans as well as sorghum and many of the other crop species that may be included in this system. Development of sorghum cultivars specifically adapted to double-cropping will be necessary, and charcoal-rot resistance is an essential quality in these genotypes. Cultural measures—such as irrigation during the post-flowering period, balanced fertilization, and the appropriate plant populations—are also important factors in reducing the incidence of charcoal rot in these areas.

Finally it is obvious that continuous and integrated efforts by pathologists and sorghum breeders working together are essential for solving sorghum disease problems.

References

Nakamura, K., and **Fernandes, N.G.** 1984. Resultados dos ensaios de mildio do sorgo realizados no ano agricola 1983/84 em colaboração com o CNPMS-EMBRAPA-Universidade Estadual Paulista, Jaboticabal, Brazil: Universidade Estadual Paulista, Jaboticabal. 5 pp.

Pinto, N.F.J.A. 1984. Virus do mosaico da cana-de-açùcar em sorgo [*Sorghum bicolor* (L.) Moench.]: Caracterização de isolados, reação de cultivares e herança da resistença. Ph.D. Thesis University of São Paulo. Piracicaba, Brazil. 136 pp.

Schaffert, R.E. 1986. Desenvolvimento de cultivares de sorgo para uso na alimentação humana. Informe Agropecuario 12(144): 13–14.

Sharma, H.C. 1983. A technique for identifying and rating resistance to foliar diseases of sorghum under field conditions. Proceedings, Indian Academy of Sciences 42(3):278–378.

Vanderplank, J.E. 1963. Plant diseases: epidemics and control. New York, NY, USA: Academic Press. 349 pp.

Vanderplank, J.E. 1982. Host-pathogen interactions in plant diseases. New York, NY, USA: Academic Press. 207 pp.

Veiga, A.C. 1986. Aspectos econômicos da cultura do sorgo. Informe Agropecuario 12(144):3–5.

Vianna, A.C., Borgonovi, R.A., and **Freire, F.M.** 1986. Alternativas de cultivo para exploração do sorgo granifero. Informe Agropecuario 12(144):28–32.

Sorghum Diseases in South America

E. Teyssandier[1]

Abstract

The relative importance of sorghum in the South American countries of Argentina, Colombia, Paraguay, Uruguay, and Venezuela is discussed. Incidence of major and minor diseases determine current comparative disease situation. The grain molds, a new race of sorghum downy mildew, and bacterial diseases in the Argentine situation are discussed. Grain molds, head blight, downy mildew, stalk rot, and virus diseases, particularly the mosaic group, are major diseases in these countries of South America. Bacterial diseases in Argentina, and perhaps elsewhere in South America, may be more important than now recognized.

Introduction

The importance of sorghum in South America can be seen by examining three statistics: 1986/87 area sown, average yield, and total production (Table 1).

It is clear that Argentina is the major producer. Chile, Peru, and Bolivia are not listed because of the small area in these nations, nor is Brazil, as Dr Casela is reporting on Brazil (this volume).

South America's sorghum production accounts for 11.6% of the world production (1979/86 average). Argentina's production is about 50% of South America's total. Of the total area devoted to sorghum in South America, grain types account for about 75% and forage about 25% of production. Both are grown predominantly for animal feed, i.e., dairy, poultry, swine, and beef production.

Nowadays, most of South America's grain-sorghum area is sown with hybrids.

Prevalent Sorghum Diseases

Sorghum in South America is subject to damage by several diseases, some limiting sorghum pro-

Table 1. Sorghum production in Argentina, Venezuela, Colombia, Uruguay, and Paraguay, 1986/87.

Country	Area sown ('000 ha)	Production ('000 t)	Yield (t ha⁻¹)
Argentina	1005	3040	3.2
Venezuela	369	850	2.2
Colombia	264	718	1.7/3.5[1]
Uruguay	31	90	2.3
Paraguay	15	22	1.4

1. 1.7 t ha⁻¹ - rainfed; 3.5 t ha⁻¹ - supplemental irrigation.

duction. Others represent potential or minor problems. Many outstanding papers have described the importance of sorghum diseases in each country of South America during the past 10 years. Information in these papers provide the basis for comparison of the current disease situation, by country, through use of a conventional rating scale from 1 to 5, where 1 represents a situation in which limitations by diseases are very severe and maximum effort to solve the problem and render sorghum production profitable is required; 5 represents few disease lim-

1. Pathologist, Cargill S.A. Alem 623, 2700 Pergamino, Buenos Aires, Argentina.

Teyssandier, E. 1992. Sorghum diseases in South America. Pages 63–66 *in* Sorghum and millets diseases: a second world review. (de Milliano, W.A.J., Frederiksen, R.A., and Bengston, G.D., eds). Patancheru, A.P. 502 324, India: International Crops Research Institute for the Semi-Arid Tropics.

itations for the crop, and relatively little effort is required to remove the problem and allow full profitability (Table 2).

Comparisons of different methods for disease screening and disease control are beyond the scope of this review.

Major Disease Problems in Argentina

Grain molds

Resistant hybrids are the most practical and economical control measure for grain molding.

At present, grain molding and grainweathering, using Castor's terminology (1981), represent major problems affecting sorghum almost every year in areas where it flowers and matures during periods of prolonged rain, heavy dew, and high relative humidities. Losses have not been estimated, but a great reduction in grain viability and grain quality occurs if panicles become infected between anthesis and harvest.

Most commercial breeding programs have devoted efforts to screen for genotypes tolerant to grain mold, grain weathering, and sprouting, probably better called "field deterioration." This

Table 2. Severity of sorghum diseases in Argentina, Venezuela, Colombia, Uruguay, and Paraguay.[1]

Disease	Pathogen	Argen-tina	Vene-zuela	Colom-bia	Urug-uay	Parag-uay
Head						
Grain molds	*Fusarium moniliforme* *Curvularia* spp, others	1	1	2	-	-
Head blight	*Fusarium moniliforme*	2	3	-	-	-
Stalk and root rots						
Fusarium stalk rot	*Fusarium* spp	1	2	1	-	-
Charcoal rot	*Macrophomina phasesolina*	1	2	1	-	-
Acremonium wilt	*Acremonium strictum*	5	4	4	-	-
Anthracnose	*Colletotrichum graminicola*	4	3	3	-	-
Mildews						
Sorghum downy mildew	*Peronosclerospora sorghi*	2	3	3	1	1
Virus						
Sugarcane mosaic group		3	1	3	2	2
Foliar						
Anthracnose	*Colletotrichum graminicola*	4	3	3	-	-
Bacterial leaf stripe	*Pseudomonas andropogonis*	2	4	4	-	-
Bacterial streak	*Xanthomonas holcicola*	3	4	4	-	-
Leaf blight	*Helminthosporium turcicum*	4	3	4	-	-
Rust	*Puccinia purpurea*	4	3	3	-	-
Gray leaf spot	*Cercospora sorghi*	3	4	5	-	-
Zonate leaf spot	*Gloeocercospora sorghi*	4	4	5	-	-

1. 1 = severe disease; 5 = minimum limitation by disease; (-) = no information available.

term, proposed by Glueck et al. (1977), is my preference as the descriptor of this complex syndrome.

Field tolerance to grain deterioration has been achieved in commercial brown-grained sorghum cultivars. This characteristic is determined by the presence of a testa and by high tannin content (see Waniska, this volume). In areas where environment has not favored panicle fungal colonization, sorghum hybrids having low tannin content normally produce high yields and find good acceptance by consumers.

High-tannin sorghums comprise about 85% of Argentina's sorghum production. Coordinated efforts of breeders and pathologists have developed new screening procedures that identify good sources of stable resistance to grain molding in low-tannin sorghums.

Head blight

Recent reports of severe incidence of head blight in central Argentina are of considerable concern because this disease had not been reported from this area, and panicle damage was serious.

Sorghum downy mildew (SDM)

Most Argentine commercial hybrids are resistant to downy mildew caused by *Peronosclerospora sorghi* (Weston and Uppal) Shaw. In 1985/86 a change in the pathogen population was detected in some areas of Chaco and Cordoba provinces. Commercial hybrids having RTx 430 or derivatives as male parents were susceptible, showing systemic and local symptoms.

Sorghum breeders have used RTx 430 as a source of sorghum downy mildew (SDM) resistance since the 1970s, developing hybrids with high levels of resistance.

The change in the SDM pathogen populations represents the major threat to sorghum production in these provinces even though many other conditions must be present for an epidemic to occur. A widespread and uniformly susceptible genotype, environmental conditions favoring reproduction of the new race, and time for disease build-up are required. Several reports (Craig and Frederiksen 1980; Pawar et al. 1985) clearly demonstrate variations in *Peronosclerospora sorghi* virulence. There is evidence that suggests that unknown factors have prevented new races from reaching their full potential, like the case of pathotype 3 in the United States of America.

Action to prevent a drastic SDM epidemic includes identification of the new race, and detection of sources of resistance. These objectives have been achieved.

Bacterial Diseases

During the last 5 years, a significant build-up of bacterial diseases has occurred in Argentina. Three aspects related with these microorganisms—leaf symptoms, grain infection, and stalk rot—occupy our attention.

Leaf symptoms

Bacterial stripe (*Pseudomonas andropogonis*) is now common in almost every sorghum crop. Usually, bacterial streak (*Xanthomonas holcicola*) is also seen, but to a lesser extent. Since 1983/84, severe bacterial leaf damage has been recorded in our breeding nursery at Pergamino, Buenos Aires Province. Some genotypes showed symptoms before flowering, rather than after bloom as expected. Something has changed. We don't know if we are dealing with a pathogen population that comprises virulent strains causing severe sorghum infection. Avezdzhanova and Sidorova (1978, pp. 63–65) compared pathogenicity of different strains of bacterial leaf spot and bacterial streak affecting sorghum.

Likewise we don't know much about time of initial infection, predisposing factors, effects on yield or grain quality, or if a vector is involved.

Grain infection. In 1984/85, a sorghum germination problem caused by *Pseudomonas* spp was identified. Large numbers of bacteria were observed in close proximity to the embryos, and black and brown lesions were seen in every case. Infected seeds emerged as poor seedlings and decayed easily.

Gaudet and Kokko (1986) mentioned a seedling disease of sorghum caused by seedborne *Pseudomonas syringae*.

Stalk rot. Last season stalk rot symptoms were observed in sorghum crops located in Cordoba Province. Plants were mature. The stalk pith was rotten in appearance and gave off a slightly sweet odor. Isolations of affected tissue in KB medium yielded white fluorescent and non-fluorescent bacterial colonies.

Conclusion

Grain mold, stalk rot, head blight, sorghum downy mildew, and virus diseases (particularly the mosaic group), are major diseases in the South American countries discussed. Bacterial diseases in Argentina represent a potentially dangerous problem that needs to be investigated more carefully. Our recommendation is to search for resistance-screening procedures and then eventual selection for genetic control.

References

Avezdzhanova, G.P., and **Sidorova, V.K.** 1978. Comparative pathogenicity of strains of the pathogens of bacterial leaf spot and bacterial streak of millet crops. Byulleten' Usesoyuznogo Nauchno Issledovatel's Kogo. Institute Zashchity Rastenil (1978) No 43.

Castor, L.L. 1981. Grain mold histopathology damage assessment, and resistance screening within *Sorghum bicolor* (L.) Moench lines. Ph.D. Dissertation, Texas A&M University, College Station, TX, USA. 170 pp.

Craig, J., and **Frederiksen, R.A.** 1980. Pathotypes of *Peronosclerospora sorghi*. Plant Disease 64:778–779.

Gaudet, P.A., and **Kokko, E.G.** 1986. Seedling disease of sorghum grown in southern Alberta caused by seedborne *Pseudomonas syringae* pv *syringae*. Canadian Journal of Plant Pathology 8:208–217.

Glueck, J.A., Rooney, L.W., Rosenow, D.T., and **Miller, F.R.** 1977. Physical and structural properties of field-deteriorated (weathered) sorghum grains. *In* Weathered Sorghum Grain. Texas Agricultural Experiment Station MP-1375 College Station, TX, USA: Texas Agricultural Experiment Station. 93 pp.

Pawar, M.N., Frederiksen, R.A., Mughogho, L.K., Craig, J., Rosenow, D.T., and **Bonde, M.R.** 1985. Survey of virulence in *Peronosclerospora sorghi* isolates from India, Ethiopia, Nigeria, Texas (USA), Honduras, Brazil, and Argentina. Phytopathology 75:1374. (Abstract).

Sorghum Diseases in Central America and the Caribbean Basin

G.C. Wall[1] and D.H. Meckenstock[2]

Abstract

Sorghum in Central America and the Caribbean Basin is the second-most food crop; production in the entire region in 1985 was 653 000 t. Of this, about 80% was produced in Central America. Sorghum diseases are outlined according to their relative importance: dwarf mosaic and sugarcane mosaic are important; in Honduras, yield in healthy vs MDMV-diseased plants differed by 52%. Gray leaf spot, the most widespread foliar disease in farmers' fields, caused 15% reduction in yield. Oval leaf spot and rust caused 6% and 4% yield reduction, respectively. Zonate leaf spot reduced grain yield by 14% in test plots. Downy mildew, acremonium wilt, anthracnose, and charcoal rot are at times severe on sorghum; the first two have reduced yields by 44% and 36%, respectively. Some widespread diseases—such as grain mold, covered smut, loose smut, and leaf blight—are usually not important.

Introduction

In Central America, 523 000 t of sorghum grain were produced in 1985. When sorghum grown on the Caribbean Islands is included, the total becomes 653 000 t (FAO 1985). In many Central American and Caribbean nations, it is second-most of the cereal crops, following maize. The sorghum plant itself may be used for forage, and the sorghum grain may be fed directly or in feed concentrates. In the home, sorghum alone or a sorghum/maize mix is used as an ingredient in tortillas, although maize is the preferred cereal for this purpose. Other traditional foods, such as *atol, alboroto,* and *chancaca* may be prepared from sorghum grain.

Certain sorghum (broomcorn) cultivars are grown for the manufacture of brooms.

Sorghum can be mixed with wheat as an extender of baking flour (Herrera and de Palomo 1984), and thus could help reduce the necessity of wheat importations for Central American and Caribbean nations.

Sorghum probably reached Central America from Africa in the 16th century, with an adjunct of the slave trade. In Guatemala, El Salvador, the Honduras, Nicaragua, and Haiti, as much as 85% of the sorghum production is from photo-period-sensitive 'native varieties' or landrace cultivars, mostly sown in association with maize.

Intercropping sorghum and maize is a common practice of resource-poor farmers. Most of these farmers sow less than 5 ha each, most of it on marginal land with slopes exceeding 8% (Hawkins et al. 1983).

In general, the sorghum crop grows with very little attention from the farmer. Diseases and pests take a high toll from an already limited potential yield. Yields may be thwarted in the first place by the lack or insufficient use of fertilizers, and by overcropping and poor soil management in general. Land that is sown by its owners is commonly overworked; for lack of space, crop rotation is not practiced frequently enough. If the land is rented or share-cropped,

1. Assistant Professor, Plant Pathology, College of Agriculture and Life Sciences, Univeristy of Guam, Mangilao, GU 96913, USA.
2. Associate Professor, Soil and Crop Sciences, Texas A&M University, INSTSORMIL, Apartado Postal 93, c/o Escuela Agricola Panamericana, Tegucipalpa, the Honduras.

Wall, G.C., and Meckenstock, D.H. 1992. Sorghum diseases in Central America and the Caribbean basin. Pages 67–73 *in* Sorghum and millets diseases: a second world review. (de Milliano, W.A.J., Frederiksen, R.A., and Bengston, G.D., eds). Patancheru, A.P. 502 324, India: International Crops Research Institute for the Semi-Arid Tropics.

soil erosion is seldom controlled. The net result of these practices adds up to a very low yield average (1.5 t ha^{-1}) for Central America and the Caribbean.

In most countries of the Central American region, there are current efforts underway to increase sorghum yields. Improved cultivars, mostly photoperiod-insensitive and early-maturing, are available and recommended for cropping under high-input systems. The farmer must rely on the availability of credit and technical help from extension agencies. Improved cultivars do not function well in association with maize.

In El Salvador and the Honduras, crop improvement efforts also include the development of higher-yielding, photoperiod-sensitive sorghums that can be used in association with maize. The practice of sowing maize with sorghum is a risk-reducing strategy, especially in areas where the maize may fail due to lack of rains at critical growth stages. The sorghum crop (maicillos) is more drought-resistant, but maize (maize, or maize grueso) is the preferred crop.

Maize-sorghum intercropping, advantageous as it is in these areas, can sometimes result in more disease problems. Maize dwarf mosaic affects maize and sorghum, and its incidence can be higher in maize-sorghum fields than in sorghum growing alone (Wall 1986).

Most of the diseases of sorghum are present in Central America. Notable exceptions are witchweed, long smut, and ergot.

Important Diseases

There are variations in the relative importance of particular sorghum diseases from one area to another, and from one season to the next. But certain diseases, such as gray leaf spot, are widespread and endemic; others are particularly important in one area. Some diseases may become important only in a particular year.

Sorghum diseases were surveyed in the Honduras between 1983 and 1985. A disease inventory was compiled, and the relative importance of these diseases was estimated from their incidence and severity. Finally, yield losses due to the most important diseases were estimated (Meckenstock and Wall 1987).

These studies show that, overall, the most important diseases on the basis of incidence times

severity were gray leaf spot, oval leaf spot, rust, ladder leaf spot, and zonate leaf spot, respectively. On a regional basis, their importance varied. Sorghum downy mildew, for instance, was totally absent in Choluteca (southern the Honduras), although it was the most important disease in the Comayaguan area of the country. Certain diseases, important at the beginning of the season, were insignificant at the time of harvest; such was the case with anthracnose. Improved cultivars (photoperiod-insensitive), which are not intercropped and are usually fertilized, had stalk rot problems, while this disease was hardly seen on intercropped, photoperiod-sensitive sorghum (maicillos). The fact that the latter are normally not fertilized may indicate that the stalk rot problem is induced; if so, it would be expected to become more important with increased use of fertilizers and improved sorghum cultivars.

Foliar diseases

The most prevalent foliar disease in Guatemala, the Honduras, and El Salvador, and perhaps in all the other countries of the region, is gray leaf spot (*Cercospora sorghi* Ellis and Everhart) (Escobedo et al. 1979; Meckenstock and Wall 1987; Wall 1980). It can be found anywhere sorghum is grown, and affects perennial weeds, such as *Sorghum halepense*, that no doubt play an important role in maintaining the inoculum of *C. sorghi* and many other pathogens throughout the year. Gray leaf spot affects sorghum plants early in their development; on intercropped susceptible varieties, young plants can be severely defoliated while still in the shade of the maize crop (Wall 1980). The fungus infects leaves and stems, with yield losses up to 15% (Meckenstock and Wall 1987).

Rust (*Puccinia purpurea*) is also found everywhere, particularly towards the end of the rainy season, becoming more severe with the onset of the dry season. It is an important disease of sorghum in Panama (Jimenez 1984). This disease can cause defoliation on susceptible varieties. A hyperparasitic fungus, *Darluca filum*, occurs on rust of sorghum (Contreras and Barahona 1974), attacking also maize rust (*P. sorghi*). Certain dipterous larvae also feed on the pustules formed by the rust fungus. Like *Cercospora sorghi*, *P. purpurea* also infects johnsongrass.

Zonate leaf spot, a disease incited by *Gloeocercospora sorghi*, may, on a regional basis, be the third most prevalent of the foliar diseases. It is a serious problem in Dominican Republic and Panama (Escobedo et al. 1979; Perez Duverge et al. 1984; Jimenez 1984; Wall 1980; Meckenstock and Wall 1987). A higher incidence of zonate leaf spot was recorded in fields where sorghum was sown alone than in sorghum intercropped with maize (Wall 1986). This foliar disease is most commonly found on the older leaves and sheaths, but it can reach epidemic proportions under favorable conditions and attack all foliage. A 14% loss in yield was measured under high disease severity levels (Meckenstock and Wall 1987).

Ramulispora sorghicola, causal agent of oval leaf spot, first reported on herbarium samples from Haiti in 1960 (Harris 1960) had not been recorded on the American mainland. Now it can be found in certain areas of El Salvador and the Honduras, reaching higher severity levels than other foliar diseases and causing defoliation (Wall 1986).

Anthracnose (*Colletotrichum graminicola*) can affect all aerial parts of the sorghum plant (Frederiksen 1986). In Guatemala, where it is considered an important disease (Escobedo et al. 1979), the pathogen can be found sporulating on infected individual grains of susceptible sorghum panicles (Frederiksen 1986). The foliar stage of the disease can be found on young plants early in the growing season. In the authors' experience, anthracnose usually becomes less important as the crop matures and the rains subside and other foliar diseases, such as gray leaf spot and rust, become more prevalent. Anthracnose is favored by hot and humid conditions. In Panama, for instance, it is one of the most important problems in sorghum production (Jimenez 1984). Johnsongrass is a wild host of this disease, as well.

Incidence of leaf blight (*Exserohilum turcicum*) usually is low but every now and then it can flare up to noticeable levels. Again, johnsongrass is one of its hosts. This disease has been serious in certain parts of Mexico in recent years. Many imported sorghum-sudan hybrids, grown for forage, are particularly susceptible to leaf blight in Central America.

Another species of *Cercospora* (*C. fusimaculans*), causing ladder leaf spot, has been described recently on sorghum (Wall et al. 1987). It

is not as widespread as gray leaf spot, but occurrence has been reported from the United States of America to Brazil. Resistance to this disease, apparently independent of that to *C. sorghi*, has been observed. Symptoms differ from those of gray leaf spot mainly in the scalariform pattern of the lesions Bacterial leaf stripe (*Pseudomonas andropogonis*), also found in many sorghum-producing areas, is of minor importance. In breeding nurseries, certain cultivars show a marked susceptibility to this disease.

Sheath blight, caused by *Sclerotium rolfsii*, has been observed in the Honduras and El Salvador, but is of minor importance.

Systemic diseases

Maize dwarf mosaic virus (MDMV) and sugarcane mosaic are aphid-transmitted virus diseases found throughout Central America. They are also spread by mechanic transmission. They can hardly be distinguished by symptoms alone; in fact, the two can occur together. The first is common on johnsongrass and other wild grasses; the second one can be found on sorghum where sugarcane is also grown. In a comparison of healthy vs diseased sorghum, MDMV-infected plants had 52% less yield than healthy plants (Meckenstock and Wall 1987). These virus diseases also occur on maize.

Maize chlorotic dwarf, another virus disease, has been reported in a remote area of the Honduras. Unlike those previously mentioned, this virus is transmitted by leaf-hoppers (*Graminella* sp), and is not known to be mechanically transmitted. At this time, this disease is of no significant importance.

Sorghum downy mildew (SDM), caused by *Peronosclerospora sorghi*, is an important disease of maize and of sorghum. It is found in southeastern Guatemala, and in several sorghum-producing areas of El Salvador, the Honduras, and Dominican Republic (Castellanos et al. 1982; Wall 1980; Fernandez and Meckenstock 1987; Perez Duverge et al. 1984). Fortunately, this disease is absent in some important sorghum production areas, such as the southern departments of the Honduras.

Oospores of *P. sorghi* are carried with sorghum seed; they can also be found in the glumes that remain attached to the few seeds occasionally formed on systemically infected plants (most systemically infected plants are barren). Infected plants also produce conidia that become airborne under favorable conditions. With certain maize-sorghum intercropping schemes, conidia produced by the maize plants may reach the sorghum plants in time to cause systemic infections. Infections can become systemic mainly during the first 3 weeks of plant growth. Systemically infected sorghum plants produce numerous oospores that are released when the dying leaves shed. These can infest the sorghum seed produced on healthy plants, and carry over in the soil. In a yield-loss study on sorghum in the Honduras, an incidence of 43% SDM reduced grain production by 44% (Wall et al. 1986).

Three different pathotypes of *P. sorghi* have been found in the Honduras so far (Fernandez and Meckenstock 1987). This development complicates implementation of effective control strategies based on resistance. Constant monitoring of the pathogen population in different production areas and deployment of the adequate resistant genotypes in each is required.

Acremonium strictum invades vascular tissues of sorghum, causing wilting and burning of the leaves. The disease has been seen all over the Honduras, on local sorghums as well as on improved cultivars. Its prevalence is very high in some areas, but its severity is usually low; only certain susceptible materials are affected seriously. Infected plants may produce no grain at all; an average of 36% grain mass reduction (Table 1) was observed on infected plants when compared to healthy ones (Meckenstock and Wall 1987). The disease can be seed-transmitted (Bandyopadhyay et al. 1987). Our own observations in the field lead us to believe that acremonium wilt can also become established on plants after they suffer insect damage, either from stalk borers or armyworms.

Stalk and panicle diseases

Covered smut (*Sporisorium sorghi*) and loose smut (*Sphacelotheca cruenta*) are common, because most farmers save part of their own harvested grain for next year's seed. When seed is treated with thiram, the problem is virtually eliminated. Head smut (*Sporisorium reilianum*) has been seen in an experiment station in El Salvador, but does not occur regularly (R. Ortiz, CENTA/MAG, San Andres, El Salvador, personal communication).

With maicillos, the problem of grain mold is minimal, as the flowering period coincides with the end of the rainy season. Photoperiod-insensitive sorghums must be sown so that their flowering times coincide with dry weather, or they can develop serious grain mold problems. Several fungi are involved, but it is mainly *Fusarium moniliforme* and *Curvularia lunata* that are involved in grain molding in Central America. They appear as pink and black discolored grains, respectively, on the panicles.

Head blight can be caused by *F. moniliforme* and *Colletotrichum graminicola*. The basal node area of the panicle is invaded, and shows a discoloration if split open. Insect damage promotes colonization by *Fusarium*. Conditions that favor grain mold are conducive to head blight.

The basal part of the sorghum stalk can also be attacked by *F. moniliforme*, leading to premature death of the plant. *F. moniliforme* infections are also associated with insect and mechanical damage.

Charcoal rot (*Macrophomina phaseolina*) can affect plants that are subjected to drought stress during grain filling. This disease is important in Panama (Jimenez 1984). It occurs mostly on high-yielding sorghums that have been heavily fertilized, particularly with high levels of nitrogen and low levels of potassium, and then subjected to drought stress (Frederiksen 1986). Charcoal rot seldom occurs on maicillos, or in maize-sorghum intercropping.

Table 1. Yield loss of grain sorghum from systemic diseases in the Honduras.[1]

Disease	Type of comparison	Yield loss (%)
MDMV	diseased vs healthy[2]	52
SDM	3% vs 43%[3]	44
Acremonium wilt	diseased vs healthy	36

1. Source: Wall (1986).
2. Comparisons between paired plants.
3. Comparisons between resistant and susceptible near-isogenic populations.

National Research Efforts

Crop improvement efforts are underway in most countries of Central America. Resource-poor farmers prefer to save seed from their own crops, rather than buy seed each year. This sustains the demand for open-pollinated varieties. In El Salvador there have been releases of improved maicillo varieties suitable for intercropping. Many improved (photoperiod-insensitive) sorghum varieties and hybrids have also been released; open-pollinated white-seeded varieties are available for making tortillas, red-seeded hybrids for grain production are intended for animal feed, and sorghum-sudan hybrids have been released for use as forage crops (Clara et al. 1984).

Photoperiod-sensitive hybrids are being developed in El Salvador in an effort to reach higher yields under the maize-sorghum intercropping system.

In the Honduras there have been releases of white-seeded open-pollinated varieties for tortillas, and a hybrid for animal feeds. Guatemala has a similar hybrid. In addition, maicillos are being improved for yield and for disease resistance in the Honduras and will soon be ready for release.

Large-scale sorghum sowings frequently make use of imported hybrids, particularly in Costa Rica, Panama, and Dominican Republic, and to an increasing extent in Guatemala. However, these must undergo local and regional evaluations that include overall adaptability, yield, and disease resistance. Field screening for SDM resistance, for instance, is done at Comayagua, the Honduras.

SDM was first reported in Central America in 1975. Since then, El Salvador's sorghum program has included greenhouse screening for SDM resistance. The disease spreads rapidly throughout the region, much to the concern of sorghum and maize workers. Because of this, a cooperative effort was made to find SDM-resistant materials in the Central American region (Fernandez and Meckenstock 1987). In 1979, the experiment station of Las Playitas in Comayagua was designated the site for all interested national programs of the region to send their maize and sorghum cultivars for field evaluation. At that time, there were no reports of different pathotypes of this fungus. Now that several pathotypes are known to occur in the

Honduras, the pathogen populations need to be monitored throughout the areas where SDM is known to exist. Once the pathotype is identified, corresponding sources of resistance can be introduced.

In the past, there has been a widespread tendency to downplay the importance of foliar diseases. There is no question now concerning the importance of these diseases in the Honduras (Table 2). Although sorghum improvement efforts in the region have included a concern for plant health in general, it is hoped that a more concentrated effort will be made in the future to develop and release materials that possess higher degrees of resistance to the most important diseases. Sources of resistance to many important diseases have already been identified (Meckenstock and Wall 1987).

Table 2. Yield losses of grain sorghum from foliar diseases in the Honduras.[1]

Disease	Extent or duration of infection	Yield loss (%)
Gray leaf spot	< 20% vs 100%[2]	15
Zonate leaf spot	310 vs 993 AUDPC[3]	14
Oval leaf spot	< 20% vs 100%	6
Rust	< 5% vs 25%	4

1. Source: Wall (1986).
2. Comparisons made between cohorts with low and with high disease-severity levels.
3. Area under the disease progress curve.

Conclusion

In Central American countries, there is an ever-increasing demand for cereals for animal and human food. Sorghum, therefore, has an important role in the region's future. Diseases constitute an important constraint to sorghum production (but not the only constraint).

In some countries there is still room for expanding the area for sorghum production. In Haiti, for example, an estimated 11% of the arable land was not yet utilized in 1979 (Coissy et. al. 1979). Guatemala has room for growth (Fuentes and Salguero 1983). In most cases, it is land tenure that limits increases in sorghum hectarage. In southern the Honduras, sorghum production is being displaced by cattle ranching (DeWalt and DeWalt 1984).

Before 1979, most sorghum farmers in Nicaragua used no fertilizers (Rizo and Obando Solis 1979); sorghum yields were low. According to FAO reports, there has been a dramatic increase in sorghum production during the last 6 years. The total sorghum-producing area in Nicaragua increased by nearly 50%; more importantly, yields were 67% higher. Total sorghum production increased by 143%.

Increases in sorghum production in Central America must be based on higher yields. Reaching higher yields requires the development and use of higher-yielding cultivars with improved disease resistance; fertilizers must also become available to the resource-poor farmer in order for him to maximize his yields. Finally, technical assistance must follow to help these farmers cope with new pest problems that are sure to come with the introduction of new cultivars or the use of fertilizers or with other changes in the cropping system.

References

Bandyopadhyay, R., Mughogho, L.K., and Satyanarayana, M.V. 1987. Systemic infection of sorghum by *Acremonium strictum* and its transmission through seed. Plant Disease 71:647–650.

Castellanos, S., Dardon, O., Ozaeta, M., Soto, G., and Cordova, H. 1982. Mildiu en maiz, descripción, incidencia, y métodos de control en Guatemala. Folleto técnico 19: ICTA, Guatemala. 24 pp.

Clara, R., Cordova, R.H., and Coty Amaya, H. 1984. El programa de mejoramiento genético del sorgo del Centro de Tecnología Agrícola. Pages 100–108 *in* Memoria III reunión anual de la Comisión Latinoamericana de Investigadores en Sorgo, 18–24 Nov 1984, San Salvador, El Salvador. San Andres, El Salvador: CENTA/ICRISAT.

Coissy, H.T., Saint Phard, J., and Prophete, E. 1979. Reporte nacional; Haiti. Pages 138–145 *in* Control integrado de plagas en sistemas de pro-

ducción para pequeños agricultores, Vol. III. Turrialba, Costa Rica: CATIE-UC/USAID-OIRAS.

Contreras, S., and Barahona, M. 1974. Evaluación del porcentaje natural de parasitismo efectuado por el micoparásito *Darluca* sp. en la roya del maíz (*Puccinia sorghi* Scha). SIADES 3(1):7–10.

DeWalt, B.R., and DeWalt, K.M. 1984. Sistemas de cultivo en Pespire, sur de the Honduras: un enfoque de agroecosistemas. Estudios antropologicos e históricos 4. Tegucigalpa, Honduras: Instituto Hondureño de Antropología e Historia e INTSORMIL. 88 pp.

Escobedo, J., Cano, M., and Gamboa, R. 1979. Reporte nacional; Guatemala. Pages 97–122 *in* Control integrado de plagas en sistemas de producción de cultivos para pequeñs agricultores. Vol. III. Turrialba, Costa Rica: CATIE-US/USAID-OIRSA.

FAO 1985. FAO Production Yearbook. Rome, Italy: Food and Agriculture Organization of the United Nations.

Fernandez, K.D., and Meckenstock, D.H 1987. Virulencia de *Peronosclerospora sorghi* en the Honduras. In Memoria de la XXXIII reunión anual del PCCMCA (Programa Cooperativo Centroamericano para el Majoramiento de Cultivos Alimenticios) 30 Mar to 4 Apr 1987. Guatemala, Guatemala:ICTA.

Frederiksen, R.A. (ed.). 1986. Compendium of Sorghum Diseases. St. Paul, MN: American Phytopathological Society. 82 pp.

Fuentes, J.S., and Salguero, E.R. 1983. Sistemas de cultivos practicados en Guatemala para la producción de maíz, fríjol, y sorgo. Pages 230–243 *in* Proceedings of the plant breeding methods and approaches in sorghum workshop for Latin America, 11–15 Apr 1983, Texcoco, Mexico: INTSORMIL-INIA-ICRISAT/CIMMYT.

Harris, E. 1960. *Ramulispora sorghicola* sp. nov. Transactions of the British Mycological Society 43(1):80–84.

Hawkins, R., Smith, M., and Arias Milla, R. 1983. Sistemas de cultivo de sorgo en Centro

America: importancia, localización y características. Pages 207–229 *in* Proceedings of the plant breeding methods and approaches in sorghum workshop for Latin America, 11–15 Apr 1983. Texcoco, Mexico: INTSORMIL-INIA-ICRISAT/CIMMYT.

Herrera, A.V., and **de Palomo, M.T.** 1984. Elaboracíon de productos a base de grano de sorgo. Pages 176–185 *in* Memoria III reunión anual de al Comisión Latinoamericana de Investigadores en Sorgo, 18–24 Nov 1984, San Salvador, El Salvador. San Andres, El Salvador: CENTA/ICRISAT.

Jimenez, D.M. 1984. Resumen de investigación en sorgo, (*Sorghum bicolor*) en Panama. Pages 160–168 *in* Memoria III reunión anual de la Comisíon Latinoamericana de Investigadores en Sorgo, 18–24 Nov 1984, San Salvador, El Salvador. San Andres, El Salvador: CENTA/ICRISAT.

Meckenstock, D.H., and **Wall, G.C.** 1987. Enfermedades de sorgo en the Honduras; su importancia y estratégias para su control. *In* Memoria del taller sobre Los maicillos criollos y otros sorgos en Meso–America: producción, utilización y mejoramiento, 7–11 Dec 1987, Tegucigalpa, the Hondoras, Secretaría de Recursos Naturales, Tegucigalpa, the Honduras: SRN-INTSORMIL-ICRISAT/CLAIS-AID/H.

Nolasco, R., Meckenstock, D.H., and **Wall, G.C.** 1984. Informe del proyecto nacional de sorgo en the Honduras. Pages 144–149 *in* Memoria III reunión anual del la Comisión Latinoamericana de Investigadores en Sorgo, 18–24 Nov 1984, San Salvador, El Salvador. San Andres, El Salvador:CENTA/ICRISAT.

Perez Duverge, R., Celado Montero, R., and

Caraballo, A. 1984. Investigaciones sobre el cultivo del sorgo en Republica Dominicana. Pages 150–159 *in* Memoria III reunión anual de la Comisión Latinoamericana de Investigadores en Sorgo, 18–24 Nov 1984, San Salvador, El Salvador. San Andres, El Salvador:CENTA/ICRISAT.

Rizo, M.P., and **Obando Solis, R.** 1979. Reporte nacional; Nicaragua. Pages 51–75 *in* Control integrado de plagas en sistemas de producción de cultivos para pequenos agricultores. Vol. III. Turrialba, Costa Rica: CATIE-UC/USAID-OIRSA.

Wall, G.C. 1980. The present status of sorghum diseases in El Salvador. Pages 18–21 *in* Proceedings of the International Workshop on Sorghum Diseases, 11–15 Dec 1978, Hyderabad, India. Patancheru, Andhra Pradesh 502 324, India: International Crops Research Institute for the Semi-Arid Tropics.

Wall, G.C., 1986. A study of sorghum diseases in the Honduras, their importance under different cropping systems, and strategies for their control. Ph.D. thesis, Texas A&M University, College Station, TX, 77843, USA. 108 pp.

Wall, G.C., Craig, J., Meckenstock, D.H., Nolasco, R., and **Frederiksen, R.A.** 1986. Efecto de *Peronosclerospora sorghi* en el rendimiento de *Sorghum bicolor* (L) Moench. In Memoria de la XXXII reunión anual del PCCMCA (Programa Cooperativo Centroamericano para el Mejoramiento de los Cultivos Alimenticios), 17–22 Mar 1986, San Salvador, El Salvador. San Andres, El Salvador: CENTA.

Wall, G.C., Mughogho, L.K., Frederiksen, R.A., and **Odvody, G. N.** 1987. A foliar disease of sorghum species caused by *Cercospora fusimaculans.* Plant Disease 71:759–760.

Sorghum Diseases in Mexico

J. Narro[1], V.A. Betancourt[2], and J.I. Aguirre[3]

Abstract

Grain sorghum (Sorghum bicolor) was introduced to Mexico in 1944; at that time, only open-pollinated varieties were sown. This crop became important in the early sixties; the first hybrids were well accepted by farmers.

In the first years, the crop was relatively free of pests and diseases but by the end of the 1970s this situation had changed, mainly in Mexico's two most important sorghum producing states: Tamaulipas and Guanajuato. Biotic and abiotic diseases occurring in Mexico are grouped according to importance, distribution, and prevalence in nine classes.

Information about control and research strategies to avoid further dissemination of the diseases is presented herein.

Introduction

Grain sorghum [*Sorghum bicolor* (L.) Moench] was introduced to Mexico in 1944 as a crop for rainfed areas, which account for 80% of the total agricultural area of Mexico. The first trials included open-pollinated varieties; but introduction, in the early sixties, of hybrid sorghums greatly increased the crop's importance. Sorghum was readily accepted by Mexican farmers because of its high yield, wide adaptation, ease of mechanization, and ease of incorporation into different production systems.

At sorghum's introduction, serious problems of diseases and pests did not occur, but now they are present throughout many of the sorghum-growing regions, and cause serious crop losses in the two major sorghum-producing states (Tamaulipas and Guanajuato). Head smut was the first serious disease problem in Tamaulipas, occurring in most commercial hybrids in the late 1960s; in the mid 1970s, downy mildew caused severe losses in the high-yielding yellow endosperm hybrids. The disease impact in Tamaulipas is critical, because 90% of the seed and 25% of Mexico's total sorghum is produced in this state. At present, crop losses occur in both rainfed and irrigated areas.

Because of disease problems, sorghum research was expanded and new hybrids, developed by private companies and Mexico's National Institute of the Forest, Agricultural, and Livestock Research (INIFAP), replaced the earlier hybrids. In 1972, Mexico's National Institute of Agricultural Research, (which became INIFAP in 1985), had released hybrids with good yield and adaptation potential, but they became susceptible, mainly to downy mildew.

Several years later, by a joint effort of breeders and pathologists using new research strategies, the hybrids RB 3030 and RB 3006 were released. These hybrids are adapted to growing conditions in northern Tamaulipas, Nuevo Leon, and Sinaloa. They exhibit good levels of resistance to downy mildew (*Peronosclerospora sorghi*) and head smut (*Sporisorium*

1. Agronomist, National Institute of Forest, Agricultural and Livestock Research (INIFAP) Apartado Postal 112, Celaya, Gto., Mexico.
2. Plant Breeder, Agricultural Research, S. of R.L., Colegio Militar 1124, Guadalajara, Jalisco, Mexico.
3. Plant Pathologist, National Insitute of Forest, Agricultural and Livestock Research (INIFAP), Apartado Postal 112, Rio Bravo, Tamaulipas, Mexico.

Narro, J., Betancourt, V.A., and Aguirre, J.I. 1992. Sorghum diseases in Mexico. Pages 75–84 *in* Sorghum and millets diseases: a second world review. (de Milliano, W.A.J., Frederiksen, R.A., and Bengston, G.D., eds). Patancheru, A.P. 502 324, India: International Crops Research Institute for the Semi-Arid Tropics.

reilianum). Both hybrids have females developed for the Bajio and their pollinator is RTx 430 (Williams 1980).

In the late 1970s and early 1980s, the sorghum diseases spread to other sorghum-producing areas. In Guanajuato, sorghum downy mildew and head blight (*Fusarium moniliforme*) were reported as prevalent (Narro et al. 1983). In Michoacan and Jalisco, leaf blight (*Exserohilum turcicum*), rust (*Puccinia purpurea*), downy mildew, stalk rot, and head blight appeared (Betancourt 1980). Areas such as Nuevo Leon were affected mainly by downy mildew, head smut,

rust, leaf blight, maize dwany mosaic virus (MDMV) and anthracnose (*Colletotrichum graminicola*) (De la Garza 1980). At present, field data on the effects of these diseases on yield have not been recorded, but it has been estimated that they are responsible for 10–15% of the crop losses, and thus are second after weeds in sorghum problems in Mexico.

The diseases of sorghum in Mexico are grouped into nine classes. Diseases are grouped according to their presence during crop development; the prevalence, importance, and distribution of each group is described in Table 1.

Table 1. Relative prevalence, importance, and distribution of sorghum diseases in Mexico, 1987.

Disease	Prevalence[1]	Importance[2]	Distribution
Seedling			
Fusarium spp, *Pythium* spp, *Rizoctonia* spp, and *Aspergillus* spp	+	1	Tamaulipas, Jalisco, Guanajuato
Downy mildews			
Sorghum: *Peronosclerospora sorghi*	++	2	Tamaulipas, Michoacan, Jalisco, Nuevo Leon, Veracruz, Nayarit, Colima, Morelos, Chiapas
Crazy top: *Sclerophthora macrospora*	+	1	Tamaulipas, Jalisco, Guanajuato, Michoacan, Veracruz, Chiapas, Nuevo Leon, Morelos
Virus			
Maize dwarf mosaic virus (MDMV)	++	2	Guanajuato, Jalisco, Tamaulipas, Chiapas, Veracruz, Morelos, Colima
Foliar			
Leaf blight: *Exserohilum turcicum*	+++	3	Jalisco, Michoacan, Guanajuato, Nayarit, Tamaulipas, Nuevo Leon, Sinaloa, Morelos, Veracruz, Campeche
Leaf anthracnose: *Colletotrichum graminicola*	++	2	Tamaulipas, Nayarit, Veracruz, Nuevo Leon, Jalisco, Guanajuato
Zonate leaf spot: *Gloeocercospora sorghi*	++	2	Jalisco, Yucatan, Tamaulipas, Veracruz, Morelos, Chiapas, Nuevo Leon, Michoacan
Gray leaf spot: *Cercospora sorghi*	++	2	Tamaulipas, Yucatan, Nayarit, Nuevo Leon, Campeche, Veracruz, Jalisco, Guanajuato, Michoacan
Bacterial leaf streak: *Xanthomonas campestris* pv. *holcicola*	+	2	Guanajuato, Sinaloa, Sonora, Jalisco, Yucatan, Morelos
Bacterial leaf stripe: *Pseudomonas andropogoni*	+	1	Tamaulipas, Chiapas, Jalisco, Yucatan, Campeche, Guanajuato, Michoacan
Sooty stripe	+	0	Jalisco, Tamaulipas, Veracruz, Campeche

Continued

Table 1. *Continued.*

Disease	Preva-lence[1]	Import-ance[2]	Distribution
Bacterial sun spot: *Pseudomonas* spp	+	0	Guanajuato, Morelos
Rust: *Puccinia purpurea*	++	2	Jalisco, Michoacan, Guanajuato, Tamaulipas, Nuevo Leon
Smuts			
Head smut: *Sporisorium reilianum*	++	3	Tamaulipas, Michoacan, Sinaloa, Guanajuato, Nuevo Leon, Nayarit, Colima, State of Mexico
Covered kernel smut: *Sporisorium sorghi*	+	0	Sinaloa, Coahuila, Guanajuato, Tamaulipas, Jalisco, Morelos
Loose kernel smut: *Sphacelotheca cruenta*	+	0	Sinaloa, Guanajuato, Jalisco, Michoacan, Tamaulipas, Morelos
Root and stalk rot			
Stalk rot: *Fusarium moniliforme*	+++	3	Guanajuato, Jalisco, Michoacan, Nayarit, Nuevo Leon, Colima
Charcoal rot: *Macrophomina phaseolina*	++	3	Tamaulipas, Michoacan, Nuevo Leon, Colima, Veracruz, Jalisco, Guanajuato
Anthracnose stalk rot: *Colletotrichum graminicola*	+	2	Tamaulipas, Nayarit, Nuevo Leon, Veracruz, Guanajuato, Michoacan
Pokkah boeng: *Fusarium mon iliforme*	+	1	Jalisco, Guanajuato, Tamaulipas, Michoacan
Acremonium wilt: *Acremonium strictum*	+	0	Guanajuato, Tamaulipas, Morelos
Panicle and seed			
Grain mold: *Curvulari* spp *Fusarium* spp and *Alternaria* spp	++	2	Tamaulipas, Nayarit, Veracruz, Colima, Michoacan, Guanajuato
Head blight: *Fusarium moniliforme*	+++	3	Guanajuato, Michoacan, Tamaulipas, Nuevo Leon
Fungi storage: *Aspergillus* spp, *Penicillum* spp, *Phoma* spp, *Alternaria* spp, *Curvularia* spp, and *Botrytis* spp	+	1	Jalisco, Tamaulipas, Guanajuato, Michoacan
Abiotic			
Chlorosis (Iron deficiency)	++	2	Guanajuato, Michoacan, Tamaulipas, Nuevo Leon
Pesticide injury	+	1	Guanajuato, Michoacan, Tamaulipas, Jalisco
Salt damage	+	1	Tamaulipas
Mycoplasm			
Yellow sorghum stunt	+	1	Guanajuato, Veracruz, Morelos

1. + = occasionally present, ++ = commonly present, and +++ = generally found on most plants in most fields.
2. 0 = causing no loss, 1 = minor importance, 2 = moderate importance, and 3 = a major deterrent at times to crop production.

Disease research conducted in Mexico during the period 1980–87 is likewise presented.

Seedling Diseases

These types of diseases are caused mainly by pathogens of the genera *Fusarium* spp, *Pythium* spp, and *Rhizoctonia* spp. In the Bajio, most farmers sow 22.5 kg hybrid seed ha^{-1}. This amount is above the average sown in other areas, as a higher seedling rate is thought advisable in view of losses to seedling decay, herbicide residues (Dual), inefficient plowing, and poor seed quality. Seedling diseases account for 20% of disease problems observed in some areas of the Bajio, and farmers are advised to verify the germination of the seed before sowing.

Downy Mildews

The most common mildews that occur in Mexico are downy mildew (*Peronosclerospora sorghi*) and crazy top (*Sclerophthora macrospora*). The latter has no economic importance.

Downy mildew (*P. sorghi*) has spread to the main grain sorghum production areas. In Tamaulipas, it has been recorded up to 80% of systemic infection on commercial hybrids (Betancourt 1980). Conidial infection is very severe at El Maluco, in the state of Michoacan, with ratings of 2.5 on a 1 to 5 scale (Narro et al. 1983). It is important to point out that this species attacks maize also, but in maize oospores are not produced, as it is commonly observed in sorghum. In commercial fields, effects on yields of susceptible hybrids may be up to 10%. In experimental fields, however, up to 50% yield loss has been recorded when systemic infection reaches 70% of incidence.

Pathotypes of downy mildew were identified by using the Downy Mildew Virulence Nursery. At Tamaulipas, pathotypes 1, 2, and 3 are present (Aguirre 1984b). In Michoacan, Guanajuato, and Jalisco only pathotype 1 has been found (Narro et al.1982; Betancourt, in press). In Mexico, pathotype 1 is the most common, averaging 40% of systemic infection in the hybrid ATx 399 × RTx 2536.

In relation to chemical control of downy mildew, the fungicide metalaxyl (Ridomil®) (C.A. 48988) was evaluated in tests conducted at Mic-

hoacan in 1981. Results indicated that a dosage of 1 g kg^{-1} of seed was very effective (Hernandez et al. 1982). These treatments were also applied to the hybrid Funk's G 766 W in commercial fields at the Bajio with excellent control (N. Avila, personal communication). For the most part though, downy mildew control has been through resistant hybrids. INIFAP released the hybrids RB 3030 and RB 3006 for Tamaulipas and Sinaloa (Williams 1980) and BJ 83 for the Bajio (Narro 1984); each of these hybrids has Tx 430 as common pollinator where a dominance to partial dominance for resistance has been found (Narro et al. 1982). So far, pathotype 3 is found only at Tamaulipas; no evidence of its presence in other areas of Mexico is known. This is fortunate, since RTx 430 shows susceptibility to this pathotype. Nor have pathotypes 4 and 5, identified at Brazil and the Honduras, been detected in Mexico (R.A. Frederiksen, personal communication).

The resistance observed in RTx 430 has been very stable; in nurseries at Beeville, Texas, hybrids with this pollinator show no more than 5% systemic infection. Thus, RTx 430 may possess several genes for disease resistance.

A program involving crosses of material developed by INIFAP in Mexico and initiated in 1985 at Texas A&M University was selecting germplasm adapted to temperate areas. Some F_4 lines from this program show promise. Although up to now we consider the sorghum downy mildew (SDM) pathogen to be under control, we need to develop new sources of resistance, since the downy mildew pathogen has the potential for change and can become very destructive in a short time.

As mentioned before, crazy top is found in very limited areas; in commercial fields it has no economic importance.

Viruses

Of the 19 viruses registered worldwide as sorghum pathogens (Toler 1980), only three are of economic importance in Mexico: sugarcane mosaic virus (SCMV), brome mosaic (BM), and maize dwarf mosaic virus (MDMV).

The latter is the most important in the sorghum production because it can infect up to 100% of susceptible material. Virus disease infections of sorghum in Mexico were first reported in 1977 (Martinez 1984). In 1981, field samples were collected and inoculations were made in one replication of the International Virus Disease Nursery (IVDN) and MDMV confirmed (Narro and Hernandez 1984). In 1982, the MDMV was identified by using the IVDN in the Jalisco area.

Toler (1980) points out that only strains A and B of the MDMV are important. Strain A infects johnsongrass; strain B does not. A study to determine the distribution of strains A and B of MDMV indicated that only strain A was present in Mexico.

Different sowing dates at the Experimental Station at the Bajio were used to study the distribution of this virus in relation to sowing date. It was observed that 35% of incidence occurred between 15 March and 15 April, with a significant reduction in later sowings (Narro and Delgadillo 1985). At the present time an evaluation of yield reduction caused by MDMV (A and B) is underway with four genotypes. In relation with genetic resistance, the best 25 inbreds of Bajio program were evaluated; 8 lines showed resistance to strain A, 15 to strain B, and 5 to both strains; the inheritance of resistance to MDMV is now being studied by using germplasm with different reaction to the virus.

Foliar Diseases

Leaf blight. (*Exserohilum turcicum* Pass.) It is causing severe damage, mainly in the states of Jalisco, Michoacan, and Guanajuato, but it is also present in seven other states.

In Jalisco it is very common to find commercial fields, rainfed and irrigated alike, highly damaged by the pathogen if sown late in the season (Distancia and Betancourt 1984). At La Barca (State of Jalisco), incidences up to 80% were found in susceptible hybrids, and few hybrids show tolerance.

Studies on chemical control of leaf blight show that preventive treatment with fungicides, such as maneb, propiconazole (Tilt®), or zineb, are not effective. In 1981, the germplasm developed at the Bajio was evaluated in three locations with high incidence of the pathogen; only one inbred (E. 13B, a kafir derivative) was resistant; in general, all kafir derivatives from BTx 3197 have shown tolerance.

Introduced material (ATx 623 × SC 0326-6 and ATx 623 × R 6956) that allowed identification of two restorer lines with good levels of resistance were identified in the TAT (Tropical Adaptation Test). As BTx 623 is susceptible, it is concluded that dominance is present in both pollinators. It should be mentioned that INIFAP has an experimental station at Ocotlan, Jalisco (near La Barca) where a breeding program for resistance to foliar diseases has been in place since 1980. Workers there have obtained the second backcross in a group of foliar resistant B lines; these are now being sterilized. Crosses of inbreds from Bajio × Texas have shown better response than those of Texas × Texas. Excellent sources for leaf-disease resistance have been developed by this program.

Bacterial streak. (*Xanthomonas campestris* pv *holcicola*.) This occurs in most of the Bajio area; at the state of Guanajuato its incidence is very high, mainly on susceptible genotypes such as Tx 623 (ratings of 4 on a 1 to 5 scale). Data on the effects of this disease on yield are not available. It probably combines with other diseases to predispose the sorghum plants to infection by others, such as *Fusarium moniliforme*. A nursery from Texas A&M Univeristy, established in 1983, contains seven lines with tolerance to this pathogen. These lines will be useful in the future as sources of resistance to foliar pathogens.

Zonate leaf spot. (*Gloeocercospora sorghi*.) It is widely distributed, mainly in the subtropical areas of Tamaulipas, the Ameca area in Jalisco, and coastal areas of Veracruz and Chiapas. Economic losses related with this disease have not been reported, however.

Gray leaf spot or **angular leaf spot.** (*Cercospora sorghi*.) It is found in almost every sorghum-growing area of Mexico, but has little effect on yield because it usually does not infect the leaves until late in the grain-filling stage, when the grain is near physiological maturity.

Leaf anthracnose. (*Colletotrichum graminicola*.) It has limited distribution in the major sorghum-growing areas of Mexico, but is more common in tropical areas where the sorghum hectarages are rather small. It is found in northern Tamaulipas and the coastal areas of Nayarit, Jalisco, and similar areas. Even though severe damage has not been observed, the disease is potentially dangerous to sorghums in the tropical areas.

Bacterial stripe. (*Pseudomonas andropogonis.*) The distribution and incidence of this disease is similar to that of bacterial streak, but *Pseudomonas* is less common than *Xanthomonas* at Guanajuato. On the other hand, there are reports that indicate a wide distribution of *Pseudomonas* in the high valleys of Jalisco and the Ocotlan area. Incidence, however, varies from year to year, and the disease is so far without economic importance

Sooty strip. (*Ramulispora sorghi.*) It is sometimes found in Tamaulipas, Jalisco, and small areas of the tropics in Mexico. The incidence of this disease is very low and is at present of little importance.

Bacterial sun spot. (*Pseudomonas* sp.) It is similar in importance and distribution to sooty stripe.

Rust. (*Puccinia purpurea* Cooke.) It is very common in tropical and subtropical areas of Mexico, particularly at the Jalisco area and in southern Tamaulipas; it can be very high in incidence and economic importance. Commercial hybrids have been severely affected in the last 5 years, with scores of 5 on a 1 to 5 severity scale. The damage, however, is not as high as expected, as the pathogen appears after flowering. Most commercial hybrids are susceptible to the pathogen. Tolerant hybrids, however, are extensively sown in areas such as Ocotlan in Jalisco.

Studies on the chemical control of rust have evaluated chlorothalonil (Bravo 500®), three sprayings—1 kg a.i. ha^{-1} applied prior to flowering, at flowering, and 15 days following—(Betancourt and Narro 1983; Frederiksen et al. 1983; Rodriguez 1983). Even though the incidence of pustules was reduced on the fungicide-treated plants, there is evidence that suggests that rust is not correlated with yield, although the presence of rust was correlated with the presence of *Fusarium*.

Smuts

Three types of smuts are present in Mexico: covered kernel, loose kernel, and head smut.

Head smut. (*Sporisorium reilianum.*) This is the only smut of economic importance in Mexico. It is found in most of the country's sorghum-growing areas, particularly in northern Tamaulipas where incidences up to 40% in commercial hybrids have been reported (Aguirre 1984a). The pathogen is highly variable and several races have been reported. In Tamaulipas, races 1, 2, 3, and 4 occur (Aguirre 1986). In Jalisco, Michoacan, and Guanajuato races 1, 2, and 3, (Hernandez et al. 1983) are common, races 1 and 3, however, are the most common with incidences from 20 to 40%. Herrera and Betancourt (1986) reported that race 3 of the pathogen is the most widely distributed in Mexico.

Control of head smut has been achieved by planting resistant hybrids such as RB 3030, RB 3006, and BJ 83 from INIFAP and others from private companies. Since the hybrids are also resistant to downy mildew, two pathological problems are solved simultaneously.

In Tamaulipas, a program for resistance to head smut and downy mildew was reinforced in 1977, with the objective of obtaining hybrids resistant to both pathogens. In 1984, several genotypes with high levels of resistance were identified; so far, the resistance to head smut and downy mildew has been stable (Aguirre 1986).

Root and Stalk Rots

Stalk rot and head blight. (*Fusarium moniliforme* Sheldon.) These are diseases widely distributed in Mexico and are economically important in the states of Jalisco, Guanajuato, and Michoacan. Most commercial hybrids are susceptible, though a few have shown moderate tolerance. At Guanajuato, incidences up to 40%, with scores of 4 on a 1 to 5 scale, have been reported (Narro and Betancourt 1983).

The effect of this disease on yield was estimated by noting yields of inbred and hybrid plants inoculated at the boot stage, anthesis, and 15 days after anthesis. Yield reductions were 33, 11, and 8%, respectively. In the case of the hybrids, when both parental lines were tolerant the yield reduction was 14% whereas in susceptible lines yield reduction accounted for 30% when plants were inoculated at the boot stage (Hernandez et al. 1987a).

In histopathological studies it was found that RTx 430, rated as tolerant, contains more lignin in the vascular tissue, less percentage of damage of the pith tissue, and higher percentage of epidermal (peripheral) tissues than the susceptible line R 28 B (Hernandez et al. 1987a).

The culture medium PDA was identified as excellent for growing the pathogen and the medium "Belay" likewise for sporulation (Delgadillo 1983). In inoculation studies using hypodermic syringe and toothpick, no significant differences on effectiveness of either were found (Hernandez et al. 1987b). At Chapingo, using the toothpick technique, trials produced 17% incidence which reduced yield by 7.3% (Vargus et al. 1985).

In general, Combine Kafir 60 has shown high levels of tolerance to this pathogen. It was also found that the incidence of rust and *Fusarium* infections are positively correlated.

In sowing-date comparisons at the Bajio, early sowing (such as 15 April) had a 60% incidence (Narro 1984). Some tolerant inbreds identified in 1982 are listed in Table 2 (Narro and Betancourt 1983).

In crosses using Tx 430 as pollinator, it was found that inheritance for resistance is recessive; for that reason resistance needs to be present in both parents.

In the INIFAP program at the Bajio, several F_4 generations from crosses between local and Texas materials have shown excellent levels of resistance.

Charcoal rot. (*Macrophomina phaseolina*.) It is very common around Apatzingan, Michoacan, and Ocotlan, Jalisco, and in other areas where severe drought occurs late in the season. In susceptible hybrids, important damage can occur from lodging. In experimental plots, we have observed that inheritance for resistance is recessive. This disease has the potential to cause yield losses in some semi-arid areas of Mexico.

Pokkah boeng (twisted top). (*Fusarium moniliforme* var *subglutinans*.) It is occasionally present in central Mexico, but is not known to be of economic importance.

Acremonium wilt. (*Acremonium strictum* W. Gams Syn. *Cephalosporium acremonium*.) It occurs occasionally in areas such as Jalisco and coastal tropical and subtropical regions of Mexico, including Guanajuato.

Table 2. Stalk rot and head blight ratings of 24 selected inbred sorghum lines showing tolerance to *Fusarium moniliforme* in sowing date studies at El Bajio Celaya, Guanajuat, and Mexico, 1982.

Inbred sorghum Line	Stalk rot rating[1]	Head blight rating
Tx 430	4.0	2.5
SC 599-11 E	2.5	2.5
SC 326-6	4.0	3.0
TAM 428	4.0	3.0
77 SC 2	4.0	4.0
77 CS 490	3.0	3.0
Tx 2536	3.5	3.5
BTx 378	4.0	4.0
BTx 623	4.0	3.5
GPR 148	4.0	2.0
E 35-1	3.5	4.0
SPV 35	3.0	3.5
UCH V1	3.0	3.5
UCH 2	3.5	3.5
IS 173	2.0	2.0
QL 3 India	3.0	3.0
CS 3541	3.5	4.0
B 198 B	2.5	1.0
B 194 B	2.5	2.0
B 214 B	2.0	2.0
B 206 B	2.0	2.0
E 15 B	2.0	2.5
R 5 B	3.0	3.0
R 27 B	3.0	3.0

1. Scored on a 1–5 scale, where 1 = resistant (R), 2 = moderately resistant (MR), 3 = moderately susceptible (MS), and 4–5 = susceptible (S).

A. strictum was first observed at Ocotlan in Jalisco by Betancourt and Frederiksen in 1981. Most hybrids are resistant, but the inbreds BTx 623 and BTx 625 are extremely susceptible to this pathogen.

Panicle and Seed Diseases

Anthracnose. Panicle and grain anthracnose (*Colletotrichum graminicola*) occur in the cloudy warm and humid sorghum-growing areas of

Mexico, particularly in Tamaulipas. Severe damage has not been reported.

Grain molds. (*Curvularia* spp, *Fusarium* spp, *Alternaria* spp.) These are important on the coastal areas of Mexico, including Tamaulipas. *Curvularia* spp, present in 90% of the molded heads, are the most common, followed by *Fusarium* spp.

Storage molds. These include several species; they occur occasionally in some areas, but are not reported to be important.

Abiotic Diseases

Chlorosis. (Iron deficiency.) It is common in the most important sorghum-growing areas of Mexico; entire fields of dwarf-yellowing plants appear. The problem has been more severe in recent years.

At Guanajuato, an estimated 10 000 ha are iron-deficient, and the chlorosis can prevent grain yields. The condition is common in calcareous soils with iron deficiencies.

In 1985, two lines were identified as tolerant to iron deficiency when grown in a greenhouse; resistance was later verified in field trials (Gonzalez and Galvan 1987). Tolerance of iron deficiency is being incorporated into elite material, and studies on the inheritance to this character are underway.

Pesticide injury. Chemical damage to sorghum plants is now a common problem in Mexico, becoming more so as chemical control of greenbugs is becoming more important. Many of the organophosphate products can burn up to 80% of the foliar area, reducing yields by an estimated 5% or more. Herbicides such as parafuat produce similar damages. Most of the RTAM 428 hybrids will show pesticide damage because the pollinator is susceptible to parathion.

Salt damage. It is caused because of salt residues from irrigation waters, occuring in large areas in northern Tamaulipas. Data on the area affected are not available, but the problem is increasingly important. Some hybrids seem to be more tolerant of salt than others. In severely affected areas, sorghum cannot be produced profitably.

Mycoplasm Diseases

Yellow sorghum stunt. It is the only sorghum disease known to be caused by a mycoplasm. It occurs occasionally in central Mexico, but incidence in susceptible lines has not been greater than 10% (Narro and Betancourt 1983).

Research Strategies in Mexico

Sorghum disease problems in Mexico are being approached mainly through the development of resistant hybrids and studies of chemical control. At Rio Bravo, Tamaulipas, research is focused on downy mildew and head smut. At Celaya, Guanajuato, the program includes studies of stalk rot and head blight, virus diseases, rust, and leaf blight. At Ocotlán, Jalisco, a program is underway for leaf blight and stalk rot resistance.

Mexico wishes to reinforce its sorghum research programs by:

1. Training personnel in sorghum pathology.
2. Determining yield losses to diseases.
3. Promoting international cooperation through establishing international disease nurseries, sharing elite resistance inbreds, and control technologies.

Acknowledgment. The authors express their gratitude to Leopoldo Mendoza Onofre for his assistance, and to agronomists Flores Gaxiola, J. Abel, A. Viscaino Guardado, Grajales Solis, M. Teniente Oviedo, Rodrigo F. J. Cruz, Ch. F. Monjaras. A., for information on sorghum diseases in southern Mexico.

References

Aguirre, R.J. 1984a. Razas fisiológicas del carbón de la panoja en el cultivo de sorgo en el norte de Tamaulipas. Page 73 *in* Proceedings XI Congreso Nacional de Fitopatologia, San Luis Potosi, 16–18 Jul 1984.

Aguirre, R.J. 1984b. Patotipos de mildiu velloso en el cultivo de sorgho—en el norte de Tamaulipas. Page 74 *in* Proceedings XI Congreso Nacional de Fitopatología, San Luis Potosí, 16–18 Jul 1984.

Aguirre, R.J. 1986. Informe parcial del grupo interdisciplinario de sorgo. Internal Report R85–1986. Rio Bravo, Tamps. National Institute of the Forest, Agricultural, and Livestock Research (Mexico).

Betancourt, V.A. 1980. Sorghum diseases in Mexico. Pages 22–28 in Sorghum diseases, a world review: proceedings of the International Workshop on Sorghum Diseases, 11–15 Dec 1978, Hyberabad, India. Patancheru, Andhra Pradesh 502 324, India: International Crops Research Institute for the Semi-Arid Tropics.

Betancourt, V.A. (In Press). Patotipos de downy mildew. Segunda Reunion Nacional sobre sorgo, 14–17 Oct 1986. Escuela Superior de Agricultura, UAS, Sinaloa, Mexico.

Betancourt, V.A., and Narro, S.J. 1983. Yellow sorghum stunt: a new disease of sorghum in Mexico. Page 77 in Proceedings, Thirteenth Biennial Grain-Sorghum Research and Utilization Conference 22–24 Feb 1983, Brownsville, Texas, USA.

de la Garza, J.L. 1980. Enfermedades del sorgo en la región de Anahuac, N. L. Page 10 in Proceedings IX Congreso Nacional de Fitopatologia, Uruapan Michoacan, 16–18 Jul 1980.

Delgadillo, J.H. 1983. Aislamiento, cultivo y esporulación de Fusarium moniliforme Sheldon, causante de la pudrición del tallo en sorgo – (Sorghum bicolor L. Moench). BSc. thesis, Universidad de Guadalajara, Escuela de Agricultura, Guadalajara, Mexico.

Distancia A., and Betancourt, V.A. 1984. Effects of sowing date on yield and disease in commercial sorghum hybrids at Jalisco, Mexico. Sorghum Newsletter 27:119.

Frederiksen, R.A., Betancourt, V.A., and Schuh, W. 1983. Disease control using fungicides in the Bajio. Pages 86–88 in Proceedings, Thirteenth Biennial Grain Sorghum Research and Utilization Conference, 22–12 Feb 1983, Brownsville, Texas, USA.

Gonzalez, M.J., and Galvan, C.F. 1987. Eficiencia de genotipos de sorgo en la utilización de Fe. Page 135 in Proceedings, XI Congreso Nacional de Fitogenetica, 25–28 Aug 1986. Facultad de Agricultura, Universidad de Guadalajara.

Hernandez, M.M., Narro, S.J., and Betancourt, V.A. 1982. Chemical control of downy mildew (Peronosclerospora sorghi) C. G. Shaw at "El Bajio" at Michoacán, Mexico. Sorghum Newsletter 25: 119–120.

Hernandez, M.M., Narro, S.J., and Betancourt, V.A. 1983. Races of head smut (Spacelotheca reiliana) Kühn Clint at the Bajio of Guanajuato and Michoacán, Mexico. Sorghum Newsletter 26: 118.45

Hernandez, M.M., Mendoza, O.L.E., and Osada, K.S. 1987a. Inoculación de sorgo en antésis con Fusarium moniliforme (Sheld). S. & H. mediante palillo y jeringa en penunculo y tercer entrenudo basal. Revista Mexicana de Fitopatologia (SC1):27–31.

Hernandez, M.M., Mendoza, O.L.E., Ortiz, C.J., and Osada, K.S. 1987b. Efecto del tizon de la panoja, Fusarium moniliforme (Sheld.) S. & H. En el rendimiento de grano en sorgo. Page 89 in Proceedings, XIV Congreso Nacional de Fitopatologiá, 15–17 Jul 1987, Morelia, Michoacán, Mexico.

Herrera, J.A., and Betancourt, V.A. 1986. Distribution of races of head smut (Sporisorium reilianum) inthe northeast and southwest areas of Mexico. Sorghum Newsletter 29:86.

Martinez, A.J. 1984. Informe anual del programa de sorgo. CAEB-CIAB-INIFAP. Celaya, Guanajuato, Mexico.

Narro, S.J. 1984. Programa de Fitopatologia del sorgo CAEB-CIAB-INIA-SARH. Pages 381–403 in Proceedings, Primera Reunión Nacional sobre Sorgo, 22–26 Oct 1984, Facultad de Agronomiá, UANL, Marin, Nuevo Leon, Mexico.

Narro, S.J., and Betancourt, V.A. 1983. Reaction of selected sorghum lines to Fusarium stalk rot and head blight at El Bajio, Mexico. Pages 78–79 in Proceedings, Thirteenth Biennial Grain Sorghum Research and Utilization Conference, 22–24 Feb 1983, Brownsville, Texas, USA.

Narro, S.J., and Delgadillo, S.F. 1985. Identificación de variantes del virus del mosaico enanismo del maíz (MDMV) en el cultivo del sorgo en el Bajió de Guanajuato. *In* Proceedings XII Congreso Nacional del la Sociedad Mexicana de Fitopatologia and XXV Annual Meeting American Phytopathological Society, Caribbean Division 11–14 Sep 1985, Guanajuato, Mexico.

Narro, S.J., and Hernandez, M.M. 1984. Evaluación de sorgos hibridos con germoplasma tropical. Page 20 *in* Proceedings XI Congreso Nacional de Fitopathologia, 16–18 Jul, San Luis Potosi, Mexico.

Narro, S.J., Hernandez, M.M., and Betancourt, V.A. 1982. Pathotypes of downy mildew (*Peronosclerospora sorghi* Weston & Uppal C. G. Shaw) at El Bajio de Michoacan. Sorghum Newsletter 25:120–121.

Narro, S.J., Hernandez, M.M., and Betancourt, V.A. 1983. Reaction of commercial sorghum hybrids to conidial infection of downy mildew (*Peronosclerospora sorghi*). Sorghum Newsletter 26:117–118.

Rodriguez, R.P.I. 1983. Prevención de la roya *Puccinia purpurea* Cooke como método indirecto para disminuir la incidencia de *Fusarium moniliforme* Sheld en sorgo (*Sorghum bicolor* L. Moench.) B.Sc. thesis, Universidad de Guadalajara, Escuela de Agricultura, Guadalajara, Mexico.

Toler, R.W. 1980. Viruses and viral diseases of sorghum. Pages 395–408 *in* Sorghum diseases, a world review: proceedings of the International Workshop on Sorghum Diseases, 11–15 December 1978, Hyberabad, India. Andhra Pradesh 502 324, India: International Crop Research Institute for the Semi-Arid Tropics.

Vargus, R.G., Leyva, Mir, G., and Romo, C.E. 1985. Etiología y resistencia varietal del sorgo (*Sorghum bicolor*) Moench a la pudrición del tallo y raíz. *In* Proceedings, XII Congreso Nacional Sociedad Mexicana de Fitopatologia and XXV Annual Meeting, American Phytopathological Society, Caribbean Division 11–14 Sep 1985, Guanajuato, Mexico.

Williams, A.H. 1980. RB Exp-182 (INIA RB-3030) y RB Exp. 607 (INIA RB-3006), nuevos sorgos para la región norte de Tamaulipas resistentes a *Peronosclerospora sorghi*. Pages 253–262 *in* Proceedings, Sociedad Mexicana de Fitogenetica, 3–7 Aug 1980, Facultad de Agrobiologia "Presidente Juarez". Uruapan, Michoacan, Mexico.

Sorghum Diseases in North America

R.A. Frederiksen[1] and R.R. Duncan[2]

Abstract

Sorghum diseases are more important in the Central, Gulf Coast, and eastern growing regions of the United States. Anthracnose, head blight, grain mold, and head smut remain as significant constraints to production in some years in these regions. Diseases are rarely important in the Great Plains, although the stalk rots and maize dwarf mosaic will affect production to some extent annually.

Introduction

Sorghum diseases in the United States of America were thoroughly reviewed by Edmunds and Zummo (1975) and Frederiksen (1986). However, these comprehensive treatises include no data as to distribution or rank in importance of each disease. Few attempts have been made to estimate the extent of damage by pathogens on sorghum in USA. Frederiksen (1980b) attempted to do so on a global scale by rating each disease as to its prevalence and importance by ecogeographic regions. Consequently, a similar tabular presentation of these estimates is made by ecological zones. There are about five ecological zones for sorghum production in USA. These, in order of production, are the Great Plains, Central, Gulf Coast, East Coast, and West (Fig. 1).

Over the past decade, U.S. sorghum area has dropped from about 5.2 to about 4.3 million hectare. Yields, however, have increased from an average of 3.6 to about 4.3 t ha^{-1}. While there is variation in the types of diseases present in each region (Table 1), much less variation is found within than among these areas. Sorghum is grown on the Great Plains (Texas, Panhandle north to South Dakota), in part because of the freedom of this area from most of the sorghum

diseases. Stalk rots and virus diseases are generally present, but losses have not been substantial (Doupnik and Frederiksen 1983). This is true, in part, because of the development of hybrids with tolerance to maize dwarf mosaic virus and because of the development of machinery to mechanically harvest lodged sorghum.

Sorghum downy mildew was widespread in Kansas and Nebraska in 1987 (L. Claflin, personal communication; Jensen et al. 1989). However, yield losses on a statewide basis were insignificant, and downy mildew is not expected to be as prevalent in 1988 because the disease has been present in Kansas for 20 years but in only 2 years has it warranted concern. This suggests that the environment for disease was unusually favorable in 1987 and that the cultivars used were susceptible to the pathotypes present. Head smut occurs in some regions of Texas and Kansas, more or less annually.

In the Central region (Missouri, Tennessee, Illinois, and Kentucky) grain sorghum production has increased over the past decade. Diseases such as crazy top, sorghum downy mildew, anthracnose, head blight, and grain mold have at times been important. Anthracnose was widely prevalent in parts of Missouri in 1986 and 1987 (J. Dale, and K. Cardwell, per-

---

1. Plant Pathologist, Texas Agricultural Experiment Station, Texas A&M Univeristy, College Station, TX 77843, USA.
2. Sorghum Breeder/Physiologist, University of Georgia, Georgia Experiment Station, Griffin, GA 30223, USA.

Frederiksen, R.A., and **Duncan, R.R.** 1992. Sorghum diseases in North America. Pages 85–88 *in* Sorghum and millets diseases: a second world review. (de Milliano, W.A.J., Frederiksen, R.A., and Bengston, G.D., eds). Patancheru, A.P. 502 324, India: International Crops Research Institute for the Semi-Arid Tropics.

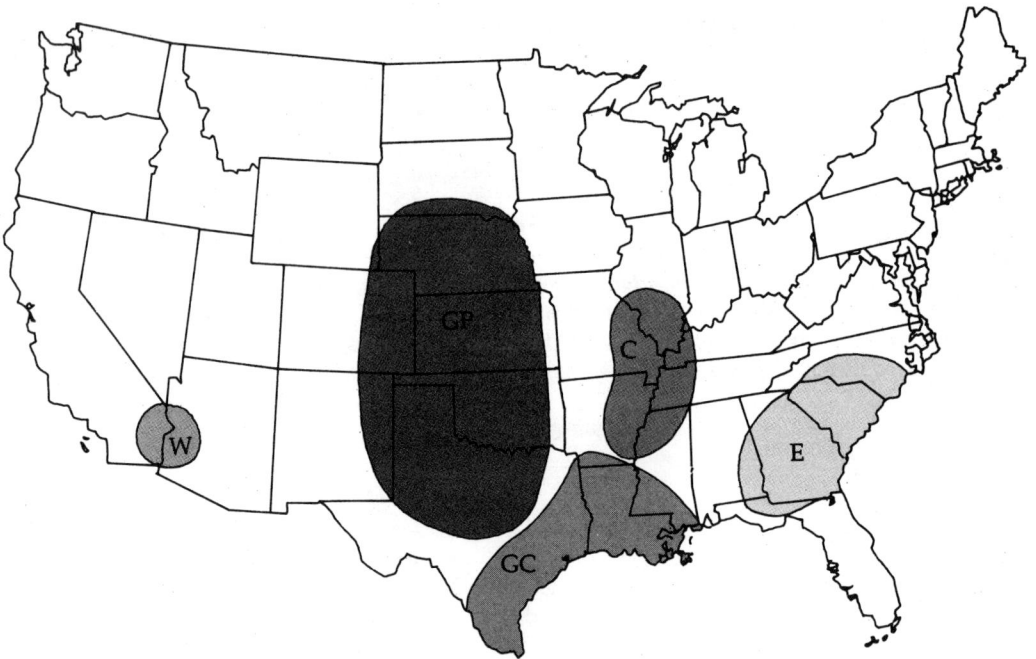

Figure 1. Agroecological sorghum-growing zones in USA. Production (million t) in 1985, by region was: GP (Great Plains) 17.5; C (Central) 6.4; GC (Gulf Coast) 5.1; W (Western) 1.9; and E (Eastern) 0.3.

Table 1. The relative prevalence and importance of sorghum diseases in the United States of America.

Disease	Region[1]				
	Great plains	Coastal plains	Central	East	West
Anthracnose	-[2] 0[3]	+++ 3	++ 2	+++ 3	- 0
Charcoal rot/ Fusarium stalk rot	++ 3	++ 2	+ 2	+ 2	+ 1
Downy mildew	+ 2	++ 3	+ 1	+ 1	- 0
Foliar diseases	+ 1	+++ 3	+ 1	+++ 2	+ 0
Grain mold/ Fusarium head blight	+ 2	++ 3	++ 3	++ 3	+ 0
Head smut	++ 2	++ 3	+ 0	+ 0	+ 0
Maize dwarf mosaic	+++ 1	+++ 1	++ 1	++ 1	+ 1

1. Sorghum-growing regions include: Great Plains, parts of Texas, Nebraska, Kansas, New Mexico, Colorado, and South Dakota. Coastal Plains: Texas, Mississippi, and Louisiana. Central: Illinois, Kentucky, Missouri, and Tennessee. East: Alabama, Georgia, North Carolina, and South Carolina. West: California and Arizona.
2. - Not reported at present, + Occassionally, ++ Commonly, +++ Generally found on most plants.
3. 0 = No loss, 1 = Probably not economically important, 2 = Occasionally important 3 = A major production constraint.

sonal communication). This region of higher rainfall (1200–1300 mm) encourages diseases such as anthracnose and fusarium head blight.

The Gulf Coast region has a history of disease problems. Downy mildew, anthracnose, and grain mold can be important in any year in this region. A major epidemic of sorghum downy mildew occurred in 1973 (Frederiksen 1980a). While the actual losses were not estimated, the gravity of the disease changed the genotypes of sorghum grown and downy mildew became an intimate part of the sorghum scientist's lexicon. Its appearance since has been sporadic due in part to the growing of susceptible hybrids, or the appearance of a new race (Craig and Fre-

deriksen 1983) or the interaction with herbicide antidotes (Craig et al. 1987). Occasionally, even seedling disease (Forbes et al. 1983) and bacterial diseases affect sorghum in the Gulf Coast area (Frederiksen et al. 1980). Head smut appears to be limited to the coastal and central sorghum-growing regions of Texas. Foliar diseases, such as zonate leaf spot, gray leaf spot, ladder spot, and rough spot become more serious along the eastern coastal areas of the Gulf.

In the Southeast, anthracnose remains the single most important problem and is crop-limiting in some years (Table 2). Many foliar diseases are present as are head blight and grain mold. In the western region, only root, stalk rot,

Table 2. Prioritization of disease problems on sorghum in the southeastern region of the United States of America.

Common name	Causal organism	Severity[1]	General symptom expression
Anthracnose	Colletotrichum graminicola (Cesati) Wilson	1	Attacks foliage, culms, panicle, seed. Lodging possibly four or five pathotypes.
Fusarium	Fusarium moniliforme Sheldon and other species	1	Panicle blight, culm rot, small seed, reduced germination, old, seedling blight, lodging.
Grain molds	Fusarium, Alternaria, Curvularia, Cladosporium, Phoma, Helminthosporium, Colletotrichum, others	2	Seed deterioration, reduced germination.
Charcoal rot	Macrophomina phaseolina (Maulb.) Ashby	2	Basal culm rot, lodging
Gray leaf spot	Cercospora sorghi Ell. & Ev.	3	Foliar (leaf blades and sheaths)
Ladder leaf spot	Cercospora fusimaculans Atk.	3	Foliar
Zonate leaf spot	Gloeocercospora sorghi Bam & Edgerton	3	Foliar
Rough leaf spot	Ascochyta sorghina Sacc.	3	Foliar
Leaf blight	Peronosclerospora sorghi (Weston & Uppal) C.G.Shaw	4	Foliar
Pokkah boeng	F. moniliforme var. subglutinans	4	Systemic, abnormal growth
Red spot	Helminthosporium rostratum Drechsler	4	Foliar-midvein
Acremonium wilt	Acremonium strictum W. Gams	4	Culm, foliar

Continued

Table 2. *Continued*

Common name	Causal organism	Severity[1]	General symptom expression
Target leaf spot	*Bipolaris sorghicola* (Lefebvre & Sherwin) Shoem.	5	Foliar
MDMV	Aphid-vectored virus	5	Systemic, foliar
Downy Mildew	*Exserohilum turcicum* (Pass.) Leonard & Suggs	5	Systemic
Bacterial leaf streak	*Xanthomonas holcicola* (Elliot) Starr & Burk	5	Foliar
Bacterial leaf stripe	*Pseudomonas andropogoni* (E.F.Sra.) Stapp	5	Foliar
Rust	*Puccinia purpurea* Cooke	5	Foliar
Yellow stunt sorghum	–	5	Systemic, foliar

1. 1 = very severe, consistent annual occurrence, 2 = sporadically severe, inconsistent annual occurrence, 3 = occasional problem, not economically damaging, 4 = minor problem, 5 = rarely observed.

and virus diseases are present. Losses are not significant although an unidentified root disease problem was reported in 1972 (Voight 1972).

References

Craig, J., and Frederiksen, R.A. 1983. Sporulation of pathotypes of *Peronosclerospora sorghi* on inoculated sorghum. Plant Disease 64:278–279.

Craig, J., Frederiksen, R.A., Odvody, G.W., and Szerszen, J. 1987. Effects of herbicide antidotes on sorghum downy mildew. Phytopathology 77: 1530–1532.

Doupnik, Ben Jr. and Frederiksen, R.A. 1983. Diseases of Major Crops. Pages 525–534 *in* Dryland Agriculture. (Dregne, H.E., and Willis, W.O., eds.). Madison, WI, USA: American Society of Agronomy.

Edmunds, L.K., and Zummo, N. 1975. Sorghum diseases in the United States and their control. Agricultural Handbook No. 468. Washington, DC, USA: United States Department of Agriculture.

Forbes, G., Collins, D., Odvody, G., and Frederiksen, R. 1983. Seedling disease in South Texas during 1983. Sorghum Newsletter 26:124–125.

Frederiksen, R.A. 1980a. Sorghum downy mildew in the United States: overview and outlook. Plant Disease 64:903–908.

Frederiksen, R.A. 1980b. Diseases of sorghum. Pages 99–109 *in* Elements of Integrated Control of Sorghum Pests. FAO Plant Production and Protection Paper 19. FAO, Rome, Italy: Food and Agriculture Organization.

Frederiksen, R.A. (ed.) 1986. Compendium of Sorghum Diseases. St. Paul, MN, USA: American Phytopathological Society. 82 pp.

Frederiksen, R.A., Rosenow, D.T., and Odvody, G.N. 1980. Bacterial diseases in Texas - 1979. Sorghum Newsletter 23:133.

Jensen, S.G., Doupnik, Ben, Jr., Wysong, D. and Johnson, B. 1989. An epidemic of sorghum downy mildew in Nebraska in 1987. Plant Disease 73:75–78.

Voight, R.L. 1972. A potential new strain of milo disease. Sorghum Newsletter 15:1.

A Short Communication on Sorghum Diseases in Europe

G.A. Forbes[1]

Abstract

Diseases of current and potential importance on sorghum in southern France, with general application to sorghum in other regions of Europe are discussed. In general, disease pressure is relatively low throughout Europe. The importance of seedling disease is not known, but it may contribute to poor stand establishment, a major problem in the area. Grain mold probably poses a potential problem for seed production, but does not appear to be a limiting factor for farmers. Other diseases or disease groups reviewed include head blight, head smut, foliar diseases, stalk rots, and those caused by viruses.

Introduction

The sources of information for this review are based on the author's observations in a few of the sorghum-growing regions in southern France, conversations with sorghum workers (generally plant breeders) in France, and a brief literature search for reports of sorghum diseases in Europe. This review is thus presented as a general view of the sorghum disease situation in southern France with application to other sorghum-growing regions of Europe. The region of sorghum production in southern France is probably representative of the other sorghum-growing regions of Europe, but particular disease situations may exist in other areas.

Sorghum in Europe is produced in the south where a Mediterranean climate prevails. It is a secondary crop of diminishing importance throughout the continent, including France, where approximately half of the European Economic Community (EEC) grain sorghum is produced. Production in France has steadily decreased, from 57 000 ha in 1984 to 39 000 ha in 1987 (EUROSTAT 1987).

Although sorghum, a drought-tolerant plant, is ecologically adapted to the dry Mediterranean summers, this advantage is now of little importance to farmers; they find the price subsidies given to other crops more attractive. Agricultural areas once used for sorghum are now being sown to soybean and sunflower.

Sorghum breeders in France list the short growing season and the lack of early-maturing materials as the greatest constraints to production. Diseases are thought to be of minor or no importance. The minor emphases placed on diseases may be partly a result of lack of close examination. Published records of systematic evaluations of sorghum diseases in Europe were not found.

If diseases were indeed of major importance, however, their incidence and effects would have to be considered in plant improvement programs. Lack of concern for sorghum pathology, therefore, probably reflects the relative good health of the crop in this region. As with sorghum in the northern USA and in Canada, sorghum in Europe faces relatively low disease pressure.

---

1. Plant Pathologist, CIP, International Potato Center, PO Box 5969, Lima, Peru.

Forbes, G.A. 1992. A short communication on sorghum diseases in Europe. Pages 89–91 *in* Sorghum and millets diseases: a second world review. (de Milliano, W.A.J., Frederiksen, R.A., and Bengston, G.D., eds). Patancheru, A.P. 502 324, India: International Crops Research Institute for the Semi-Arid Tropics.

Even so, certain diseases have been cited or reported as suspect on the basis of climatic and agricultural particularities of the region. Below is a brief discussion of some of the current and potential disease problems.

Fungal Diseases

The majority of diseases occurring on sorghum in southern France are of fungal origin (Table 1).

Table 1. Potential importance of sorghum diseases in the sorghum-growing regions of Europe.

Disease	Potential importance[1]
Fungal	
Seedling	+
Grain mold and head blight	+
Head smut	+
Foliar	-
Stalk rots	+
Viral	
MDMV	+
Bacterial	-

1. (+) = potentially important, (-) = not potentially important.

Seedling Disease

Poor stand establishment constitutes a major problem for European farmers. The development of cultivars which emerge and grow well in cool, wet soils is one of the primary concerns of sorghum improvement programs in France. To date, low percentage emergence and reduced seedling vigor have been attributed to low soil temperatures for extended periods after sowing.

Poor stand establishment has also been a problem in the northern USA and Canada, restricting the area where the crop can be produced. As in Europe, the condition has generally been attributed to low soil temperatures. Recent studies, however, have implicated seedling dis-

ease as a contributing factor of potential importance (Gaudet and Major 1986). Based on this North America research, one could easily suspect that seedling disease is involved with stand establishment problems occurring in Europe.

Grain mold

At least two studies on grain mold of sorghum in France have been reported (Sifaw 1978; Ruiz 1979). Both are basically grain mycofloral inventories and provide little indication of the importance of grain mold in the region. Nonetheless, these studies demonstrate an awareness of concern for grain mold by sorghum scientists in France. These studies were perhaps initiated by seed production. It does not appear that grain mold poses a threat to farmers in the Mediterranean area, but can be considered a potential constraint to seed production.

Head blight

At a recent meeting of sorghum workers held in southwestern France, several participants expressed concern over a panicle disorder referred to as "Fusariosis." From the descriptions given, the disorder appears to resemble head blight (Castor and Frederiksen 1981), but the causal role of *Fusarium moniliforme* has not been confirmed. Differences in susceptibility among cultivars have been noted. This disorder is apparently increasing in incidence and may warrant closer examination.

Head smut

Head smut is not now a problem on sorghum in Europe. The disease was on the increase in the late 1970s and early 1980s, but is now controlled by the use of resistant hybrids. Head smut occurs regularly on maize in the Mediterranean area, but apparently with little economic significance.

Stalk rots and wilts

There is little information on stalk rots of sorghum in Europe, yet this is one group of diseases that could be expected to cause problems when climatic conditions are appropriate. Inoculum levels of *Macrophomina phaseolina* are probably high in most soils in southern France, where it is an important pathogen of several crops, including sunflower and soybean (Davet et al. 1986).

Fusarium spp have also been reported causing stalk rots of maize throughout France (Rouhani et al. 1979). Acremonium wilt appears to be present, at least in breeders' nurseries, but there is no evidence of an important incidence in production fields.

Foliar Disease

The Mediterranean summer is characterized by warm days, cool nights, and low rainfall. Consequently, one could easily assume that foliar diseases are not a problem. Last year the author, searching for foliar diseases, visited several sorghum fields just before harvest. None were found.

Viral Diseases

Maize downy mosaic virus (MDMV) is prevalent throughout the sorghum-growing regions of southern Europe (P. Signoret, personal communication), and has been reported in France, Italy, and Yugoslavia. MDMV-B probably occurs in Yugoslavia (R. Toler, personal communication). Viruses are currently controlled by resistance found in materials coming from major international seed companies, many of which are active in the region.

Bacterial Diseases

There are no recent records of bacterial diseases occurring on sorghum in Europe.

References

Castor, L.L., and **Frederiksen, R.A.** 1981. Fusarium head blight occurrence and effects on sorghum yield and grain characteristics in Texas. Plant Disease 64:1017–1019.

Davet, P., Herbach, M., Rabat, M., and **Piquemal, G.** 1986. Effet de quelques facteurs intrinsèques ou extrinsèques d'affaiblissement des tournesols sur leur sensibilité au dessèchement précoce. Agronomie 6:803–810.

EUROSTAT. 1987. Crop production quarterly statistics. Brussels, Belgium: CECA-CEE-CEEA.

Gaudet, D.A., and **Major, D.J.** 1986. Factors affecting seedling emergence of sorghum for short-season areas. Plant Disease 70:572–575.

Rouhani, H., Davet, P., Poinso, B., Beyries, A., and **Messiaen, C.M.** 1979. Inventaire et évaluation du pouvoir pathogène des composants de la microflore fongique sur racines de maïs en France. Annales de Phytopathologie 11:69–93.

Ruiz, M. 1979. Recherche d'une technique d'inoculation et du comportement du grain de sorghum vis-à-vis de quelques moisissures. ENSA, 34060 Montpellier, France: DAA Chaire de Biologie et Pathologie Végétales. 43 pp.

Sifaw, M. 1978. Etude de la mycoflore des semences du sorgho en France. ENSA Montpellier, France: D.E.S. 57 pp.

Part 2

Regional and Country Reports
(Pearl Millet)

World Review of Pearl Millet Diseases: Knowledge and Future Research Needs

S.B. King[1]

Abstract

*About 95% of the world's pearl millet (*Pennisetum glaucum*) growing area is located in Sahelian and sub-Sahelian Africa and in the Indian subcontinent. The most important disease of pearl millet is downy mildew, which is widespread. Second in importance is witchweed, especially* Striga hermonthica *in African fields. Smut, ergot, and rust are also widespread, but of considerably less importance. Bacterial and viral diseases are of minor importance and certainty of their identity is often lacking. Nematodes are likewise probably widespread, but their importance in pearl millet production is virtually unknown. Information on biology and epidemiology of the major diseases is incomplete, and precise information on the environmental factors that promote disease is often lacking. Field-screening techniques for downy mildew, smut, and ergot are available and stable resistances have been identified. The importance of these three diseases in India has increased considerably since the adoption of hybrids by Indian farmers. Differences in virulence have been confirmed in pathogen populations of downy mildew and rust, but not in smut or ergot. Successful resistance breeding has been achieved for downy mildew and smut, but it is problematic for ergot, especially in hybrids. Confirmed resistance to* S. hermonthica *has not been found in pearl millet; screening techniques need refinement.*

Introduction

Pearl millet [*Pennisetum glaucum* (L.) R. Br.] is the staple cereal crop best suited to the harsh climate of the seasonally hot, drought-prone, semi-arid regions of Africa and the Indian subcontinent. Average grain yields in the more than 27 million ha of pearl millet cultivated in farmers' fields are estimated to be about 500 to 550 kg ha[-1], although reported yields vary considerably from one region to another and from one year to another. The crop is grown for grain mainly in Africa (more than 15 million ha, 95% of which are in the Sahelian and sub-Sahelian zones spanning from Sudan to Senegal), and the Indian subcontinent (more than 11 million ha) (ICRISAT 1987a). About 0.5 million ha are grown

in the southern USA, primarily for forage (ICRISAT 1987b).

Smut and rust are widely distributed across the pearl millet growing areas of the world, although rust rarely occurs at more than a low level in the Sahelian zone of Africa. Downy mildew and ergot are widespread in the pearl millet growing areas of Asia and Africa alike; but do not occur on pearl millet in the Americas. The downy mildew pathogen, *Sclerospora graminicola*, has been reported, however, to occur naturally on *Setaria viridis*, a common weed of corn fields in parts of midwestern USA (Melhus et al. 1928). The distribution of a number of less important leaf diseases of pearl millet is not well defined. Two species of witchweed, *Striga hermonthica* and *S. asiatica*, parasitize pearl millet;

---

1. Principal Plant Pathologist, Cereals Program, ICRISAT Center, Patancheru, Andhra Pradesh 502 324, India.

King, S.B. 1992. World review of pearl millet diseases: knowledge and future research needs. Pages 95–108 *in* Sorghum and millets diseases: a second world review. (de Milliano, W.A.J., Frederiksen, R.A., and Bengston, G.D., eds). Patancheru, A.P. 502 324, India: International Crops Research Institute for the Semi-Arid Tropics. (CP 743).

S. *hermonthica* predominates in Sahelian and sub-Sahelian Africa and S. *asiatica* predominates in northwestern India, southern Africa, and parts of eastern Africa.

Yield losses in pearl millet due to specific diseases have been estimated, but this information is scanty and usually very localized, generally applying only to specific years or experiments. The relative importance, on a global basis, of the major pearl millet diseases is not known. Therefore, ICRISAT's pearl millet pathologists (S.B. King, S.D. Singh, R.P. Thakur, and J.Werder) have attempted to estimate global yield losses to diseases based on a general understanding of yield losses in the major pearl millet production areas. Our estimate of the relative importance of pearl millet diseases (excluding nematodes) are as follows: downy mildew 45%, *Striga* spp 32%, smut 9%, ergot 7%, rust 3%, viruses, >1%, and other diseases 3%. Downy mildew is especially important in India and to a somewhat lesser degree in western Africa, and *Striga* is especially important in western Africa.

This review presents information in some detail on only the more important diseases of pearl millet, worldwide. I have attempted to identify important gaps in knowledge that should receive research attention in the future, and I have reported very little on what I consider to be the less important diseases, although some of these may be important in some locations in some years.

Downy Mildew

Sclerospora graminicola was first reported on pearl millet in India by Butler (1907), but downy mildew (DM) did not become a serious problem on a national level until after the widespread cultivation in India of hybrids in the late 1960s. The first of several major epidemics occurred in 1971 (Safeeulla 1976; Singh et al. 1987a). These hybrids were evidently bred under little downy mildew pressure and were based mainly on the male-sterile line from Georgia, Tift 23A, that was bred in the absence of downy mildew and was highly susceptible to it. In 1970, India recorded a record grain production in the pearl millet crop exceeding 8 million t, attributed largely to the widespread cultivation of hybrids. However, in the following year a drop in grain production of more than 40% was attributed mostly to downy

mildew and to the increased vulnerability of the crop resulting from the greater genetic uniformity introduced by hybrids (Pokhriyal et al. 1976; Safeeulla 1977).

There is little question about the taxonomy of the downy mildew pathogen; *Sclerospora graminicola* (Sacc.) Schroet. is the accepted name. It was first described as *Protomyces graminicola* on *Setaria verticillata* by Saccardo in 1876 but a few years later was renamed S. *graminicola* by Schroeter. There appear to be two distinct pathotypes of S. *graminicola*, one that attacks *Setaria* spp and another that attacks pearl millet, although they are morphologically the same (Williams 1984a).

Considerable information is available on the biology, epidemiology, and control of S. *graminicola* and pearl millet downy mildew (Safeeulla 1976; Nene and Singh 1976; Williams 1984a; Shetty 1987; Singh et al. 1987b). The fungus produces both sexual spores (oospores) and asexual spores (sporangia, zoospores). It is heterothallic, requiring the presence of two mating types for production of oospores, though such combinations are not required for infection and subsequent asexual spore production (Michelmore et al. 1982). Only two mating types have been found, and there is compatibility between those from western Africa and those from the Indian subcontinent (Idris and Ball 1984). The nuclear condition of S. *graminicola*, particularly of sporangia and their production, has been described (Safeeulla 1976; Shetty and Ahmed 1981), but further clarification is desirable.

There is little doubt that S. *graminicola* can be transmitted as oospores on the seed surface or as mycelium in embryonic tissues of seed. However, there seems to be no agreement on whether mycelia-infected seed is capable of giving rise to downy mildew infection on pearl millet plants (Sundaram et al. 1973; Shetty et al. 1977; Williams 1979; Williams et al. 1980). There is ample evidence of virulence differences in the pathogen in India and Africa (Girard 1975; Shetty and Ahmed 1981; Ball 1983; Singh and Singh 1987), with isolates from Nigeria and Niger being the most aggressive (Ball et al. 1986).

Field-screening techniques involving a sick plot, infector rows, or a combination of both have been developed and are being used with varying degrees of success at a number of locations in India and in several countries in western Africa. The success of the infector row system depends heavily on sporangial production, dissemination, zoospore release, and infection, especially during early growth of test materials (Williams et al. 1981; Singh et al. 1987a). Sporadic rains accompanied by high soil temperatures are often characteristic of early-season conditions in pearl millet cultivation areas of the semi-arid tropics; supplemental irrigation at this time may help alleviate the problems of high temperature and low moisture, thus increasing the reliability of screening. In extreme cases of drought and high temperature, irrigation may fail to assure conditions necessary for effective field screening.

Many control measures for downy mildew have been suggested. These include removing diseased plant debris at the end of the season and roughing diseased plants as they appear during the season (Kenneth 1977; Thakur 1980). Reports are available on the influence of soil nutrients on downy mildew (Deshmukh et al. 1978; Singh and Agarwal 1979), but the soil nutrient/ DM relationship is not well defined. The systemic fungicide metalaxyl has been shown to be effective as a seed treatment against early infection; as a foliar spray it causes symptom remission (Williams and Singh 1981; Dang et al. 1983; Singh et al. 1984). However, the most practical control measure is host resistance.

Numerous sources of resistance have been identified among accessions of germplasm and in breeding lines in India and in some countries of western Africa (Singh et al. 1987a). However, there seems to have been relatively little direct use of accessions selected for their downy mildew resistance. ICRISAT screening indicates that resistance is most frequently found in accessions from central to east-central western Africa (Andrews et al. 1985a; Williams 1984b). Multilocational testing in India and in countries in western Africa have indicated stable resistance in a number of lines (ICRISAT 1986). It is interesting to note, however, that in multilocational testing in western Africa, local landraces growing at their center of origin often appear to be more susceptible than landraces introduced from elsewhere in the region (J. Werder, personal communication). The variability for downy mildew reaction among plants within lines has allowed selection for resistance in susceptible lines and the conversion of susceptible lines into resistant lines (Singh 1983; Singh et al. 1988). The phenomenon of recovery resistance, whereby an infected plant outgrows downy mildew, has recently been reported (Singh and King 1988); the phenomenon appears to be common and widespread in accessions of germplasm and breeding lines. A number of studies have been made on inheritance of resistance to downy mildew (Appadurai et al. 1975; Gill et al. 1978; Singh et al. 1978; Basavaraj et al. 1980; Shinde et al. 1984). The picture is not entirely clear, but reports generally suggest that resistance is polygenic and involves one or more dominant genes.

Because downy mildew is of major importance in western Africa and the Indian subcontinent—the world's two main areas of pearl millet cultivation—it is essential that research be conducted to fill gaps in our knowledge. One research area involves the oospore. It plays an important role as a primary inoculum, but its biology is little understood. There is a need to know the manner in which oospores germinate and infect plants, and the factors (environment, host) that influence germination and infection. Development of reliable tests for oospore viability and ways to influence viability, longevity, and germination could lead to more effective screening and control of this disease.

There should be a constant effort to improve the efficiency and reliability of downy mildew screening techniques both in the field and in the greenhouse or laboratory, with a view to also increase flexibility to better meet the needs of breeding programs. In this regard, we have initiated mass screening of breeding materials at the seedling stage at ICRISAT Center.

It is likely that *S. graminicola* must invade meristematic tissue for expression of downy mildew symptoms (Williams 1984a), but relatively little is known relating infection with symptom expression or the reasons for the great variability in symptoms. The phenomenon of recovery resistance is, in the broad sense, related to this matter of symptom expression. Studies that will elucidate what happens to *S. graminicola* at the tissue level during recovery of resistance, as well as the possible influence of environment, especially temperature, on symptom expression, are needed.

Knowledge of the genetics of resistance in pearl millet and the genetics of virulence in *S. graminicola* is incomplete. A more complete understanding could lead to the development of a successful strategy of resistance breeding and gene deployment to control the disease. For such studies, it is necessary to first increase the level of inbreeding in so-called inbred lines, possibly through doubled haploids. An in vitro technique for maintenance of sporangia over time is desirable. Tissue culture may be a possibility for maintenance of sporangial cultures. For studies on genetics of pathogen virulence, a technique to induce oospore germination is also needed.

At present, there seems to be little need for pathologists to identify more sources of downy mildew resistance in accessions of germplasm. Sufficient resistance is probably available, in most breeding programs, in agronomically superior lines. Research efforts could better be directed to reliable identification of resistances in such material. In India, where downy mildew is an especially critical problem in hybrids, a joint effort by breeders and pathologists becomes essential for rapid and effective progress. In view of the recent experiences of resistance breakdown in hybrids in India (Singh et al. 1987a), it seems advisable that a strategy be developed whereby the same hybrid is not grown in an area for more than 4 successive years. This might be achieved through production, distribution, and sale of seed of specific hybrids. There is an urgent need in India to increase the genetic diversity of parent lines, especially female parents, so that hybrids grown by farmers may be genetically more diverse. A concerted effort to increase the number of good hybrid parents is under way at ICRISAT, and the possibility of using topcross hybrids to reduce vulnerability to downy mildew is being investigated.

Breeding programs in Africa can certainly learn a lesson from the experience of resistance breakdown to downy mildew in India, even though the problem has not yet presented itself, possibly because of the current minimal use of improved cultivars, and of almost no hybrids, by farmers in Africa.

Rust

Rust is generally considered to be a moderately important disease in most pearl millet production areas; although in Sahelian and sub-Sahelian Africa it is usually of little consequence. It reportedly attacks pearl millet at all stages of plant growth (Ramakrishnan 1963), but severe infection normally does not occur in the field until late in plant development. This has little effect on grain yield, though forage and fodder value may be reduced considerably.

Four rust pathogens have been reported on pearl millet: *Puccinia penniseti*, *P. substriata* var *indica*, *P. substriata* var *penicillariae*, and *P. stenotaphri* (Ramachar and Cummins 1965; Sathe 1969). The complete life cycle of pearl millet rust is well known (Kulkarni 1958; Ramakrishnan 1963; Ramakrishnan and Soumin 1948). The uredinial and telial stages occur on pearl millet and the spermagonial and aecial stages occur on the alternate host, *Solanum melongena*. Reports on period of viability of urediniospores vary, but generally they can be maintained for several weeks at moderately low temperature (10°C) and low humidity. Incubation period for symptom expression is 6 to 8 days.

Control of rust by use of fungicides (Kapooria 1972; Bhadur et al. 1975; Parambaramani et al. 1971) and cultural methods such as sowing date and fertilizer application (Kandasamy et al. 1971; Sivaprakasham and Pillayarsamy 1975) have been attained with varying degrees of success, and disease inhibition by biological control agents occuring in nature. However, none of these methods currently hold much promise as a practical control measure for farmers.

A number of sources of rust resistance have been identified, although the extent of their use in breeding seems limited. Resistance in some cases is likely based on slow rusting traits (Singh and Sokhi 1983). However, two cases of resistance based on a hypersensitive reaction controlled by single, and likely different, dominant genes have been reported (Hanna et al. 1985; Andrews et al. 1985b). The resistance identified in Georgia, USA, is effective in India, but the reverse does not seem to be true (Hanna and King, unpublished). It seems unlikely, however, that extensive use of rust resistance based on a single dominant gene would be long-lasting, especially since the alternate host, egg plant, is commonly grown in most millet-production areas.

Although rust may be a relatively unimportant disease in most locations and years, there are many areas of research that merit attention. An understanding of the taxonomic identities and relationships among the pathogens reported to cause rust on pearl millet is needed. Research is also needed to identify and quantify environmental factors that promote epidemics, and to better establish the relationship between physiological maturity of the host and susceptibility to rust. A reliable method to preserve urediniospores for at least several months would facilitate the development and implementation of more reliable field, and possibly greenhouse, identification of resistance. Almost all disease rating scales now being used for pearl millet rust are based on percentage leaf area affected. Rating systems that assess pustule type would also be useful. Evaluation of existing breeding materials for reaction to rust is advisable before launching into new, major programs to identify rust resistance in accessions of germplasm. With the development of more precise screening techniques, studies should be undertaken on the inheritance of resistances other than those based on a hypersensitive reaction.

Virus and Virus-like Diseases

Virus-like symptoms on pearl millet have been reported from western Africa, southern Africa, the Indian subcontinent, and the USA. Symptoms are of two types, streak and mosaic. Streak at up to 80% frequency was noted in some pearl millet lines at New Delhi, India, in the 1970 off-season (Seth et al. 1972a). This disease was solely vector-transmitted (*Cicadulina mbila*) and designated as the *Pennisetum* strain of maize streak virus. A similar disease characterized by intermittent streaking, dwarfing, and sometimes excessive tillering was observed in 50% of the plants (estimated 30% yield loss) in a pearl millet seed multiplication field in southern India during the 1986 dry season (S.B. King, unpublished). Similar symptoms have been observed (S.B. King) in isolated pearl millet plants at Patancheru, India, and in Mali, Nigeria, Niger, and Zambia.

Mosaic symptoms associated with dwarfing have been reported on pearl millet in northern and southern Côte d'Ivoire (Kukla et al. 1984). The disease was called pearl millet mosaic; based on host range, physiochemical properties, and serology, it was closely related to the mechanically transmissible, aphid-borne guinea grass mosaic virus. Seth et al. (1972b) reported a mosaic on pearl millet that was readily transmitted mechanically and by *Rhopalosiphum maidis*. The virus infected several other grass hosts. In the USA, Jensen et al. (1983) greenhouse-tested pearl millet genotypes for reaction to maize dwarf mosaic virus strains A and B. Many pearl millet lines were infected to at least a low frequency by one or both strains. I have observed mosaic incidence exceeding 80% in a late-sown isolation of pearl millet near Lusaka, Zambia.

The whole area of virus and virus-like disease in pearl millet requires sorting out. At present these diseases appear not to be very important, but our present lack of knowledge places production of pearl millet in a rather vulnerable position. Initially, research is needed on virus identification and characterization, host range, transmission, and potential damage to affected plants. This should be followed by resistance identification, and then perhaps breeding for resistance.

Ergot

Ergot was first reported on pearl millet in the early 1940s in India (Thomas et al. 1945). There were epidemics in the late 1950s in the area now known as Maharashtra state (Bhide and Hegde 1957; Shinde and Bhide 1958). However, ergot did not draw serious attention in India until after the release of the first hybrid in the mid-1960s. Subsequent epidemics occurred, primarily in the early 1970s (Thakur 1984). Epidemics have not been reported in Africa. Ergot causes losses in grain yield, because infected florets do not produce grain. It reduces grain quality, because ergot sclerotia contain several alkaloids and contaminate grain. The alkaloids are generally of the agroclavine group, and are toxic to man and livestock.

The extent of pearl millet ergot toxicity in man is not clearly understood. Symptoms include nausea, vomiting, giddiness, and somnolence (Krishnamachari and Bhat 1976; Bhat et al. 1976). Although death of humans has been reported in India, mainly in the popular press, there seems to be some question whether sclerotial ingestion at an expected level would in fact cause death.

The ergot pathogen on pearl millet was generally recognized as *Claviceps microcephala*, until Loveless (1967) named it *C. fusiformis*, based on assessment of specimens in Africa. *C. fusiformis* was confirmed in India (Siddiqui and Khan 1973a) and is now the accepted nomenclature.

Many aspects of the disease cycle are well understood (Thakur 1984; Thakur et al. 1984).

Sclerotia in the soil germinate to produce ascospores that become airborne and infect individual florets through fresh stigmas. The ovary becomes infected, and within 5 to 7 days a sticky viscous liquid (honeydew) containing micro- and macroconidia exudes from diseased florets. Secondary spread is believed to occur by splashing rain and by physical contact among panicles. Insects can carry inoculum from diseased to healthy panicles (Sharma et al. 1983; Verma and Pathak 1984), but evidence suggesting that insects are important in disease epidemiology seems circumstantial.

Infected florets produce sclerotia that reach maturity at about the same time as grain. Several alternative grass hosts have been reported, including *Panicum antidotale* (Thakur and Kanwar 1978) and *Cenchrus ciliaris* (Singh et al. 1983), but the importance of these inoculum sources in disease epidemiology is not clearly understood. There is no confirmation of virulence differences among isolates of *C. fusiformis*. It is believed that moderate temperature and high humidity favor disease development (Ramaswamy 1968; Siddiqui and Khan 1973b), but precise information is not available. In India, epidemics have been associated with rainy weather; in western Africa ergot generally occurs more frequently in the higher rainfall areas. Factors influencing sclerotial germination are not well understood.

Cultural control methods, including intercropping, early sowing, adjusting soil fertility, and deep plowing to bury sclerotia have been suggested (Thakur 1984), but the practicality of these methods at the farmer's level is not known. Fungicidal control, demonstrated experimentally by Thakur (1984), is not economical for farmer use. Avoiding sowing of sclerotia-contaminated seed is likely to have some merit. Considerable evidence indicates that high availability of pollen during protogyny helps to control the disease (Thakur et al. 1983c). Control of ergot through pollen management might be possible in farmers' fields, but the technique has not been tested fully.

Disease resistance may hold some promise as a means of ergot control, but there are definitely some problems, especially for hybrids. Resistance could not be assessed adequately until the development of a reliable field screening technique that restricted pollen-based escape to a low level (Thakur et al. 1982). Resistance was not found in accessions of germplasm, but crosses between less susceptible plants followed by pedigree breeding and selection under high disease pressure produced numerous lines and sib-bulks with very high levels of resistance (<1% ergot when controls have 70% ergot). Many of these lines have shown stable resistance in several years of multilocational testing in India and some African countries (Thakur et al. 1985). Some of these materials, compare favorably for yield with released varieties in India, although they are generally late, especially in northern India (ICRISAT 1985). Attempts to utilize these sources of resistance in population breeding and hybrid breeding at ICRISAT Center and elsewhere have met with only limited success. Resistance is polygenic and recessive (Thakur et al. 1983a), and it appears that producing a resistant hybrid will require that both parents have resistance of a similar type.

There is some evidence that cytoplasm may influence resistance. F_1 hybrids based on sterile cytoplasm are generally more susceptible than hybrids based on normal cytoplasm (R.P. Thakur, unpublished). Poor fertility restoration may be involved.

Resistance appears to depend to a large extent on rapid pollination of stigmas (Thakur and Williams 1980; Willingale and Mantle 1985). Constriction of stylodial tissue occurs in response to pollination and to infection by *C. fusiformis* (Willingale and Mantle 1985; Willingale et al. 1986). This constriction may be an important resistance mechanism, but the evidence is only circumstantial as it is not known how long the ovary remains susceptible, following fertilization, to invasion by the pathogen. Constriction can also be the result of natural aging.

There are several areas still requiring research, and we are pursuing a number of these at ICRISAT Center. These include the relative abilities of ascospore and conidial inocula to incite disease, as possible differences in rates of spore germination between the two spore types may be critical in expression of ergot reaction in the pearl millet. As pollen appears to be very important in resistance, an in vitro test for pollen viability is needed. Likewise, a method for rapid assessment of stigmatic constriction due to aging is required.

Successful breeding of hybrids is probably possible, but whether it is possible to breed hybrids that are both ergot resistant and high yielding is questionable. Yields of ergot-resistant hybrids will probably always lag behind those of other hybrids in a breeding program. Will farmers accept resistance with lower yield in a situation where epidemics are generally infrequent? Seed producers are not likely to be interested in producing seed of hybrids that may not sell well or that have a very limited market. Another question requiring clarification is the actual seriousness of ergot toxicity. Finding answers to these questions is critical, because any program for breeding of ergot-resistant hybrids or even one for breeding ergot-resistant varieties is very resource consuming. For areas where ergot is a likely problem perhaps other methods, such as pollen management or using existing resistant lines directly as varieties, should be considered before additional resources are expended in breeding for resistance. Breeding for ergot resistance from scratch is not practical for a resource-limited program; it would be better to examine the direct use of already identified resistances as varieties.

There is some need to determine if genotype × environment × pathogen interactions affect quantitative or qualitative differences in the alkaloid content of ergot sclerotia. Host reaction to ergot depends considerably on speed of flowering events, and these are influenced, likely, considerably by temperature, with cooler temperatures generally prolonging the protogyny period.

Smut

Smut was first reported on pearl millet in the early 1930s in Senegal (Chevalier 1931) and India (Ajrekar and Likhite 1933). As with downy mildew and ergot, smut gained greater significance as a pearl millet disease in India, especially in northern India, with the adoption of hybrids. In the Sahelian Zone of western Africa, smut is generally considered to be more important than ergot.

Tolyposporium penicillariae Bref. seems well accepted as the name of the causal organism, although *Moesziomyces* has been suggested as a more appropriate genus name (Knower 1977). Like ergot, infection is believed to be through stigmas (Nary 1946); however, unlike ergot, highest levels of smut infection are obtained if inoculation is done before newer (boot-leaf stage) rather than at newer (Thakur et al. 1983b). Infected florets produce sori that are usually top-shaped and somewhat larger than grain. They are at first green, but as panicles mature they turn brown and by harvest rupture relatively easily. Sori contain teliospores in aggregates called 'sporeballs', each consisting of up to several hundred teliospores (Subba Rao and Thakur 1983). Soilborne teliospores germinate and produce windborne sporidia that infect florets. It is possible that teliospores also become windborne and germinate to produce sporidia on the pearl millet plant. Secondary spread of inoculum may also occur from early-flowering panicles to late-flowering ones (Vasudeva and Iyengar 1950).

Tarr (1962) reports having observed sori on pearl millet that were typically cylindrical and elongate, resembling sori of long smut of sorghum. The fungus concerned was morphologically identical with *Tolyposporium ehrenbergii*, the sorghum long smut pathogen.

Teliospores of pearl millet smut have been observed to germinate only at a low frequency on artificial media and in water (Bhatt 1946; Subba Rao and Thakur 1983). The fungus can be readily cultured on several media, including potato dextrose agar, potato agar, and carrot agar. Growth is by budding. Precise information on environmental conditions associated with natural infection has not been reported, although it is generally believed that a warm, humid environment accompanied by some wind promotes the disease.

Some fungicides have been found to reduce smut infection when sprayed onto panicles at the boot stage or shortly thereafter. However, this is not considered to be a practical method of control for pearl millet farmers. Little information is available on effective control through cultural practices.

An effective screening technique to identify smut resistance has been developed (Thakur et al. 1983b). It involves inoculating panicles with a sporidial suspension at the boot-leaf stage, covering the boot with parchment paper bags, and maintaining high relative humidity with some sprinkler irrigation on rain-free days.

Many resistant lines have been bred, and many of these have shown stable resistance to smut in multilocational testing in India and western Africa (Thakur et al. 1986). In limited testing, resistance identified in India is effective in western Africa (ICRISAT, unpublished). Smut resistance is more readily available and more easily identified than ergot resistance. Resistance is probably dominant and probably controlled largely by one or a few genes (Yadav 1974; Phookan 1987). The *tr* gene that removes styler branches is reported to confer a high level of resistance to smut (Wells et al. 1987), but the same gene apparently confers greater susceptibility to rust. At ICRISAT Center, several composites have moderate to high levels of resistance, and success has been achieved in breeding smut-resistant varieties and synthetics. There is a high probability for success in the breeding of smut-resistant hybrids (Thakur and Chahal 1987). Under experimental field conditions, dwarfs seem to have more smut than do the tall or medium-tall types, and panicles with good exertion generally have less smut than those with poor exertion. Ergot resistance almost assures smut resistance, but the reverse is not true. A pollen-based mechanism is reported to be operative for smut resistance (Thakur et al. 1983b), but it appears that in comparison with ergot, the smut fungus requires more time from inoculation to infection of the ovary. Therefore, smut-resistant genotypes can probably tolerate a longer protogyny period than can ergot-resistant genotypes. Multilocational testing in India and western Africa, and additional tests at ICRISAT SAT Center, have not given evidence for the existence of pathotype differences in *T. penicillariae*.

More research is needed to clarify the infection process, including the nuclear status of infective mycelium, and to precisely understand the environmental conditions that favor the disease. Additional information on the inheritance of resistance would be useful, although current breeding for resistance stands a reasonable chance of success without this knowledge. Because of the lack of evidence for pathotype differences, resistances identified at one location are likely to be effective at other locations. A screening technique that reduces the amount of labor required would be useful. In this regard, at Samaru, Nigeria, and Bengou, Niger, simply bagging heads, without inoculation, generally gives at least moderate levels of disease pressure for screening.

Nematodes

Although the more commonly reported genera of nematodes that attack pearl millet are *Melo-*
idogyne, *Pratylenchus*, and *Tylenchorhnchus*, at least 25 genera of plant parasitic nematodes are reported to be associated with pearl millet (Sharma 1985). However, the extent of damage caused by nematodes in pearl millet is not well documented. In Senegal studies conducted in farmers' pearl millet fields demonstrated yield reduction due to nematodes to be as high as 40% (J.D. Prot, personal communication). Surveys and related research are needed to better document the importance of nematodes in reducing pearl millet yields.

Striga spp

Striga hermonthica is a very important problem on pearl millet in Sahelian and sub-Sahelian Africa. *S. asiatica* occurs on pearl millet in India and southern Africa, and to some extent in eastern Africa, but it is of far less importance than *S. hermonthica*. Very little has been reported with respect to these parasites on pearl millet, although both attack sorghum and maize on which three has been considerabe research. Studies in Africa indicate that there are strains of *S. hermonthica* that are more or less specific to pearl millet or sorghum, and possibly also to maize (Jones 1955; King and Zummo 1977). Resistance to *S. hermonthica* or *S. asiatica* in pearl millet has not been confirmed. Although relatively little resistance-identification work has been conducted on pearl millet, it may be that high levels of resistance do not exist in this crop. Local varieties in western Africa generally show some tolerance to *S. hermonthica*. However, tolerance is not a very useful strategy for handling the *Striga* problem, because tolerant lines support the growth of relatively large numbers of *Striga* plants that in turn produce considerable quantities of *Striga* seed that remain in the soil and become troublesome in succeeding years.

A concerted effort should be made to determine if resistance to *Striga* occurs in pearl millet. However, an improvement in the reliability of screening techniques may be necessary before successful resistance identification is achieved. It is likely that control of *Striga* in pearl millet will depend heavily on cultural practices, but resistance, if found, could play an important role in an integrated approach to the control of this important parasite.

Other Diseases

Numerous other diseases, to be mentioned here only briefly, have been reported for pearl millet: [Luttrell (1954); Ramakrishnan (1963); Ferrais (1973); Wells and Hanna (1987)]. Some are important, but generally only on a localized basis; most seem to be of little concern.

Fungal pathogens reported to cause seedling blights or leaf diseases include one or more species of the following genera: *Beniowskia, Cercospora, Curvularia, Dactuliophora, Fusarium, Gloeocercospora, Helminthosporium* (a group of several genera and species), *Phoma, Placosphaeria, Pyricularia, Ramulispora, Rhizoctonia,* and *Sclerotium.* False mildew (*Beniowskia sphaeroidea*) seems limited to parts of southern and eastern Africa, and circular leaf spot (*Dactuliophora elongata*) may be limited to the higher rainfall areas of pearl millet cultivation in western Africa.

Several bacteria have been reported as leaf pathogens on pearl millet. These diseases are generally of minor importance, but there is a need for accurate identification of the bacteria involved. Recent research on two bacterial diseases in Africa, bacterial leaf stripe (Claflin et al. 1987) and bacterial leaf streak (Qhobela and Claflin 1987) may help to clarify the situation.

Preharvest molding of grain can occur and even cause considerable damage when unusually wet conditions prevail at the time grain approaches physiological maturity, or when harvesting is delayed. However, grain molding is not a common problem of pearl millet in its major production areas. Some fungi associated with grain molding are species of *Cladosporium, Curvularia, Fusarium, Oidium,* and *Helminthosporium.*

Charcoal rot (*Macrophomina phaseolina*) does occur in pearl millet, and *Fusarium* spp have been found associated with pearl millet stalks. However, from the lack of information available on stalk and root rots of pearl millet, it is perhaps reasonable to assume that these diseases are of less importance to pearl millet production.

Conclusion

Considerable progress has been achieved in understanding the biology and epidemiology of at least the major pathogens and diseases of pearl millet. Large-scale field screening techniques, having a moderate to high degree of reliability, have been developed for most of the major diseases, and resistances have been identified. Some networks for multilocational testing have been established and others are being developed. Breeding for resistance to many diseases is progressing.

These achievements are significant, but much research remains to be done along these and other lines. Improvement of screening methodologies to provide efficiency, reliability, and suitability for breeding is an urgent need. Hotspot locations alone are not sufficient for screening. Speed of progress in breeding for resistance will depend to a large degree on the extent of collaboration between breeders and pathologists.

Human and material resources for pathology research in pearl millet are often very limited, and care should be taken that these resources are directed toward research that has the best chance of success in benefiting the farmer. These resources are in shortest supply in the major geographic areas of pearl millet cultivation, in the zone spanning from Sudan to Senegal. Williams' survey of journal articles on pearl millet diseases published in the Review of Plant Pathology (1984b) (formerly the Review of Applied Mycology) from 1971 to 1980 is worth noting. He reported that 8024 articles were published on diseases of five major cereals (wheat, rice, maize, sorghum, pearl millet) during this 10–year period; of these, only 149 (2%) dealt with pearl millet. Furthermore, only 5 of these originated in Africa, whereas 129 were from India and the remaining 15 were from other countries.

References

Ajrekar, S.L., and Likhite, V.N. 1933. Observations on *Tolyposporium penicillariae* Bref. (the bajra smut fungus). Current Science 1:215.

Andrews, D.J., King, S.B., Whitcombe, J.R., Singh, S.D., Rai, K.N., Thakur, R.P., Talukdar, B.S., Chavan, S.B., and Singh, P. 1985a. Breeding for disease resistance and yield in pearl millet. Field Crops Research 11:241–258.

Andrews, D.J., Rai, K.N., and Singh, S.D. 1985b. A single dominant gene for rust resistance in pearl millet. Crop Science 25:565–566.

Appadurai, R., Parambaramani, C., and Natarajan, U.S. 1975. Note on the inheritance of susceptibility of pearl millet to downy mildew. Indian Journal of Agricultural Sciences 45:179–180.

Ball, S.L. 1983. Pathogenic variability of downy mildew (Sclerospora graminicola) on pearl millet. I. Host cultivar reactions to infection by different pathogen isolates. Annals of Applied Biology 102:257–264.

Ball, S.L., Pike, D.J., and Burridge, C.Y. 1986. Characterization of populations of Sclerospora graminicola. Annals of Applied Biology 108:519–526.

Basavaraj, R., Safeeulla, K.M., and Murty, B.R. 1980. The role of gene effects and heterosis for resistance to downy mildew in pearl millet. Indian Journal of Genetics and Plant Breeding 40:537–548.

Bhadur, P., Singh, S., Sinha, V.C., Goel, L.B., and Sharma, S.K. 1975. Note on the laboratory evaluation of some systemic fungicides against cereal rusts. Indian Journal of Agriculture Sciences 45:274–275.

Bhat, R.V., Roy, D.N., and Tulpule, P.G 1976. The nature of alkaloids of ergoty pearl millet or bajra and its comparison with alkaloids of ergoty rye and ergoty wheat. Toxicology and Applied Pharmacology 36:11–17.

Bhatt, R.S. 1946. Studies in the Ustilaginales. 1. The mode of infection of the bajra plant (Pennisetum typhoides Stapf. & Hubbard) by the smut, Tolyposporium penicillariae Bref. Journal of Indian Botanical Society 25:163–186.

Bhide, V.P., and Hegde, R.K. 1957. Ergot of bajra Pennisetum typhoides (Burm.) Stapf. and Hubb. in Bombay State. Current Science 26:116.

Butler, E.J. 1907. Some diseases of cereals caused by Sclerospora graminicola. Memoirs of the Department of Agriculture in India. Botanical Series 2:1–24.

Chevalier, A. 1931. Une maladie de pénicillaire au Sénégal. Revue de Botanique Appliquée 11:49–50.

Clafiln, L.E., Ramundo, B.A., Leach, J.E., and Erinle, I.D. 1987. Pseudomonas avenae, the causal agent of bacterial leaf stripe of pearl millet in Nigeria. Phytopathology 77:1766. (Abstract)

Dang, J.K., Thakur, D.P., and Grover, R.K. 1983. Control of pearl millet downy mildew caused by Sclerospora graminicola with systemic fungicides in an artificially-contaminated plot. Annals of Applied Biology 102:99–106.

Deshmukh, S.S., Mayee, C.D., and Kulkarni, B.S. 1978. Reduction of downy mildew of pearl millet with fertilizer management. Phytopathology 68:1350–1353.

Ferrais, R. 1973. Pearl Millet (Pennisetum typhoides). Review Series 1/1973. Commonwealth Bureau of Pastures and Field Crops, Hurley, Maidenhead, Berks SL6 5LR, UK. 70 pp.

Gill, K.S., Phul, P.S., Chahal, S.S., and Singh, N.B. 1978. The role of gene effects and heterosis for resistance to downy mildew in pearl millet. Indian Journal of Genetics and Plant Breeding 40:537–548.

Girard, J.C. 1975. Downy mildew of pearl millet in Senegal. Pages 59–73 in Proceedings of Consultants' Group Meetings on Downy Mildew and Ergot of Pearl Millet, 1–3 Oct. 1975, ICRISAT, Hyderabad, India. Patancheru, A.P. 502 324, India: International Crops Research Institute for the Semi-Arid Tropics.

Hanna, W.W., Wells, H.D., and Burton, G.W. 1985. Dominant gene for rust resistance in pearl millet. Journal of Heredity 76:134.

ICRISAT. 1985. Pearl millet. Pages 81–103 in Annual Report 1984. Patancheru, A.P. 502 324, India: International Crops Research Institute for the Semi-Arid Tropics.

ICRISAT. 1986. Report of the 1985 International Pearl Millet Downy Mildew Nursery (IPM-DMN). Progress Report MP 9.46. Patancheru, A.P. 502 324, India: International Crops Research Institute for the Semi-Arid Tropics. pp. 16. (Limited distribution.)

ICRISAT. 1987a. Looking ahead: a 10–year plan. Patancheru, A.P 502 324, India: International

Crops Research Institute for the Semi-Arid Tropics

ICRISAT. 1987b. Proceedings of the International Pearl Millet Workshop, 7–11 Apr 1986, ICRISAT Center, India. Patancheru, A.P. 502 324, India: International Crops Research Institute for the Semi-Arid Tropics

Idris, M.P., and **Ball, S.L.** 1984. Inter- and intra-continental sexual compatibility in *Sclerospora graminicola*. Plant Pathology 33:219–223.

Jensen, S.G., Andrews, D.J., and **DeVries, N.E.** 1983. Reactions of pearl millet germplasm from the world collection to maize dwarf mosaic virus strains A and B. Plant Disease 67:1105–1108.

Jones, W.K. 1955. Further experiments on witch-weed control: II. The existence of physiological strains of *Striga hermonthica*. Empire Journal of Experimental Agriculture 23:206–213.

Kandasamy, D., Sivaprakasam, K., and **Krishnamurty, C.S.** 1971. An observation of the different levels of nitrogen on the incidence of rust disease of pearl millet (*Pennisetum typhoides* Stapf. and Hubb.). Plant and Soil 34:757–760.

Kapooria, R.G. 1972. On the chemical control of *Puccinia penniseti*. Ghana Journal of Science 12:67–68.

Kenneth, R. 1977. *Sclerospora graminicola* (Sacc.) Schroet. Pages 96–99 *in* Diseases, Pests, and Weeds in Tropical Crops (Kranz, J., Schutterer, H., and Koch, W., eds.). Berlin, Federal Republic of Germany: Verlag, Paul Parey.

King, S.B., and **Zummo, N.** 1977. Physiologic specialization in *Striga hermonthica* in western Africa. Plant Disease Reporter 61:770–773.

Krishnamachari, K.A.V. R., and **Bhat, R.V.** 1976. Poisoning of ergoty bajra (pearl millet) in man. Indian Journal of Medical Research 64:1624–1628.

Kukla, B., Thourenel, J.C., and **Fauquet, C.** 1984. A strain of guinea grass mosaic virus from pearl millet in the Ivory Coast. Phytopathologische Zeitschrift 109:65–73.

Kulkarni, U.K. 1958. Studies on the development and cytology of *Puccinia penniseti*. Zimm. Transactions of the British Mycological Society 41:65–67.

Loveless, A.R. 1967. *Claviceps fusiformis* sp. nov. the causal agent of agalactia of sows. Transactions of the British Mycological Society 50:15–18.

Luttrell, E.S. 1954. Diseases of pearl millet in Georgia. Plant Disease Reporter 38:507–514.

Melhus, I.E., Vanttaltera, F.H., and **Bliss, D.E.** 1928. A study of *Sclerospora graminicola* (Sacc.) Schroet. on *Setaria viridis* (l.) Beauv. and *Zea mays* L. Research Bulletin 111, Agricultural Experiment Station, Iowa State College of Agriculture and Mechanic Arts, Ames, Iowa.

Michelmore, R.W., Pawar, M.N., and **Williams, R.J.** 1982. Heterothallism in *Sclerospora graminicola*. Phytopathology 72:1368–1372.

Nene, Y.L., and **Singh, S.D.** 1976. Downy mildew and ergot of pearl millet. PANS 22:366–385.

Parambaramani, C., Vidhyasekharan, P., and **Kandaswamy, T.K.** 1971. Preliminary studies on the use of new oxathin fungicides for control of some rusts. Madras Agricultural Journal 58:705–706.

Phookan, A.K. 1987. Studies on pearl millet smut with special reference to pathogenic variability, inheritance of resistance and chemical control. Ph.D. thesis, Haryana Agricultural University, Hisar, Haryana, India. 116 pp.

Pokhriyal, S.C., Unnikrishnan, K.V., Singh, B., Dass, R., and **Patel, R.R.** 1976. Combining ability of downy mildew resistant lines in pearl millet. Indian Journal of Genetics and Plant Breeding 36:403–409.

Qhobela, M., and **Claflin, L.E.** 1987. Identification of the causal agent of bacterial leaf streak of pearl millet (*Pennisetum americanum* (L.) Leeke). Phytopathology 77:1767. (Abstract)

Ramachar, P., and **Cummins, G.B.** 1965. The species of *Puccinia* on the *Paniceae*. Mycopathologia et Mycologia Applicata 25:7–60.

Ramakrishnan, T.S. 1963. Disease of millets. New Delhi, India: Indian Council of Agricultural Research. 152 pp.

Ramakrishnan, T.S., and Soumin, C.K. 1948. Studies on the cereal rust *Puccinia penniseti* Zimm. and its alternate host. Indian Phytopathology 1:97–103.

Ramaswamy, C. 1968. Meteorological factors associated with the ergot epidemic of bajra (*Pennisetum*) in India during the Kharif season-1967– a preliminary study. Current Science 37:331–335.

Safeeulla, K.M. 1976. Biology and control of the downy mildews of pearl millet, sorghum and finger millet. Mysore, Karnataka, India: University of Mysore. 304 pp.

Safeeulla, K.M. 1977. Genetic vulnerability: the basis of recent epidemics in India. Annals of the New York Academy of Sciences 287:72–85.

Sathe, A.V. 1969. A new rust record on *Pennisetum typhoides*. Mycologia 61:198–200.

Seth, M.L., Raychaudhuri S.P., and Singh, D.V. 1972a. Bajra (pearl millet) streak, a leaf hopper borne cereal virus in India. Plant Disease Reporter 56:424–428.

Seth, M.L., Raychaudhuri S.P., and Singh, D.V. 1972b. A mosaic disease of Bajra (*Pennisetum typhoides* Berm. F.) Stapf. and Hubb. in India. Indian Journal of Agricultural Sciences 42:322–325.

Sharma, S.B. 1985. A world list of nematode pathogens associated with chickpea, groundnut, pearl millet, pigeonpea, and sorghum. ICRISAT Pulse Progress Report 42. Patancheru, A.P. 502 324, India: International Crops Research Institute for the Semi-Arid Tropics. pp 36. (Limited distribution.).

Sharma, Y.P., Singh, R.S., and Tripathi, R.K. 1983. Role of insects in secondary spread of the ergot disease of pearl millet (*Pennisetum americanum*). Indian Phytopathology 36:131–133.

Shetty, H.S. 1987. Biology and epidemiology of downy mildew of pearl millet. Pages 147–160 *in* Proceedings of the International Pearl Millet Workshop, 7–11 Apr 1986, ICRISAT, Hyderabad, India. Patancheru, A.P. 502 324, India: International Crops Research Institute for the Semi-Arid Tropics.

Shetty, H.S., Neergaard, P., and Mathur, S.B. 1977. Demonstration of seed transmission of downy mildew or green ear disease, *Sclerospora graminicola*, in pearl millet, *Pennisetum typhoides*. Proceedings of the Indian National Science Academy 43:201–205.

Shetty, S., and Ahmed, R. 1981. Physiological specialization in *Sclerospora graminicola* attacking pearl millet. Indian Phytopathology 34:307–309.

Shinde, P.A., and Bhide, V.P. 1958. Ergot of Bajra (*Pennisetum typhoides*) in Bombay State. Current Science 27:499–500.

Shinde, R.B., Patil, F.B., and Sangave, R.A. 1984. Resistance to downy mildew in pearl millet. Journal of Maharashtra Agriculture University 9:337–338.

Siddiqui, M.R., and Khan, I.D. 1973a. Renaming *Claviceps microcephala*, ergot fungus on *Pennisetum typhoides* in India as *Claviceps fusiformis*. Transactions of the Mycological Society of Japan 14:195–198.

Siddiqui, M.R., and Khan, I.D. 1973b. Dynamics of inoculum and environment in relation to ergot incidence on *Pennisetum typhoides* (Burm.) Stapf. and Hubbard. Transactions of the Mycological Society of Japan 14:280–288.

Singh, B.B., and Sokhi, S.S. 1983. Effect of rust of pearl millet on yield components. Indian Phytopathology 36:89–91.

Singh, F., Singh, R.K., Singh, R.M., and Singh, R.B. 1978. Genetic analysis of downy mildew (*Sclerospora graminicola*) resistance in pearl millet (*Pennisetum typhoides* (Brum.) S. & H.). Zeitschrift für Pflanzenzüchtung 81:54–57.

Singh, G., Vyas, K.L., and Bhatt, B.N. 1983. Occurrence of pearl millet ergot on *Cenchrus ciliaris* Pers. in Rajasthan. Indian Journal of Agricultural Sciences 53:481–483.

Singh, P., and Agarwal, R.K. 1979. Effects of zinc and phosphatic fertilizers on the incidence of downy mildew and the nutrient contents in pearl millet. Indian Journal of Agricultural Sciences 49:459–462.

Singh, S.D. 1983. Selection for downy mildew resistance in the parental lines of the susceptible hybrid BJ 104. International Working Group on Graminaceous Downy Mildews (IWGGDM) Newsletter 5:2

Singh, S.D., Gopinath, R., Luther, K.D.M., Reddy, P.M., and Pawar, M.N. 1984. Systemic remissive property of metalaxyl against downy mildew in pearl millet. Plant Disease 68:668–670.

Singh, S.D., and Singh, G. 1987. Resistance to downy mildew in pearl millet hybrid NHB–3. Indian Phytopathology 40:178–180.

Singh, S.D, Ball, S., and Thakur, D.P. 1987a. Problems and strategies in the control of downy mildew. Pages 161–172 in Proceedings of the International Pearl Millet Workshop, 7–11 Apr 1986, ICRISAT, Hyderabad, India. Patancheru, A.P. 502 324, India: International Crops Research Institute for the Semi-Arid Tropics.

Singh, S.D., Gopinath, R., and Pawar, M.N. 1987b. Effects of environmental factors on asexual sporulation of Sclerospora graminicola. Indian Phytopathology 40:186–193.

Singh, S.D., and King, S.B. 1988. Recovery resistance to downy mildew in pearl millet. Plant Disease 72:425–428.

Singh, S.D., Williams, R.J., and Reddy, P.M. 1988 Isolation of downy mildew resistant lines from a highly susceptible cultivar of pearl millet. Indian Phytopathology 41:450–456.

Sivaprakasham, K., and Pillayarsamy, K. 1975. Effect of nitrogen on the incidence of rust disease of pearl millet caused by Puccinia penniseti: Zimm. Madras Agricultural Journal 62:221–223.

Subba Rao, K.V., and Thakur, R.P. 1983. Tolyposporium penicillariae, the causal agent of pearl millet smut. Transactions of the British Mycological Society 81:597–603.

Sundaram, N.V., Sastry, D.V.R., and Kanwar, S.K. 1973. Note on the seed-borne infection of

downy mildew Sclerospora graminicola (Sacc.) Schroet. of pearl millet. Indian Journal of Agricultural Sciences 43:215–217.

Tarr, S.A.J. 1962. Diseases of sorghum, sudan grass, and broom corn. Kew, Surrey, UK: The Commonwealth Mycological Institute. 380 pp.

Thakur, D.P. and Kanwar, Z.S. 1978. Ability of natural incitent Claviceps microcephala from Panicum antidotale to produce ergot symptoms in Pennisetum typhoides. Indian Journal of Agricultural Sciences 48:540–542.

Thakur, D.P. 1980. Utilization of downy mildew resistant germplasm and other resources for increased productivity of pearl millet under arid and semi arid regions of India. Annals of Arid Zone 19:265–270.

Thakur, D.P. 1984. Ergot disease of pearl millet. Review of Tropical Plant Pathology 1:297–328.

Thakur, R.P., and Williams, R.J. 1980. Pollination effects on pearl millet ergot. Phytopathology 70:80–84.

Thakur, R.P., Williams, R.J., and Rao, V.P. 1982. Development of ergot resistance in pearl millet. Phytopathology 72:406–408.

Thakur, R.P., Talukdar, B.S., and Rao, V.P. 1983a. Genetics of ergot resistance in pearl millet. Page 737 in Abstracts of contributed papers of the XV International Congress of Genetics (Part 2), 12–21 Dec 1983, New Delhi, India, New Delhi, India: Oxford and IBH Publishing Co.

Thakur, R.P., Subba Rao, K.V., and Williams, R.J. 1983b. Effects of pollination on smut development in pearl millet. Plant Pathology 32:141–144.

Thakur, R.P., Williams, R.J., and Rao, V.P. 1983c. Control of ergot in pearl millet through pollen management. Annals of Applied Biology 103–31–36.

Thakur, R.P., Rao, V.P., and Williams, R.J. 1984. The morphology and disease cycle of ergot, caused by Claviceps fusiformis, in pearl millet. Phytopathology 74:210–205.

Thakur, R.P., Rao, V.P., Williams, R.J., Chahal, S.S., Mathur, S.B., Pawar, N.B., Nafade, S.D., Shetty, H.S., Singh, G., and Bangar, S.D. 1985. Identification of stable resistance to ergot in pearl millet. Plant Disease 69:982–985.

Thakur, R.P., Subba Rao, K.V., Williams, R.J., Gupta, S.C., Thakur, D.P., Sundaram, N.V., Frowd, J.A., and Guthrie, E.J. 1986. Identification of stable resistance to smut in pearl millet. Plant Disease 70:38–41.

Thakur, R.P., and Chahal, S.S. 1987. Problems and strategies in the control of ergot and smut in pearl millet. Pages 173–182 in Proceedings of the International Pearl Millet Workshop, 7–11 Apr 1986, ICRISAT, Hyberabad, India. Patancheru, A.P. 502 324, India: International Crops Research Institute for the Semi-Arid Tropics.

Thomas, K.M., Ramakrishnan, T.S., and Srinivasan, K.V. 1945. The occurrence of ergot in south India. Proceedings of the Indian Academy of Sciences B 21:93–100.

Vanky, K. 1977. *Moesziomyces*, a new genus of Ustilaginales. Botaniska Notiser 130:131–135.

Vasudeva, R.S., and Iyengar, M.R.S. 1950. Secondary infection in bajra smut disease caused by *Tolyposporium penicillarie* Bref. Current Science 19:123.

Verma, O.P., and Pathak, V.N. 1984. Role of insects in secondary spread of pearl millet ergot. Phytophylactica 16:257–258.

Wells, H.D., Hanna, W.A., and Burton, G.W. 1987. Effects of inoculation and pollination on smut development in near-isogenic lines of pearl millet. Phytopathology 77:293–296.

Wells, H.D., and Hanna, W.A. 1987. Inheritance of a severe bipolaris leafspot of pearl millet. Phytopathology 77:1772. (Abstract)

Williams, R.J. 1979. Position paper on seed transmission of some graminaceous downy mildews. Presented at the International Conference on Graminaceous Downy Mildew Diseases, 28 Nov–4 Dec 1979, Beliagio, Italy. (Limited distribution.)

Williams, R.J., Pawar, M.N., and Huibers-Govaert, I. 1980. Factors affecting staining of *Sclerospora graminicola* oospores with triphenyl tetrazolium chloride. Phytopathology 70:1092–1096.

Williams, R.J., and Singh, S.D. 1981. Control of pearl millet downy mildew by seed treatment with metalaxyl. Annals of Applied Biology 97:263–268.

Williams, R.J., Singh, S.D., and Pawar, M.N. 1981. An improved field screening technique for downy mildew resistance in pearl millet. Plant Disease 65:239–241.

Williams, R.J. 1984a. Downy mildews of tropical cereals. Advances in Plant Pathology 2:1–103.

Williams, R.J. 1984b. Disease resistance in pearl millet. Pages 245–296 in Review of Tropical Plant Pathology Vol. 1 (Raychaudhuri, S.P., and Verma, J.P., eds.). New Delhi, India: Today and Tomorrow's Printers and Publishers.

Willingale, J., and Mantle, P.G. 1985. Stigma constriction in pearl millet, a factor influencing reproduction and disease. Annals of Botany 56:09–115.

Willingale, J. Mantle, P.G., and Thakur R.P. 1986. Postpollination stigmatic constriction, the basis of ergot resistance in selected lines of pearl millet. Phytopathology 76:536–539.

Yadav, R.P. 1974. Inheritance of smut resistance in pearl millet. Agra University Journal of Research 23:37–39.

Pearl Millet Diseases in Western Africa

J. Werder[1] and S.K. Manzo[2]

Abstract

Downy mildew (DM) is an important disease, followed in importance by smut. Ergot is found, but this disease is generally confined to pearl millet in higher rainfall areas. Several leaf diseases occur, including leaf blast, rectangular leaf spot, circular leaf spot, and three bacterial diseases, but these are of minor importance.

Downy mildew infection is often characterized by green ear symptoms. Virulence differences have been demonstrated for the DM pathogen; nurseries are currently in use to study virulence and resistance stability of western African materials. Smut is especially common on panicles with poor exertion, and infection is highest when the boot-leaf stage coincides with rainfall and high humidity.

Striga hermonthica is especially serious in areas of low soil fertility. Local varieties show some tolerance, but good evidence for resistance is lacking. Control of this parasite will require integrated efforts of scientists of several disciplines.

Introduction

Pearl millet [*Pennisetum glaucum* (L.) R. Br.] is the principal staple cereal grown in the dry semi-arid Sahelian regions of western Africa. The major pearl millet growing areas of the world are Africa, with 15.09 million ha (12.85 million in western Africa) (Table 1), and southern Asia with 11.34 million ha (ICRISAT 1987a).

More than 50 fungal, bacterial, and viral pathogens of pearl millet have been reported (Ferraris 1973). Only a few pathogens and weeds, however, cause major damage to pearl millet in western Africa. The main constraints of pearl millet production in western Africa are physical stresses like drought, poor crop establishment, and low soil fertility, and biological stresses: insect pests, diseases, and weeds.

Downy mildew [*Sclerospora graminicola* (Sacc.) Schroet.] and *Striga hermonthica* are the most important pathogens and weed parasites in western African pearl millet sowings. Other diseases such as smut (*Tolyposporium penicillariae*), ergot (*Claviceps fusiformis*), and leaf spots—though of minor importance can cause yield losses in certain areas or on certain germplasm.

Table 1. Area sown to pearl millet and average yield of pearl millet in eight western African countries.

Country	Area[1] (million ha)	Average yield (kg ha[-1])
Burkina Faso	0.90	424
Cameroon	0.50	761
Chad	1.15	404
Ghana	0.14	NA[2]
Mali	1.36	692
Niger	3.08	423
Nigeria	4.72	565
Senegal	1.00	449
Western Africa	12.85	531
Africa	15.09	564

1. Source: ICRISAT 1987b, 2. NA = Not available.

1. Principal Millet Pathologist, ICRISAT Sahelian Center, B.P. 12404, Niamey (via Paris), Niger.
2. Head, Plant Protection Department, Institute of Agricultural Research, Ahmadu Bello University, Zaria, Nigeria.

Werder, J., and **Manzo, S.K.** 1992. Pearl millet diseases in western Africa. Pages 109–114 *in* Sorghum and millets diseases: a second world review. (de Milliano, W.A.J., Frederiksen, R.A., and Bengston, G.D., eds). Patancheru, A.P. 502 324, India: International Crops Research Institute for the Semi-Arid Tropics. (CP 734).

Downy Mildew

The downy mildew organism induces systemic and rarely local infection on pearl millet. Systemic symptoms can appear at any growth stage from seedling to flowering. However, except in highly susceptible genotypes, infection of seedlings is rarely seen in western Africa. Infection by *S. graminicola* generally produces malformed inflorescences with various degrees of transformation into leafy structures called green ears. This prevents inflorescence development and thus is detrimental to grain yield (Girard 1976; Girard and Delassus 1978).

In a survey of farmers' fields in Niger, downy mildew (DM) incidence varied from nil (low-rainfall area) to 30% (high-rainfall areas) (ICRISAT 1985). In the north of Cameroon, DM incidence was from 10 to 29% (IRA/ICRISAT 1987) and in Nigeria from nil to 15% (Selvaraj 1979). An unpublished crop-loss study in Gaya, Niger, showed a yield loss of 50% due to DM attack; in Nigeria losses were from 4 to 22% (Selvaraj 1979).

Pathogen variation has been demonstrated with various procedures by many researchers (Ball and Pike 1983, 1984; Shetty and Ahmad 1981; Girard 1976; Selvaraj 1979). The differential pathogenic reaction of DM isolates in India and in western Africa is a major reason why millet cultivars improved and produced in India are not adapted to most countries of western Africa.

The WADMVN

ICRISAT's Western African Downy Mildew Variability Nursery (WADMVN), initiated in 1986, provides useful information on the pathogenic variability of DM within the western African region. This nursery includes varieties, hybrids, and breeding lines from India and western Africa (10 entries) and is sown as a "differential." The same seed sources are to be used for 3 years. Results from the 1st year have demonstrated striking differences between Burkina Faso, Nigeria, Niger, and Senegal for pathogenicity of *S. graminicola* (ICRISAT 1987b). Evidence for differences in virulence among the four locations was especially noticeable for NHB 3, 81 B, and MBH 110 (Table 2). To test stability of resistance, breeding products must be exposed to a range of DM pathotypes in different environments.

In view of these requirements, ICRISAT has initiated the Western African Downy Mildew Observation Nursery (WADMON). This is a regional cooperative activity through which pearl

Table 2. Downy mildew (DM) incidence (%) of the Western African Downy Mildew Variability Nursery (WADMVN), at Bambey, Bengou, Kamboinse, and Samaru, rainy season 1987.

Entry	Designation	Origin	Mean[2]	Bambey[3]	Bengou[4]	Kam-boinse[5]	Samaru[6]
			DM incidence[1] (%) at dough stage				
700651	Breeding line	IAR, Nigeria	1	0	1	0	4
		INERA/ICRISAT					
IKMV 8201	Variety	Burkina Faso	3	0	3	1	6
INMV 8220	Variety	IAR/ICRISAT, Nigeria	3	5	4	3	0
MBH 110	Hybrid	India	27	4	30	74	0
NHB 3	Hybrid	India	30	1	21	58	38
81 B	Maintainer	ICRISAT, India	43	4	69	0	100
Trial mean (10 entries)			14	5	15	15	20

1. Based on a mean number of 69 to 90 plants per entry in two replications and two rows per replication.
2. Mean of all locations ranked.
3. Institut Sénégalais de Recherches Agricoles, Senegal.
4. ICRISAT Sahélian Center, Niger.
5. Institut d'Etudes et de Recherches Agricoles, Burkina Faso.
6. Institute for Agricultural Research, Nigeria.

millet lines and cultivars from national programs in western African and from ICRISAT's Sahelian Center are tested for stability of DM resistance. In 1986, cultivars from four national programs were sent for testing to sites in five countries in western Africa (ICRISAT 1986).

Economic resources of the Sahelian farmer are very limited, and chemical control of DM is non-affordable. The most practical means of control of *S. graminicola* in Sahelian farmers' fields will be crop varieties resistant to or with relatively low levels of susceptibility to downy mildew.

Panicle Diseases

The two main panicle diseases of pearl millet are smut (*Tolyposporium penicillariae* Bref.) and ergot (*Claviceps fusiformis* Loveless). In smut infections, some grains, in the head are replaced by green smut sori that eventually turn dark brown and break off, releasing balls of teliospores. Highest smut severity usually occurs on heads with poor exertion. That part of the head covered by the flag leaf is often ideal for the development of the smut fungus.

Smut is found in all millet-growing areas of western Africa, but yield losses are of less importance than those caused by downy mildew. Selvaraj (1979) estimated yield losses of 2–3% due to smut in Nigeria. In late-sown millet at Bengou, Niger, boot-leaf stage coincided with high and well-distributed rainfall and high humidity contributing to much higher smut incidence on BJ 104 (Table 3).

Table 3. Influence of sowing date on smut (*Tolyposporium penicillarie*) severity at INRAN station, Bengou, Niger, rainy season 1987.

Date of sowing[1]	Smut severity[2](%)	Date of leaf-boot stage
15 Jun	20 (99)[3]	21 Jul
24 Jun	40 (89)	3 Aug
6 Jul	67 (46)	18 Aug
18 Jul	40 (50)	29 Aug

1. On sowing date, 10 rows sown in a nonreplicated plot.
2. Based on a scale of 0 to 100%, where 0 = no symptoms, and 100 = totally covered.
3. Number of plants scored is indicated in parentheses.

Some breeding materials and certain varieties showing resistance in India also show resistance in western Africa. A large number of varieties and lines of western African origin were found to have resistance or at least the ability to endure infection.

Yield losses in pearl millet due to ergot are of minor importance in western Africa (Selvaraj 1980), but can be devastating on short-duration photoperiod-insensitive cultivars. Observations in farmers' fields in the north of Cameroon showed only 3% ergot severity (IRA/ICRISAT 1987). The sclerotia of this pathogen contain alkaloids that can be hazardous if eaten.

Leaf Diseases

Leaf diseases, although currently less important than DM, are potentially damaging in western Africa. Leaf blast and rectangular leaf spot can be found in nearly all millet-growing areas of western Africa, whereas circular leaf spot (*Dactuliophora elongata*) is found only in the high rainfall zones (600 mm). Circular leaf spot appears as concentric or zonate spots studded with black sclerotia (Tyagi 1985).

Rust (*Puccinia penniseti*) is rarely found in the Sahelian region. It is mainly observed in zones with 800 mm rainfall. The fungus appears very late in the season and usually has little influence on yield (Girard and Delassus 1978). In some wetter parts of Nigeria (e.g., the Guinea Savannah) rust can have some effect on yield.

Phoma leaf spot (*Phoma sorghina = Phyllosticta penicillariae*), described by Saccas (1954), appears in areas with higher rainfall.

A midrib infection on millet in Nigeria, was identified as *Curvularia eragrostidis* (Bindawa 1988). *Placosphaeria* sp (Jouan and Delassus 1971), another leaf disease found rarely on millet, was observed in Gongola state, Nigeria, in 1986.

Bacterial leaf streak has been described as *Pseudomonas* sp whereas bacterial streak symptoms on samples from Niger have been identified as *Xanthomonas* sp. The symptoms of this bacterial disease look very similar to those reported on proso millet (Elliott 1923). Bacterial disease on pearl millet was reported (Zummo 1976) as yellow leaf blotch (*Pseudomonas* sp).

Ekukole (1985) found a few infections of maize streak virus on pearl millet in the Samaru area.

Striga hermonthica

The root parasite *Striga hermonthica* is a major constraint to millet production and is the most costly weed species in western African cereals production (Laycock 1986). *S. hermonthica* in the sorghum and millet fields of subsistence farmers is said to be the most important pest-control problem in western Africa (Ogborn 1984). *Striga hermonthica* can substantially reduce yield in nearly all millet-growing areas of the western African Sahel. Yield losses up to 65% are recorded in Burkina Faso (Ramaiah 1985). At the subsistence level, the *Striga* problem will in-

crease with growing population and consequent cultivation of millet on unsuitable land.

In general, *Striga* infestations in farmers' fields are more severe in areas with low fertility. Pearl millet varieties of western African origin tested for *Striga* reaction in pots and in a sick plot in Niger in 1987 did not show resistance (Table 4). The susceptible control, Sadoré local, had the highest *Striga* infestation, but the best yield as well. This might indicate that local varieties are tolerant to *Striga hermonthica* if the infestation pressure is not too high. In general, progress in identification of resistance sources is slow, and future research strategies to address the *Striga* problem will require integration of efforts of agronomists, physiologists, breeders, soil scientists, and plant pathologists. A regional approach, coordinated and clearly focused within the western African region, is essential.

Table 4. Reaction of entries from the ICRISAT Pearl Millet Zone A Trial (IMZAT) to *Striga hermonthica* in *Striga*-infested pots and in a *Striga*-sick plot, ISC, Sadoré, Niger, rainy season 1987.

Entry	Origin	Mean[1] *Striga* plants pot[-1] at 66 DAS[2]	Survival[3] (%) at 107 DAS	Mean[1] *Striga*[4] plants pot[-1] (9 M[2])	Grain yield (t ha[-1])
ICMV IS 85 333	ISC, Niger	29.7 (2.80)	100	3.8 (1.29)	0.78
CT 2	INRAN, Niger	34.2 (2.68)	83	13.8 (2.65)	0.76
IKMC 1	INERA/ICRISAT, Burkina Faso	36.8 (3.45)	100	7.5 (1.99)	0.57
SE 2124	IAR, Nigeria	39.5 (3.50)	67	4.5 (1.48)	0.24
IKMV 8201	INERA/ICRISAT, Burkina Faso	58.2 (3.67)	83	6.0 (1.65)	0.28
SE 361	IAR, Nigeria	78.0 (4.10)	83	3.5 (0.97)	0.22
Control (Sadoré local)		52.3 (3.90)	67	55.3 (3.78)	1.14
SE		(±0.11)		(±0.08)	±0.04
Trial mean (15 entries)					0.55
CV (%)		(32)		(38)	40

1. Entries ranked by mean *Striga* plants per pot and logarithmic transformation values (log x + 1.1) are shown in parentheses.
2. DAS = Days after sowing.
3. Number of millet plants in pots surviving at 107 DAS as percentage of total plants counted at 45 days.
4. *Striga* plants counted 90 DAS.

References

Ball, S.L., and Pike, D.J. 1983. Pathogenic variability of downy mildew (*Sclerospora graminicola*) on pearl millet. II. Statistical techniques for analysis of data. Annals of Applied Biology 102:256–273.

Ball, S.L., and Pike, D.J. 1984. Intercontinental variation of *Sclerospora graminicola*. Annals of Applied Biology 104:41–51.

Bindawa, A.A. 1988. Studies on anthracnose on sorghum and maize caused by *Colletotrichum graminicola* (Ces.) Wils and a midrib spot on millet. M.Sc. thesis, Faculty of Agriculture, Department of Plant Protection, Ahmadu Bello University, Zaria, Nigeria. 85 pp.

Ekukole, G. 1985. Studies on the maize streak virus and its vectors. M.Sc. thesis, Faculty of Agriculture, Department of Plant Protection, Ahmadu Bello University, Zaria, Nigeria. 99 pp.

Elliott, C. 1923. A bacterial disease of proso millet. Journal of Agricultural Research 26(4): 151–160.

Ferraris, R. 1973. Pearl millet (*Pennisetum typhoides*). Review Series No. 1/1973. Hurley, Maidenhead, Berks SL6 5LR, UK: Commonwealth Bureau of Pastures and Field Crops. 70 pp. 343 refs.

Girard, J.C. 1976. Downy mildew of pearl millet in Senegal. Pages 59–73 *in* Proceedings of the Consultants' Group Meetings on Downy Mildew and Ergot of Pearl Millet. 1–3 Oct 1975. ICRISAT, Hyderabad, India. Patancheru, A.P. 502 324, India: International Crops Research Institute for the Semi-Arid Tropics.

Girard, J.C., and Delassus, M. 1978. Les maladies parasitaires des mils et des sorghos au Sénégal. Presented at SODEVA (Société de Développement et de Vulgarisation Agricole) training course, 7 Apr 1976, Kaolack, Sénégal. Dakar, Sénégal: ISRA (Institut Senegalais de Recherches Agricoles).

ICRISAT. 1985. Pearl Millet Improvement Program. Pages 15–35 *in* Annual Report, 1985. Sahelian Center, Niamey, Niger: International Crops Research Institute for the Semi-Arid Tropics.

ICRISAT. 1986. Report of the western African Downy Mildew Observation Nursery (WADMON). Sahelian Center, Niamey, Niger: International Crops Research Institute for the Semi-Arid Tropics. 20 pp.

ICRISAT. 1987a. Looking ahead: A 10–year plan. Patancheru, A.P. 502 324, India: International Crops Research Institute for the Semi-Arid Tropics. (Limited distribution.)

ICRISAT. 1987b. Cereals, pearl millet. Pages 69–126 *in* Annual Report 1986. Patancheru, A.P. 502 324, India: International Crops Research Institute for the Semi-Arid Tropics.

IRA/ICRISAT. 1987. Technical report: Sorghum and pearl millet diseases in the North of Cameroon. Institut de la Recherche Agronomique/ International Crops Research Institute for the Semi-Arid Tropics. Sahelian Center, Niameny, Niger: International Crops Research Institute for the Semi-Arid Tropics. 19 pp.

Jouan, B., and Delassus, M. 1971. Principales maladies des mils et sorghos observées au Niger. L'Agronomie Tropicale 8:830–860.

Laycock, D. 1986. Recent results on *Striga* investigations in the CILSS Integrated Pest Management Project, with special reference to Niger. Presented at the FAO/OAU Pan-African Consultation on *Striga* and its Control, 20–24 Oct 1986, Maroua, Cameroon.

Ogborn, J. 1984. *Striga*: Research priorities with specific reference to agronomy. Pages 195–212 *in Striga* Biology and Control (Ayensu E.S., et al. eds.). Paris, France: ICSU Press.

Ramaiah, K.V. 1985. Hand pulling on *Striga hermonthica* in pearl millet. Tropical Pest Management 31:326–327.

Saccas, A.M. 1954. Les champignons parasitaires des sorghos (*Sorghum vulagare*) et des penicillaires (*Pennisetum typhoideum*) en Afrique Equatoriale Française. L'Agronomie Tropicale 2,3,6:1954. 117 pp.

Selvaraj, J.C. 1979. Research on pearl millet diseases in Nigeria. Samaru Conference Paper 28, Samaru, Zaria, Nigeria: Ahmadu Bello University. 20 pp.

Selvaraj, J.C. 1980. Ergot disease of pearl millet in Nigeria. FAO Plant Protection Bulletin 28(4): 129–132.

Shetty, H.S., and Ahmad, R. 1981. Physiologic specialisation in *Sclerospora graminicola*. Indian Phytopathology 34(3):307–309.

Tyagi, P.D. 1985. Identity of *Dactuliophora* leaf spot of pearl millet. Proceedings of the Indian Academy of Sciences (Plant Sciences) 94(2–3): 407–413.

Zummo, N. 1976. Yellow leaf blotch, a new bacterial disease of sorghum, maize, and millet in western Africa. Plant Disease Reporter 60(9): 798–799.

Pearl Millet Diseases in Southern Africa

W.A.J. de Milliano[1]

Abstract

Pearl millet is thought to have been in southern Africa for more than 2000 years. In the last 15 years, the region had occupied about 2% of the world's pearl millet area. Ten of the 12 countries grew pearl millet and produced 3 to 4% of the grain in Africa annually, and about 1% of the world production. National grain yields were low and fluctuated below 1 t ha-1. Tanzania, Zimbabwe, Angola, and Namibia had the largest areas under production. These were no indications that fluctuations in areas under production, low grain yield, or low total production could be attributed to high disease incidence. Varieties began to replace landraces, but hybrids were used in research only.

There is still a need to identify pearl millet diseases and study their occurrence and incidence in southern Africa. Data about losses caused by diseases were not available. Grain losses because of ergot and smut, and losses of dry matter because of fungal foliar diseases occurred regularly. Loss of grain and dry matter losses because of downy mildew occurred occasionally. Witchweeds and grain molds were of importance only in some regions or during some seasons. Disease research in relation to pearl millet improvement was stimulated by the initiation of the SADCC/ICRISAT Sorghum and Millet Improvement Program in 1984. The present status of information on diseases and research results is reported here and areas for further research are indicated.

Introduction

It is believed that pearl millet reached southern Africa by 900 to 800 BC (Robinson 1966). In the past 15 years, pearl millet was grown in at least 9 of the 12 countries of southern Africa. It is grown for its grain, stems, and total biomass. Lesotho and Swaziland recently introduced pearl millet for experimental forage studies, but the crop is not (yet) grown by farmers in these nations. From 1969 to 1985, the whole region produced approximately 5–6% of the millet grain in Africa annually, approximately 2% of the world production [FAO tapes, 1986 (1988)]. Pearl millet and finger millet (SADCC/ICRISAT 1985) are lumped together in the FAO statistics, but on the basis of estimated ratios of the two crops, southern Africa probably produced 3–4% of Africa's pearl millet grain annually, and about 1% of the world production.

Angola, Namibia, Tanzania, and Zimbabwe had the largest areas under production (Table 1). There are no indications that fluctuations in the areas sown could be attributed to high disease incidence. The area under production appeared stable in Angola, Botswana, Mozambique, the Republic of South Africa, and Zaire. Tanzania and Namibia appeared to have a steady increase in the area under production, possibly as a result of economic factors, successful extension, and response of farmers to successive droughts. In Zambia and Zimbabwe, pearl millet areas were reduced because (1) maize is preferred, is hardly damaged by birds, and is easier to process for

1. Principal Cereals Pathologist, SADCC/ICRISAT Regional Sorghum and Millet Improvement Program, PO Box 776, Bulawayo, Zimbabwe. Present address: Section Leader, Phytopathology, Zoaduniz Westeinde 62, PO Box 26, 1600 AA Enklhuizen, the Netherlands.

de Milliano, W.A.J. 1992. Pearl millet diseases in southern Africa. Pages 115–122 *in* Sorghum and millets diseases: a second world review. (de Milliano, W.A.J., Frederiksen, R.A., and Bengston, G.D., eds). Patancheru, A.P. 502 324, India: International Crops Research Institute for the Semi-Arid Tropics. (CP 732).

Table 1. Area ('000 ha) sown to pearl millet and proportion (%) of the total area sown to pearl millet in 11 southern African countries and Africa, 1969–71, 1979–81, and 1984–86.[1]

Country	1969–71 Area ('000 ha)	1969–71 Proportion (%)	1979–81 Area ('000 ha)	1979–81 Proportion (%)	1984–86 Area ('000 ha)	1984–86 Proportion (%)
Angola	92.3	10	80.0	7	76.7	9
Botswana	19.2	2	11.7	1	9.3	1
Lesotho	0	0	0	0	0	0
Mozambique	19.0	2	20.0	2	20.0	2
Namibia	47.6	5	58.0	5	84.9	10
Swaziland	0	0	-	0	_[2]	0
South Africa	28.6	3	19.3	2	18.9	0.2
Tanzania	210.7	22	450.0	42	334.3	39
Zaire	26.6	3	31.4	3	41.0	5
Zambia	127.5	13	53.3	5	20.1	2
Zimbabwe	380.0	40	352.9	33	250.0	30
Total Eleven African countries	951.1		1 076.6		855.2	
Africa	15 225.4		13 784.0		15 831.3	

1. Source: FAO tapes, 1986 (1988); for Namibia and the Republic of South Africa, FAO 1981 (pp. 106–109) and 1986 (pp. 120–121).
2. No information.

food purposes, (2) economic factors, including subsidies on maize (but not on pearl millet) in Zambia; (3) food-aid programs in Zimbabwe; and (4) shortage of seed stocks after several years of drought.

Grain Yield

Grain yield, i.e., grain harvested per hectare sown, was generally low, although generally in line with world pearl millet yields. It fluctuated at levels less than 1000 kg ha[-1]. Tanzania appeared to have a steady increase in yield over the last 15 years (Fig. 1). In Botswana, with its extended droughts, however, there were very drastic fluctuations in yield. Tanzania appeared to have a yield higher than the world mean yield and the Africa mean yield. Yields tend to be low because the crop is grown in low-rainfall and drought-prone areas, on soils of low fertility, by farmers who cannot afford inputs other than family manpower. There are no indications,

however, that low yields or fluctuations in yield have occurred because of high disease incidence.

Regular and Occasional Losses

In southern Africa, pearl millet has received even less research attention than sorghum. Data about pearl millet yield losses because of diseases are not available. Consequently I can only venture suggestions based on disease lists, discussions, and personal observations (Riley 1960; Angus 1965, 1966. pp 65–66; Ebbels and Allen 1978; FAO 1972; Peregrine and Siddiqi 1972; Plumb-Dhindsa and Mondjane 1984; Rothwell 1983; and Botswana: Ministry of Agriculture, 1987).

1. There is little evidence that areas under production (Table 1) were reduced because of high disease incidences.
2. National grain yields appear not to have been reduced because of a specific disease, but

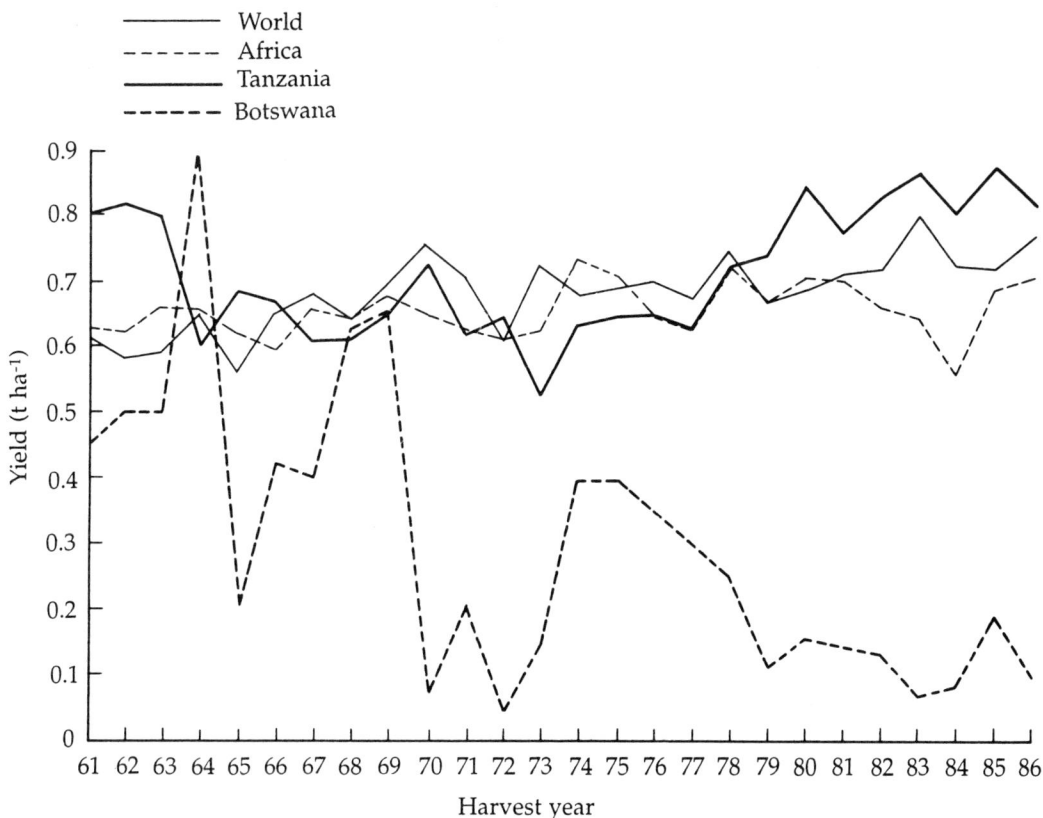

Figure 1. Millet yields (t ha⁻¹), 1961-86. Source: FAO tapes,1986 (1988).

losses may have occurred at the household level, outside of the national marketing system.

3. Loss of grain and other dry matter occasionally occurred because of downy mildew (*Sclerospora graminicola*) in Malawi, Tanzania, Zambia, and Zimbabwe, and possibly also in some other countries of southern Africa.

4. Loss of dry matter as a result of fungal foliar diseases such as leaf spots (*Beniowskia sphaeroidea, Cochliobolus bicolor, Colletotrichum* sp, *Gloeocercospora* sp), and rust (*Puccinia penicillariae*) did occur regularly.

5. Regular losses of grain due to ergot (*Claviceps* sp) and smut (*Tolyposporium penicillariae*) occurred, particularly in Malawi, Tanzania, Zambia, and Zimbabwe, and possibly in other countries.

6. Loss of grain and other dry matter occurred regularly because of *Striga asiatica* in Zimbabwe.

7. Loss of grain quality after infection by grain molds occurred only rarely, as pearl millet grain tends to ripen during a dry period.

Pearl Millet Diseases

Seed rots and seedling blight

Information on seed rots and seedling blight is not available, but stand establishment (emergence and seedling survival under low moisture and high temperature stresses) is known to be of high importance (ICRISAT 1988). In the western areas of Zambia, seedling blight occurred when the conditions were favorable for infection. The role of pathogens (including nematodes) in poor stand establishment requires study.

Foliar diseases

Foliar diseases in southern Africa are caused by viruses and fungi. No information is available on bacterial diseases (Table 2). Page et al. (1985) reported chlorosis and stunting caused by nematodes (*Pratylenchus, Paralongidorus,* and *Xiphinema* spp) in Zimbabwe.

Viral diseases

Virus-like symptoms were observed in pearl millet in Lesotho, Swaziland, Zambia, and Zimbabwe. Serological tests indicated that pearl millet in Zambia, including the 1987 release WC-C75, was affected by virus (Reddy 1987). Several viruses, such as maize dwarf virus(es) and sugarcane mosaic virus(es), may be involved. The causal agents of the disease symptoms are still to be identified in the other countries.

Fungal diseases

Downy mildew (*Sclerospora graminicola*), false mildew (*Beniowskia sphaeroidea*), leaf spots (*Cochlioblus bicolor, Colletotrichum* sp, *Gloeocercospora*

Table 2. Pearl millet diseases in the countries of southern Africa.[1]

Pearl millet disease	ANG	BOT	LES	MAL	MOZ	NAM	RSA	SWA	TAN	ZAM	ZIM
Seed rots and seedling blight											
Fungal	N	N	N	N	N	N	N	N	N	O	O
Foliar											
Viral	N	N	O	N	N	N	N	O	O	O	O
Bacterial	N	N	N	N	N	N	N	N	N	N	N
Nematodes	N	N	N	N	N	N	N	N	N	R	R
Fungal											
Downy mildew	N	O	N	R	R	N	R	N	R	R	R
False mildew	N	N	N	R	N	N	N	N	R	R	R
Leaf spots	O	O	O	R	N	O	O	O	R	R	R
Rust	O	R	N	R	R	O	N	N	R	R	R
Stalk and root rots											
Fungal											
Charcoal rot	N	O	N	O	N	N	O	N	N	N	O
Fusarium spp	N	N	N	O	N	N	N	N	O	N	O
Inflorescence											
Viral	N	N	N	N	N	N	N	N	N	N	N
Fungal											
Ergot	O	R	N	R	N	N	N	N	R	R	R
Grain molds	O	O	N	R	N	N	N	N	O	O	O
Smut	O	R	N	R	R	O	N	N	R	R	R
Witchweeds											
Striga asiatica	N	O	A	O	N	N	N	N	O	O	O
Striga hermonthica	N	A	A	A	N	N	A	N	R	A	A

1. O = observed, R = officially reported, A = absent, N = no information; ANG = Angola, BOT = Botswana, LES = Lesotho, MAL = Malawi, MOZ = Mozambique, NAM = Namibia, RSA = Republic of South Africa, SWA = Swaziland, TAN = Tanzania, ZAM = Zambia, ZIM = Zimbabwe.

sp, *Cercospora fusimaculans*, *Phyllachora* sp, and *Bipolaris urochloae*) and rust (*Puccinia penicillariae*) caused foliar diseases (Riley 1960; Angus 1965; FAO 1972; Peregrine and Siddiqi 1972; Kenneth 1975; Rothwell 1983; and Plumb-Dhindsa and Mondjane 1984.)

Downy mildew

In Tanzania, Malawi, Mozambique, Zambia, and Zimbabwe, local germplasm can become heavily affected with downy mildew (DM). However, when downy mildew differentials were tested under natural disease pressure, BJ 104, BK 560, MBH 110, WC-C75, and ICMS 7703 were resistant (less than five infected plants per 4.5-m row) and 7042 DMS and Impala Seed (Zimbabwe local) were susceptible at three sites—Ngabu (Malawi), Lusitu (Zambia), and Wedza (Zimbabwe). This suggests a weak pathogenicity of the downy mildew at the sites of testing. Ball (1987), using isolates from the same sites, had similar results in experiments in plastic tunnels in the UK.

The majority of the germplasm developed at ICRISAT Center and in western Africa appeared to have a good level of resistance. However, there appears to be a need to identify susceptible germplasm and confirm resistance with controlled disease pressure.

False mildew

The disease was recorded on klits grass (*Setaria verticillata*) in 1926 and reported on pearl millet in 1952, but high severities on pearl millet were observed for the first time in Zimbabwe in the 1985/86 cropping season (Mtisi and de Milliano 1991). The disease appeared to be favored by moist conditions (i.e., wet years). The host range appears to be limited to genera of the Graminea family. Genetic resistance was found in introductions (e.g., 700516, 7042 DMR, ICMPES 29, and ICMPES 1500-7-3-2) and in entries of the Regional Cooperative Variety Trial (e.g., IVS-A82, and IBVM 8402 SN). Most entries from the Pearl Millet Dwarf Variety Trial including RMP 1 and GAS from Zimbabwe were moderately susceptible in Zimbabwe. Very little is known concerning the influence of false mildew on yield, but the indications are that yield is affected only

to a minor extent (de Milliano as reported in SADCC/ICRISAT 1987). Confirmation is required.

Leaf spots

At certain locations in Malawi (e.g., at Ngabu in 1987), Tanzania (e.g., at Hombolo in 1988) Zambia (e.g., at Kaoma in 1986 and 1987), and Zimbabwe (at Henderson in 1986 and 1987; at Kadoma, 1986; at Matopos, 1986; and at Wedza, 1986), pearl millet was affected by leaf spot diseases. Highly susceptible lines, such as 7042 DMR, occasionally failed to produce seed (e.g., at Henderson, 1987). Genetic resistance appeared to be present, at least at certain locations in certain introductions (e.g., 700516 and ICMPES 28). The causal organisms still are not all known and identification is ongoing with assistance of the International Mycological Institute, Kew, UK.

With assistance of scientists from ICRISAT Center, it was found that *Bipolaris urochloae* affected pearl millet in Zimbabwe (Singh 1987; Sivanean 1987). Disease identification, determination of yield losses, and elimination of very susceptible genotypes need further attention.

Rust

Rust may be of some importance in a few countries, such as Tanzania. From 1986 to 1988, the gene for rust resistance in 7042 RR appeared to be effective in Malawi, Zimbabwe, and Tanzania. P 1564 was immune in Zimbabwe (at Mzarabani) but moderately susceptible in Tanzania (at Ilonga). Similarly P 24-1 was immune in Malawi (Ngabu) but susceptible in Tanzania (Ukiriguru). BJ 104 DMR was susceptible in Zimbabwe (Mzarabani) but resistant in Tanzania (Ukiriguru). This may indicate that there are differences in rust populations in southern Africa.

Additional sources from ICRISAT Center and elsewhere will be screened for resistance to rust, in search of resistant lines that are better agronomically than 7042 RR.

Stalk and root rots

Charcoal rot (*Macrophomina phaseolina*) and possibly *Fusarium* spp caused stalk and root rots.

Charcoal rot was observed in drought-stricken areas of Botswana (Botswana: Ministry of Agriculture 1987), Zimbabwe, the Republic of South Africa, and possibly Malawi, in particular at research stations. Page et al. (1985) did not report charcoal rot damage in pearl millet fields in the communal areas of Zimbabwe. It appears that this disease is of minor importance in the region. Some work to test sources of resistance is ongoing in Botswana. Many genotypes were susceptible to charcoal rot at Sebele (Botswana) in 1987 and 1988, but Botswana Serere 7A was relatively resistant.

Growth deformations are occasionally found in research plots and in farmers' fields, but the disease is thought to be of minor importance only.

Inflorescence diseases

Ergot (*Claviceps* sp) and smut (*Tolyposporium penicillariae*) are major diseases.

Ergot. Many local germplasm collections late-sown in 1987–88 at Henderson and at Panmure in Zimbabwe were heavily affected under natural disease pressure. Most of the germplasm, including introductions tested at Ilonga in Tanzania, were heavily affected in 1987. Local millet in farmers' fields was heavily affected at Gairo in 1987 and in the Singida area in 1988. Pearl millet in farmers' fields in the Shire Valley was affected in 1987. Several of the lines developed for ergot resistance (e.g., ICMPES 28) at ICRISAT Center showed low severities under natural disease pressure at several locations in Malawi, Tanzania, and Zimbabwe during several years. WC-C75 was moderately susceptible in Zambia.

In 1987, with an ergot pathologist from ICRISAT Center assisting, screening with artificial inoculation began. Five out of 100 ICRISAT's ergot-resistant entries had low severities (less than 10% of inoculated head area affected) after artificial inoculation. Immunity did not occur. ICMPES 28 and ICMPES 29 had severities exceeding 30% after inoculation. However, it is possible that flowering was prolonged because of out-of-season sowing. Additional utilization of the ergot-resistant germplasm received from ICRISAT Center is planned.

In 1988, sclerotia collection began; these may be found almost anywhere ergot occurs. Studies to determine if toxicity is related to genotype and environment are required. Scientists at Imperial College (UK) have received sclerotia samples from many places in southern Africa for future studies.

Smut. Smut is important in many pearl millet growing areas of southern Africa. At certain locations, e.g., at Ngabu in Malawi; at Hombolo in Tanzania; and at Panmure in Zimbabwe, smut appears to compete with ergot for infection sites in the pearl millet head.

In 1986 and 1987, genotypes developed at ICRISAT Center had low smut severities under natural disease pressure in Malawi and Zimbabwe, in particular ICMPES 28, ICMPES 29, ICMPS 100-5-1, and ICMPS 900-9-3.

In 1988, in cooperation with Wye College, UK, testing of artificial methods of inoculation and studies of the smut life cycle were initiated.

Witchweeds

Pearl millet is affected by *Striga asiatica* and *S. hermonthica* (Riches et al. as reported in SADCC/ICRISAT 1987).

Striga asiatica. Pearl millet is generally said to be resistant to this witchweed in southern Africa. Only on rare occasions will a single plant be found in a field of millet in Botswana (Riches et al. as reported in SADCC/ICRISAT 1987); the same is reported from Hombolo in Tanzania. In the Masvingo region of Zimbabwe, pearl millet fields are commonly affected. Little is known of the extent of the damage caused by *S. asiatica* in pearl millet.

Striga hermonthica. In the northern districts of Tanzania, this parasite is of importance on sorghum and possibly on millets. At least two strains of the parasite, one specific to pearl millet and the other specific to sorghum have been reported in Sudan (Wilson-Jones 1955). At Ukiriguru Research Station in Tanzania, it was observed that sorghum was heavily affected, while on Serere 17A hardly a *Striga* plant was found. It must be determined if such physiological strains occur in the region, in particular in Tanzania.

It is planned to develop *Striga* sick plots (*S. asiatica* and *S. hermonthica*) at a number of locations in southern Africa to screen for *Striga* resistance in pearl millet, sorghum, and maize.

Conclusion

Disease resistance appears to be the most appropriate method to obtain stable yields, as pearl millet farmers usually cannot afford inputs other than family manpower. Some complex diseases, such as ergot, form a major challenge to research. Considering the importance of food security in the low-rainfall areas where pearl millet is the most appropriate crop, it appears appropriate to secure the pearl millet yields. However, given the difficulties in developing resistance to some pearl millet diseases, outstanding efforts will be required and expected. The importance of pearl millet for the semi-arid tropics of Africa cannot be over emphasized.

Acknowledgment. D. Rohrbach assisted with the preparation of the introduction and the material on grain yields.

References

Angus, A. 1965. Annotated list of plant pests and diseases in Zambia (Parts 1–7). Mount Makulu, Zambia: Mount Makulu Research Station, Plant Pathology Laboratory. 384 pp. (Limited distribution).

Angus, A. 1966. Annotated list of plant pests and diseases in Zambia (Supplement). Mount Makulu, Zambia: Mount Makulu Research Station, Plant Pathology Laboratory. (Limited distribution).

Ball, S.L. 1987. Pearl millet downy mildew. Annual Report 1986. ODA/University of Reading. Research Project 3492H. 16 pp. (Limited distribution).

Botswana: Ministry of Agriculture. 1987. Annual Report for the Division of Arable Crops Research 1985–1986. Gabarone, Botswana: Ministry of Agriculture. 274 pp. (Limited distribution).

Ebbels, D.L., and Allen, D.J. 1978. A supplementary and annotated list of plant diseases, pathogens and associated fungi in Tanzania. Phytopathological Paper No. 22. 89 pp.

FAO. 1972. Report to the Government of Botswana on plant diseases based on the work of G. I. Nilsson. UNDP Report No. TA 3057. Rome, Italy: Food and Agricultural Organization of the United Nations. 34 pp.

FAO. 1981. FAO Production yearbook 1980. Rome, Italy: Food and Agricultural Organization of the United Nations.

FAO. 1986. FAO Production yearbook 1985. Rome, Italy: Food and Agricultural Organization of the United Nations.

FAO tapes, 1986. 1988. Rome, Italy: Food and Agricultural Organization of the United Nations.

ICRISAT. 1988. Proceedings of the International Pearl Millet Workshop, 7–11 Apr 1986, ICRISAT Center, India. Patancheru, Andhra Pradesh 502 324, India: International Crops Research Institute for the Semi-Arid Tropics. 354 pp.

Kenneth, R. 1975. CMI Descriptions of pathogenic fungi and bacteria No. 457. *Sclerospora graminicola*. Kew, Surrey, England: Commonwealth Mycological Institute. 2 pp.

Mtisi, E., and de Milliano, W.A.J. 1991. Occurrence and host range of false mildew, caused by *Beniowskia sphaeroidea*, in Zimbabwe. Plant Disease 75:1.

Page, S.L., Mguni, C. M., and Sithole, S.Z. 1985. Pests and diseases of crops in communal areas of Zimbabwe, 1986/85 growing season. Overseas Development Administration Technical Report. Harare, Zimbabwe: Plant Protection Research Institute. 203 pp. (Limited distribution).

Peregrine, W.J.H., and Siddiqi, M.A. 1972. A revised and annotated list of plant diseases in Malawi. CAB Phytopathological Papers 16:29.

Plumb-Dhindsa, P., and Mondjane, A.M. 1984. Index of Plant Diseases and associated organisms of Mozambique. Tropical Pest Management 30(4): 407–429.

Reddy, D.V.R. 1987. Report on a visit to Malawi and Zambia, 5–22 Mar 1987. Patancheru, Andhra Pradesh 502 324, India: International Crops Research Institute for the Semi-Arid Tropics. 36 pp. (Limited distribution).

Riley, E.A. 1960. A revised list of plant diseases in Tanganyika Territory. Mycological Papers 75:42.

Robinson, K.R. 1966. A preliminary report of the recent archeology of Ngonde, Northern Malawi. Journal of African History 7:169–188.

Rothwell, A. 1983. A revised list of plant diseases occurring in Zimbabwe. Kirkia 12 (II): 275.

SADCC/ICRISAT. 1985. Proceedings of the First Regional Workshop on sorghums and millets for southern Africa. Harare, Zimbabwe, 23–26 Oct 1984. Bulawayo, Zimbabwe: SADCC/ICRISAT Sorghum and Millet Improvement Program. 89 pp. (Limited distribution).

SADCC/ICRISAT. 1987. Proceedings of the Third Regional Workshop on Sorghums and Millets for southern Africa. Lusaka, Zambia, 6–10 Oct 1986. Bulawayo, Zimbabwe: SADCC/ICRISAT Sorghum and Millet Improvement Program. 476 pp. (Limited distribution).

Singh, S.D. 1987. Report on a trip to Zimbabwe, 26 Oct 1986 to 30 Apr 1987. Patancheru, Andhra Pradesh 502 324, India: International Crops Research Institute for the Semi-Arid Tropics. 36 pp. (Limited distribution).

Sivanean, A. 1987. Bipolaris urochloae (Putterill) Shoem on pearl millet in Zimbabwe. International Mycological Institute. Report No. 4524/87/Y25. 1 pp. (Limited distribution).

Wilson-Jones, K. 1955. The witchweeds of Africa. World Crops 5:263–266.

Pearl Millet Diseases in India: Knowledge and Future Research Needs

K.A. Balasubramanian[1]

Abstract

Pearl millet, Pennisetum glaucum *(P. typhoides) is an important grain and forage cereal of India. It is grown during the rainy s͘ ͘son. Introduction of hybrids using cytoplasmic male sterility produced dramatic increases in yield on the subcontinent, and more than 2 million ha were sown to such hybrids when the crop was hit by severe epidemics of downy mildew* (Sclerospora graminicola). *The hybrids were more susceptible than landraces to ergot, as well. Other pearl millet diseases of importance are smut and rust. The importance of rust became obvious when it was learned that the pathogen can affect seedling plants.*

Satisfactory methods of control of pearl millet diseases in India seem to be limited to host-plant resistance, although chemical control is practiced in research efforts of high input. Farmers do not, in general, use chemical control. With practically each disease of pearl millet in India, scientists feel that information on plant and pathogen and plant-pathogen interactions is incomplete, and additional study should benefit the effort to develop resistant plants. Computers, demonstrated to be of great value in studies of diseases of other crops, have not been fully utilized. Multiple-disease resistance is considered to be an achievable goal.

Introduction

Pearl millet, *Pennisetum glaucum* (L.) R. Br. (*P. typhoides* (Burm.) Stapf and Hubb.) is grown for grain and forage. It occupies an area of about 26 million ha distributed in the tropical and subtropical areas of Africa and the Indian subcontinent (FAO 1978). Of the total millet crop produced in India, 71% is contributed by four states—Gujarat, Rajasthan, Maharashtra, and Uttar Pradesh. Another 28% of the total production is contributed by Haryana, Andhra Pradesh, and Punjab. The millet crop in India is produced during the rainy season. Pakistan, Myanmar, and Sri Lanka together produce 4% of the millet crop in South Asia (Sivakumar et al. 1984).

In India, the first pearl millet hybrids using cytoplasmic male sterility provided a dramatic grain yield increase of about 70% or more over the existing varieties (ICAR 1966). By 1971, more than 2 million ha were sown to hybrids, when the first of several major epidemics of downy mildew occurred (Safeeulla 1977). The hybrids as a class were found to be more susceptible than open-pollinated varieties to ergot.

Downy mildew (*Sclerospora graminicola* Sacc. Schroet), ergot (*Claviceps fusiformis* Lov.), smut (*Tolyposporium penicillariae* Bref.), and rust (*Puccinia penniseti* Zimm.), are important diseases of pearl millet in India.

Downy Mildew

Downy mildew has been reported from all countries growing pearl millet except South America

1. Professor and Head, Department of Plant Pathology, Andhra Pradesh Agricultural Univeristy, Bapatla, Andhra Pradesh 522 101. India.

Balasubramanian, K.A. 1992. Pearl millet diseases in India: knowledge and future research needs. Pages 123–127 *in* Sorghum and millets diseases: a second world review. (de Milliano, W.A.J., Frederiksen, R.A., and Bengston, G.D., eds). Patancheru, A.P. 502 324, India: International Crops Research Institute for the Semi-Arid Tropics.

and Australia. In India, the disease has been reported in all areas that grow pearl millet.

The downy mildew pathogen, *S. graminicola*, produces large numbers of sporangia at temperatures between 15 and 25°C. At cardinal temperatures of 5–7°C, 18°C, and 30–33°C the sporangia produce oospores. The number of oospores released from a sporangium varies from 3 to 13 (Suryanarayana 1965; Bhat 1973). Products of sexual reproduction, the oospores, pose serious problems in determining their viability and their ability to germinate. Great variation, anywhere from 8 months to 10 years, appears in reports on the survival and dormancy of oospores. The main reason for such a variation seems to be the differences in conditions under which the survival and dormancy tests were conducted. Many workers have claimed successful germination of oospores (Nene and Singh 1976). But no one method appears to be reproducible. The degradation of oospores in soil is also highly conjectural. In the absence of a reliable method to germinate oospores, the survival and viability of oospores is indirectly inferred from their infectivity on pearl millet seedlings in artificially inoculated sick plots.

That the inoculum is associated with the seed is generally accepted. The debatable point is whether the inoculum is seedborne internally or externally. It is generally accepted that the seeds carry oospores on their surface. The role of sporangia in inciting the disease has been proven, and there is sufficient evidence that young seedlings of pearl millet become systemically affected with sporangial inoculum in the presence of adequate humidity (Williams et al. 1981).

Information on physiological specialization in the fungus is far from complete. The differential interactions of pearl millet varieties at different locations are, in the absence of data from well-defined experiments, discussed in terms of environmental differences or of physiological races.

The host range of *S. graminicola* is considerable, embracing the tribes Maydae, Andropogonae, Paniceae, and Agrostidae of Graminaceae. With this wide host range, it is not surprising when downy mildew appears on pearl millet in areas where it has not before been grown.

Among the several types of inoculation used in screening varieties for reaction to downy mildew, the combination of sick plot method containing oospores in the soil with sporangial inoculation appears to be more dependable. This suggests the possibility of two types of resistance in pearl millet, one type to soilborne oospores, and the other type to airborne sporangia. *S. graminicola*, being a biotroph (strict or physiologically obligate parasite), refuses axenic culture. A close scrutiny of the physiology of the fungus is required to bring this fungus from the tissue-culture stage, as it is today, to the axenic-culture stage. The achievement could perhaps unravel the mystery surrounding the germination of *S. graminicola* oospores.

Among agronomic management practices for avoiding downy mildew, early sowing is generally observed to be better than other methods.

Metalaxyl fungicide treatment holds good promise to keep downy mildew under check, providing sustained protection to the millet crop. However, this fungicide is usually not available in the free market in developing countries.

In the absence of a good and easily available fungicide for seed treatment, the alternative strategy becomes the use of resistant varieties. Sources of resistance have been found in several national and international programs. Genes involving resistance have to be systematically identified, especially those providing durable resistance.

Once identified, these genes must be incorporated in adequate levels to perceptibly change the pearl millet/downy mildew profile. It is doubtful that this can be achieved until we have strong and adequate supporting data on the physiology and biochemistry of pearl millet plants affected by downy mildew.

Ergot

Apart from a 1956 epidemic in India (Bhide and Hegde 1957; Shinde and Bhide 1958), ergot of pearl millet incited by *C. fusiformis* assumed importance only with large-scale commercial cultivation of F_1 hybrids (Thakur and Williams 1980). In addition to reducing yields, the adverse effect of ergot is mainly contamination of grain with toxic alkaloids. Sclerotia of the fungus contain the toxic substances.

The infection takes place mainly through the stigma of the flower, but entry through the ovary wall prior to fertilization has been observed (Sundaram 1969). The inoculum consists of airborne ascospores produced by germinating sclerotia lying in the soil and also by conidia from the same season's crop (Thakur et al. 1984). Although cross-inoculation studies have been carried out by Small (1922) and Reddy et al. (1969), the role of collateral hosts in initiation of the disease, during the same season, is not clear. In cross-inoculation studies, Reddy et al. (1969) observed that the pearl millet ergot pathogen was unable to colonize sorghum, but ergot sorghum on the sugary-disease pathogen was able to infect pearl millet. This can be a potential threat to pearl millet where ergot of sorghum is endemic.

It is generally observed that ascosopores of *C. fusiformis* do not infect fertilized flowers. Artificial pollination before the application of inoculum has been found to prevent infection (Thakur and Williams 1980). This poses an obvious hurdle in screening varieties for resistance. Spraying heavy (10^6 conidia mL^{-1}) suspensions of conidia into the previously protected protogynous inflorescence (Thakur et al. 1982) is a significant advancement in creating ergot infection artificially. Gill et al. (1979) demonstrated low levels of susceptibility at pre-protogyny, high levels of susceptibility during protogny, and high levels of resistance during anthesis.

An approach to dealing with ergot would be to exploit host-plant resistance. Unfortunately, there are few well-documented sources of resistance, in spite of screening large numbers of germplasm accessions in the national and international programs. Development of resistant lines by intermating moderately resistant lines has been suggested by Gill et al. (1979). Reaction of developed lines can be scored using a standard rating scale (Thakur and Williams 1980). Lines resistant to ergot have been bred through pedigree selection from accessions available at ICRISAT (Thakur and Chahal 1987). This could be the starting point of the long journey to find and incorporate resistance to ergot into pearl millet. Chemistry of ergot-affected earheads can contribute significantly to the quest for resistance. Kannaiyan et al. (1973) reported tryptophan content of spikelets was maximum before fertilization. They also observed that in vitro, asparagine and proline supported growth of the fungus, while tryptophan impeded growth. Variations in the morphology of the ergot fungus and its capacity to attack pearl millet cultivars, have been reported by Chahal et al. (1985). These workers grouped their isolates, based on pathogenicity, into highly virulent (Kovilpatti, Jobner, Ludhiana, and Aurangabad) and weakly virulent (Coimbatore, ICRISAT, and Mysore) groups. These results can be confirmed by the study of protein isozymes by electrophoresis.

Future studies should focus on the epidemiology of this disease and the mechanisms and genetics of resistance.

Smut

Smut incited by *Tolyposporium penicillariae* Bref. has become a major disease, especially since the introduction of high-yielding varieties. The infection is confined to the individual spikelet; smut sori are produced in place of grains. Infection is by sporidia produced by soilborne teliospores, acting through the stigmas (Bhatt 1946). Of the several inoculation methods tried, injection of artificially cultured sporidia into the leaf sheath at the boot stage appears to be dependable (Gill et al. 1979; Thakur et al. 1983). Earheads can be graded for their resistance or susceptibility on a standard severity index (Thakur and Williams 1980). Although application of the fungicides oxycarboxin, carboxin, and captafol as sprays helped to reduce smut significantly (Gill et al. 1979), this type of management has not made an impact and host-plant resistance is the sought-after method of managing smut in pearl millet.

The International Pearl Millet Workshop in 1986 emphasized the exploitation of host-plant resistance to contain the disease. In the absence of adequate information the aim of future research should be the understanding of the biology of the pathogen and the epidemics of the disease. Mechanisms of resistance and in-depth studies of variation in the pathogen population are the other important items for concentrated study.

Rust

Once considered a minor disease, rust (*P. penniseti*) has in recent years gained considerable importance in pearl millet production. A little more than a decade ago, rust would not appear until the boot-leaf or grain-filling stage of the pearl millet plants. In the late 1970s, however, young pearl millet seedlings were found to be susceptible to rust.

Speakers at the 1977 International Pearl Millet Workshop acknowledged that rust received the lowest priority and attracted the least attention in the millet-growing states of Punjab, Haryana, Rajasthan, and Andhra Pradesh. Failure to recognize the importance of this disease was further demonstrated when only a single rust paper was presented during the 1986 International Pearl Millet Workshop 9 years later.

An acceptable level of rust resistance has not yet been found in pearl millet cultivars. Of the 614 entries tested in the All India Coordinated Pearl Millet Improvement Project (AICPMIP), only 32 were found to have some resistance to rust. In the field, Cobb's modified rust-rating scale is being used to assess rust incidence. There is a need to develop a rating scale that includes seedling infection.

Resistance based on a single dominant gene, designated *Rppl*, has been discovered recently at ICRISAT (Andrews et al. 1985), opening the possibility of incorporating adequate rust resistance.

Recommendations

A major thrust in the present management of pearl millet diseases appears to depend on the exploitation of host-plant resistance. Development of lines with multiple-disease resistance has attracted the attention of several workers and attempts have been made to identify such lines since 1985 (Andrews et al. 1985). Although computer-aided prediction programs are available for a few crop diseases, pearl millet pathology programs are seriously deficient of such technology.

Genotype deployment should be looked into seriously, instead of relegating susceptible lines permanently to the background.

A combination of computer-aided multiple-regression analysis, genotype deployment, and multiple-disease resistance will give pearl millet pathology a new impetus and take the management of pearl millet disease through a new realm.

References

Andrews, D.J., Rai, K.N. and **Singh, S.D.** 1985. A single dominant gene for rust resistance in pearl millet. Crop Science 25:565–566.

Bhat, S.S. 1973. Investigations on the biology and control of *Sclerospora graminicola* on bajra. Ph.D. thesis, University of Mysore, India.

Bhatt, R.S. 1946. Studies in Ustilaginales. I. The mode of infection of the bajra plant (*Pennisetum typhoides* Stapf and Hubb.) by the smut *Tolyposporium penicillariae* Bref. Journal of the Indian Botanical Society 25:163–186.

Bhide, V.P., and **Hegde, R.K.** 1957. Ergot of bajra (*Pennisetum typhoides* (Burm.) and Stapf and Hubb.) in Bombay state. Current Science 26:116.

Chahal, S.S., Rao, V.P., and **Thakur, R.P.** 1985. Variation in morphology and pathogenicity in *Claviceps fusiformis*, the causal agent of pearl millet ergot. Transactions of the British Mycological Society 84:325–332.

FAO. 1978. Production Year Book 32. FAO, Rome, Italy: Food and Agricultural Organization of the United Nations.

Gill, K.S., Phul, P.S., Chahal, S.S., Singh, N.B., Bharadwaj, B.L., and **Jindla, L.N.** 1979. Studies on disease resistance for the improvement of bajra. Project report 1974–1979. Ludhiana, India: Punjab Agricultural University.

ICAR. 1966. Progress report of the All India Coordinated Millet Improvement Project 1965–66. New Delhi, India: Indian Council of Agricultural Research.

Kannaiyan, J., Vidhyasekaran, P., and **Kandaswamy, T.K.** 1973. Amino acid content of bajra in relation to ergot disease resistance. Indian Phytopathology 26:358–359.

Nene, Y.L., and **Singh, S.D.** 1976. Downy mildew and ergot of pearl millet. Pest Articles and News Summary 22:366–385.

Reddy, K.D., Govindswamy, C.V., and **Vidyasekharan, P.** 1969. Studies on ergot disease of cumbu (*Pennisetum typhoides*). Madras Agricultural Journal 56:367–377.

Safeeulla, K.M. 1977. Genetic vulnerability. The basis of recent epidemics in India. *In* The Genetic Basis of Epidemics in Agriculture (Day, R.P., ed.) Annals of The New York Academy of Sciences 287:72–85.

Shinde, P.A., and Bhide, V.P. 1958. Ergot of bajra in Bombay state. Current Science 27:499–500.

Sivakumar, M.V.K., Huda, A.K.S., and Virmani, S.M. 1984. Physical environment of sorghum and millet growing areas in South Asia. Proceedings, International Symposium on Agrometeorology of Sorghum and Millet in the Semi-Arid Tropics. Patancheru, Andhra Pradesh 502 324, India: International Crops Research Institute for the Semi-Arid Tropics. 469 pp.

Small, W. 1922. Diseases of cereals in Uganda Circular, Department of Agriculture, Uganda Protectorate.

Sundaram, N.V. 1969. Current plant pathological problems in new agricultural strategy with special reference to sorghum and millets. Indian Phytopathological Society Bulletin 5:67–72.

Suryanarayana, D. 1965. Studies on the downy mildew diseases of millets in India. Indian Phytopathological Society Bulletin 3:72–78.

Thakur, R.P., and Williams, R.J. 1980. Pollination effects on pearl millet ergot. Phytopathology 70:80–84.

Thakur, R.P., Rao, V.P., and Williams, R.J. 1984 The morphology and disease cycle of ergot caused by *Claviceps fusiformis* in pearl millet. Phytopathology 74:201–205.

Thakur, R.P., Subba Rao, K.V., and Williams, R.J. 1983. Evaluation of a new field screening technique for smut resistance in pearl millet. Phytopathology 73:1255–1258.

Thakur, R.P., Williams, R.J., and Rao, V.P. 1982. Development of resistance to ergot in pearl millet. Phytopathology 72:406–408.

Thakur R.P., and Chahal, S.S. 1987. Problems and strategies in the biology and control of ergot and smut of pearl millet. Pages 173–182 *in* Proceedings, Pearl Millet Workshop, 7–11 Apr 1986. ICRISAT Center, India, Patancheru, A.P. 502 324, India: International Crops Research Institute for the Semi-Arid Tropics.

Williams, R.J., Singh, S.D., and Pawar, M.N. 1981. An improved field screening technique for downy mildew resistance in pearl millet. Plant Disease 65:239–241.

Ergot Disease of Pearl Millet: Toxicological Significance and a Mechanism for Disease Escape

P.G. Mantle[1]

Abstract

Ergot disease of pearl millet, caused by Claviceps fusiformis, *not only causes crop losses but potentially adds a toxic component. Extensive reproductive failure in pigs fed ergotised millet in southern Rhodesia in the 1950s was manifest, in the absence of any other symptom, as agalactia. This was subsequently shown in the mouse to be due to the central stimulatory action of the principal ergot alkaloids agroclavine and elymoclavine. The peripheral pharmacology of the classical C. purpurea alkaloids that cause gangrenous ergotism is absent in the C. fusiformis alkaloids, rendering the millet ergot sclerotia a less toxic hazard to man and farm animals than might be assumed.*

Millet disease is exacerbated by pronounced protogyny that is common in some commercial hybrids. Ergot-resistant pearl millets selected at ICRISAT Center have very short protogyny and escape infection as rapid pollination quickly precludes the pathogen's entry via the stigma by the constriction of a region of the fused stylodia. Control of ergot disease in pearl millet can thus be achieved directly by efficient pollen management exploiting postpollination stigmatic constriction.

Introduction

Ergot disease is one of the two ovary diseases of pearl millet, the other being the smut caused by *Tolyposporium penicillariae*. Ergot is prevalent in parts of western and eastern Africa and in India. It became particularly recognized in both continents during the 1950s as a significant cause of reduced grain yield. The loss of photosynthate in the release of honeydew and associated encouragement of grain molds are factors which compound the simple loss of potential grain by ergot pathogen substitution. It is reasonable to suppose that the disease is much more ancient, in keeping with the crop plant itself. It is not clear whether the fungus arose separately in the two continents or has spread from a single source as pearl millet has been grown more widely in the semi-arid tropics. Ergot became even more significant as a disease with the intro-

duction of high-yielding hybrid millets in India in the 1960s (Thakur and Chahal 1987).

In spite of some initial controversy, it appears that the fungi-causing ergot disease of pearl millet can be classified within the description of *Claviceps fusiformis* made by Loveless (1967). The only important omission in Loveless' description concerns recognition of two types of asexual conidia, which occur usually as a mixture of the fusiform macrospores and the much smaller spherical microspores.

Toxicology

Interest in defining this plant pathogen was also stimulated in southern Rhodesia concomitantly in the 1950s following several extensive outbreaks of agalactia in sows fed rations containing ergotized millet (Shone et al. 1959).

---

1. Professor, Biochemistry Department, Imperial College, London, SW7 2AY, UK.

Mantle, P.G. 1992. Ergot disease of pearl millet: toxicological significance and a mechanism for disease escape. Pages 129–132 *in* Sorghum and millets diseases: a second world review. (de Milliano, W.A.J., Frederiksen, R.A., and Bengston, G.D., eds). Patancheru, A.P. 502 324, India: International Crops Research Institute for the Semi-Arid Tropics.

Experimental feeding of the suspected toxigenic material to pregnant and lactating sows clearly implicated the ergot sclerotia, contaminating the grain at the rate of approximately 1–2%, in reproductive failure due not to any detectable effect on the foetus in utero, but to inhibited mammary function leading quickly to starvation of normal piglets born at normal term. Evidence of peripheral vascular disturbance, commonly associated with the classical ergot fungus of temperate cereals and grasses (*Claviceps purpurea*) was never expressed.

Most of the pharmacological properties of ergot sclerotia (*Claviceps* spp) can be attributed to the ergolene alkaloids derived biosynthetically from a common pathway of secondary metabolism commencing with the condensation of tryptophan with the isoprene precursor dimethylallyl prophosphate. *Claviceps purpurea* elaborates the most complex derivatives of lysergic acid, but *C. fusiformis* completes only part of the pathway, the clavine alkaloids, agroclavine and elymoclavine accumulating instead as endproducts occurring at about 0.3% (w/w) of the sclerotial tissue and as trace amounts even in the honeydew exudate associated with sphacelial fructification. Pure alkaloids of this type, extracted from sclerotia grown on pearl millet or produced by the fungus in fermentation, exhibit negligible peripheral pharmacology but are potent central stimulators (Mantle 1968a). For example, mice given daily approximately 0.5 mg agroclavine in food became obsessively hyperactive, tearing blotting paper on which they were bedded into minute shreds. Sheep given 40 mg agroclavine subcutaneously persistently licked the walls of the pen for 2 to 3 hours, an expression of quite atypical behavior. In contrast, at a dose of 350 mg day^{-1} (which elicited no overt signs), pregnant mice failed to prepare for lactation and pups died of starvation. At a lower dose, lactation was almost switched off, being sufficient only to support one grossly malnourished pup whose eyes were still closed at 4 weeks of age and whose weight was less than one fifth of the weight of a normal animal at 3 weeks of age. When the higher dose was given to lactating mice, lactation was switched off and the pups ceased to thrive. When a similar dose was given to mice on days 3, 4, and 5 after insemination blastocysts failed to implant and pregnancy did not establish (Mantle 1969).

These findings pointed to an endocrine-mediated mechanism which logically results from central (hypothalamic) stimulation causing, in the mouse or rat, a diminution of prolactin release from the pituitary gland and the consequent paucity of lactogenic stimulus. Prolactin being also a luteotropic hormone in these experimental animals, its absence leads in turn to insufficient progesterone to prepare the uterus wall for blastocyst implantation.

Striking though these effects are in the mouse or rat, without eliciting any overt signs of altered behavior, it was not possible to produce similar agalactia in either the goat or the cow where relatively small changes in lactation performance are comparatively easy to monitor. Further, the known differences in mechanisms of endocrine control of pregnancy in laboratory rodents and the human do not allow confident prediction that any notable reproductive hazard could be ascribed to lightly ergotised millet as a human food. Although apparently no experiments on the effect of clavine alkaloids on lactation in the pig have been reported, the close mimicking of the Rhodesian porcine agalactia by the mouse implies that the pig is at least sensitive to the inhibition of prolactin release with respect to its role in initiating and maintaining lactation. The conclusion, therefore, is that while ergotised millet may be undesirable for human consumption, any effects are not likely to involve serious hazard. Nevertheless, what might be assumed to be a potentially reasonable outlet for substandard grain, namely as pig feed, is certainly contraindicated for pregnant (near term) or lactating sows. Since species other than pigs (e.g., buffalo, goat, donkey) may be more usually available in millet-growing areas subject to ergot epidemics, there is a clear need for the results of research on their tolerance of ergot to be published in the most relevant scientific literature. An even wider market concerning feed for poultry or fish could well be explored to advantage. Ergot has acquired a toxigenic image which is partly well justified (Mantle 1978) but which must not be allowed to be a basis for unreasonable and unscientific extrapolation across the whole range of *Claviceps* spp. While *Claviceps purpurea* deserves the maximum caution, *C. fusiformis* on millet commands less concern unless contamination exceeds about 1%. There are grounds to believe that ergot disease of sorghum, the other cereal important in the semi-arid tropics, is essentially nontoxic (Mantle 1968b, and unpublished).

Disease Escape

Research during the 1970s, by ICRISAT scientists, to develop ergot resistant cultivars of pearl millet has been successful and the topic has been reviewed recently by Thakur and Chahal (1987). Concurrently it had been recognized that improved pollen management, facilitating early and efficient capture by stigmas as they are exerted, minimized disease expression. Complementary studies, supported by UK's Overseas Development Administration, sought to explain the mechanism operating in the ergot-resistant cultivars and revealed for the first time a fundamental botanical event which had until then, surprisingly, escaped notice. Within about 5 hours of compatible pollen reaching the surface of stigmatic lobes on exerted stigmas, a group of cells in a region of the fused stylodia begin to lose turgor and progressively collapsed (Willingale and Mantle 1985). The collapsed region unsuitable for the passage of further pollen tubes and soon leads to desiccation and abscission of the distal stigmatic tissue. Although the postpollination constriction of fused stylodia is possibly the most dramatic instance of a localized response to potential fertilization, some analogous events were being observed independently and concurrently by Heslop-Harrison et al. (1985) in maize and some grasses. Clearly, such a quick response, either to the passage of pollen tubes through the potential abscission zone or by some hormonal response to invasion of the upper ovary wall, will exclude ovary pathogens for which the normal route of entry is via the stigma, mimicking the behavior of pollen tubes. Stigmatic constriction is an important factor in escape from ovary disease in pearl millet and should be exploited. This could best be achieved by reducing protogyny within the inflorescence to a minimum and if possible arranging synchrony of anthesis with exertion of stigmatic hairs from within the confines of the floral cavity, as is evident in cultivars ICMPE 13-6-30 and 134-6-9 (Willingale et al. 1986). The resulting economy of pollen could also allow selection for minimum size of exerted stigmas to reduce the effective receptive area for pathogen inoculum. Stigmatic constriction is initiated also following the passage of C. fusiformis hyphae down the stylodial transmission tracks in the absence of pollination (Willingale and Mantle 1987). Since the pathogen is much slower than a pollen tube, taking about 16 h to induce the constriction, it seems likely that it is not just a response to invasion of the potential constriction region. Otherwise, there might be a response before the hyphal tips had passed through the group of cells. Constriction also occurs simply as an aging response (notable in the commercial susceptible hybrid BJ 104) in inflorescences enclosed in the conventional bag. If the bag is carefully removed shortly before anthesis, long white turgid stigmas will be seen as expected but, since the fused stylodia have already constricted, the stigmas can be wiped away with the slightest touch. They are in effect already useless as an effective route of pollination.

This mechanism for ergot disease escape does not exclude the possibility that other forms of resistance may exist, but it would be in itself sufficient to ensure that ergot disease in pearl millet becomes insignificant if, by arranging efficient pollen management, the natural rapid postpollination stigmatic constriction phenomenon in pearl millet is exploited to the full.

References

Heslop-Harrison, Y., Heslop-Harrison, J., and **Reger, B.J.** 1985. The pollen-stigma interaction in the grasses. 7. Pollen-tube guidance and the regulation of tube number in *Zea mays* L. Acta Botanica Neerlandica 34: 193–211.

Loveless, A.R. 1967. *Claviceps fusiformis* sp. nov., the causal agent of an agalactia of sows. Transactions of the British Mycological Society 50(1): 15–18.

Mantle, P.G. 1968a. Inhibition of lactation in mice following feeding with ergot sclerotia [*Claviceps fusiformis* (Loveless)] from the bulrush millet [*Pennisetum typhoides* (Staph and Hubbard)] and an alkaloid component. Proceedings of the Royal Society, Series B, 170: 423–434.

Mantle, P.G. 1968b. Studies on *Sphacelia sorghi* McRae, an ergot of *Sorghum vulgare* Pers. Annals of Applied Biology 62: 443–449.

Mantle, P.G. 1969. Interruption of early pregnancy in mice by oral administration of agroclavine and sclerotia of *Claviceps fusiformis* (Loveless). Journal of Reproduction and Fertility 18:81–88.

Mantle, P.G. 1978. Ergotism in cattle; sheep; pigs and poultry. Pages 145–151; 207–213; 273–275; 307–308 *in* Mycotoxic Fungi, Mycotoxins, Mycotoxicosis, Vol. 2 (Wyllie, T.D., and Morehouse, L.G., eds.) New York, USA, and Basel, Switzerland: Marcel Dekker.

Shone, D.K., Phillip, J.R., and Christie, G.J. 1959. Agalactia of sows caused by feeding the ergot of the bulrush millet, *Pennisetum typhoides*. Veterinary Record 71(7): 129–132.

Thakur, R.P., and Chahal, S.S. 1987. Problems and strategies in the control of ergot and smut in pearl millet. Pages 173–182 *in* Proceedings of the International Pearl Millet Workshop, 7–11 Apr 1986, ICRISAT Center, Patancheru, Andhra Pradesh 502 324, India: International Crops Research Institute for the Semi-Arid Tropics.

Willingale, J., and Mantle, P.G. 1985. Stigma constriction in pearl millet, a factor influencing reproduction and disease. Annals of Botany 56: 109–115.

Willingale, J., and Mantle, P.G. 1987. Stigmatic constriction in pearl millet following infection by *Claviceps fusiformis*. Physiological and Molecular Plant Pathology 30: 247–257.

Willingale, J., Mantle, P.G., and Thakur, R.P. 1986. Post-pollination stigmatic constriction, the basis of ergot resistance in selected lines of pearl millet. Phytopathology 76: 536–539.

Part 3

Current Status of Sorghum Diseases

Bacterial Diseases of Sorghum

L.E. Claflin[1], B.A. Ramundo[2], J.E. Leach[3], and M. Qhobela[4]

Abstract

Bacterial diseases on sorghum have been diagnosed in all major sorghum-producing areas. Bacterial stripe (Pseudomonas andropogonis), bacterial streak (Xanthomonas campestris pv holcicola) and bacterial spot (Pseudomonas syringae) are the principal diseases. This review summarizes previous research efforts and reports research activities in progress or completed, but yet unpublished. Various facets of the etiology and epidemiology of the major diseases are discussed, as are the effects of stage of growth on leaf stripe and leaf blight development. Dot-immunobinding assay and restriction endonuclease analysis, recently developed techniques for identifying causal agents, are explained. The bacterial pathogen responsible for a disease of pearl millet observed in northern Nigeria in 1987 was identified as caused by Pseudomonas avenae.

Introduction

Numerous bacteria have been reported as parasitic to sorghum (Table 1). Of these, the three most commonly reported are bacterial stripe, (*Pseudomonas andropogonis*), bacterial streak (*Xanthomonas campestris* pv *holcicola*), and bacterial spot (*Pseudomonas syringae* pv *syringae*). Bacterial diseases have been reported from all sorghum-producing areas in the world and may reduce yields where climatic conditions favor disease development. Reports of research on these sorghum diseases are limited, with fewer than 40 literature citations since 1929; most appearing in state or country reports mentioning bacterial diseases.

Etiology and epidemiology of bacterial diseases in sorghum is poorly understood, again because reports are few and most are primarily observational in nature. For example, it is often stated that bacterial diseases are favored by a warm and moist environment, but there is no experimental evidence to support such a statement. Likewise, *P. andropogonis* and *X. campestris* pv *holcicola* are reported to be seedborne (Noble and Richardson 1968); however, experimental data was not presented. *P. syringae* pv *syringae* has been shown to be seedborne (Gaudet and Kokko 1986).

In this review, previous research is summarized; data that are being published by our laboratories are also included. Descriptions of methods and materials, usually absent in publications of this nature, are included to assist those with limited access to scientific literature.

Bacterial Leaf Stripe

Causal organism

P. andropogonis is an aerobic, gram negative, nonspore-forming rod, measuring 0.5×1.5 μm, with one large, sheathed polar flagellum. A fluo-

---

1. Professor, Department of Plant Pathology, Kansas State University, Manhattam, KS 66506, USA.
2. Research Assistant at the above address.
3. Associate Professor at the above address.
4. Former graduate student at the above address. Present address: Agricultural Research Station, Box 829, Maseru, Lesotho.

Claflin, L.E., Ramundo, B.A., Leach, J.E., and Qhobela, M. 1992. Bacterial diseases of sorghum. Pages 135–151 *in* Sorghum and millets diseases: a second world review. (de Milliano, W.A.J., Frederiksen, R.A., and Bengston, G.D., eds). Patancheru, A.P. 502 324, India: International Crops Research Institute for the Semi-Arid Tropics.

Table 1. Bacterial diseases of sorghum.

Common name	Pathogen	Symptom	Reference
Bacterial stripe	*Pseudomonas andropogonis*	Linear, intervenial lesions of various colors depending upon host genotype	Elliot 1929
Bacterial leaf blight	*Pseudomonas avenae*	Irregular shaped lesions with tan centers and dark red margins	Bradbury 1973a
Bacterial spot	*Pseudomonas syringae* pv *syringae*	Small, cylindrical lesions with light colored centers	Claflin, L.E., and Ramundo, B.A., unpublished; Kendrick 1926
Bacterial streak	*Xanthomonas campestris* pv *holcicola*	Linear, intervenial lesions, very similar to bacterial stripe	
Bacterial leaf stripe	*Pseudomonas rubrisubalbicans*	Irregular dark red lesions on leaves and sheaths	Hale and Wilkie 1972; Bradbury 1967b
Red stripe	*Pseudomonas rubrulineans*	Same as bacterial leaf blight	Claflin, L.E., and Ramundo, B.A., unpublished; Bradbury 1967b
Gummosis disease of sugarcane	*Xanthomonas campestris* pv *vasculorum*	Yellow, corotic stripes on leaves	Zummo 1984; Bradbury 1973b
Bacterial soft rot	*Erwina chrysanthemi* pv *chrysanthemi*	Necrotic or heavily pigmented stripes or blotches on upper leaves; stalk and leaf tissue rotted in whorls	Jenson et al. 1986
Bacterial sun spot	*Pseudomonas* spp	Circular to elliptic, dark red or brown spots with tan or light centers, found in eastern Senegal	Zummo and Freeman 1975
Yellow leaf blotch	*Pseudomonas* spp	Scattered cream-yellow to light beige lesion; symptoms resemble *Striga* spp; throughout western Africa	Zummo 1976

rescent pigment is not produced on King's Medium B (KB) (King et al. 1954). Smooth, round, and cream-colored colonies are formed on yeast extract-glucose-calcium carbonate media (YDCA). Acid is produced from arabinose, fructose, galactose, glucose, and myo-Inositol. The bacterium can utilize citrate (weak) and malonate as sole sources of carbon. Hydrolysis of gelatin and starch are negative. The organism fails to grow at 4°C and 41°C. Arginine dihydrolase, lipase production, nitrate reduction, and oxidase production are negative. *P. andropogonis* elicits a hypersensitive response on tobacco (Claflin and Ramundo, unpublished).

Host range

Hosts, other than sorghum, reported to be susceptible to *P. andropogonis* include Johnson grass and maize (Vidaver and Carlson 1978); velvet bean (Burkholder 1957); clover (Hayward 1972); teosinte, soybean, and chickpea (Caruso 1984); white clover (Gitaitis et al. 1983); red clover, alfalfa, snapbean, red kidney bean, broadbean, lespedeza, vetch, bougainvillea (Rothwell and Hayward 1964); blueberry, carob, *Gypsophila paniculata*, and statice (Moffett et al. 1986), and sugarcane (Claflin and Ramundo, unpublished). Bacterial leaf stripe has been diagnosed in sorghum areas of North and South America, Africa, Europe, USSR, China, India, southeast Asia, and the South Pacific (Commonwealth Mycological Institute 1973).

Symptoms

Bacterial stripe symptoms consist of linear interveinal lesions that are purple, tan, red, or yellow, depending upon the host reaction (Fig. 1). Bacterial exudate is often observed on the abaxial side of the leaf. Lesions may also be found on the peduncle, rachis branches, seeds, and in the interior of the stalk. If warm weather and high relative humidity persists, lesions may extend

Figure 1. Bacterial stripe on sorghum cv B 35-6 caused by *Pseudomonas andropogonis*.

upwards of 20 cm with the leaf and leaf sheath nearly encompassed with symptoms. Bacterial stripe symptoms are often similar to those of bacterial streak, and both closely mimic those of several fungal diseases.

Although research data on field dissemination are not available, it is probably by wind and rain. Infested seed or debris probably account for long-distance dissemination. Weed hosts, volunteer plants, and infested seed are likely sources of overseasoning bacterial organisms. Sources of tolerant and susceptible germplasm are listed in Table 2.

Table 2. Reaction of selected sorghum lines to bacterial stripe (*Pseudomonas andropogonis*) and bacterial streak (*Xanthomonas campestris* pv *holcicola*).

| Line | Rating[1] | |
	Stripe	Streak
80 B 3039-5	6.0	5.8
Tx 2783	5.8	2.2
B 35-6	5.6	2.6
81 BH 5359	5.6	5.1
81 BH 5426	5.6	3.3
Tx 2536	5.6	4.2
77 CS 1	5.6	4.1
76 CS 478	5.0	5.6
76 CS 256 (TS)	4.8	5.5
GR 2-14-1	3.7	5.4
81 L-13688	4.6	5.4
81 BH 5496	2.5	2.3
R 6956	2.5	2.5
R 3338	2.5	1.0
SC 326-6	2.5	2.2
81 BH 5559	2.4	2.5
BTX 378	1.7	2.2
QL 3 (India)	1.5	2.1
81 BH 5646	2.0	2.2
CV 223-4-1-1	3.1	2.2
SC 170-6-17	5.0	2.0
9 L-3510	3.0	1.5

1. Ratings based on the percentage of leaf area with symptoms defined as follows: 1 = trace <1%, 2 = 2–10%, 3 = 11–25%, 4 = 26–50%, 5 = >56%, and 6 = all plants dead.

Source: Clafin and Rosenow (1983)

Temperature and relative humidity

A susceptible genotype of *P. andropogonis* (80 B 3039-5) from the Texas A&M University All Disease and Insect Nursery (ADIN) was used to determine the optimal temperature and relative humidity for bacterial stripe development in sorghum (Claflin and Machtmes, unpublished). Plants were inoculated with a Hagborg device (Hagborg 1970) at the four-leaf stage of growth and then placed in growth chambers at relative humidity levels of 90, 75, and 50%. Temperature combinations were 24/18 (16 h day/8 h night), 30/24, and 36/24°C. Plants were rated, 2 weeks after inoculation, on the percentage of leaf area with symptoms. Ratings were defined as follows: 1 = trace –1%; 2 = 2 –10%; 3 =11 –25%; 4 = 26 –50%; 5 = 50%; and 6 = death of plant.

Significant differences in disease expression of *P. andropogonis* were observed between the three RH levels, with maximum ratings at the 90% level. Disease expression was maximum (overall rating of 3.3) at 30/24°C. Significant differences were also observed at temperatures of 36/24°C and 24/18°C (ratings of 2.5 and 2.2, respectively).

Bacterial Leaf Blight

Causal organism

P. avenae (Syn. *P. alboprecipitans*) is the causal agent of bacterial leaf blight disease of maize, oats, barley, wheat, Italian millet, barnyard millet, proso millet, foxtail millet, finger millet, rice, and rye (Rosen 1922; Schaad et al. 1975; Shakya et al. 1985; Bradbury 1973a). *P. avenae* cells are aerobic, gram-negative, with dimensions averaging around 0.6×1.8 μm. Acid is produced from arabinose, fructose, galactose, glucose, glycerol, and sorbitol. Citrate and malonate are utilized as sole sources of carbon. Starch and gelatin are not hydrolyzed. Growth occurs at 41°C, but not at 4°C. Lipase and oxidase are produced and nitrates are reduced. Colonies growing on YDCA are smooth, round, and viscid, and are cream-colored with a brownish tinge. A halo may be detected around colonies when the pH of the medium is 6.8 or less and contains beef extract (Claflin and Ramundo, unpublished).

P. avenae has been reported from North America, Asia, South America, and Africa (Commonwealth Mycological Institute 1976). *P. rubrilineans* is synonymous to *P. avenae* (Claflin and Ramundo, unpublished data) and the latter has been reported from numerous countries in Africa, Asia, Australia and Oceania, and Central America (Commonwealth Mycological Institute 1987)

Bacterial Leaf Stripe of Pearl Millet

While surveying in 1984 for pearl millet diseases in northern Nigeria, nearly every sowing observed while traveling the almost 700 km from Sokoto to Jos showed a bacterial leaf disease (Claflin et al. 1987). Lesions, 3–25 cm long, were usually confined within the leaf veins (Fig. 2). Isolations on YDCA and KB yielded cream-colored, nonfluorescing colonies, closely resembling those of *P. andropogonis*, within 3 days.

Physiological and biochemical tests, however, indicated the two to be quite different. The millet isolate differed from *P. rubrilineans*, the causal agent of red stripe disease in sugarcane (Christopher and Edgerton 1930), only in that *P. rubrilineans* is negative for production of acid from salicin. Few differences were observed when comparing the millet isolate and *P. rubrilineans* strains to those of *P. avenae* and *P. rubrisubalbicans*. The millet strains were capable of acid production from salicin, whereas strains of the other species tested negative. *P. avenae* and *P. rubrisubalbicans* utilized malonate as the sole carbon source whereas millet strains and *P. rubrilineans* were negative. Most strains of *P. avenae*, *P. rubrilineans*, *P. rubrisubalbicans*, and the millet pathogen produced acid from arabinose, fructose, galactose, glucose, glycerol, and sorbitol. All of the strains tested except those of *P. andropogonis* grew at 41°C and, except for one strain of *P. rubrisubalbicans*, no growth occurred at 4°C. All of the bacterial

Figure 2. Bacterial leaf stripe on pearl millet caused by *Pseudomonas avenae*.

strains produced a hypersensitive response in tobacco.

The average doubling times for *P. avenae, P. andropogonis, P. rubrilineans, P. rubrisubalbicans*, and the millet pathogen were 103, 148, 100, 110, and 108 minutes, respectively.

Pathogenicity tests

Pathogenicity of various *P. avenae, P. rubrilineans*, and *P. rubrisubalbicans* strains was determined on maize: Pioneer 3195 (Pioneer Hi-Bred International, Johnston, IA) Gold Cup (Harris-Moran Seed Co., Rochester, NY); grain sorghum (*Sorghum bicolor*): 80B3039 (D. T. Rosenow, Texas A&M University, Lubbock, TX); pearl millet: Sidney dwarf and Serere 3A (D. J. Andrews, University of Nebraska, Lincoln, NE); and sugarcane (*Saccharum officinarum*): cultivar 67500 (N. Zummo, Mississippi State University, Mississippi State, MS). Cultures were grown at 28°C overnight in NBY, centrifuged at 13 000 g for 10 min; resuspended and then centrifuged and washed with 12.5 mM PO_4 buffer two additional times. The cell concentration was adjusted to 100 Klett units. Leaves were inoculated, 2 weeks after emergence, using a Hagborg device. Inoculated plants were maintained in a greenhouse (22–28°C) with automated misting (6 min h^{-1} for 6 h day^{-1}). Symptoms were recorded 2 weeks after inoculation. All *P. avenae* isolates tested were pathogenic to maize, sorghum, millet, and sugarcane.

On maize, lesions were translucent with light tan centers and dark brown margins. The margins of the lesions were usually water-soaked and confined to interveinal tissue. Lesions on sorghum had tan centers with dark red margins (Fig. 3). Leaf veins in proximity to the margin were dark red. Lesions on millet were light brown in the centers and dark brown at the margins, up to 10 cm in length, and found only on tissue between leaf veins. Water-soaking was observed only at the edge of the margin. On sugarcane, elongated dark red lesions up to 45 cm long were confined to interveinal tissue.

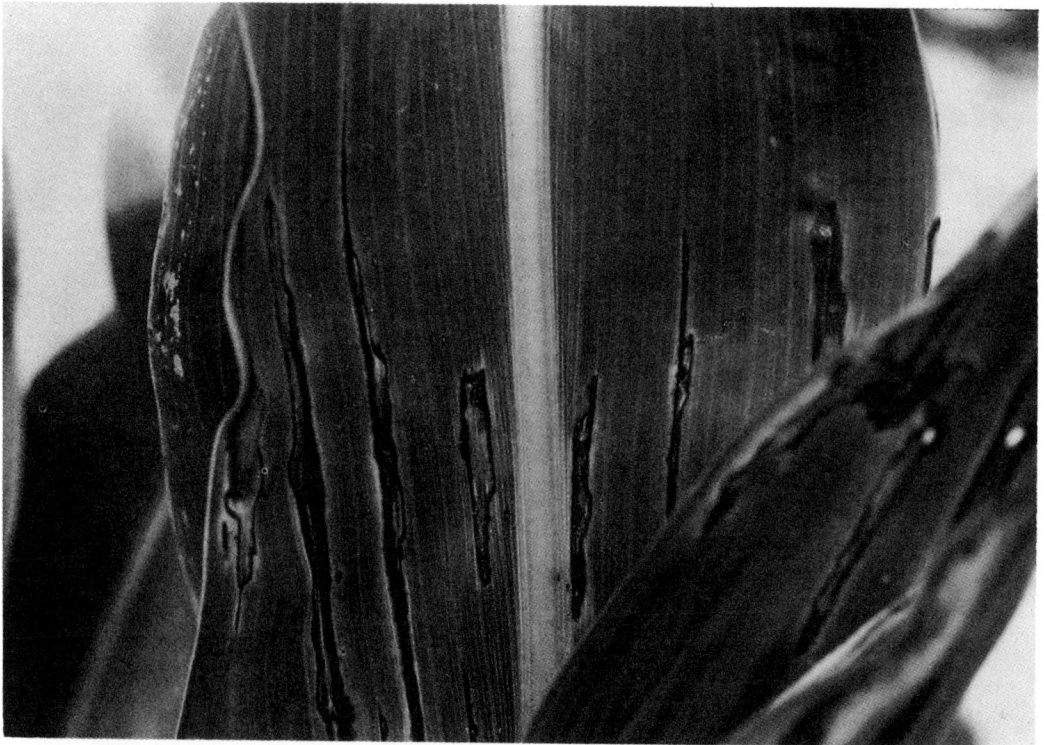

Figure 3. Bacterial leaf blight on sorghum cv B 35-6 caused by *Pseudomonas avenae*.

All *P. rubrilineans* strains tested were pathogenic to maize, sorghum, and millet and, except for strain NCPPB 3112, to sugarcane. Lesions on those plants were indistinguishable from those of *P. avenae*. With some strains, a hypersensitive-like reaction was observed on maize, sorghum, and millet. Within 24 to 48 h after inoculation with the Hagborg device, the tissue subjected to inoculation became necrotic, assuming a translucent appearance. Within 3 or 4 days, the translucent-like areas turned light brown but failed to enlarge, remaining confined to the tissue infiltrated by the Hagborg. Inoculations with PBS showed only the injury caused by the Hagborg, without necrosis.

Most strains of *P. rubrisubalbicans* were pathogenic to sugarcane, although several provided weak reactions. With the exception of strain PD-DCC 3109, the other strains were negative or caused hypersensitive reactions on maize, sorghum, and millet. The hypersensitive reactions appeared to be identical to those for *P. rubrilineans*, described above. Strain PDDCC 3109 on maize produced circular to rectangular water-soaked areas around the light brown and/or translucent necrotic areas. Water-soaking consisted of small areas resembling freckles, usually with a yellow halo around the periphery.

Sorghum lesions were small (1–3 mm) circular to rectangular dark brown areas, with light brown zones or streaks radiating from the primary lesion. Lesions were also characterized by light yellow areas around the periphery. On millet, tan water-soaked areas were most noticeable as narrow streaks and were not necessarily vein delimited. On older lesions, light brown necrotic areas surrounded by tannish water-soaked areas were characteristic.

The pearl millet isolate incited symptoms on maize, sorghum, millet, and sugarcane and symptoms were identical to those incited by *P. rubrilineans* and *P. avenae*.

Strains of *P. andropogonis* were most virulent on maize and sorghum, mildly virulent on sugarcane, and provoked a weak response on millet.

Antisera production of *P. andropogonis* and *P. avenae*

Pseudomonas avenae (syn *P. alboprecipitans*) strains ATCC 19860 and ICPB PA134, *P. an-*dropogonis strains ATCC 23061 and ATCC 23062, *P. rubrilineans* strain ATCC 19307, *P. rubrisubalbicans* strain ATCC 19308, and *P. syringae* pv *syringae* strains ICPB PS 146 and PS 296 were utilized as antigens in antisera production. The bacteria were grown on YDCA for 96 h at 28°C and washed with sterile 10 mM PO_4 buffered saline (0.85% NaCl, pH 7.2) [PBS] by centrifuging three times at 17 000 g. Cells were resuspended in PBS and fixed by dialyzing against a 2% glutaraldehyde solution at room temperature for 3 h. The cells were then dialyzed for 24–36 h against 100-fold volumes of PBS (with five or six changes). Equal volumes of bacterial suspensions (ca. 2×10^{10} cfu mL^{-1}) and Freund's incomplete adjuvant (Difco) were emulsified with the aid of a Spex mixer-mill for 2 min at high speed.

New Zealand white rabbits were injected intramuscularly with one mL of the emulsfied suspension at weekly intervals. Rabbits were bled from the marginal ear vein after the fourth injection. Injections continued until serial two-fold agglutination titers exceeded 1:2048. Sera were collected 3 to 4 h after bleeding and stored in serum bottles without preservatives at −20°C.

Dot-immunobinding assay, *P. avenae* and *P. andropogonis*

The dot-immunobinding assay (DIA) was performed (DeBlas and Cerwinski 1983; Leach et al. 1987) using Schleicher and Schuell (Keene, NH) nitrocellulose membranes (0.2 mm, BA83) which were divided into 1×1 cm squares by marking with a ballpoint pen, washed in distilled water for 5 min and air dried before use. Four mL of serial 10-fold dilutions of the bacterial cultures were individually spotted on the grids. Distilled water was used as a control. Each nitrocellulose membrane was cut into strips (4– × 6–cm) in which the serial dilutions (normally four) were arranged in a top to bottom descending order of dilution. The bacterial cells were fixed to membranes by soaking in 10% acetic acid and 25% ethanol solution for 15 min. This was followed by a rinse in distilled water and then another rinsing in three or four changes of 50 mM Tris-HC1 (pH 7.2) containing 200 mM NaCl and 0.1% Triton X-100 (TBS-T100). Antiserum was diluted in TBS-T100 to 1:2000 (v/v). The membranes were incubated in the antiserum dilution for 2 h at room temperature, or overnight at 4°C, and then rinsed in four changes of TBS-T100 with 5 min for each rinse. The strips were then incubated for 2 h in protein A-alkaline phosphatase conjugate (Sigma, St. Louis, MO) diluted (1:1000; v/v) in TBS-T100. The membranes were rinsed in TBS-T100 followed by two rinses in Tris-buffered saline. Following washing, the strips were placed in a substrate solution [12 mg fast violet B salt (Sigma), 2 mL Napthol AS-MX phosphate alkaline solution (Sigma), and 48 mL distilled water] for 30 min on a tabletop shaker in the dark. The substrate was prepared immediately before use. The strips were then rinsed with running tap water for 5 to 10 min and air-dried. Positive reactions gave a violet color (Fig. 4).

P. avenae (ATTC 19860) antiserum used in DIA tests did not distinguish *P. rubrilineans*, *P. avenae*, and the millet strain at dilutions of 10^5 (Fig. 4). Comparable results were obtained with *P. rubrilineans* antiserum. Faint cross-reactions were observed at a dilution of 10^8 for *P. syringae* pv *syringae* and at dilutions of 10^7 and 10^8 for

P. rubrisubalbicans. In several experiments, *P. rubrilineans*, *P. avenae*, and the millet strain provided faint confirmation readings at dilutions of 10^5 (approximately 800 colony-forming units per application). Therefore, we concluded that *P. avenae*, *P. rubrilineans*, and the millet strain are synonymous.

The bacterial pathogen affecting millet production in northern Nigeria was identified as *P. avenae* on the basis of biochemical, physiological, and immunological tests. The millet pathogen and *P. rubrilineans* were synonymous to *P. avenae* strains and were clearly distinct from the closely related pathogen *P. rubrisubalbicans* (Claflin and Ramundo, unpublished). Zummo (1976) described a yellow leaf blotch of maize, sorghum, and pearl millet in western Africa with symptoms partially resembling those we observed on pearl millet. His photographs reveal a more oval-shaped lesion, closely resembling symptoms of *Striga* spp, although it was described as vein-limited. Bacterial stripe symptoms that we observed were always vein-limited

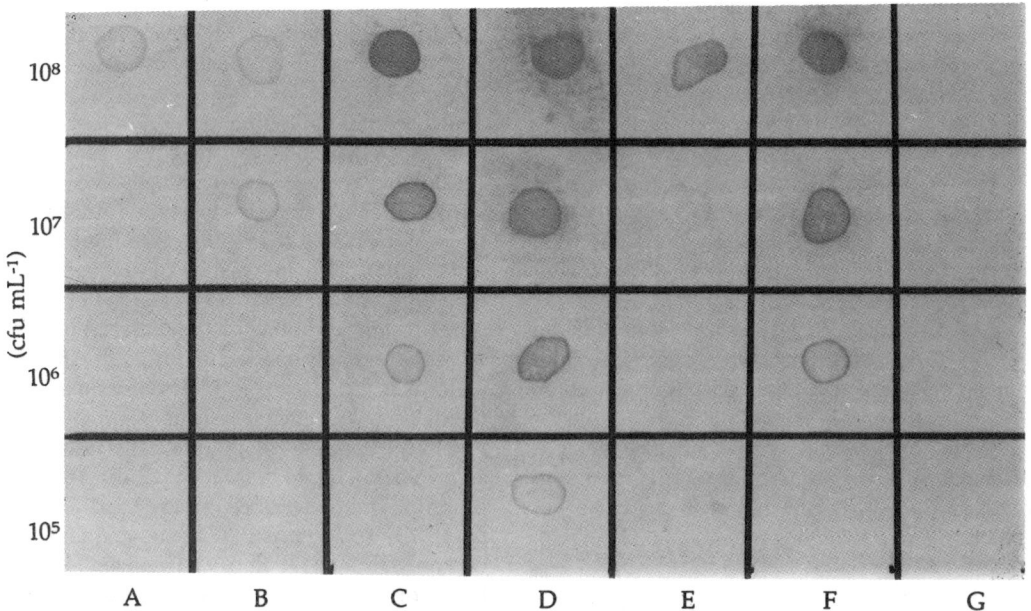

Figure 4. Nitrocellulose membrane from a dot-immunobinding assay of various bacterial causal agents including *P. syringae* pv *syringae* (A), *P. andropogonis* (B), *P. avenae* (C), *P. rubrilineans* (D,E), *P. rubrisubalbicans* (F), and water control (G). *P. rubrilineans* antiserum was used at a dilution of 1:1000.

and occurred more often at the leaf tips, suggesting that hydathodes may be a portal of entry for the bacteria. We did not detect bacterial leaf blight symptoms on maize or sorghum in Nigeria, even when these crops were growing adjacent to or intersown with pearl millet.

Epidemiology of *P. avenae*

In growth chamber studies, bacterial leaf blight was favored by hot (36/24°C) and moist conditions (90% RH) (Akhtar and Claflin 1986a). In Kansas, *P. andropogonis* and *P. avenae* were capable of overwintering in infested plant debris (Akhtar and Claflin 1986b). Leaf and stalk tissue and seeds were examined by fluorescent antibody staining (FAS) for bacteria every 2 weeks from February through May. More bacteria were observed in leaf tissue than in stalk tissue or in seed samples. Neither of the bacteria species was recovered from seed samples plated on YDCA medium, whereas the pathogens were easily recovered from stalk and leaf tissue.

Effect of sorghum growth stage on bacterial stripe and bacterial leaf blight development

Three plants from each row of 26 grain sorghum hybrids (four replications) were inoculated with *P. andropogonis* and *P. avenae* at 30, 40, 50, and 90 days after sowing in 1982 and 30 to 40 days in 1983 (Akhtar and Claflin, unpublished). This approximated, respectively, Stage 3 (growing point differentiation: seven to ten leaves), Stage 4 (flag leaf visible in whorl), Stage 5 (boot), and Stage 9 (physiological maturity). Although our data (Table 3) showed Stage 9 to have the highest reading, it is probably important in most years that plants be inoculated prior to Stage 5. Climatic conditions in 1982 were optimal for disease development. In 1983, however, periods of hot and dry conditions persisted and limited disease was observed. Timely rainfall, warm temperatures, overcast skies, and frequent dews after inoculation appear vital for extensive disease development.

Table 3. Effect of inoculating grain sorghum plants at various growth stages with *Pseudomonas andropogonis* and *P. avenae*.[1]

Growth stage[2]	*P. andropogonis*		*P. avenae*	
	1982	1983	1982	1983
3	2.45[3] b[4]	1.76 a	1.39 a	1.09 a
4	1.97 c	1.11 b	1.01 b	0.20 b
5	1.51 d		1.04 b	
9	2.84 a		1.25 a	

1. Source: Akhtar and Claflin, unpublished.
2. Growth stages: 3 = growing point differentiation (7–10 leaves); 4 = flag leaf visible; 5 = boot stage; 9 = physiological maturity.
3. Rating scale of 0–5, where 0 = no symptoms; 1 = 1–10; 2 = 11–25; 3 = 26–50; 4 = 51–75; 5 = 76–100% of leaf area affected. Each number represents an average of 3 plants of 26 hybrids inoculated at each stage of growth.
4. Numbers within a column not followed by a common letter are significantly different at the 0.05 level as determined by Duncan's Multiple Range Test.

Bacterial Spot

Causal organism

P. syringae cells are aerobic, gram-negative rods measuring $0.5–0.7 \times 0.8–2.2$ µm. The cells are motile with one or more polar flagella. *P. syringae* produces a green fluorescent pigment on KB. The bacterium does not hydrolyze starch; catalase and aesculin tests are positive whereas oxidase and indole production are negative. Acid is produced from glucose, fructose, sucrose, galactose, mannose, arabinose, xylose, mannitol, glycerol, and sorbitol. Optimum growth temperature is near 28°C; maximum is 35°C and minimum is 1°C (Hayward and Waterston 1965).

Host range

Numerous monocot and dicot hosts, including many genera from the Graminae family, are susceptible. *P. syringae* is disseminated over long distances by seed and plant debris. Bacterial spot incidence is increased by cool weather, rain, and high humidity. The bacteria are presumed to gain entry into the plant through wounds, stomata, or hydathodes. Gaudet and Kokko (1986) detected *P. syringae* on 8 to 67% of the sorghum seeds from different lots, and reported *P. syringae* to be the cause of poor sorghum seedling emergence in Alberta, Canada. They found large numbers of bacteria inside the seed coat, distal to the coleorhiza and on the exterior of the seed in the region of the embryo, after 3 days of germination. After 7 days, bacteria were observed throughout the seed, including embryo and endosperm.

Symptoms

Bacterial leaf spot may be difficult to diagnose. Symptoms resemble pesticide injuries, those of several fungal diseases, and particularly of older lesions.

Control recommendations include crop rotation, destruction of debris, and selecting seed from disease-free plants. Bacterial leaf spot has not been reported to be of wide occurrence; therefore, control measures are not likely to be warranted.

A recently reported semiselective medium (KBC) for *P. syringae* (Mohan and Schaad 1987) consists of King's B medium supplemented with boric acid (1.5 mg mL^{-1}), Cephalexin® (80 µg mL^{-1}), and Cycloheximide® (200 µg mL^{-1}). This semiselective medium should be advantageous for isolating *P. syringae* from plant tissues, seed, soil, and water, and thus invaluable in epidemiological and etiological research (Figs. 5,6,7).

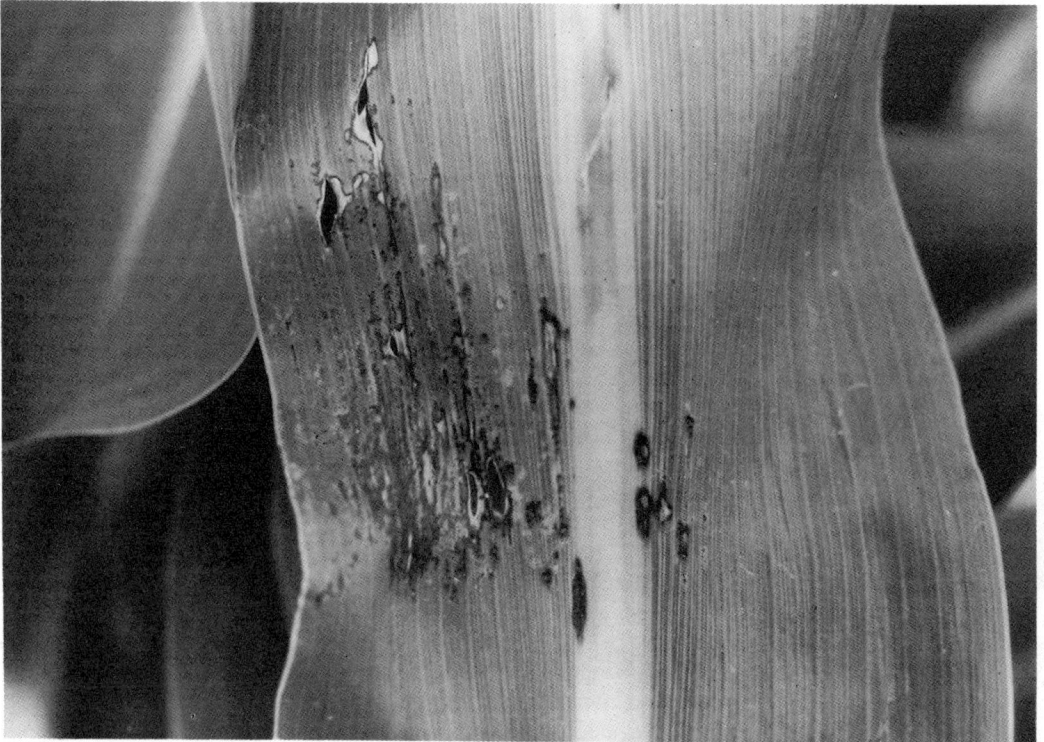

Figure 5. Bacterial spot on sorghum cv B 35-6 caused by *Pseudomonas syringae* pv *syringae*.

Figure 6. Bacterial spot on sorghum cv BTx 2755 caused by *Pseudomonas syringae* pv *syringae*.

Figure 7. Onion-leaf phase of bacterial spot on sorghum cv Tx 7078 caused by *Pseudomonas syringae* pv *syringae*.

Bacterial Streak

Causal organism

X. campestris pv *holcicola* cells are gram-negative, about 0.4–0.9 × 1.0–2.4 μm in size, and have one or two polar flagella. Acid is produced from fructose, galactose, glucose, mannose, and sucrose. Casein and Tween 80 are hydrolyzed, but starch is not. Gelatin is not liquified, oxidase is not produced, nitrates are not reduced, and malonate can be the sole source of carbon. *X. c.* pv *holcicola* will not grow at 4° or 41°C. Bacterial growth is slow; colony development requires 4 to 5 days. Colonies on YDCA are yellow, circular, convex, and very viscous (Qhobela and Claflin, unpublished).

Host range

Hosts for *X. campestris* pv *holcicola* include Sudangrass, maize, sorghum, broomcorn, and Johnsongrass. Pathogenicity tests were negative on sugarcane, pearl millet, finger millet, and proso millet (Qhobela and Claflin, unpublished). Bacterial leaf streak has been reported from South America, North America, Mexico, Australia, Africa, New Zealand, the Philippines, and the USSR (Commonwealth Mycological Institue 1987).

Epidemiology

Growth chamber studies at 50, 75, and 90% RH showed that bacterial streak is not influenced by high relative humidity levels (as is bacterial stripe), but optimal temperatures for disease expression were the same (30/24°C). (Claflin and Machtmes, unpublished).

Symptoms of bacterial streak are often very similar to those of bacterial stripe. Small interveinal water-soaked lesions are initial symptoms. With favorable climate, the lesions may extend to 30 cm and often coalesce to produce blotches covering most of the lower portion of the leaf (Fig. 8). Color of lesions corresponds to

Figure 8. Bacterial streak on sorghum cv B 35-6 caused by *Xanthomonas campestris* pv *holcicola*.

host genotype. Copious amounts of bacterial exudate are usually apparent on both leaf surfaces.

Widespread dissemination of the bacteria is probably attributable to seed or infested debris. Entry into the plant is possibly through wounds, stomata, or hydathodes. Tolerance or susceptibility of the germplasm tested is listed in Table 2.

Restriction endonuclease analysis of the genome of *Xanthomonas campestris* pv *holcicola*

Xanthomonas campestris pv *holcicola* cannot be differentiated from other pathovars of *X. campestris* by physiological or biochemical tests. Serological probes are useful, but unless monoclonal antibodies are used, the serological probes often cross-react with other pathovars. Development of monoclonals is expensive, a constraint that prohibits their use with pathogens of limited economic importance. Restriction endonuclease analysis (REA) has accurately differentiated other *X. campestris* pathovars (Leach et al. 1987; Lazo et al. 1987). Plant assays require several weeks to complete; DNA restriction analyses of many isolates can be completed within a week. These features of the REA procedure make it attractive for specific purposes.

REA is based on the identification of specific DNA fragmentation patterns. The number and locations of endonuclease restriction sites along the DNA strand are unique for each genome. Separation of the fragments by digestion with specific endonucleases on agarnose gels reveals fragment-size classes unique for each genome. Such unique fragment classes form the specific restriction patterns, or fingerprints, of individual isolates.

To identify unique DNA banding patterns which correlate with the pathovar *holcicola* and differentiate it from all other pathovars of *X. campestris*, *X. campestris* pv *holcicola* isolates from USA (Kansas, Nebraska, and Texas), Mexico, Lesotho (Africa), and Australia were screened by restriction analysis of their DNA. The DNA restriction patterns were compared with those of 24 *X. campestris* pathovars. Total DNA was extracted (Shepard and Polisky 1979, pp. 503–506) and digested to completion with various restriction enzymes (Maniatis et al. 1982). The DNA fragments were separated by electrophoresis in

0.75% agarose gels and stained with ethidium bromide. *X. campestris* pv *holcicola* was then characterized by visual assessment of unique fragment subsets (or bands) within the total genomic profile.

An *Eco*R1 restriction pattern, consisting of a broad space (spanning about 4.8 to 5.0 kb) and a pair of bands at about 3.7 and 3.9 kb, differentiates *X. campestris* pv *holcicola* from the 24 different pathovars of *X. campestris* tested (for example, see Fig. 9). Fifteen *X. campestris* pv *holcicola* isolates were screened; the pattern was consistent in all isolates (Fig. 10). While isolates of other pathovars may have bands in the same positions, not one had the complete pattern. Thus the subset pattern was unique to *X.campestris* pv *holcicola* and REA is useful in confirming identity of this pathovar.

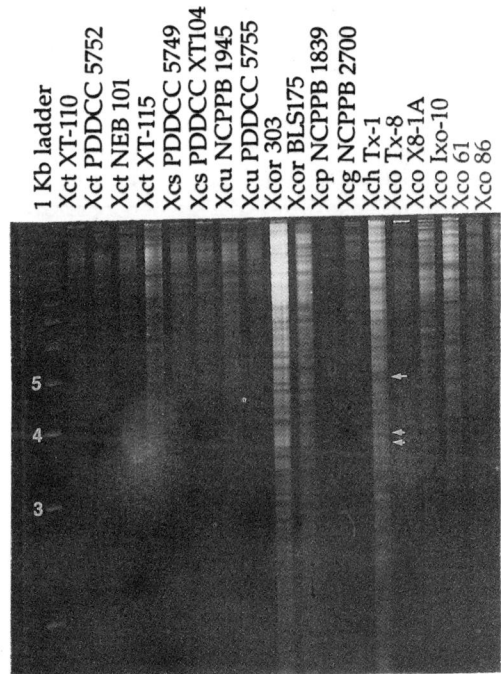

Figure 9. *Xanthomonas campestris* pv *holcicola* DNA *Eco* R1 restriction pattern. Pattern which differentiates *X. c.* pv *holcicola* from other pathovars is shown by arrows. Lane 1, 1 kb ladder size marker; 2-5, *X. c.* pv *translucens;* 6-7, pv *secalis;* 8-9, pv *undulosa;* 10-11, pv *oryzicola;* 12, pv *phleipratensis;* 13, pv *graminis;* 14, pv *holcicola;* 15-19, pv *oryzae*.

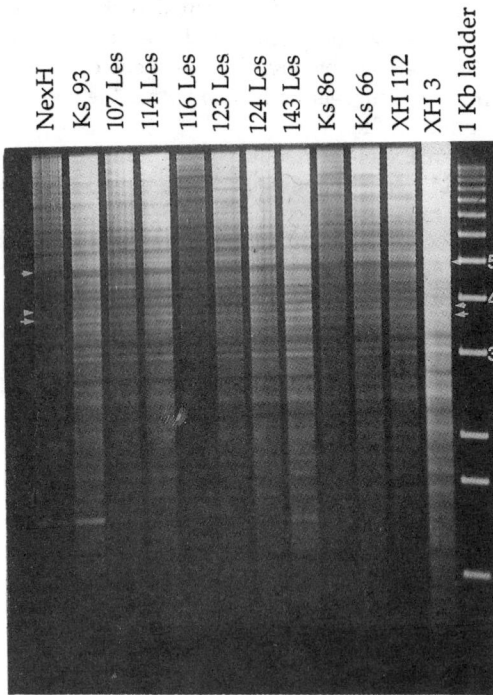

Figure 10. *Xanthomonas campestris* pv *holcicola* DNA *EcoR*1 restriction pattern. Pattern which differentiates *X. c.* pv *holcicola* from other pathovars is designated by arrows. Lane 1, *Xch* isolate from Nebraska; 2, Kansas; 3-8, Lesotho; 9-10, Kansas; 11, Australia; 12, Texas; 13, 1 kb ladder size marker.

Identification of races of *X. campestris* pv *holcicola*

Pathogenic specialization has been demonstrated between other *X. campestris* pathovars and their specific host cultivars. For example, *X. campestris* pv *oryzae* exhibits race-specificity on a differential set of rice cultivars containing specific genes for resistance (Mew 1987). Other examples include *X. campestris* pv *malvacearum* on cotton and *X. campestris* pv *vesicatoria* on pepper (Stall et al. 1986). To determine if *X. campestris* pv *holcicola* exhibited pathogenic specialization on sorghum lines, 70 genotypes of sorghum from Texas A&M University's All Disease and Insect Nursery (ADIN) were screened for resistance or susceptibility to six *X. campestris*

pv *holcicola* isolates [Xh 3 (Texas), Xh 112 (Australia), Les 114, Les 143 (Lesotho), KS 93 (Kansas), and Tx 4501 (Texas)].

Sorghum plants were grown in a greenhouse (25–27°C day, 22–24°C night) in square pots (4 inches on a side), one plant per pot, in bactopotting soil. Plants were inoculated at 4 to 7 weeks after sowing (approximately the four- to six-leaf stage of growth). Bacteria were grown overnight at 28°C in nutrient broth shake cultures. Bacterial cells were pelleted by centrifugation at 13 000 g for 10 min and then resuspended in water to approximately 5×10^9 cfu mL^{-1} (50 Kletts). One leaf on each sorghum plant was infiltrated with the bacterial suspension by a Hagborg apparatus. The trial utilized a split-plot design with one plant per replication and four replications per treatment. Symptoms (watersoaking vs no water-soaking) were determined at 4 and 6 days postinfiltration.

Initially, the 70 sorghum genotypes were screened for their response to six *X. campestris* pv *holcicola* isolates. Eight genotypes responded differentially to one or more of the isolates; they were screened again with a larger number of *X. campestris* pv *holcicola* isolates. The pattern of host responses is shown in Table 4. The recognition of race specificity in *X campestris holcicola* is important to sorghum breeders because it indicates that more than one pathogen race is required in screening for resistance to bacterial streak.

Our data suggest that isolates of *X. campestris* pv *holcicola* can be differentiated by their reaction with different sorghum cultivars. Therefore, it is likely that *X. campestris* pv *holcicola* exhibits pathogenic specialization to sorghum. This is significant in that (1) different sorghum genes for resistance to *X. campestris* pv *holcicola* must exist, (2) the potential for a large number of *X. campestris* pv *holcicola* races exists, and (3) identification of sources of resistance to *X. campestris* pv *holcicola* will require screening with a mixture of pathogen isolates.

Semiselective medium for *X. campestris* pv *holcicola*

A semiselective medium has been adapted for use in recovering *X. campestris* pv *holcicola* from bacterial streak-infested plant tissue. The medium, MXP (Claflin et al. 1987) was initially de-

Table 4. Reaction of six *Xanthomonas campestris* pv *holcicola* isolates on eight sorghum lines.[1]

Line	Reaction to *Xanthomonas campestris* pv *holcicola* isolates from					
	Nebraska	Kansas	Texas	Mexico	Lesotho	Australia
BTx 623	R[2]	S[3]	S	S	S	S
BTx 630	R	R	R	R	R	R
B 8106	R	R	S	S	R	R
QL 3	R	R	R	R	R	R
TAM 428	R	S	S	S	S	S
Tx 434	R	S	S	S	S	S
84 C-7730	R	S	S	S	S	S
85 L-21905	R	S	S	S	S	S

1. Source: Leach et al., unpublished.
2. R = resistant reaction.
3. S = susceptibility.

veloped for recovering *X. campestris* pv *phaseoli* from common blight-infested bean debris. Reducing the concentration of gentamycin from 2 mg L^{-1} to 1 mg L^{-1} provided an excellent medium for *X. campestris* pv *holcicola*. The medium consisted of 0.8 g K_2HPO_4, 0.6 g KH_2PO_4, 0.7 g yeast extract, 8 g soluble potato starch, 10 g potassium bromide, 1 g glucose, and 15 g agar in 1 L of distilled water. After autoclaving and after the medium cooled to 48°C, add 1 mL of Daconil 2787 (chlorothalonil®; to prepare stock suspension, add 1.2 mL of Daconil® 2787 to 38.8 mL of H_2O); 10 mL of cephalexin (for stock solution, add 1.0 g cephal-exinmonohydrate to 500 ml H_2O); 10 mL of kasugamycin [stock solution, add 1.0 g kasugamycin hemisulfate to 500 µl H_2O), and 5 mL of gentamycin sulfate (stock solution: 100 mg in 500 mL H_2O) per liter of media. The antibiotics should be filter sterilized prior to use. To enhance starch hydrolysis zones, add 30 µl L^{-1} of methyl violet 2B (1% solution in 20% ethanol), and 60 µl L^{-1} of methyl green (1% aqueous solution).

The medium was utilized to characterize *X. campestris* pv *pennamericanum*, the causal agent of bacterial leaf streak of pearl millet (Qhobela and Claflin 1987).

References

Akhtar, M.A., and Claflin, L.E. 1986a. Immunofluorescent detection of *Pseudomonas avenae* in plant tissue. Pakistan Journal of Agricultural Research 7:238–240.

Akhtar, M.A., and Claflin, L.E. 1986b. Detection of *Pseudomonas andropogonis* in plant tissue employing fluorescent antibody stain. Pakistan Journal of Agricultural Research 7:319–322.

Bradbury, J.F. 1967a. *Pseudomonas rubrilineans.* Descriptions of pathogenic fungi and bacteria, No. 127. Kew, Surrey, UK: Commonwealth Mycological Institute.

Bradbury, J.F. 1967b. *Pseudomonas rubrisubalbicans.* Descriptions of pathogenic fungi and bacteria, No. 128. Kew, Surrey, UK: Commonwealth Mycological Institute.

Bradbury, J.F. 1973a. *Pseudomonas alboprecipitans.* Descriptions of pathogenic fungi and bacteria, No. 371. Kew, Surrey, UK: Commonwealth Mycological Institute.

Bradbury, J.F. 1973b. *Xanthomonas vasculorum.* Descriptions of pathogenic fungi and bacteria, No. 380. Kew, Surrey, UK: Commonwealth Mycological Institute.

Burkholder, W.H. 1957. A bacterial disease of clover and velvet beans. Phytopathology 47:48–50.

Caruso, F.L. 1984. Bacterial blight of chickpea

incited by *Pseudomonas andropogonis*. Plant Disease 68:910–913.

Christopher, W.M., and Edgerton, C.W. 1930. Bacterial stripe diseases of sugarcane in Louisiana. Journal of Agricultural Research 41:259–267.

Claflin, L.E., Ramundo, B.A., Leach, J.E., and Erinle, I.D. 1987. *Pseudomonas avenae*, the causal agent of bacterial leaf stripe of pearl millet in Nigeria. Phytopathology 77:1766.

Claflin, L.E., and Rosenow, D.T. 1983. Screening grain sorghum lines for resistance to bacterial stripe and streak. Biennial Grain Sorghum Research and Utilization Conference. Brownsville, TX. 13:101–103.

Claflin, L.E., Vidaver, A.K., and Sasser, M. 1987. MXP, a semiselective medium for *Xanthomonas campestris* pv *phaseoli*. Phytopathology 77: 730–734.

Commonwealth Mycological Institute. 1987. Distribution Maps of Plant Diseases. Map No. 395, Edition 3. Kew, Surrey, UK: Commonwealth Mycological Institute.

Commonwealth Mycological Institute. 1987. Distribution Maps of Plant Diseases. Map No. 39, Edition 5. Kew, Surrey, UK: Commonwealth Mycological Institute.

Commonwealth Mycological Institute. 1976. Distribution Maps of Plant Diseases. Map No. 511, Edition 1. Kew, Surrey, UK: Commonwealth Mycological Institute.

Commonwealth Mycological Institute. 1973. Distribution Maps of Plant Diseases. Map No. 495, Edition 1. Kew, Surrey, UK: Commonwealth Mycological Institute.

DeBlas, A.L., and Cerwinski, H.M. 1983. Detection of antigens on nitrocellulose paper immunoblots with monoclonal antibodies. Analytical Biochemistry 133:214–219.

Elliott, C. 1929. A bacterial stripe disease of sorghum. Journal of Agricultural Research 38:1–22.

Elliott, C. 1930. Bacterial streak disease of sorghums. Journal of Agricultural Research 40:963–976.

Gaudet,D.A., and Kokko, E.G. 1986. Seedling disease of sorghum grown in Southern Alberta caused by seedborne *Pseudomonas syringae* pv *syringae*. Canadian Journal of Plant Pathology 8:208–217.

Gitaitis, R.D., Miller, J., and Wells, H.D. 1983. Bacterial leaf spot of white clover in Georgia. Plant Disease 67:913–914.

Hagborg, W.A.F. 1970. A device for injecting solutions and suspensions into thin leaves of plants. Canadian Journal of Botany 48:1135–1136.

Hale, C.N., and Wilkie, J.P. 1972. Bacterial leaf stripe of sorghum caused by *Pseudomonas rubrisubalbicans*. New Zealand Journal of Agricultural Research 15:457–460.

Hayward, A.C. 1972. A bacterial disease of clover in Hawaii. Plant Disease Reporter 56:446–450.

Hayward, A.C., and Waterston, J.M. 1965. *Pseudomonas syringae*. Descriptions of pathogenic fungi and bacteria, No. 46. Kew, Surrey, UK: Commonwealth Mycological Institute.

Jensen, S.G., Mayberry, W.R., and Obrigawitch, J.A. 1986. Identification of *Erwinia chrysanthemi* as a soft-rot-inducing pathogen of grain sorghum. Plant Disease 70:593–596.

Kendrick, J.B. 1926. Holcus bacterial spot on species of *Holcus* and *Zea mays*. Phytopathology 16:236–237.

King, E.O., Ward, M.K., and Raney, D.E.1954. Two simple media for the demonstration of pyocyanin and fluorescin. Journal of Laboratory Medicine 44:301–307.

Lazo, G.R., Roffey, R., and Gabriel, D.W. 1987. Pathovars of *Xanthomonas campestris* are distinguishable by Restriction Fragment Length Polymorphism. International Journal of Systematic Bacteriology 37:214–221.

Leach, J.E., Ramundo, B.A., Pearson, D.L., and Claflin, L.E. 1987. Dot-immunobinding assay for detection of *Xanthomonas campestris* pv *holcicola* in sorghum. Plant Disease 71:30–33.

Maniatis, T., Fritsch, E.F., and Sambrock, J. 1982. Molecular Cloning Laboratory Manual. Cold Spring Harbor, NY, USA: Cold Spring Harbor Laboratory.

Mew, T.W. 1987. Current status and future prospects of research on bacterial blight of rice. Annual Review of Phytopathology 25:359–382.

Moffett, M.L., Hayward, A.C., and Fahy, P.C. 1986. Five new hosts of *Pseudomonas andropogonis* occurring in Eastern Australia: host range and characterization of isolates. Plant Pathology 35:34–43.

Mohan, S.K., and Schaad, N.W. 1987. An improved agar plating assay for detecting *Pseudomonas syringae* pv *syringae* and *P. s.* pv *phaseolicola* in contaminated bean seed. Phytopathology 77:1390–1395.

Noble, M., and Richardson, M.J. 1968. An annotated list of seed-borne diseases. 2nd. edn. Commonwealth Mycological Institute. Phytopathological Papers 8:138.

Qhobela, M., and Claflin, L.E. 1987. Identification of the causal agent of bacterial leaf streak of pearl millet (*Pennisetum americanum*). Phytopathology 77:1767.

Rosen, H.R. 1922. A bacterial disease of foxtail (*Chaetochloa lutescens*). Annuals of Missouri Botanical Garden 9:333–388.

Rothwell, A., and Hayward, A.C. 1964. A bacterial disease of bougainvillea. Rhodesian Journal of Agricultural Research 2:97–99.

Schaad, N.W., Kado, C.I., and Sumner, D.R. 1975. Synonymy of *Pseudomonas avenae* Manns 1905 and *Pseudomonas alboprecipitans* Rosen. 1922. International Journal of Systemic Bacteriology 25:133–137.

Shakya, D.D., Vinther, F., and Mathur, S.B. 1985. Worldwide distribution of a bacterial stripe pathogen of rice identified as *Pseudomonas avenae*. Phytopathologische Zeitschrift 114: 256–259.

Shepard, H., and Polisky, B. 1979. Methods in Enzymology. Wu, R. (ed.). New York, NY, USA: Academic Press.

Stall, R.E., Loschke, D.C., and Jones, J.B. 1986. Linkage of copper, resistance and avirulence loci on a self-transmissible plasmid in *Xanthomonas campestris* pv *vesicatoria*. Phytopathology 76:240–243.

Vidaver, A.K., and Carlson, R.R. 1978. Leaf spot of field corn caused by *Pseudomonas andropogonis*. Plant DIsease Reporter 62:213–216.

Watson, D.R.W. 1971. A bacterial pathogen of sorghum in New Zealand. New Zealand Journal of Agricultural Research 14:944–947.

Zummo, N. 1976. Yellow leaf blotch: A new bacterial disease of sorghum maize, and millet in West Africa. Plant Disease Reporter 60:798–799.

Zummo, N. 1984. Sorghum diseases in West Africa: An illustrated text. Washington, DC, USA: USDA Animal and Plant Health Inspection Service and US Agency for International Development.

Zummo, N., and Freeman, K.C. 1975. Bacterial sun spot, a new disease of sugarcane and sweet sorghum. Sugarcane Pathology Newsletter 13/14:15–16.

Detection and Identification of Viruses and Virus Diseases of Sorghum

R.W. Toler[1] and L.M. Giorda[2]

Abstract

Plant virology is the study of viruses that infect plants. Sorghum viruses are distributed world-wide and cause economic damage to the host. A wide range of abnormalities is caused by these viruses. Some of the most noticeable effects are changes in color, death of tissue, and changes in growth habit. Most if not all sorghum viruses infect the plant systemically. Symptomology may play a role in identification, but symptomology alone is not definitive. Diagnosis of a sorghum virus disease on a routine basis is required for the development of control measures. Indicator host and host-range studies are important in initial diagnosis. Early diagnosis may include microscopic studies of cellular inclusions. These may be followed by quick-dip electron microscopy (EM). Serological tests may follow. Combined EM and serology studies, e.g., immunosorbent electron microscopy, may be in order. The EM studies should provide information on particle morphology, serology, and on the relationship of known viruses and strains. For finite criteria of a virus, purification is in order. Antiserum specific to the virus being studied can be prepared to the purified virus. After purification, information including sedimentation coefficient, molecular weight, isoelectric point, extinction coefficient, 260/280 ratio, and buoyant density can be obtained. Molecular studies should identify protein species, terminal amino acids, and nonparticle proteins. Lipids and polyamines present are characterized. The nucleic acid is analyzed for type and strandedness. Number and size of nucleic acids are determined, as well as base composition and ratios. Other distinguishing characteristics of the nucleic acid include presence of a methylated cap, amino acid-accepting ability, or polyadenine at the 3' end. Subgenomic RNA may or may not be present. Some viruses, such as the rhabdoviruses and reoviruses, may contain polymerases within the particle. Virus identification encompasses evaluation of the sum total of all information available, demonstrating relatedness or nonrelatedness to known sorghum viruses, or establishing the identity of a new virus.

Symptomology

The earliest detection of sorghum viruses in the field is made by noting their observable effects on the host. However, observable changes induced by plant viruses may be similar to those associated with other diseases and disorders in the field or area and cause misidentification.

Usually, the first symptom observed is some type of leaf discoloration. Discolored plants are compared to healthy plants of the same cultivar in the same sowing. General chlorosis or yellowing may indicate infection by viral or other pathogens. However, one must eliminate possible nutritional effects, particularly those resulting from minor element deficiency or excess (Clark 1986). Viral chlorosis must also be differentiated from genetic chlorosis or anomalies caused by biotic or abiotic agents (Jordan et al. 1986). Chlorotic effects caused by herbicides,

1. Professor of Plant Pathology and Microbiology, Texas A&M University, College Station, TX 77843, USA.
2. Sorghum Research Leader, INTA, Cordova, Argentina 5988.

Toler, R.W., and Giorda, L.M. 1992. Detection and identification of viruses and virus diseases of sorghum. Pages 153–159 *in* Sorghum and millets diseases: a second world review. (de Milliano, W.A.J., Frederiksen, R.A., and Bengston, G.D., eds). Patancheru, A.P. 502 324, India: International Crops Research Institute for the Semi-Arid Tropics.

153

insecticides, fungicides, or other chemicals must likewise be cataloged and separated (Merkle 1986). Another type of discoloration is chlorotic streaking or striping, appearing as wide or narrow bands of chlorosis running parallel to the midvein. This type of chlorosis may appear in younger leaves and in the sheaths. Streaking and striping caused by a virus or viruses must be differentiated from those caused by genetic anomalies or chemicals (Rosenow 1986).

Red coloration produced by anthocyanin developing in spots or stripes is often associated with virus infection. This is sometimes referred to as red-leaf reaction (Toler 1986). In the case of sorghums with tan pigments, the spots appear as tan or brown instead of red. The red areas usually appear early as spots that enlarge, coalesce, and finally become necrotic. As a precaution, this symptom must be distinguished from chemical damage and from nonviral pathogens causing similar symptoms.

Sorghum viruses may also induce ring spots. These are usually chlorotic, but may be necrotic. Rings may appear in conjunction with other symptoms, such as mosaic. Ring spots are not commonly diagnostic on sorghum. Discoloration also includes necrosis. Some hypersensitive sorghums infected with virus react by producing local necrotic spots that may eventually encompass the entire leaf and cause death of the plant (Toler 1986). Again, agents such as insects and other pathogens must be eliminated as incitants.

Mosaics or mottling, either chlorotic or necrotic, are often found on leaves and sheaths of virus-infected sorghums. Mosaics may occur on the entire leaf area or in defined patterns on the leaf. This symptom is more often associated with younger leaves or tillers. Mosaic, and particularly necrosis, may be confused with chemical damage or with symptoms caused by other pathogens, and the latter must be recognized and eliminated as possible inciters. Similarly, symptom patterns induced by mixed infections of viruses or virus strains must be distinguished from symptoms produced by either strain separately (Giorda and Toler 1985).

Stunting and dwarfing often accompany virus diseases of sorghum. With these symptoms, the affected plant must be compared to healthy plants of the same cultivar growing in the same location. Stunting occurs when the internodes in a particular portion of the stalk are noticeably shortened. This occurs most often in the upper portion of the plant, and may be associated with a particular growth phase. Stunting at a particular growth stage may be indicative of the time of virus infection. Dwarfing connotes reduced size of the entire plant, usually uniformly. The degree of dwarfing or stunting may be measured by comparing the height of the diseased plant with associated healthy plants (Alexander et al. 1983; Giorda 1983). Drought stress, excessive moisture, malnutrition, genetic root rots, and nematodes also cause stunting and dwarfing and must be eliminated as causes (Jordan and Peacock 1986; Rosenow 1986; Starr 1986).

Delayed flowering is a symptom associated with virus disease in sorghum. Flowering in diseased plants may be delayed from 1 to 12 days. Delay in heading may contribute to an increased incidence of other diseases and damage by midge (Toler 1985). Grain yield losses are reflected in lower total grain weight and test weight (Alexander et al. 1984; Alexander et al. 1985; Giorda 1983; Henzell et al. 1979; Toler 1985). Other factors associated with yield loss include smaller heads and seeds on the diseased plants. Confusion of virus symptoms with those caused by other pathogens or parasites can usually be systematically eliminated by visual inspection and light microscopic examination of the plants for nematodes, midge, and dodder. Facultative fungal causal agents can be identified by inspection or by light microscope, and by culturing on media. Bacterial infections can be identified by culturing and examination by light microscope. Pathogens such as mycoplasma and rickettsia can be separated from viral pathogens by axenic culture, electron microscopy, and immunofluorescent microscopy (Rocha et al. 1986). Viroids consist of naked single-stranded RNA (Diener 1983) and do not produce nucleocapsids or virions (have no protein or lipoprotein). Thus they can be separated from viruses by nucleic acid hybridization (Owens and Diener 1981).

Transmission and Host Range

Plant viruses are obligate pathogens, so the identification procedure usually begins with modified Koch's postulates: transmission of the causal agent from a diseased plant to a healthy plant of the same cultivar (genome) with the resulting production of the symptoms previously observed. Testing for potential vectors is a critical step, especially if the virus is not sap-inoculable. The following means of transmission should be tested: grafting, arthropod, nematode, fungus, parasitic seed plant (dodder), seed, and pollen. Following identification of a vector, the mechanism of transmission must be ascertained. The next identification step is usually the establishment of a host range, using both dicotyledonous and monocotyledonous test plants. In this phase the identification of diagnostic, propagation, and assay host species is desirable for identifying, multiplying, and quantifying the virus, respectively (Hamilton et al. 1981).

At this point, one can deviate somewhat from empirical procedures and conduct serological tests (Von Wechmar et al. 1983) without purification of the virus, provided antisera to known viruses are available. Selecting the antisera is a crucial step, as the selection must include as many and the types needed to identify the unknown virus and determine its interrelationships with known virus groups, viruses, and virus strains. If the serological test is positive to a known virus, additional testing can identify similarities to or deviations from the type virus. If the serology tests are negative, two quick tests may identify the unknown. One is the quick-dip test, designed to demonstrate the presence of rod-shaped virus particles as well as their relative lengths and morphologies (rigid, flexuous, or bacilliform) by electron microscopy (EM). However, the test is not reliable for small polyhedral viruses, because of confusion with ribosomes. If the quick-dip test can determine morphology of the virus particle, this alone will serve to eliminate many virus groups.

Immunosorbent electron microscopy (ISEM) provides both serological and particle morphology data (Derrick and Brlansky 1976). In this procedure, antiserum-treated grids are floated on a drop of infected crude sap, washed, stained, and viewed by EM. If the antiserum is homologous, the virus particles will adhere to the grid and their morphology can be noted (Derrick 1975; Giorda et al. 1986a). In virus identification, the greater the number of criteria used and tests applied, the greater are the chances of positive identification of a known virus, or placement of an unknown virus in an established virus group. If serology, quick-dip, and ISEM are negative, the next step is to check for possible viral inclusions (Christie and Edwardson 1977). This involves staining and examination by light or electron microscope. Viral inclusion may be specific for a single virus, or general for a particular virus group. Detection of ultramicroscopic inclusions, such as pinwheels, requires electron microscopy.

Purification and Molecular Characteristics

Study of the ultrastructure of infected host tissue is required before purification. Such study will confirm presence or absence of inclusions, types of tissue infected, location of virus in the host cell, structural or cellular changes in the host, and morphology of the virus particle (Kurstak 1981). At this point, if evidence is insufficient or the virus appears to be one not yet described, purification is necessary. Purification schemes are available for viruses of all types (Giorda and Toler 1986a; Giorda et al. 1987; Langham 1986; Van Regenmortel 1982a). Preliminary information on nonspecific criteria such as thermal inactivation point, dilution-end point, effect of pH, longevity in vitro, and the effect of diethyl ether is helpful (Noordam 1973). After purification, the sedimentation coefficient, molecular weight, isoelectric point, extinction coefficient, 260/280 ratio of unfractionated and separated components, and buoyant density in CsCl or Cs_2SO_4 can be measured (Matthews 1981; Noordam 1973). Purity is critical. If artifacts produced by nonviral material remain, the results are likely to be misread. Also, one must determine if more than a single component is present in the virus preparation. Analytical ultracentrifugation and Schlieren optics are employed to separate viral components sedimenting at different rates. If purified preparations contain more than one component, each component can be categorized by number or name (Trautman and Hamilton 1972).

Molecular characterization is another step in the identification of viruses. Protein is separated from the nucleic acid and the molecular weight determined. Next, the number of protein species in the particles are counted and described by size and number (Langham 1986; Smith 1984). Terminal amino acids are identified, and then the entire amino acid composition is sequenced (Langham 1986). The presence of viral encoded nonparticle proteins may also be determined. The nonparticle proteins are named and sized, and if possible, functionally characterized. Other components of viruses, such as polyamines, may be identified. Lipids are found in enveloped viruses. At least two virus groups contain envelopes of 10 to 30% lipids. Virus proteins may have other molecules attached. For example, some viruses are glycosylated and others may be phosporylated (Matthews 1981).

Nucleic acid can be analyzed for type—deoxyribonucleic acid (DNA) or ribonucleic acid (RNA)—and strandedness. Both RNA and DNA may be either single- or double-stranded (ssRNA, ssDNA, dsRNA, dsDNA) (Dodds et al. 1984; Harrison 1985; Matthews 1981). To date, unusual or minor bases have not been reported in plant virus nucleic acids. The number of nucleic acid species is another identifying criterion, since the genome of a virus may be comprised of one or more separate RNAs (Goldbach 1986; Matthews 1981). The molecular weight of the nucleic acid and its base composition and ratios are distinguishing characteristics. Another distinguishing genomic characteristic is the presence or absence of a covalently linked protein or methylated cap at the five prime (5′) end. At the three prime (3′) end, the virus may or may not have an amino acid-accepting ability. Also, the 3′ end may have a polyadenine or tRNA-like structure with a CCA (anti)condon. Infectivity may or may not require a protein capsid (nucleocapsid). Subgenomic RNA may or may not be present (Goldbach 1986; Matthews 1981; Rowlands et al. 1987). Also, some viruses (e.g., rhabdoviruses and reoviruses) contain polymerases within the particle (Joklik 1981). The absorbance of its viral nucleic acid per unit weight at 260 μm varies, depending o base composition, and may therefore serve as a distinguishing factor.

Antiserum production, testing

After purification, antiserum can be produced. The reaction of virus with polyclonal or monoclonal antibodies is useful in virus identification and study of virus relationships. Many assay techniques are available for studying antigen antibody reactions (Hill et al. 1984; Van Regenmortel 1982b). The two most useful for virus identification are gel-diffusion tests and enzyme-linked immunosorbent assay (ELISA)

(Kerlan et al. 1982). The gel-diffusion test most commonly used is the Ouchterlony double-diffusion test. The position on the plate where antigen and antibody meet in optimum proportions produce precipitation zones characteristic of the particular antigen-antibody system used. In the gel-diffusion technique, crude or purified virus preparations may be used with any type of antibody. An additional advantage is that a number of dilutions of antigen or antiserum samples can be tested at the same time. The ELISA technique employs several systems, but the most commonly used is the double antibody sandwich (Koenig and Paul 1982). In this procedure, the inside walls of a polystyrene plate are coated with gamma globulin from the antiserum and the plate is then washed. The virus sample is added and incubated. After washing, the antibody linked to an enzyme with gluteraldehyde is added, the plate washed, and an enzyme substrate added. The colorless substrate, if positive to the virus, converts to a colored product that can be observed visually or measured spectrophotometrically. This test is very sensitive and can provide quantitative data for study of host-pathogen interaction, epidemiological investigations, and characterization studies (Clark 1981; Giorda et al. 1986a). Virus samples may be crude or purified preparations, and the technique is particularly suitable for processing large numbers of samples. It is highly specific and can be used where mixed infections, or more than one virus strain, is involved (Clark 1981; Koenig and Paul 1982).

Host response may also be used as a diagnostic tool. The symptom response of a host to an unknown virus can be compared to that of a known virus and the symptom homology or heterology established (Alexander et al. 1985; Giorda et al. 1986b; Persley et al. 1985; Toler 1985). Genotypes resistant to specific viruses can be inoculated with the unidentified virus, and observed during the growing season. The observations provide corroborative evidence based on resistance (Alexander et al. 1983; Giorda et al. 1986b; Toler 1985).

Traditional approaches to detection and diagnosis of plant viruses include: symptoms; transmissibility; host range; vector relationships; immunological, physical, and chemical properties of the virus particle; morphology; its capsid protein; and genome. With the development of biotechnology, new methods based on nucleic acid hybridization, cloning, and sequencing (Dodds et al. 1984; Gabriel 1986; Maule et al. 1983; Sela et al. 1984), as well as viral specific dsRNA (Morris et al. 1983), are being used for detection and diagnosis.

Virus identification encompasses evaluation of the sum total of all evidence, positive or negative, demonstrating relatedness or nonrelatedness of the virus in question to known sorghum viruses, and their strains or variants.

References

Alexander, J.D., Toler, R.W., and Miller, F.R. 1983. Reaction of grain sorghum accessions to maize dwarf mosaic virus strain B. Biennial Grain Sorghum Research and Utilization Conference 13:103–104.

Alexander, J.D., Toler, R.W., and Miller, F.R. 1984. Effects of maize dwarf mosaic virus strain B infection on growth and yield of susceptible sorghum. Sorghum Newsletter 27:124–126.

Alexander, J.D., Toler, R.W., and Giorda, L.M. 1985. Correlation of yield reductions with severities of disease symptoms in sorghum infected with sugarcase mosaic or maize dwarf mosaic virus. Biennial Grain Sorghum Research and Utilization Conference 14:109–112.

Christie, R.G., and Edwardson, J.R. 1977. Light and electron microscopy of platn virus. Florida Agricultural Experiment Station Monograph 9. Gainesville, FL, USA: Florida Agricultural Experiment Station.

Clark, M.F. 1981. Immunosorbent assays in plant pathology. Annual Review of Phytopathology. Vol. 19 (Baker, K.F., ed.) Palo Alto, CA, USA: George Banta Co.

Clark, R.B. 1986. Nutrient deficiencies and toxicities. Pages 57–59 in Compendium of Sorghum Diseases, (Frederiksen, R.A., ed.) St. Paul, MN, USA: American Phytopathological Society 2:42 (Abstract).

Derrick, K.S. 1975. Serological relationships among strains of sugarcane mosaic. Proceedings of the American Phytopathological society 2:42 (Abstract).

Derrick, K.S., and Brlansky, Jr., H. 1976. Assay for viruses and mycoplasma using serologically specific electron microscopy. Phytopathology 66:815–820.

Diener, T.O. 1983. Viroids. Advances in Virus Research 28:241–279.

Dodds, J.A., Morris, T.J., and Jordan, R.L. 1984. Plant viral double-stranded RNA. Pages 151–168 in Annual Review of Phytopathology, Vol 22 (Baker, K.F., ed.) Palo Alto, CA, USA: George Banta Co.

Gabriel, C.J. 1986. Detection of double-stranded RNA by immunoblot electrophoresis. Journal of Virological Methods 13:279–283.

Giorda, L.M. 1983. Identification and evaluation of an isolate of sugarcane mosaic virus. M.S. thesis. Texas A&M University College Station, TX, USA.

Giorda, L.M., and Toler, R.W. 1985. Effect of single and mixed infection of maize dwarf mosaic virus strain A, sugarcane mosaic virus strain H and an isolate of sugarcane mosaic virus strain H on 27 accessions of grain sorghum. Sorghum Newsletter 38:103–105.

Giorda, L.M., and Toler, R.W. 1986a. Use of virus concentration, symptomatology, and yield as a measure of "resistance" to MDMV-A in different sorghum cultivars. Phytopathology 76:1072. (Abstract)

Giorda, L.M., Toler, R.W., and Miller, F.R. 1986b. Identification of sugarcane mosaic virus strain H isolate in commercial grain sorghum. Plant Disease 70:624–628.

Giorda, L.M., Toler, R.W., and Odvody, G.N. 1987. Purification and serology of a newly recognized virus disease in sorghum × sudangrass hybrids. Phytopathology 77:17. (Abstract).

Goldbach, R.W. 1986. Molecular evolution of plant RNA virus. Pages 289–310 in Annual Review Phytopathology, Vol. 24 (Baker, K.F., ed.). Palo Alto, CA, USA: George Banta Co.

Hamilton, R.I., Edwardson, J.R., Francki, R.I.B., Hsu, H.T., Hull, R., Koening, R., and Koenig,

R., and **Milne, R.G.** 1981. Guidelines for the identification and characterization of plant viruses. Journal of General Virology 54:223–241.

Harrison, B.D. 1985. Advances in gemini virus research. Pages 55–82 *in* Annual Review of Phytopathology, Vol 23 (Baker, K.F., ed.) Palo Alto, CA, USA: George Banta Co.

Henzell, R.G., Persley, D.M., Fletcher, D.S., Greber, R.S., and **von. Slobbe, M.** 1979. The effect of sugarcane mosaic on the yield of eleven grain sorghum (*Sorghum bicolor*) cultivars. Australian Journal of Experimental Agriculture and Animal Husbandry 19:225:232.

Hill, E.K., Hill, J.H., and **Durand, D.P.** 1984. Production of monoclonal antibodies to viruses in the potyvirus group: Use in radioimmunoassay. Journal of General Virology 65:525–532.

Joklik, W.K. 1981. Structure and function of the Reovirus genome. Microbiology Review 45:483–501.

Jordan, W.R., and **Peacock, J.M. 1986.** Water and temperature stresses. Pages 59–60 *in* Compendium of Sorghum Diseases. (Frederiksen, R.A., ed.). St. Paul, MN, USA: American Phytopathological Society.

Kerlan, C., Mille, B., Detienne, G., and **Dunez, J.** 1982. Comparison of immunoelectronmicroscopy, immunoenzymology (ELISA) and gel diffusion for investigating virus strain relationships. Annales de Virologie 133E:1–14.

Koenig, R., and **Paul, H.L.** 1982. Variants of ELISA in plant virus diagnosis. Journal of Virological Methods 5:113–125.

Kurstak, E. (ed.), 1981. Handbook of Plant Virus Infections and Comparative Diagnosis. Amsterdam, The Netherlands: Elsevier/North-Holland Biomedical Press. 945 pp.

Langham, M.A.C. 1986. An analysis of the strain relationship of maize dwarf mosaic virus. Ph.D. Dissertation, Texas A&M University, College Station, TX, USA.

Matthews, R.E.F. 1981. Plant Virology. New York, USA: Academic Press.

Maule, A.J., Hull, R., and **Donson, J.** 1983. The application of spot hybridization to the detection of DNA and RNA viruses in plant tissues. Journal of Virological Methods 6:215–224.

Merkle, M.G. 1986. Pesticide injury. Pages 60–61 *in* Compendium of Sorghum Diseases. (Frederiksen, R.A., ed.). St. Paul, MN, USA: American Phytopathological Society.

Morris, T.J., Dodds, J.A., Hillman, B., Jordan, R.L., Lommel, S.A., and Tamaki, S.T. 1983. Viral specific dsRNA: Diagnostic value for plant virus disease identification. Plant Molecular Biology Reporter 1:27–30.

Noordam, D. 1973. Identification of plant viruses. Methods and experiments. Wageningen, The Netherlands: Centre for Agricultural Publishing and Documentation.

Owens, R.A., and **Diener, T.O.** 1981. Sensitive and rapid diagnosis of potato spindle tuber viroid disease by nucleic acid hybridization. Science 213:670–671.

Persley, D.M., Henzell, R.G., Greber, R.S., Teakle, D.S., and **Toler, R.W.** 1985. Use of a set of differential sorghum inbred lines to compare isolates of sugarcane mosaic virus from sorghum and maize in nine countries. Plant Disease. 69:1046–1049.

Rocha, Ada, Ohki, S.T., and **Hiruki, C.** 1986. Detection of mycoplasmalike organisms in situ by indirect immunofluorescence microscopy. Phytopathology 76:864–868.

Rowlands, D.J., Mayo, M.A., and **Mahy, B.W.J.** 1987. The molecular biology of the positive strand RNA viruses. New York, USA: Academic Press.

Rosenow, D.T. 1986. Genetic disorders. Page 60 *in* Compendium of Sorghum Diseases. (Frederiksen, R.A., ed.). St. Paul, MN, USA: American Phytopathological Society.

Sela, I., Reidman, M., and **Weissabach, A.** 1984. Comparison of dot molecular hybridization and enzyme-linked immunosorbent assay for detecting tobacco mosaic virus in plant tissues and protoplasts. Phytopathology 74:385–389.

Smith, B.J. 1984. SDS Polyacrylamide gel electrophoresis of proteins. Pages 41–55 *in* Methods in molecular biology. Vol 1, Proteins, (Walker, J.M., ed.). Clifton, NJ, USA: Humana Press. 365 pp.

Starr, J.L. 1986. Nematodes: *Pratylenchus* spp. Page 53 *in* Compendium of Sorghum Diseases. (Frederiksen, R.A., ed.). St. Paul, MN., USA: American Phytopathological Society.

Toler, R.W. 1985. Maize dwarf mosaic, the most important virus disease of sorghum. Plant Disease 69:1011–1015.

Toler, R.W. 1986. Diseases caused by viruses and virus-like organisms. Pages 42–29 *in* Compendium of Sorghum Diseases. (Frederiksen, R.A., ed.). St. Paul, MN, USA: American Phytopathological Society.

Trautman, R., and Hamilton, M.G. 1972. Analytical ultracentrifugation. Pages 491–522 *in* Principles and Techniques in Plant Virology. (Clarence, I.K., and Agrawal, H.O., eds.). New York, NY, USA: Van Nostrand Reinhold Company.

Van Regenmortel, M.H.V. 1982a. Virus purification. Pages 25–36 *in* Serology and Immunochemistry of Plant Viruses. Mayor, NY, USA: Academic Press.

Von Wechmar, M.B., Rybicki, E.P., and Erasmus, D.S. 1983. Some serological techniques for detecting plant viruses. Pages 219–228 *in* Proceedings of the International Maize Virus Disease Colloquium and Workshop, 2–6 August 1982. (Gordon, D.T., Knoke, J.K., Nault, C.R., and Ritter, R.M., eds.). The Ohio State University, Wooster, OH, USA: The Ohio Agricultural Research and Development Center.

Biotic and Abiotic Factors Associated with Seedling Diseases of Sorghum

G.N. Odvody[1]

Abstract

Soil moisture and temperature exert major influences on seed germination and growth and on biotic and abiotic factors affecting seedling emergence and establishment. The affected factors include soil microflora, arthropods, nematodes, other fauna, and physical and chemical soil components. Seed rots, seedling diseases, and the soil environment can interact to cause retarded germination, low seedling vigor, stunting, seedling blights, and seedling death. Seedborne fungi, including storage molds (Aspergillus and Penicillium spp), Rhizopus spp, grain mold fungi (Fusarium moniliforme, F. semitectum, and Curvularia lunata), and foliar pathogens (Colletotrichum graminicola and Gloeocercospora sorghi) may rot unprotected seed in cold wet soils. Seedborne and soilborne Fusarium and Penicillium spp are reported to sometimes cause seedling blights. Soilborne Pythium arrhenomanes and P. graminicola are the primary causes of pre- and postemergent damping-off and seedling blight of sorghum in cold wet soil. Seed treatments with standard contact fungicides (captan or thiram) control seed rot fungi, but not Pythium spp New systemic fungicides (metalaxyl) and resistant germplasm may provide protection against Pythium seedling diseases. Cropping practices and resistant germplasm may be used to minimize seed damage caused by grain mold and weathering fungi.

Introduction

Germinating seeds, emerging, and emerged seedlings of grain sorghum are exposed to widely variable soil environments and soil microflora in the diverse areas of the world where sorghum is grown. Of the multitude of soil environmental factors that interact and affect the establishment of sorghum seedlings, temperature and moisture are considered to be the major physical factors. They directly affect the growth processes of sorghum seedlings, and also may affect and interact in a major way with other factors influencing sorghum-seedling emergence and establishment— namely soil physics and chemistry, insects, nematodes, and soil microflora.

The soil microflora consists of many interacting organisms, including bacteria, fungi, actinomycetes, viruses, algae, and protozoa. The bacteria and fungi can be strict saprophytes, others are primarily pathogens, and another group (the facultative parasites) are capable of both pathogenic and saprophytic activity on sorghum tissues. These latter two groups of organisms, especially fungi, are the most active against sorghum seedling tissue. Their deleterious effect, however, is highly dependent on the soil environment (including other soil microflora) and susceptibility of the sorghum tissue.

The soil environment for seed germination and growth is in constant flux; the changes in soil microflora and physical conditions have

1. Associate Professor, Plant Pathology, Texas A&M Univeristy Research and Extension Center, Route 2, Corpus Christi; TX 78410, USA.

Odvody, G.N. 1992. Biotic and abiotic factors associated with seedling diseases of sorghum. Pages 161–166 *in* Sorghum and millets diseases: a second world review. (de Milliano, W.A.J., Frederiksen, R.A., and Bengston, G.D., eds). Patancheru, A.P. 502 324, India: International Crops Research Institute for the Semi-Arid Tropics.

direct effects on sorghum seedling growth and vulnerability to soilborne organisms.

The major biotic and abiotic diseases affecting sorghum seedlings, from germination through the establishment of the permanent root system are discussed here.

Seed Germination and the Primary Root System

Freeman (1970) reviewed the root-system morphology of sorghum. The radicle emerges from the coleorhiza of the germinating seed and develops into the primary root that then forms several lateral roots. The coleoptile emerges from the seed and extends toward the soil surface by elongation of the mesocotyl (first internode). The upper limit of the mesocotyl is delimited by formation of the coleoptilar node near the soil surface, where the plumule emerges through the coleoptile.

The mesocotyl produces lateral (adventitious) roots similar to the primary root and together they form the primary, transitory or temporary root system of sorghum. Most of these are small in comparison to later roots and are nearly uniform in diameter throughout their length. This transitory root system is progressively displaced by the permanent root system which is entirely adventitious; whorls of roots develop sequentially from the coleoptilar and higher nodes of the crown. Later, buttress roots are initiated from nodes above ground. Depending upon plant-growth rate and soil environment (including soil microflora activity), the temporary (primary) root system may cease to function within a few weeks.

Seedling diseases and the soil environment can delay or impair establishment of the permanent root system, causing pre- and postemergent damping-off of sorghum seedlings, reduced stands, and poor seedling growth.

Abiotic Factors

Soil temperature and moisture

Many soils have physical properties that are not conducive to emergence of sorghum seedlings, and those factors may be exacerbated by moisture and temperature relationships related to rainfall patterns and subsequent solar radiation and surface drying that forms hard crusts on the soil surface. Such conditions are common to many sorghum-growing regions. Other soil factors that can affect or influence seedling growth include pH, salinity, texture, and pore size; chemical composition; and organic content. These factors also indirectly affect seedling establishment by influencing the composition and numbers of the resident soil microflora. The latter, including the facultative parasites and pathogens of sorghum seedlings, in turn influence one another.

Soil temperature directly affects the growth of sorghum; temperatures cooler than 18 °C are generally deleterious to germination of sorghum seed and emergence (Stoffer and Riper 1963). In the field. Germination response varies widely (4.6–16.5 °C) in sorghum germplasm (Miller 1982). These temperatures are not favorable for seedling establishment and growth. Cool soil temperatures can inhibit seed germination and emergence, but cold itself (if not below 0 °C) does not damage the sorghum seed.

Hot temperatures at the soil surface can damage emerging seedlings and cause them to grow parallel to, but beneath, the soil surface instead of emerging. Optimal soil temperatures for germination are from 21 to 35 °C. Temperatures of 40–48 °C at germination may be lethal to sorghum (Peacock 1982).

Soil moisture must be spatially adequate to allow germination and subsequent seedling growth and establishment of the temporary root system. In many areas the rapid depletion of soil moisture after sowing either disallows germination, or germination proceeds but is arrested by lack of moisture and the seedling dies prior to emergence. An excess of soil moisture depletes soil oxygen, damaging roots of emerging and established seedlings. The other negative effect of high soil moisture is through its effect on soil microflora activity against sorghum seedlings (Leukel and Martin 1943; Forbes 1984).

Organic matter and allelopathy

Organic material affects soil structure and chemistry through its direct physical attributes and through the dynamic microfloral communities active in its decomposition in soil. Some of the decomposition products of certain weed and crop plants are inhibitory to sorghum growth through a process called allelopathy (Mohamed-Saleem and Fawusi 1983; Shahid Shaukat et al. 1985). Conversely, sorghum residues are reported to have allelopathic effects on some weeds, including one genus (*Digitaria*) reported to be allelopathic to sorghum (Defrank and Putman 1979). The quinones produced as hydrophobic root hair exudates may prove interesting in this regard (Netzly and Butler 1986). The other important effect of soil organic material is its function as a site for survival and colonization by microorganisms pathogenic to sorghum seedlings and or by organisms inhibitory to these pathogens.

Pesticides

Soil-applied pesticides, especially herbicides, are often phytotoxic to sorghum and can inhibit seedling growth. McLaren (1983) demonstrated that four herbicides, including atrazine, inhibited secondary root formation and promoted pre- and postemergent damping-off of sorghum seedlings.

Biotic Factors

Seed-related

Many organisms have been isolated from sorghum seed, but their relationship to seedling diseases or reduction in seedling growth differ and are often unknown. Some of the fungi are passively associated with the seed and are of no significance, but others, like *Rhizopus* spp, can cause seed rots in cold, wet soil (Leukel and Martin 1943). Others, such as *Colletotrichum graminicola* (Basu Chaudhary and Mathur 1979) and *Gloeocercospora sorghi* (Watanabe and Hashimoto 1978), are seedborne pathogens that later attack seedlings and may or may not also damage the seed. Some fungi deleteriously affect seed and subsequent seedlings through direct damage to the seed before sowing.

Seedborne fungi damaging the seed are usually divided into two groups, one based on biological behavior and species and the other on where the attack on the seed occurs. The storage fungi (primarily *Aspergillus* spp and *Penicillium* spp) usually will damage sorghum seed in storage under moisture conditions lower than those in which field fungi attack seed before maturity (grain mold) or after maturity (grain weathering fungi) (Castor 1981). The most common grain mold fungi are *Fusarium moniliforme*, *F. semitectum*, *Curvularia lunata*, and *Phoma sorghina* (Bandyopadhyay 1986); of these, *F. moniliforme* and *C. lunata* may be the most important (Frederiksen 1982). The most common fungi associated with grain weathering are species of *Fusarium*, *Alternaria*, and *Cladosporium* (Frederiksen 1982). Gaudet and Kokko (1986) reported that *Pseudomonas syringae* pv *syringae* was seedborne in sorghum seed produced in southern Alberta, Canada, and was responsible for stunting and discoloration of roots and coleoptiles of seedlings.

The cumulative effect of damage by these organisms can be loss of seed viability, seed rotting in soil (by these or other fungi), retarded germination, low seedling vigor, and sometimes blighting of the seedlings (*Fusarium*) (Williams and Rao 1980).

Soilborne organisms

Some of the seedborne organisms causing damage to seed are also soilborne and under conducive soil environments can attack ungerminated and germinating seed (Leukel and Martin 1943). The primary seed rotting fungi in soil are soilborne and seedborne species of *Fusarium* and *Penicillium* and seedborne species of *Rhizopus* and *Aspergillus*.

The primary organisms associated with pre- and postemergent damping-off and seedling diseases of sorghum are species of *Pythium* (Leukel and Martin 1943; Pratt and Janke 1980; Forbes et al. 1985). Odvody and Forbes (1984) reviewed *Pythium* root and seedling rots of sorghum. The *Pythium* species most commonly associated with seedling root loss are *P. arrhenomanes* and *P. graminicola*.

In the field, *Pythium* predominantly infects and damages the roots and mesocotyls of sorghum seedlings in cool, wet soils before or after emergence. Symptoms on seedlings are either brown or gray water-soaked roots or root tips, or lesions on roots that become flaccid and necrotic (Forbes 1984). The mesocotyl produces greater pigmentation response to the pathogen than do the roots. Most of the seedlings succumbing to preemergent damping-off die prior to emergence and those that do emerge may wilt rapidly and die (postemergent damping-off). Often development of the permanent adventitious root system will proceed at different rates in plants stunted by *Pythium* attack, and this becomes apparent in affected fields by wide variations in plant height and spacing.

Soil moisture and temperature are the most important factors associated with *Pythium* attack of sorghum. Oospores of *Pythium* spp probably supply the initial inoculum for infection, but it is not known if they germinate directly or if they produce zoospores. Free water would be required for zoospore production and infection. Low temperatures (15°C) in association with high soil moisture has historically been associated with *Pythium* attack of sorghum (Leukel and Martin 1943) but the optimal temperature for growth of *Pythium* spp is normally much higher (Hendrix and Campbell 1973). The transitory root system of sorghum seems especially vulnerable in cool wet soils, because seedling growth slows and development of the permanent root system is delayed. However, with field soil in a controlled environment, Forbes (1984) demonstrated that high soil moisture was the key requirement for most damage caused to sorghum seedlings by *Pythium arrhenomanes* although there were some important temperature × moisture interactions.

Any stress factor that delays establishment of the permanent root system probably increases the potential for *Pythium* attack in wet (and especially) cool soils. Lines and varieties, because of their reduced vigor, may be more vulnerable than hybrids. Seedlings from old seed are more susceptible than those plants of younger seed, because the older-seed seedlings show reduced vigor, especially in cool, wet soils. Other factors, such as seed damage, may reduce seedling vigor or allow other organisms to reduce vigor, and increase vulnerability to *Pythium* spp.

Controls

Control of all negative effects of some soil environmental factors against sorghum seedlings is probably not possible, but cropping and tillage practices should be utilized, when possible, to rerduce or avoid deleterious soil environments.

Seed rots caused by seedborne and soilborne fungi are easily controlled by fungicide seed treatments. Fungicide treatment also provides needed extra protection for low-vigor seedlings emerging from seed damaged by storage and grain-weathering fungi and by grain mold. Traditional contact fungicides (like captan and thiram) do not provide protection against attack by *Pythium* spp, partly because *Pythium* attacks at root and mesocotyl sites distal to the fungicide, but also because these fungicides do not act against *Pythium*. New systemic fungicides, such as the acylanilides (metalaxyl) and phosphonates (fosetyl-Al), are active against *Pythium* in other crops and may provide some control in sorghum.

Differences in response to temperature and moisture stress may be found in sor-ghum germplasm, so cultivars may be chosen or avoided to minimize the deleterious effects of these and other soil abiotic factors. Seed of these sorghums also differ in susceptibility to grain mold and weathering fungi. Cultivars can be selected to reduce seed damage associated with these fungi, or cropping sequences that will allow seed maturation in environments unfavorable for grain mold and weathering may be possible. Sources of resistance to *Pythium* seedling disease have been identified (Forbes et al. 1987), but more rapid and consistent screening techniques are needed.

Research Needs

1. Major studies of abiotic and biotic factors affecting sorghum growth, especially as they function in areas where seedling diseases are unrecognized and little understood.
2. Better understanding of *Pythium* and its occurrence on sorghum in field environments, including incidence, severity, parameters of conducive environments, frequency of those environments, and interactions with host development.
3. Determine the value of fungicide treatments to protect seed against fungal attack in areas where sorghum is grown under conditions believed to be nonconducive to their development.

References

Bandyopadhyay, R. 1986. Grain mold. Pages 36–38 *in* Compendium of Sorghum Diseases, (Frederiksen, R.A., ed.). St. Paul, MN, USA: American Phytopathological Society. 82 pp.

Basu Chaudhary, K.C., and **Mathur, S.B.** 1979. Infection of sorghum seeds by *Colletotrichum graminicola* 1. Survey, location in seed and transmission of the pathogen. Seed Science and Technology 7:87–92.

Castor, L.L. 1981. Grain mold histopathology, damage assessment, and resistance screening within *Sorghum bicolor* (L.) Moench lines. Ph.D. Dissertation, Texas A&M University, College Station, Texas, USA.

Defrank, J., and **Putman, A.R.** 1979. Efficacy of rotational crop residue for weed control. Pages 84–85 *in* Abstracts, 1979 Meeting of the Weed Science Society of America.

Forbes, G.A. 1984. Development of a technique for screening seedlings of *Sorghum bicolor* (L.) Moench for resistance to seedling disease. M.Sc. Thesis, Texas A&M University, College Station, Texas, USA.

Forbes, G.A., Collins, D.C., Odvody, G.N., and **Frederiksen, R.A.** 1985. A seedling epiphytotic of sorghum in South Texas caused by *Pythium arrhenomanes*. Plant Disease 69:726.

Forbes, G.A., Ziv, O., and **Frederiksen, R.A.** 1987. Resistance in sorghum to seedling disease caused by *Pythium arrhenomanes*. Plant Disease 71:145–148.

Frederiksen, R.A. 1982. Disease problems in sorghum. Pages 263–271 *in* Volume I: Sorghum in the Eighties. Proceedings of the International Symposium on Sorghum, 2–7 Nov 1981, ICRISAT Center, India. Patancheru, Andhra Pradesh 502 324, India: International Crops Research Institute for the Semi–Arid Tropics.

Freeman, J.E. 1970. Development and structure of the sorghum plant and its fruit. Pages 28–72 *in* Sorghum Production and Utilization. (Wall, J.S., and Ross, W.M., eds.). Westport, CN, USA: AVI Publishing Company. 702 pp.

Gaudet, D.A., and **Kokko, E.G.** 1986. Seedling disease of sorghum grown in southern Alberta caused by seedborne *Pseudomonas syringae* pv *syringae*. Canadian Journal of Plant Pathology 8:208–217.

Hendrix, F.F., Jr. and **Campbell, W.A.** 1973. Pythiums and plant pathogens. Annual Review of Phytopathology 11:77–97.

Leukel, R.W., and **Martin, J.H.** 1943. Seed rot and seedling blight of sorghum. USDA Technical Bulletin 839. Washington, DC, USA: United States Government Publishing Office. 26 pp.

McLaren, N.W. 1983. The effect of herbicides, cultivars and fungicides on pre- and post-emergence damping-off and seedling blight of sorghum (*Sorghum bicolor* (L.) Moench). Crop Production 12:101–103.

Miller, F.R. 1982. Genetic and environmental response characteristics of sorghum. Pages 393–402 *in* Sorghum in the Eighties. Proceedings of the International Symposium on Sorghum, 2–7 Nov 1981, ICRISAT Center, India. Patancheru, Andhra Pradesh 502 324, India: International Crops Research Institute for the Semi-Arid Tropics.

Mohamed-Saleem, M.A., and **Fawusi, M.O.A.** 1983. A note on the effects of tropical weed decomposition on seed germination and seedling growth of some agricultural crops. Agriculture, Ecosystems and Environment 10:347–352.

Netzly, D.H., and **Butler, L.G.** 1986. Roots of sorghum exude hydrophobic droplets containing biologically active components. Crop Science 26:775–778.

Odvody, G.N., and **Forbes, G.A.** 1984. Pythium Root and Seedling Rots. Pages 31–35 *in* Sorghum Root and Stalk Rots. A Critical Review. Proceedings of the Consultative Group Discussion on Research Needs and Strategies for Control of Sorghum Root and Stalk Diseases, 27 Nov–2 Dec 1983, Bellagio, Italy. Patancheru, Andhra Pradesh 502 324, India: International Crops Research Institute for the Semi-Arid Tropics.

Peacock, J.M. 1982. Response and tolerance of sorghum to temperature stress. Pages 143–159 *in*

Sorghum in the Eighties. Proceedings of the International Symposium on Sorghum, 2–7 Nov 1981, ICRISAT Center, India. Patancheru, Andhra Pradesh 502 324, India: International Crops Research Institute for the Semi-Arid Tropics.

Pratt, R.D., and Janke, G.D. 1980. Pathogenicity of three species of *Pythium* to seedlings and mature plants of grain sorghum. Phytopathology 70:766–771.

Shahid Shaukat, S., Ghazala, P., Khan, D., and Ahmad, M. 1985. Phytotoxic effects of *Citrullus colocynthis* (L.) Schrad on certain crop plants. Pakistan Journal of Botany 17:235–246.

Stoffer, R.V., and Riper, G.E. 1963. Effect of soil temperature and soil moisture on the physiology of sorghum. Agronomy Journal 55:447.

Watanabe, T., and Hashimoto, K. 1978. Recovery of *Gloeocercospora sorghi* from sorghum seed and soil, and its significance in transmission. Annals of the Phytopathological Society of Japan 44:633–640.

Williams, R.J., and Rao, K.N. 1980. A review of sorghum grain mold. Pages 79–92 *in* Sorghum diseases, a world review: proceedings of the international workshop on sorghum diseases, 11–15 Dec 1978, Hyderabad, India. Patancheru, Andhra Pradesh 502 324, India: International Crops Research Institute for the Semi-Arid Tropics.

Foliar Diseases of Sorghum

G.N. Odvody[1] and P.R. Hepperly[2]

Abstract

The major fungal pathogens of sorghum leaf blades are leaf blight (Exserohilum turcicum), target leaf spot (Bipolaris sorghicola), gray leaf spot (Cercospora sorghi), ladder spot (C. fusimaculans), rust (Puccinia purpurea), zonate leaf spot (Gloeocercospora sorghi), sooty stripe (Ramulispori sorghi) leaf anthracnose (Colletotrichum graminicola), rough leaf spot (Ascochyta sorghina), and tar spot (Phyllachora sacchari). The leaf sheath diseases are southern sclerotial rot (Sclerotium rolfsii), banded leaf and sheath blight (Rhizoctonia spp), and zonate leaf spot on sheaths. Many pathogens survive as mycelia, spores, or sclerotia within sorghum host residues, on or in soil. The sclerotia also exist freely in soil and, depending on the fungus, germinate to produce initial inoculum of mycelia or conidia. Initial inoculum of the foliar pathogens is disseminated primarily by wind, rain and soil splash, and soil contact. Many pathogens survive on living weed hosts or their residues. E. turcicum, C. graminicola, and G. sorghi are pathogens of hosts other than Sorghum spp. Rhizoctonia spp and S. rolfsii are facultative parasites of several hosts. Economic impact of foliar pathogens should be assessed before controls are pursued. Host plant resistance is the primary control of foliar pathogens, but other controls may be effective and also complement and conserve existing resistance sources. Monogenic and polygenic resistance to most foliar pathogens has been identified and may be used individually or in combination to provide adequate disease control.

Introduction

Foliar diseases of sorghum are among the most common and recognizable diseases of sorghum [*Sorghum bicolor* (L.) Moench]. Extensive knowledge of some diseases and their causal agents is on record, and only a minimal amount is available on others.

This discussion deals only with sorghum foliar diseases caused by fungi. The fungal pathogens, common names of the diseases they cause, initial inoculum, and other basic facts are listed in Table 1. Foliar diseases caused by bacteria, the foliar phase of sorghum downy mildew, and nonfoliar phases of anthracnose are discussed in other papers. The many minor and other reported foliar pathogens of sorghum are enumerated in Tarr (1962) and the recent compendium of sorghum diseases (Frederiksen 1986).

Symptomology

Most foliar diseases of sorghum have very distinct symptoms, especially in mature lesions. Factors like multiple diseases on the same leaf, host pigment reactions, and other complex situations can obscure otherwise typical disease symptoms. Host pigmentation can dramatically affect the appearance of disease symptoms and influence disease assessment. On tan genotypes, the symptoms of disease are not so obvious and

1. Associate Professor, Plant Pathology, Texas A&M University Research and Extension Center, Route 2, Corpus Christi, TX 78410, USA.
2. Tropical Research Station, PO Box 70, Mayaguez, Puerto Rico 00709-0070.

Odvody, G.N., and Hepperly, P.R. 1992. Foliar diseases of sorghum. Pages 167–177 *in* Sorghum and millets diseases: a second world review. (de Milliano, W.A.J., Frederiksen, R.A., and Bengston, G.D., eds). Patancheru, A.P. 502 324, India: International Crops Research Institute for the Semi-Arid Tropics.

Table 1. Foliar pathogens of sorghum caused by fungi.[1]

Causal agent	Common name of the disease	Primary survival mechanisms	Initial inoculum	Inoculum dispersal	Non-sorghum hosts
Leaf blade diseases					
Exserohilum turcicum	Leaf blight	M+C in S+R, WH	C		Y/HS
Bipolaris sorghicola	Target leaf spot	M+C in R, WH	C		N
Cercospora sorghi	Gray leaf spot	M in R, WH	C	Wind	N
Cercospora fusimaculans	Ladder spot	M in R, WH?	C		N
Puccinia purpurea	Rust	WH	Urediospores		N
Gloeocercospora sorghi	Zonate leaf spot	SC in S+R	C		Y
Ramulispora sorghi	Sooty stripe	SC in S+R	C from SC		N
Ramulispora sorghicola	Oval leaf spot	SC in S+R	C from SC	Rain and	N
Colletotrichum graminicola	Leaf anthracnose	M in R, WH, SB	C from SC	wind	Y/HS
Ascochyta sorghina	Rough leaf spot	M+C? in R, WH	C from pycnidia		N
Phyllachora sacchari	Tar spot	WH	Ascopores		N
Leaf sheath diseases					
Gloeocercospora sorghi	Zonate leaf spot	SC in S+R	C from SC	Soil and	Y
Sclerotium rolfsii	Southern sclerotial rot	SC in S+R	M	soil	Y
Rhizoctonia spp	Banded leaf and sheath blight	SC+M in S+R	M	Splash	Y

1. C = conidia, HS = host specific, M = mycelia, N = no, R = host residue, S = soil, Sb = seedborne, SC = sclerotia, WH = weed host, Y = yes.

may lead to underestimation of disease incidence; conversely, lesions on red and purple genotypes are in greater contrast, and could lead to overestimates. Some foliar diseases (target leaf spot and gray leaf spot) have symptoms or symptom progression similar to those produced by other pathogens or other biotic and abiotic factors.

Most foliar pathogens of sorghum primarily attack the leaf blade but under epidemic and extended disease-conducive environments they may also attack leaf sheaths. Puccinia purpurea Cooke also attacks peduncles and Colletotrichum graminicola (Ces.) G.W. Wilson also attacks stems and seed. There are a few pathogens that primarily attack only the leaf sheath (Sclerotium rolfsii Sacc.) and secondarily attack the leaf blade (Rhizoctonia spp). Gloeocercospora sorghi D. Bain and Edg. has historically been observed principally on leaf blades but it may be more common on leaf sheaths where it goes unnoticed on basal leaf sheaths or is mistaken for charcoal rot (Odvody and Madden 1984).

New Diseases

The only new disease described since 1978 that is recurrent and has extensive geographic distribution is ladder leaf spot caused by Cercospora fusimaculans Atk. (Wall et al. 1987). Despite similarities in lesion size and shape, the scleriform lesions caused by *C. fusimaculans* (pale-brown with darkly-pigmented borders and ladder-like markings) are distinct from those of *Cercospora sorghi* Ell. & Ev. (discolored throughout with the host's pigmentation) (Tarr 1962). These two *Cercospora* spp also differ in morphology and pathogenicity (Wall et al. 1987).

Pathogen Survival

Many foliar pathogens of sorghum survive, in the absence of the host, as mycelia or spores within sorghum host residues on or in the soil. In some environments these and other pathogens are perpetuated on living weed sorghum hosts that provide a readily available source of inoculum for newly established susceptible sorghums.

Exserohilum turcicum (Pass.) Leonard and Suggs is known to form chlamydospores within cells of the conidium; these chlamydospores can survive in soil without host tissue, but their function as inoculum on sorghum has not been verified. A related pathogen, *Bipolaris sorghicola* (Lefebvre and Sherwin) Shoem, also produces chlamydospores and, although pathogenic, they were thought to represent a minor contribution to survival (Odvody and Dunkle 1975). Overwintered lesion residue of *B. sorghicola* often produced mycelium from the open ends of leaf veins incubated under high humidity in the laboratory (Odvody and Dunkle 1975).

The pathogens *Ramulispora sorghi* (Ell. and Ev.) L.S. Olive and Lefebvre, *Ramulispora sorghicola* Harris, and *G. sorghi* survive primarily as sclerotia on or in soil; the sclerotia can be within or free from host residue (Girard 1980). Coley-Smith and Cooke (1971) classified these sclerotia as sporogenic because they germinate by producing a mass of conidia similar to that later produced on foliar lesions. The other two sclerotial classifications of Coley-Smith and Cooke, based on type of germination, are myceliogenic (mycelia production) and carpogenic (production of sexual fruiting structure). The common leaf sheath blight organisms, *S. rolfsii* and *Rhizoctonia* spp can survive in soil as mycelia in tissue residue or as free myceliogenic sclerotia.

Phyllachora sacchari P. Henn. and *P. purpurea* survive within lesions on living weed sorghum hosts. *Ascochyta sorghina* Sacc., *C. sorghi*, and probably *C. fusimaculans* survive primarily as mycelia in lesions on living or dead tissue of crop and weed host sorghums (Dalmacio 1986; Tarr 1962).

Initial Inoculum and Dispersal

All known leaf blade pathogens of sorghum are dependent upon wind-dissemination of their initial and secondary inoculum, but some appear to be better-adapted to long-distance dispersal than those whose dispersal is more closely linked with the presence of free water. The conidia of *E. turcicum* are easily wind-disseminated, with most spores released in the morning hours (Meredith 1965). Leach (1980a, 1980b) demonstrated that release of conidia of *E. turcicum* (and *Bipolaris maydis*) from conidiophores was a "spore-discharge" influenced by infrared irradiation and changes (usually a reduction) in relative humidity and electrostatic forces. This phenomenon probably has implications for other foliar pathogens, e.g., *B. sorghicola*, *C. sorghi*, and *C. fusimaculans*, whose conidia are freely borne on conidiophores in an aerial environment. Though not aerially-borne on conidiophores, urediospores of *P. purpurea* are highly dependent on wind dispersal from erumpent uredia on living host plants that provide the source of initial inoculum.

Many of the other foliar pathogens produce inoculum in some kind of protective structure, probably in association with an external, viscous water soluble matrix. The initial inoculum of *P. sacchari* is ascospores produced within locules of stromatic fungal tissue containing paraphyses described as "slimy," which could indicate such a matrix (Tarr 1962). Pycnidiospores like those produced by *A. sorghina* are often associated with viscous liquids that may protect spores from dessication. This is evident when pycnidiospores are extruded from the pycnidium in a slimy cirrhus. The acervuli of *Colletotrichum graminicola* (Ces.) G.W. Wilson also produce a protective mucilaginous matrix in which conidial masses are borne (Ramadoss et al. 1985). The matrix functions to protect conidia from dessication, and with free water, wind, and rain-splash allows rapid dispersal of conidia to other infection sites. Conidial spore masses of *G. sorghi*, *R sorghi*, and *R. sorghicola* produced from sclerotia or on leaf lesions are borne in a similarly functioning water-soluble matrix (Olive et al. 1946; Bain and Edgerton 1943). Under dry conditions, single conidial masses of *G. sorghi* and *R. sorghi* are held so tightly together that their removal from lesions is possible only en masse, but when the mass is placed into free water, the conidia immediately disperse as individual spores (Odvody, unpublished observation).

The spore-dispersal mechanisms are consistent with the epidemiology of these pathogens, especially the latter group that form spores in protective structures and matrices. They have high free water requirements for both inoculum dispersal and initial infection.

The initial inoculum of *S. rolfsii* and *Rhizoctonia* spp causing blights of the leaf sheaths is primarily mycelial, and derives probably from sclerotia and other colonized substrate, because infection usually begins on basal sheaths near the soil line (Odvody and Madden 1984; O'Neill and Rush 1982). *Rhizoctonia* spp may also produce basidiospores as initial and secondary inoculum (O'Neill and Rush 1982).

Host Range and Pathology Variability

Exserohilum turcicum and *C. graminicola* can be pathogenic on sorghum, maize (*Zea mays*), and other grasses, but naturally occurring isolates from host crops are generally genus-specific (Frederiksen 1980, 1984). *Gloeocercospora sorghi* is reported to attack sorghum, maize, and some other grasses, but more information concerning host specificity (Tarr 1962) is needed.

Sclerotium rolfsii and the *Rhizoctonia* spp (probably *R. solani*) causing sheath blights on sorghum are pathogenic to a wide variety of hosts, and might be described as facultative parasites (O'Neill and Rush 1982; Odvody and Madden 1984).

The other pathogens listed (Table 1) are generally regarded as occurring only on *Sorghum* spp. Isolated reports of some of these pathogens on other hosts could be the result of error in identification of the pathogen, saprophytic instead of parasitic attack, or extremely unusual conditions allowing infection. The large number of minor, seldom-reported diseases of sorghum could be due to similar factors, especially as senescing sorghum leaves become vulnerable to weak pathogens and saprophytes (Frederiksen 1986).

Of the pathogens listed in Table 1, only *C. graminicola* (Frederiksen 1984) and *P. purpurea* (Bergquist 1974) currently have pathotypes (races) described on sorghum.

Geographic Distribution

Frederiksen (1982), S.B. King, and N.V. Sundaram classified the major diseases of sorghum on the basis of prevalence and severity in temperate (outside 34° latitude), subtropical (between 23°15′ and 34° latitude), and tropical (within 23°15′ latitude) regions. According to their classification, the foliar pathogens *E. turcicum, C. graminicola, C. sorghi, P. purpurea, G. sorghi, R. sorghi,* and *A. sorghina* are all commonly or generally found on sorghum grown in the subtropics and on sorghum grown in the tropical lowland during summer. Occurrence of these pathogens was less consistent in the cooler temperate and tropical highland or tropical winter environments. The pathogens *C. sorghi, E. turcicum,* and *P. purpurea,* easily wind-disseminated, are apparently the most consistent in their occurrence and incidence across all these diverse sorghum-growing environments. *Bipolaris sorghicola* also occurs, but only occasionally, in all these environments. The potential misidentification of *B. sorghicola* lesions as those of *C. sorghi* could partially account for its lower reported occurrence.

The reported observations of *C. sorghi* may also include occurrences of the newly described *C. fusimaculans. Ramulispora sorghicola* seems to be restricted to warmer conditions in the subtropics and tropical lowland summers, and it usually does not have a high disease incidence. *Phyllachora sacchari* is limited to high-rainfall tropical areas (Dalmacio 1986), possibly due in part to requirements for a living weed-host reservoir for significant pathogen survival and the need for wet conditions for infection, disease development, and inoculum dispersal.

Host : Parasite Interaction

Initial infection of sorghum by foliar pathogens requires high moisture conditions for specific periods to allow spore germination and penetration of host tissue. As wind and rain-splash are the important modes of dispersal of *G. sorghi*, *R. sorghi*, *R. sorghicola*, *A. sorghina*, *C. graminicola*, and *P. sacchari*, host proximity is more important than with *E. turcicum*, *B. sorghicola*, and the *Cercospora* spp.

These pathogens can occur on younger plants, but in environments favoring disease most of them cause greater damage to older plants because of a greater susceptibility in postanthesis foliage and the concomitant increase in inoculum for infection. For most of the pathogens, individual lesion development is less restricted on older foliage of susceptible sorghum. The latent periods between infection, lesion appearance, and sporulation are shorter than on younger tissue of the same cultivar. Lesion size also increases on older leaves; delimitation by leaf veins is reduced and lateral and longitudinal coalescence of lesions is more common.

Often, the spread of a pathogen from one or a few initial lesions is evident through the occurrence of younger lesions on the same leaf or surrounding leaves of the same or other plants. Those pathogens disseminated by wind and rain-splash, like *G. sorghi* and *R. sorghi*, often produce a series of coalescing leaf-margin lesions beneath an older lesion on the upper part of the same leaf.

After initial infection, hyphae of *E. turcicum* invade leaf vessels and cause a localized wilt as the lesion extends first longitudinally within vessels then laterally to give a fusiform appearance (Frederiksen 1980). Susceptibility to *E. turcicum* is reported to decrease with sorghum maturity (Frederiksen 1980), but the occurrence of leaf blight on leaves of all ages in the field suggest that this host : pathogen interaction is more complex.

Zonate leaf spot can occur as a seedling disease, but the typically small rectangular lesions, nonsporulating and highly pigmented, are indicative of greater resistance in young foliar tissue in comparison to older leaves. If economic damage occurs on young seedlings, it may be due in part to an overwhelming number of lesions incited by initial inoculum from soilborne sclerotia (Odvody, unpublished observation).

The type of damage that pathogens cause is somewhat related to their preferred colonization site on the host plant. Since it is the foliage of the plant that produces the energy for all other parts of the plant, foliar dysfunction at critical stages of plant growth can inhibit roots, stalks, and seed heads. The most direct effect of foliar pathogens might be from reduced photosynthesis, but foliar infections could increase leaf transpiration, reduce carbohydrate translocation to other organs, and/or increase uptake of carbohydrate and other nutrients from the other plant organs to dysfunctional leaves. Premature foliar and plant senescence is a logical consequence.

One of the sites of earliest and progressive dysfunctions of leaves by foliar pathogens involves the stomata, because many of the foliar pathogens produce large numbers of specialized fungal structures (conidiophores, fungal stromata, and sclerotia) directly beneath and extruded through the stomata. Rust causes a physical disruption of the epidermis during lesion development and may be more damaging per unit of lesion area than are some other pathogens.

Foliar diseases may dramatically affect other disease incidence and severity (e.g., increased stalk rots) on sorghum, possibly through direct effects on host physiology.

Sorghums with a high susceptibility to one or more foliar pathogens usually develop a higher disease incidence and severity much earlier in their maturity than the more resistant sorghums growing in the same environments. Differences in susceptibility may be unnoticed, however, if evaluations are conducted too late in the maturity of the sorghum. In some cases, the factors of environment and inoculum can be so overwhelming that differences in disease reaction are obscured.

Economic Impact Assessment

The economic impact of foliar pathogens is a topic of constant debate. Yield losses attributed to a particular pathogen will usually vary with sorghum-growing region and with environment (Frederiksen 1982). The direct effect of foliar pathogens can easily be related to loss of forage value in terms of killed foliage, but it is harder to demonstrate that foliar diseases cause a loss in other forage components (stalk, sugars, etc.) or reductions in grain yield. Meaningful loss assessment values require a standard assessment procedure, e.g., same or an equivalent cultivar either protected from or more resistant to the pathogen(s) being evaluated. Isogenic lines and fungicides have been used to evaluate the economic impact of several sorghum diseases (Craig 1982). For foliar pathogens, fungicides have generally been used to demonstrate variable loss effects (Hepperly, in press; Hepperly et al. 1987). These types of measurements are useful as they indicate potential disease loss but the results are often locational and environment specific, or valid only with cultivars having similar reactions to the pathogen(s).

Commonly, evaluation of the severity of foliar diseases involves an estimation of the leaf area visibly affected by the disease. There is often a lack of correlation between leaf area affected by disease and expected yield loss; foliar diseases of sorghum are no exception (Hepperly, in press). A pathogen at a low disease incidence may cause economic loss, but another occurring at a much higher incidence may cause no measurable loss. It is always more convenient to do single-disease evaluations at crop maturity, but periodic evaluations would be more meaningful. The occurrence of several foliar pathogens in the later growth stages of the plant make it more difficult to distinguish between cultivars differing in susceptibility or resistance at their earlier growth stages. This also makes it imperative to consider differences in cultivar maturities even when they have been sown on the same date and spatially exposed to the same pathogen and environmental stresses.

In Puerto Rico, multiple levels of control with specific chemicals have been utilized to create multiple disease levels in plants of the same genetic background (Hepperly et al. 1987; Hepperly, in press). Levels of chemical control, especially using the systemic fungicide oxycarboxin, can be related by regression analysis to yield levels. These responses can be used to make comparisons between genotypes with differing resistance levels. However, some fungicides, including systemic ones, give variable results and can affect other diseases or factors that influence yields.

Isogenic lines resistant and susceptible to a disease or diseases can be used to measure disease losses without the pleiotropic effect of fungicides; but they take several years to develop and are most practical with resistance governed by single major genes.

A modification of the isoline strategy uses F_3 families developed from susceptible by resistant

crosses and then separated by their disease reactions (Burton and Wells 1981). This strategy was successfully utilized in assessing loss associated with systemic sorghum downy mildew in Texas (Craig and Odvody 1985) and the Honduras (Wall, personal communication). The technique seems particularly adapted to evaluation of foliar diseases and should allow greater utilization of polygenic resistance factors. The populations can be developed rapidly, but choices for the parental crosses should consider other factors that may affect the assessment of disease loss—such as yield potential, vulnerability to other pest factors, and local adaptability.

Evaluation of Host-Plant Resistance

Several control approaches will minimize damage by foliar pathogens, but each must be viewed in the context of complementing host-plant resistance. The methods by which we evaluate host-plant resistance depend upon several interrelated factors. Some of these factors are (1) type of resistance desired or available in germplasm, (2) growth stage at which host expresses susceptibility or resistance to the pathogen, (3) regularity of occurrence of the disease(s) or a disease-conducive environment, (4) interference of environmental and pest factors with evaluation, (5) ease of inoculum production in culture and its pathogenicity and efficacy as inoculum in the field and in controlled environments, (6) representative nature of cultural inoculum to at least the local naturally occurring local pathogen populations, (7) efficacy of introduced or resident natural inoculum sources, and (8) opportunity for multilocational evaluation to determine effects of variable environments and variable pathogen populations.

The type of resistance sought against a pathogen is ultimately dictated by that existing in available germplasm, influenced in part by other desirable characters in such resistance sources, and by the degree of need for the resistance. With foliar diseases of sorghum there is often a greater need to develop moderate resistance and avoid high susceptibility, rather than develop high levels of resistance (Frederiksen and Franklin 1980). Sources of resistance to most foliar pathogens of sorghum have been identified and at least partially characterized (Williams et al. 1980; Frederiksen and Rosenow 1986). General leaf-disease resistance (polygenic) perhaps coupled, where needed, with pathogen-specific (monogenic or oligogenic) resistance has been successful in other crops and might be a valuable approach with sorghum (Hooker and Perkins 1980). Sorghums with the "stay-green" or nonsenescent characters (Duncan 1984; Rosenow 1984) may be valuable sources for general leaf-disease resistance.

Sorghum breeders often treat sorghum as a self-pollinating crop which is most amenable to identification of monogenic resistance, while population breeding—being dependent on both self- and cross-pollination—is useful in identifying monogenic and polygenic sources of resistance. Frederiksen and Rosenow (1986) suggest that resistance traits should be present at levels that can be readily evaluated and they must be heritable and stable across environments. They state that progress in disease-resistance breeding can be made using several methods, including pedigree line breeding, backcross breeding, and population breeding.

Resistance (monogenic or polygenic) that reduces the rate of disease development against all races of a pathogen effectively is especially needed for pathogens with known races (Frederiksen and Rosenow 1986). This type of resistance may also reduce selection pressure for variants of other foliar pathogens. General or durable resistance conferred by many genes can have several additive and synergistic effects in host response to pathogens but their collective response under disease-conducive environments is the important factor and more easily measured.

The world collections of landrace and other sorghums represent a vast and diverse resource containing all types of resistances to many foliar and other pathogens. Scientists have already identified a large number of elite disease-resistant materials and they are being used and evaluated in sorghum programs around the world. Yet, this represents only a small proportion of what is available.

There are a multitude of germplasm-screening techniques that range from fully natural disease development in the field to strictly controlled pathogen and environment conditions in the greenhouse and laboratory (Sharma 1980; Williams et al. 1980; Frederiksen and Franklin 1980). Mughogho (1982) stated that no screening techniques were available for sooty stripe and gray leaf spot. This was due largely to the lack of laboratory techniques to mass produce conidia for greenhouse or field inoculation. Odvody (1981) developed a sporulation technique for *Cercospora sorghi* that was successfully utilized in germplasm screening by Wall (1983). *Ramulispora sorghi* sporulates in almost a yeast-like fashion on potato dextrose agar if incubated under a continuous source of longwave UV light (Odvody, unpublished). The conidia can be serially transferred and streaked, as with bacteria, onto media for mass production.

Other Disease Controls

Destruction of infested host debris

The ecology of foliar pathogens is of considerable influence on how cultural practices and farming systems affect the incidence of specific diseases. Most foliar pathogens of sorghum are favored by tillage that permits pathogen survival in surface host-residue lesions. Control by minimalization of host residues is more effective with some pathogens than others. Pathogens disseminated primarily by wind, like *E. turcicum*, can more easily overcome distance limitations imposed by such practices (Frederiksen 1986). Those pathogens surviving as sclerotia in soil, e.g., *R. sorghi*, may be long-lived in the soil but the more rapid destruction of lesion debris increases the exposure of sclerotia to the deleterious soil microenvironment. *Colletotrichum graminicola* survives only in host debris, but its ability to attack more than one plant organ compensates somewhat for that dependency.

Destruction of collateral hosts

Puccinia purpurea, a wind-disseminated pathogen, must survive on a living host. Destruction of collateral hosts may at times be an effective control for *P. purpurea* and several other pathogens, but this practice is often impractical except in the proximity of field areas.

Crop rotation and disease escape

Crop rotations are especially effective where pathogen survival is dependent on debris, but the length of rotation necessary to substantially reduce soilborne pathogens surviving as sclerotia and other long-lived propagules may make this practice impractical. Selection of appropriate sowing dates is successfully used in many areas of the world to avoid or minimize foliar disease losses. The plants are still fully susceptible but the environment or the lack of inoculum may limit disease incidence. Examples include sowing in the dry season or when air temperatures are unfavorable for disease development and spread. Using taller genotypes or reducing plant populations can reduce disease spread, because the lower leaves (where most initial infections occur) are less exposed to environments favorable for infection.

Exclusion or elimination of the pathogen

Foliar pathogens, due to diverse survival mechanisms, are very hard to eradicate from sorghum-growing regions, especially if weed sorghums provide survival sites away from field locations. Only *C. graminicola* is known to be seed-transmitted, although others are sometimes reported as seed-surface contaminants. Certainly, most of the foliar pathogens are easily transported on infected sorghum forage. Such dissemination is usually prevented through routine quarantine procedures, but within geographic regions where diseases are already established there may be significant pathogen spread and introduction in this manner.

Chemical control

Foliar application of chemicals to control foliar diseases of sorghum is usually impractical, because control is either unwarranted or the cost exceeds benefits, especially with sorghum grown in stressful environments with low-cost inputs. Chemicals applied to seed to avoid the transmission of seedborne pathogens, such as *C. graminicola*, may be quite beneficial.

Biological control

The efficacy of many disease-control practices such as tillage, crop rotation, and host-residue destruction is dependent on biological processes that may accurately be described as biological controls. Without understanding their specific mechanisms, we manipulate, exploit, and enhance these natural processes. There is always a potential for the use of one organism to control another. There are many naturally occurring soilborne and aerial parasites of pathogenic fungi, but their isolation and use apart from specific natural habitats may not be feasible. *Eudarluca australis* Speg. is a fungal hyperparasite of rust fungi, including *P. purpurea*, but its occasional abundant appearance in rust pustules probably comes too late to have a practical effect on rust epidemiology (Frederiksen 1980). The potential of using low-virulence or avirulent strains of a pathogen to induce resistance in susceptible genotypes should be investigated for pathogens with known variability like *C. graminicola* (Dean and Kuc 1986).

Future Research Needs

1. Determine economic impact of foliar diseases on sorghum in all geographic regions and environments.
2. Develop foliar-disease resistance that is (a) effective across environments, (b) durable through reduced selection pressure for physiological variants within a pathogen population, and (c) active against several foliar pathogens, especially during vulnerable growth stages of the host.
3. Increase international cooperation in multi-locational evaluation and characterization of germplasm response to foliar pathogens and exchange of elite germplasm.
4. Increase efforts to locate, identify, and collect landrace and other sorghum germplasm sources endangered by adoption of introduced elite germplasm or other factors.
5. Develop locally-effective control alternatives to complement and conserve existing host-plant resistance.
6. Integrate field and controlled-environment screening and research techniques to simultaneously obtain correlative information concerning (a) host-pathogen interactions, (b) characterization of germplasm response to foliar pathogens, and (c) identity of sources of resistance.
7. Employ multidisciplinary team research to evaluate the physical and physiological fac-

tors associated with expression of host-plant susceptibility or resistance, pathogen responses to the host, and pathogen ecology apart from the host.

8. Study intensively the biological mechanisms influencing survival and function of pathogen propagules in soil and host debris.

9. Determine factors influencing natural and introduced epiphytic and other leaf microflora on host, pathogen, and host : pathogen interaction.

References

Bain, D.C., and Edgerton, C.W. 1943. The zonate leaf spot: A new disease of sorghum. Phytopathology 33: 220–226.

Bergquist, R.R. 1974. The determination of physiologic races of sorghum rust in Hawaii. Proceedings of the American Phytopathological Society 1:67

Burton, G.W., and Wells, H.D. 1981. Use of near-isogenic host populations to estimate the effect of three foliage diseases on pearl millet forage yield. Phytopathology 71:331–333.

Coley-Smith, J.R., and Cooke, R.C. 1971. Survival and germination of fungal sclerotia. Annual Review of Phytopathology 9:65–92.

Craig, J. 1982. Factors reducing sorghum yields: diseases. Pages 283–287 in Sorghum in the Eighties: proceedings of the International Symposium on Sorghum, 2–7 Nov 1981, ICRISAT Center, India. Patancheru, Andhra Pradesh 502 324, India: International Crops Research Institute for the Semi-Arid Tropics.

Craig, J., and Odvody, G.N. 1985. Assessment of downy mildew loss with near-isogenic sorghum populations. Phytopathology 75:1300. (Abstract).

Dalmacio, S.C. 1986. Tar spot. Pages 15–16 in Compendium of Sorghum Diseases, (Frederiksen, R. A., ed.). St. Paul, MN, USA: American Phytopathological Society.

Dean, R.A., and Kuc, J. 1986. Induced systemic protection in cucumber: time of production and movement of signal. Phytopathology 76:966–970.

Duncan, R.R. 1984. The association of plant senescence with root and stalk diseases in sorghum. Pages 99–110 in Sorghum root and stalk rots, a critical review: proceedings of the Consultative Group Discussion on Research Needs and Strategies for Control of Sorghum Root and Stalk Rot Diseases, 27 Nov–2 Dec 1983, Bellagio, Italy. Patancheru, Andhra Pradesh 502 324, India: International Crops Research Institute for the Semi-Arid Tropics.

Frederiksen, R.A. 1980. Sorghum leaf blight. Pages 243–248 in Sorghum diseases, a world review: proceedings of the International Workshop on Sorghum Diseases, 11–15 Dec 1978, Hyderabad, India. Patancheru, Andhra Pradesh 502 324, India: International Crops Research Institute for the Semi-Arid Tropics.

Frederiksen, R.A. 1982. Disease problems in sorghum. Pages 263–271 in Sorghum in the Eighties: proceedings of the International Symposium on Sorghum, 2–7 Nov 1981, ICRISAT Center, India. Patancheru, Andhra Pradesh 502 324, India: International Crops Research Institute for the Semi-Arid Tropics.

Frederiksen, R.A. 1984. Anthracnose stalk rot. Pages 37–42 in Sorghum root and stalk rots, a critical review: proceedings of the Consultative Group Discussion on research Needs and Strategies for Control of Sorghum Root and Stalk Rot Diseases, 27 Nov–2 Dec 1983, Bellagio, Italy. Patancheru, Andhra Pradesh 502 324, India: International Crops Research Institute for the Semi-Arid Tropics.

Frederiksen, R.A. (ed.) 1986. Compendium of Sorghum Diseases. St. Paul, MN, USA: American Phytopathological Society. 82 pp.

Frederiksen, R.A., and Franklin, D. 1980. Sources of resistance to foliar disease of sorghum in the International Disease and Insect Nursery. Pages 265–268 in Sorghum diseases, a world review: proceedings of the International Workshop on Sorghum Diseases, 11–15 Dec 1978, Hyderabad, India. Patancheru, Andhra Pradesh 502 324, India: International Crops Research Institute for the Semi-Arid Tropics.

Frederiksen, R.A., and Rosenow, D.T. 1986. Controlling sorghum diseases. Pages 65–67 in The Compendium of Sorghum Diseases, (Frederiksen, R.A, ed.). St. Paul, MN, USA: American Phytopathological Society.

Girard, J.C. 1980. A review of sooty stripe and rough, zonate, and oval leaf spots. Pages 229–239 in sorghum diseases, a world review. Proceedings of the International Workshop on Sorghum Diseases, 11–15 Dec 1978, Hyderabad, India. Patancheru, Andhra Pradesh 502 324, India: International Crops Research Institute for the Semi-Arid Tropics.

Hepperly, P.R. (In press). Sorghum rust I. Journal of Agriculture University of Puerto Rico 72(1).

Hepperly, P.R., Feliciano, C., and Sotomayor-Rios, A. 1987. Influence of panicle fungicides and harvest schedules on sorghum seed quality under tropical conditions in Puerto Rico. Journal of Agriculture University of Puerto Rico 71(1):75–83

Hooker, A.L., and Perkins, J.M. 1980. Leaf Blight—the state of the art. Pages 68–87 in Proceedings of the 35th Annual Corn Research Conference, Washington, D.C., USA: American Seed Trade Association.

Leach, C.M. 1980a. Vibrational releases of conidia by Drechslera maydis and D. turcica related to humidity and red-infrared radiation. Phytopathology 70:196–200.

Leach, C.M. 1980b. Evidence for an electrostatic mechanism in spore discharge by Drechslera turcica. Phytopathology 70:206–213.

Meredith, D.S. 1965. Violent spore release in Helminthosporium turcicum. Phytopathology 9: 65–92.

Mughogho, L.K. 1982. Strategies for sorghum disease control. Pages 273–282 in Sorghum in the Eighties: proceedings of the International Symposium on Sorghum, 2–7 Nov 1981, ICRISAT Center, India. Patancheru, Andhra Pradesh 502 324, India: International Crops Research Institute for the Semi-Arid Tropics.

Odvody, G.N. 1981. Sporulation techniques for Cercospora sorghi. Pages 80–81 in Proceedings of the Twelfth Biennial Grain Sorghum Research and Utilization Conference, 25–27 Feb 1981, Lubbock, TX. Texas Grain Sorghum Producers Board, Box R, Abernathy, TX 79311, USA.

Odvody, G.N., and Dunkle, L.D. 1975. Occurrence of Helminthosporium sorghicola and other minor pathogens of sorghum in Nebraska. Plant Disease Reporter 59:120–122.

Odvody, G.N., and Madden, D.B. 1984. Leaf sheath blights of Sorghum bicolor caused by Sclerotium rolfsii and Gloeocercospora sorghi in South Texas. Phytopathology 74:264–268.

Olive, L.S., Lefebvre, C.L., and Sherwin, H.S. 1946. The fungus that causes sooty stripe of Sorghum spp. Phytopathology 36:190–200.

O'Neill, N.R., and Rush, M.C. 1982. Etiology of sorghum sheath blight and pathogen virulence on rice. Plant Disease 66:1115–1118.

Ramadoss, C.S., Uhlig, J., Carlson, D.M., Butler, L.G., and Nicholson, R.L. 1985. Composition of the mucilaginous spore matrix of Colletotrichum graminicola, a pathogen of corn, sorghum and other grasses. Journal of Agricultural and Food Chemistry 33:728–732.

Rosenow, D.T. 1984. Breeding for resistance to root and stalk rots in Texas. Pages 209–218 in Sorghum root and stalk rots, a critical review: proceedings of the Consultative Group Discussion on Research Needs and Strategies for Control of Sorghum Root and Stalk Rot Diseases, 27 Nov–2 Dec 1983, Bellagio, Italy. Patancheru, Andhra Pradesh 502 324, India: International Crops Research Institute for the Semi-Arid Tropics.

Sharma, H.C. 1980. Screening for leaf-disease resistance in India. Pages 249–265 in Sorghum diseases a world review: proceedings of the International Workshop on Sorghum Diseases, 11–15 Dec 1978, Hyderabad, India. Patancheru, Andhra Pradesh 502 324, India: International Crops Research Institute for the Semi-Arid Tropics.

Tarr, S.A.J. 1962. Diseases of sorghum, sudangrass and broomcorn. Kew, Surrey, England: Commonwealth Mycological Institute. 380 pp.

Wall, G.C. 1983. Development of a technique to screen for resistance to *Cercospora sorghi* in sorghum. M.Sc. thesis. Texas A&M University, College Station, TX, USA. 65 pp.

Wall, G.C., Mughogho, L.K., Frederiksen, R.A., and Odvody, G.N. 1987. Foliar disease of sorghum species caused by *Cercospora fusimaculans*. Plant Disease 71:759–760.

Williams, R.J., Rao, K.N., and Dange, S.R.S. 1980. The International Sorghum Leaf Nursery. Pages 269–283 *in* Sorghum diseases, a world review: proceedings of the International Workshop on Sorghum Diseases, 11–15 Dec 1978, Hyderabad, India. Patancheru, Andhra Pradesh 502 324, India: International Crops Research Institute for the Semi-Arid Tropics.

Nematode Pathogens of Sorghum

J.L. Starr[1]

Abstract

Plant-parasitic nematodes have only recently received recognition as pathogens of sorghum. The impact of nematodes on sorghum yield potential relative to other pest groups, however, is poorly documented. Growth suppression and sorghum yield loss due to nematode parasitism has been documented for nine species of nematodes in seven genera; an additional 22 species are known to parasitize sorghum. The associations of nematodes with sorghum are reviewed and emphases suggested for future research.

Introduction

Plant-parasitic nematodes are ubiquitous in environments that support vascular plants. All plant species examined to date are hosts (support nematode development and reproduction) to one or more species of nematodes. Similarly, nearly every plant species studied has been shown to suffer damage from the feeding activities of nematode parasites. However, not all nematodes parasitic (able to reproduce) on a given host are pathogenic (able to cause disease and suppress growth) to that host. Despite the universal occurrence of plant-parasitic nematodes, they are still frequently considered not to be of major importance for many crop species. Until recently sorghum was not considered to suffer significant damage due to plant-parasitic nematodes. Evidence documenting the host status of sorghum to numerous nematode species is accumulating; many of these nematodes are also pathogenic. Evidence of parasitism and pathogenicity by different nematode species on sorghum is reviewed.

Sedentary Endo- and Semiendo-Parasites

Eight species from four genera of sedentary nematodes with endo or semiendo feeding habits have been shown to be parasitic on sorghum (Table 1). Pathogenicity, however, has been demonstrated for only one species. *Meloidogyne incognita* suppressed dry weight accumulation of sorghum in greenhouse tests by 15%, field-test yields were suppressed by 33% (Orr 1967). Similarly, Thomas and Murray (1987) observed significant yield increases when field microplots infested with *M. incognita* at 490 eggs 500 cm^{-3} soil were treated with 1,3-dichloropropene or aldicarb.

Data on pathogenicity are lacking for the other *Meloidogyne* spp known to parasitize sorghum. It is known, however, that only selected populations of *M. naasi* are able to parasitize sorghum. Mitchell et al. (1973) separated *M. naasi* into five races based on differential parasitism of four indicator hosts, race 5 populations reproduced on sorghum.

1. Associate Professor, Department of Plant Pathology and Microbiology, Texas Agricultural Experiment Station, College Station, TX 77843 USA.

Starr, J.L. 1992. Nematode pathogens of sorghum. Pages 179–185 *in* Sorghum and millets diseases: a second world review. (de Milliano, W.A.J., Frederiksen, R.A., and Bengston, G.D., eds). Patancheru, A.P. 502 324, India: International Crops Research Institute for the Semi-Arid Tropics.

Similarly, several species of *Heterodera, Rotylenchulus,* and *Senegalonema* are known to parasitize sorghum (Table 1) but data are lacking on actual pathogenicity.

Migratory Endoparasites

Within the genus *Pratylenchus,* seven species have been reported as parasites of sorghum (Table 1). *Pratylenchus brachyurus* (Sharma and Medeiros 1982), *P. crenatus* (Motalaote et al. 1987), *P. hexincisus, P. neglectus,* and *P. scribneri* (Dickerson et al. 1978) are reportedly unable to parasitize sorghum. Thus, for *P. brachyurus* and

P. hexincisus there is some confusion because sorghum has been reported both as host and nonhost. Although Motalaote et al. (1987) did not detect significant variation among 10 sorghum cultivars with respect to parasitism by *P. brachyurus,* it is possible that variation among sorghum genotypes and/or *Pratylenchus* populations could explain the apparent discrepancies regarding parasitism of sorghum by some species.

Pathogenicity of two species of *Pratylenchus* has been documented for sorghum. Norton (1958) reported that initial populations of 65 *P. hexincisus* individuals L^{-1} of soil suppressed sorghum dry weight accumulation by 20% in

Table 1. Genera and species of nematodes known to be parasitic or pathogenic on *Sorghum bicolor.*[1]

Nematode	Parasitism	Pathogenicity	Citations/References
Sedentary endoprasites			
Meloidogyne incognita	+	+	Birchfield (1983), Orr (1967)
M. javanica	+		Sharma and Medeiros (1982)
M. naasi	+		Coetzee and Botha (1965)
Heterodera gamiensis		+	Merny and Netscher (1976)
H. sorghi	+		Jain et al. (1982)
Rotylenchulus borealis		+	Vovlas and Inserra (1982)
R. parvus	+		Page et al. (1985)
Senegalonema sorghi	+		Germani et al. (1984)
Endoparasites			
Pratylenchus brachyurus	+	-	Motalaote et al. (1987), Endo (1959)
P. coffeae	+		Motalaote et al. (1987)
P. crenatus	-		
P. hexincisus	+	+	Norton (1958)
P. penetrans	+		Marks and Townsend (1972)
P. sudanensis	+		Yassin (1973)
P. thornei		+	O'Brien (1982)
P. zeae	+	+	Motalaote et al. (1987), Endo (1959)
Ectoparasites			
Criconemella sphaerocephal	+	+	Kinlock (1987)
Quinisulcius acutus	+	+	Cuarezam-Teran and Trevathan (1985)
Trichodorus allius	+	+	Smolik (1977)
Longidorus africanus	+	+	Lamberti (1969)
Tylenchorhynchus martini	+	+	Hafez and Claflin (1982)
T. nudus	+	+	Smolik (1977)

1. (+) Parasitism indicates reproduction on sorghum while (+) Pathogenicity indicates that growth or yield suppression due to parasitism has been observed.

greenhouse tests. Furthermore, he observed a disease complex between *Macrophomina phaseolina* and *P. hexincisus*; the effects of the two pathogens on sorghum were additive. Both Cuarezam-Teran and Trevathan (1958) and Motalaote et al. (1987) reported that at initial populations of ca. 500 nematodes L^{-1} of soil, *P. zeae* would suppress growth of sorghum in greenhouse tests. Motalaote et al. (1987) also showed that the relationship between initial nematode populations and sorghum growth response fits a quadratic model (Fig. 1).

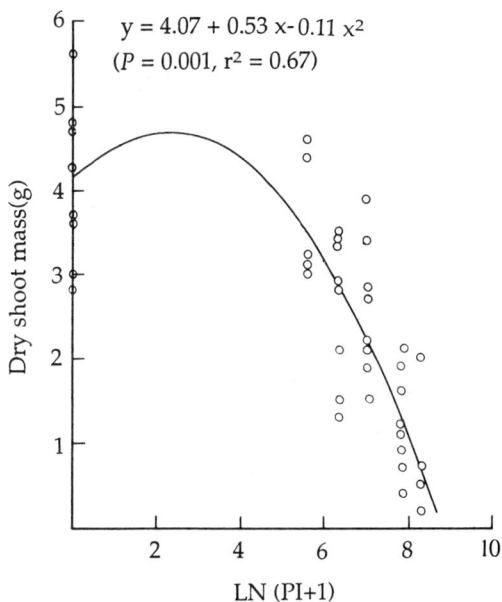

Figure 1. Growth response of sorghum cv Pioneer 8222 to infection by *Pratylenchus zeae* in greenhouse tests (reprinted with permission from Matalote et al. 1987).

In a comparison of sorghum response to infection by *P. brachyurus* and *P. zeae*, these workers noted that sorghum growth was unaffected by initial populations of *P. brachyurus* as high as 8000 individuals L^{-1} of soil in greenhouse tests. They also noted that sorghum was a relatively poor host for *P. brachyurus* and a relatively good host for *P. zeae*. They suggested that the rapid and extensive necrosis of sorghum seedling roots was a hypersensitive-type reaction which limited reproduction by *P. bra-*

chyurus. *Pratylenchus zeae*, in contrast, induced little necrosis of sorghum roots. In a separate study with different sorghum cultivars, Starr et al. (1984) reported that *P. zeae* did cause significant necrosis of sorghum roots, and that the hydrogen cyanide potential of infected tissues was reduced. Motalaote et al. (1987) also reported evidence for variation among sorghum lines in ability to support reproduction of *P. zeae*; even though all sorghum lines tested were hosts, there were significant differences in *P. zeae* reproduction among the different lines.

Johnson and Burton (1973, 1977) observed highly significant forage yield increases of pearl millet and sorghum/sudangrass hybrids in response to nematicide treatments. These tests were conducted in fields infested with a polyspecific nematode community, so it was not possible to identify which of the nematode species were actually responsible for the damage to the crops. However, the authors did suggest, based on observed nematode population dynamics, that *P. brachyurus* and *P. zeae* were causally involved in the suppression of crop growth.

Ectoparasites

As with the other nematode groups, several species of nematodes with ectoparasitic feeding habits have been associated with sorghum. In some cases it is difficult to assess the role of the ectoparasites because the data are from field tests involving polyspecific nematode communities, as was the case with the work of Johnson and Burton (1973, 1977). *Trichordus christiei*, *Criconemoides ornatus*, *Belonolaimus longicaudatus* and *Xiphinema americanum*, in addition to endoparasitic *Pratylenchus* species, infested the weed-free plots in these tests. All nematode species exhibited population increases, showing sorghum to be a host for these species. However, based on the relationships between nematicide treatments, yield responses, and nematode populations, Johnson and Burton (1973) concluded that *B. longicaudatus* was the only ectoparasite present that was pathogenic on the sorghum. The *Pratylenchus* species present were also pathogenic.

There is more concrete evidence of pathogenicity for other ectoparasitic species. Three species of stunt nematodes have been reported to be pathogens of sorghum. In 1977, Smolik reported significant growth suppression of sorghum in response to populations of *Tylenchorhynchus nudus* of ca. 400 individuals 100 cm⁻³ of soil in greenhouse tests. Furthermore, he noted that the effects of the nematode were greater under conditions of low versus high soil fertility. *Tylenchorhynchus martini* suppressed yield of sorghum at populations of 2000 to 5000 individuals 250 cm⁻³ of soil in field tests (Hafez and Claflin 1982) but not at initial populations of 750 to 1750 individuals 250 cm⁻³ soil (Todd and Claflin 1984). In both tests *T. martini* did reproduce on sorghum. Cuarezam-Teran and Trevathan (1985) reported that the stunt nematode *Quinisulcius acutus* caused significant root pruning and root necrosis (Fig. 2) and was thus parasitic and pathogenic on sorghum in greenhouse tests. They also examined the combined effects of *Q. acutus* and *P. zeae* on sorghum; the combined affects of the two species were less than additive. Their experimental design, however, was

inadequate to determine if a true negative (antagonistic) interaction occurred.

Of the ring nematodes, to date only *Criconemella sphaerocephala* has been shown to be pathogenic to sorghum. Kinlock (1987) reported a significant negative linear correlation between initial populations of *C. sphaerocephala* and sorghum height and grain yield in field tests. In these studies initial populations of approximately 300 nematodes 100 cm⁻³ of soil were required to cause a 10% yield loss.

Among the Adenophora two species have been shown to be pathogenic to sorghum. In greenhouse tests, *Trichodorus allius* at 125 individuals 100 cm⁻³ of soil suppressed growth of sorghum at 15–25°C but not at 10°C or 30–35°C (Smolik 1977). *Longidorus africanus*, at populations as low as 50 nematodes 500 cm⁻³ of soil, suppressed sorghum growth in greenhouse tests (Lamberti 1969). After 10 weeks populations had increased to as many as 750 *L. africanus* 500 cm⁻³

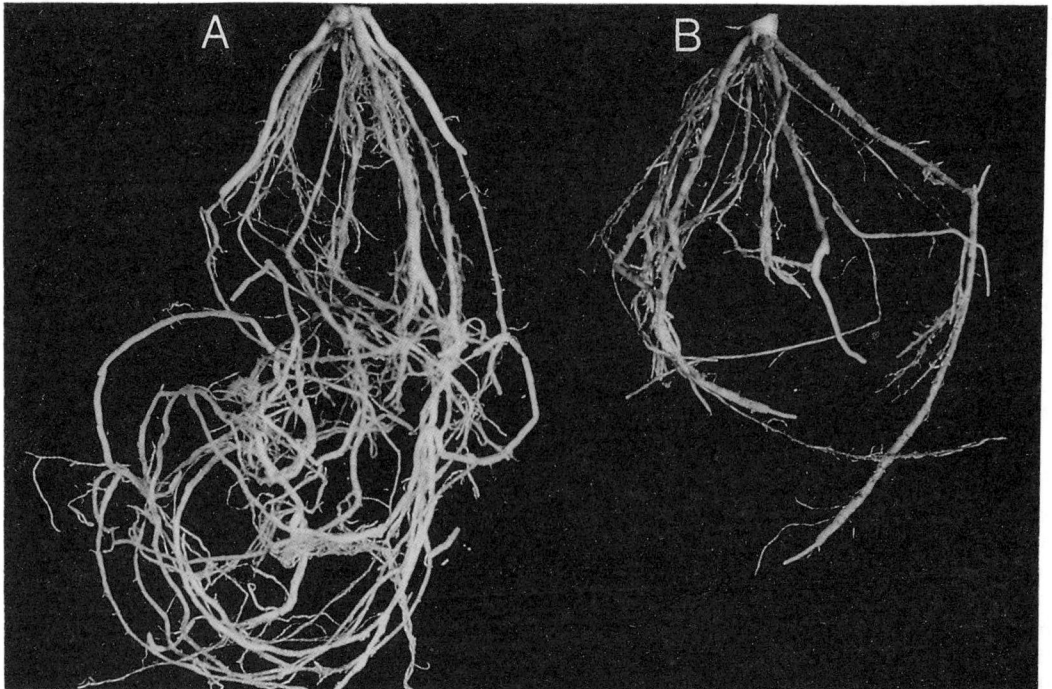

Figure 2. Root symptoms of *Quinisulcins acutus* on sorghum. A: noninoculated control. B: inoculated with 1000 nematodes L⁻¹ of soil.

of soil. Typical symptoms of *L. africanus* on sorghum included swollen and distorted roots.

Management Alternatives

Management of nematode populations to reduce crop losses in sorghum production systems may be difficult. Dramatic yield responses to nematicide treatments have been observed, but it is not likely that a chemical approach will ever be a practical or economically viable option for sorghum. Biological control of nematode populations does occur in nature (Kerry 1981), but the ability to manipulate such systems to practical degree is not likely for some time. Host resistance and cultural control, on the other hand, holds promise for alleviating damage by nematode pathogens to sorghum.

Resistance to the highly specialized sedentary endoparasites is likely to be relatively easy to identify; numerous crop species are known to contain sources of resistance to *Meloidogyne*, *Heterodera*, and *Rotylenchulus* species (Anonymous 1978). Resistance to the less specialized migratory ectoparasites will be more difficult to identify and may be more difficult to manipulate in traditional plant breeding programs. However, as indicated previously, the data available indicate that some variability for reaction to different nematode species does exist within the sorghum germplasm pool. Recent successes in identifying resistance to similarly nonspecific fungal pathogens and arthropod pests (Harris 1980) suggest that similar success awaits us with regard to nematodes if appropriate effort is put forth.

Crop rotation as a cultural control system can help alleviate nematode damage to sorghum. Successful and practical rotations require knowledge of nematode population dynamics on all crops in the rotation and works best if a single nematode species predominates. In some instances crop rotation can significantly reduce the impact of nematodes on crop productivity while in other instances the rotation may simply serve to maintain diversity in the ecosystem and perhaps avoid major problems that frequently accompany monoculture systems. Development of effective rotational systems is difficult due to the constraints of the environment, cultural biases, and availability of suitable alternate crops.

Because nematode damage is often related to plant stresses from other sources, minimization of these stresses will help reduce nematode damage. Improvements in soil fertility and reduction of water stress will enable sorghum to escape some nematode damage. Cultural practices to increase the organic content of soils are beneficial in that they can lessen the impact of nematodes.

Conclusion

These few reports provide ample evidence of many nematode species parasitic and pathogenic on sorghum. Additional species are likely to be identified with further study. The *Heterodera* and *Rotylenchulus* species now recognized as parasitic on sorghum will undoubtedly prove to be pathogenic in view of the aggressiveness of members of these genera on other plant species.

Because most nematode-related crop productivity problems lack diagnostic symptoms and the pest itself is not readily visible, nematodes are frequently an overlooked component of the pest complex attacking a particular crop. Although nematodes may not be a problem in every field in which crops are grown, their frequency distribution is sufficient to make nematodes, as a group, pests of significant impact for nearly all crop species.

At present there is a critical need to examine the parasitism of sorghum by numerous species of nematodes associated with sorghum in field surveys. The relative pathogenicity of these species must also be determined. Determination of the relationship between nematode populations and sorghum yield potentials under field conditions and the importance of particular nematode species relative to other pest groups that limit sorghum yields is of high priority.

In summary, plant-parasitic nematodes are suppressing sorghum yield potentials and probably to a greater degree than is currently recognized. Additional efforts are needed to identify the truly important nematode pathogens of sorghum and to document their impact on sorghum yields. Lastly, reduction of nematode-related yield losses will probably best be achieved by development of resistant or tolerant sorghum cultivars, by crop rotations, and by cultural practices which eliminate other stress factors.

References

Anonymous. 1978. Indexed bibliography of nematode-resistance in plants. Station Bulletin 639; Corvallis, Oregon: Agricultural Experiment Station, Oregon State University.

Aytan, S., and **Dickerson, O.J.** 1969. *Meloidogyne naasi* on sorghum in Kansas. Plant Disease Reporter 53:737.

Birchfield, W. 1983. Wheat and grain sorghum varietal reaction to *Meloidogyne incognita* and *Rotylenchulus reniformis*. Plant Disease 67:41–42.

Coetzee, V., and **Botha, H.J.** 1965. A redescription of *Hypsoperine acronea* (Coetzee, 1956) Sledge and Golden 1964. (Nematoda: Heteroderidae), with a note on its biology and host specificity. Nematologica 11:480–485.

Cuarezam-Teran, J.A., and **Trevathan, L.E.** 1985. Effects of *Pratylenchus zeae* and *Quinisulius acutus* alone and in combination on sorghum. Journal of Nematology 17:169–174.

Dickerson, O.J., Frantz, J.J., and **Lash, L.D.** 1978. Influence of crop rotation on nematode populations in Kansas. Journal of Nematology 10:284 (Abstract.)

Endo, B.Y. 1959. Response of root lesion nematodes, *Pratylenchus brachyurus* and *P. zeae* to various plants and soil types. Phytopathology 49:417–421.

Germani, G., Luc, M., and **Baldwin, J. G.** 1984. A new Rotylenculinae, *Senegulonema sorghi* n. gen., n. sp. (Nematoda: Tylenchida). Revue de Nematologie 7:49–56.

Hafez, S.L., and **Claflin, L.E.** 1982. Control of plant parasitic nematodes on grain sorghum and yield responses, 1981. APS Fungicide and Nematicide Tests 37:198.

Harris, M.K. (ed.). 1980. Biology and breeding for resistance to arthropods and pathogens in agricultural plants. MP-1451. College Station, Texas: Texas Agricultural Experiment Station. 603 pp.

Jain, K.K., Sethi, C.L., Swarup, G., and **Srivastava, A. N.** 1982. *Heterodera sorghi* n. sp., a new cyst forming nematode parasitizing sorghum. Revue de Nematologie 5:201–204.

Johnson, A.W., and **Burton, G.W.** 1973. Comparison of millet and sorghum-sudangrass hybrids grown in untreated soil and soil treated with two nematicides. Journal of Nematology 5:54–59.

Johnson, A.W., and **Burton, G.W.** 1977. Influence of nematicides on nematodes and yield of sorghum-sudangrass hybrids and millets. Plant Disease Reporter 61:1013–1017.

Kerry, B. 1981. Fungal parasites: a weapon against cyst nematodes. Plant Disease 65:390–393.

Kinlock, R.A. 1987. The population dynamics of *Criconemella sphaerocephala* and the nematode's relationship to sorghum yield. Journal of Nematology 19:536. (Abstract).

Lamberti, F. 1969. Pathogenicity of *Longidorus africanus* on selected field crops. Plant Disease Reporter 53:421–424.

Marks, C.F., and **Townsend, J.L.** 1972. Multiplication of the root lesion nematode, *Pratylenchus penetrans*, under orchard cover crops. Canadian Journal of Plant Science 52:187–188.

Merny, G., and **Netscher, C.** 1976. *Heterodera gambiensis* n. sp. (Nematoda: Tylenchida) parasite du mil et du sorgho en Gambie. Cahiers Office de la Recherche Scientifique et Technique d'Outre-Mer Série Biologie 11:209–218.

Mitchell, R.E., Malek, R.B., Taylor, D.P., and **Edwards, D.I.** 1973. Races of the barley rootknot nematode, *Meloidogyne naasi*. I. Characterization by host preference. Journal of Nematology 5:41–44.

Motalaote, B., Starr, J.L., Frederiksen, R.A., and **Miller, F.R.** 1987. Host status and susceptibility of sorghum to *Pratylenchus* species. Revue de Nematologie 10:81–86.

Norton, D.C. 1958. The association of *Pratylenchus hexincisus* with charcoal rot of sorghum. Phytopathology 48:355–358.

O'Brien, P.C. 1982. A study of the host range of *Pratylenchus thornei*. Australian Plant Pathology 11:3–5.

Orr, C.C. 1967. Observations on cotton root-knot nematode in grain sorghum in West Texas. Plant Disease Reporter 51:29.

Page, S.L.J., Mguni, C.M., and Sithole, S.Z. 1985. Pests and diseases of crops in communal areas of Zimbabwe. United Kingdom Overseas Development Administration Technical Report. 203 pp.

Sharma, R.D., and Medeiros, A.C. de S. 1982. Reacoes de alguns genotypos de sorgho sacarino aos nematoides, *Meloidogyne javanica Pratylenchus brachyurus*. Pesquisa Agropecuaria Brasileira 17:697–701.

Smolik, J.D. 1977. Effects of *Trichodorus allius* and *Tylenchorhynchus nudus* on growth of sorghum. Plant Disease Reporter 61:855–858.

Starr, J.L., Newton, R.J., and Miller, F.R. 1984. Presence of dhurrin in sorghum root tissue and the effect of pathogenesis on hydrogen cyanide potential. Crop Science 24:739–742.

Thomas, S.H., and Murray, L. 1987. Yield reductions in grain sorghum associated with injury by *Meloidogyne incognita* race 3. Journal of Nematology 19:559 (Abstract).

Todd, T.C., and Claflin, L.E. 1984. Evaluation of insecticide-nematicides for control of stunt nematodes on grain sorghum, 1983. APS Fungicide and Nematicide Tests 39:96.

Vovlas, N., and Inserra, R.N. 1982. Biological relationship of *Rotylenchulus borealis* on several plant cultivars. Journal of Nematology 14: 373–377.

Yassin, A. M. 1973. A root lesion nematode parasitic to cotton in Gezira. Cotton Growers Review 50:161–168.

Striga (Witchweeds) in Sorghum and Millet: Knowledge and Future Research Needs

A. T. Obilana[1] and K.V. Ramaiah[2]

Abstract

Striga spp (witchweeds), are notorious root hemiparasites on cereal and legume crops grown in the semi-arid tropical and subtropical regions of Africa, the southern Arabian Peninsula, India, and parts of the eastern USA. These weed-parasites cause between 5 to 90 % losses in yield; total crop-loss data have been reported. Immunity in hosts has not been found.

Past research activities and control methods for Striga are reviewed, with emphasis on the socioeconomic significance of the species. Striga research involving biosystematics, physiological biochemistry, cultural and chemical control methods, and host resistance are considered. We tried to itemize research needs of priority and look into the future of Striga research and control. In light of existing information, some control strategies which particularly suit subsistence and emerging farmers' farming systems with some minor adjustments are proposed. The authors believe that a good crop husbandry is the key to solving the Striga problem.

Introduction

Striga species (witchweeds) are parasitic weeds growing on the roots of cereal and legume crops in dry, semi-arid, and harsh environments of tropical and subtropical Africa, Arabian Peninsula, India, and a small part of USA. In some parts of Africa the profusion of witchweeds have serious impact on the socioeconomic life of farmers. Heavily infested farms are abandoned and occasional migrations of farming communities because of Striga has been reported. The statement that Striga is a new threat to Africa's food crops is not so. It is endemic in Africa's cereal and legume food crop production.

The weed is parasitic to cereals and legumes, but there is significant variation in the reaction of different crop species. Because Striga has evolved, parallel with sorghum, over the centuries, the indigenous crop has developed the whole spectrum of tolerance (on average about 60%), susceptibility (in about 30%), and resistance (in about 10%). On the other hand, in maize, susceptibility has been the common reaction as resistant varieties are still being identified and confirmed. The reaction of millet is complex, with ecological zone implications. Resistance to Striga has not yet been found in millet, even though millet coexists with sorghum in some environments. Reaction of rice to Striga is not well known, but indications are that susceptibility to Striga parasitism is normal in rice.

The cowpea, Vigna unguiculata, is the legume most affected by Striga in dry areas of Africa, where it is a common food plant.

Being an obligate hemiparasite, Striga causes tremendous damage to the host plants before Striga emerges from the soil. Persistent drought worsens the situation. Little attention is paid to it or its control. As a result the Striga problem is one of the most serious production problems on cereals and cowpeas south of the Sahara.

1. Principal Sorghum Breeder, SADCC/ICRISAT Regional Sorghum and Millets Improvement Program, PO Box 776, Bulawayo, Zimbabwe.
2. Principal Cereals Breeder, ICRISAT West African Sorghum Improvement Program, B.P. 320, Bamako, Mali.

Obilana, A.T., and Ramaiah, K.V. 1992. Striga (witchweeds) in sorghum and millet: knowledge and future research needs. Pages 187–201 in Sorghum and millets diseases: a second world review. (de Milliano, W.A.J., Frederiksen, R.A., and Bengston, G.D., eds). Patancheru, A.P. 502 324, India: International Crops Research Institute for the Semi-Arid Tropics. (CP 741).

African farmers on impoverished soils in the marginal areas are subsistence-oriented and less resourceful, possibly for survival reasons. They recognize the *Striga* problem, but have no simple control measures to solve it. Costly control methods are available, and are used by farmers in countries like USA. In developing countries, progressive farmers who can afford to apply fertilizers and follow good agronomic practices have fewer problems with *Striga*. Even in this situation, use of excess fertilizers in very dry weather may be impracticable. Therefore, the task ahead, to improve the overall productivity of the farm land and transform subsistence and emerging farmers into entrepreneurs, is tremendous. We believe a good crop husbandry is the key to solving the *Striga* problem.

We review historical developments of formal and informal work on *Striga* reemphasize its socioeconomic implications in terms of its continuous spread and its cost in crop loss, and discuss the latest developments in *Striga* research. The research needs of priority are itemized and the future of *Striga* research and control is considered.

Finally, control strategies particularly suited to the subsistence and emerging farmers' farming systems with some minor adjustments are proposed.

Socioeconomic Implications

Why research on *Striga*? The answer to this question lies in the damage done by *Striga* in terms of the farming systems, crop losses, and farm losses (abandonment) in regions of semi-arid Africa and India where the parasitic weed is endemic. In these *Striga*-infested regions, the environment is so harsh and marginal for crop productivity that only a few drought-resistant staple crops are grown—sorghum, millets, cowpeas, and some maize—and they are extremely parasitized. Food shortages are routine.

The socioeconomic impact of *Striga* infestation on cereal crops in western Africa was assessed by Obilana (1983a) and Ramaiah (1984).

The success of *Striga* as a parasitic weed is due to several of its characteristics, related somehow with the farming systems in semi-arid areas where its hosts are grown. *Striga* seeds survive in arid soils for 15 years. The number of seeds produced per plant ranges from 40 000 to

500 000 for *S. asiatica* (several authors) and 25 000 for *S. forbesii* (Obilana et al. 1988, pp. 342–364). Seeds, small in size, are efficiently dispersed by man (in use, transport by machinery, and seed movement), by animals (in droppings), and by water (in field erosion). Extensive longevity, together with ability to form "biotypes," "ecotypes," and "crop-specific" types and its ease of dispersal has made necessary serious and significant farming systems changes; several small African tribes and family groups have migrated from location to location because of *Striga*. Several cases of farm abandonment or change in cropping patterns have been reported in southern Africa (Obilana et al. 1988, pp. 342–364). Some 30 to 40% of total farmlands have been abandoned to sorghum or maize cultivation in some countries in western and southern Africa.

In terms of crop yields, *Striga* damage has been most significant. In the eastern African region, grain yield loss for susceptible sorghum varieties was estimated to be 59% (Doggett 1965). For western Africa, Obilana (1983b) and Ramaiah (1984) recorded actual yield losses in sorghum due to *Striga* damage. Obilana reported 5% loss of potential yield in resistant cultivars, 95% in susceptible varieties; and 45–63% in tolerant sorghums. Ramaiah (1984) reported yield-loss estimates of 10–35% in experimental plots. Where *Striga* infestation is intense and varieties are susceptible, 100% crop losses in farmers' fields are common. However, in most farmers' plots with mostly tolerant sorghums and millets, these crops coexist with *Striga*. In terms of actual monetary values, *Striga*-caused losses between U.S. $28 million and $87 million annually, are suffered by western African farmers (Obilana 1983a). Although actual values have not been reported for southern Africa, estimated grain-yield losses due to *Striga* could reach between 15 and 95% in sorghum, millet, and maize varieties and hybrids.

Sorghum, millets, and maize are principal staple foods in most countries of Africa. Considering that cultivation of these three cereals occupies 54.6 million ha and produces 54.3 million t of grain the approximate value of these crops (in U.S. dollars) is about $12.438 billion (Ramaiah 1984). As these three crops are major hosts of *Striga* in Africa, and it is known that *Striga* can cause yield losses of 100%, this figure could represent the cost of *Striga* infestation in all African maize, sorghum, and millets fields. Recently, Vasudeva Rao et al. (In Press) estimated mean grain yield loss in sorghum in India to range from 14.7 to 32.0% in the rainy season and 21.9 to 84.5% in the postrainy season. They write that potential loss could be total for the crop, with average of 19.7% grain yield loss in hybrid sorghums due to *S. asiatica*. Assuming only 10% of the hybrid sorghum crop is affected, losses of about 75 000 tons, valued at about U.S. $7.5 million occur annually.

It is possible that *Striga* could spread to additional areas, especially with the persistent occurrence of drought. The parasitic weed is not yet present everywhere. Unless effective integrated control measures are taken as soon as possible, *Striga* could become a serious threat to the cultivation and productivity of all rainfed cereals in the semi-arid tropics (SAT) of Africa.

Similarly alarming is the prospect for legumes—especially cowpeas. The threat posed to cowpea by *Striga* is most serious in the semi-arid savannas of western Africa and the dry arid lands of southern Africa. Although exact values of economic losses due to *Striga* in cowpeas have not been recorded, guestimates and visual observations indicate 50–100% yield loss in severe infestations (Obilana 1987). Spread of the cowpea *Striga* into wetter grassland and veld areas has been rapid, confounding the situation, and causing much concern. Economic implications of *Striga* damage on cowpeas becomes more significant in view of the role of the crop in mixed-cropping systems used in the semi-arid and arid savanna and veld regions of Africa.

Striga Research and Control

Early research

Several species of *Striga* were recognized as serious parasitic weeds as early as 1900 in India, Africa, and parts of USA. Parker (1983) traced the research history of this parasitic weed, and summarized part of the problem.

As early as 1905, Burtt-Davy in southern Africa described witchweeds in the Botanical Notes of the Transvaal Agricultural Journal, recognizing the species as parasitic. Within a decade, experiments with *Striga* were conducted in the region. Work on control methods—including

agronomic and cultural practices, e.g., fertilizer use and crop rotations, were reported by Pearson (1913). The earlier works concluded that crop rotations, catch cropping using sorghum and sudangrass (Timson 1931), and nitrate fertilizer were valuable in controlling *Striga*.

Saunders (1933) classified the biology of *Striga asiatica*, studied the use of catch crops (including the trap cropping), and pioneered the selection of resistant varieties in sorghum. This last activity identified several "resistant" varieties, including "Radar" (Saunders 1942). The resistance in "Radar" was complex, as it was reported to break down, and then again *Striga*-free in recent studies (Riches et al. 1987, pp. 358–372).

The use of chemicals to control witchweeds was an early objective. Inorganic herbicides, like sodium chlorate, were found to selectively kill *S. asiatica* in maize (Timson 1934).

Present research

Current research into *Striga* has sought using improved techniques to update and confirm earlier findings; come up with the biosystematics and classification of all types of *Striga* in Africa; continue surveying spread of known types and possible new types; and continue studies seeking environment-specific integrated *Striga*-management procedures.

Following the findings of earlier workers, after a lull of about two decades, Visser and Botha (1974), in their chromatographic investigations on *Striga* seed germination, found that crop-root exudates stimulate germination of *Striga* seeds, and that the stimulant substances can be separated readily by high-pressure chromatography. Their work illustrated the involvement of several stimulant substances, which need to be identified and characterized.

Latest Achievements

Musselman's recent book on *Striga* (1987) provides "state of art" knowledge on various aspects of the parasitic weed. The attempt here is only to highlight the latest developments, rather than to undertake an extensive review.

189

Taxonomy and biosystematics

In spite of the economic importance of this genus, relatively little is known about its taxonomy and biosystematics. The number of species is not known. According to Raynal-Roques (1987) there are about 36 species. Musselman (1987) described 30 species, of which 24 are found in western and central Africa alone (Raynal-Roques 1987). Earlier Musselman (1987) indicated 23 *Striga* species, suggesting that the African region may be its center of diversity. Lesser centers are the southern Arabian Peninsula and India. In a recent summary of herbarium documents, Riches et al. (1987, pp. 358–372) and Obilana et al. (1988, pp. 342–364) found reports of nine of the African *Striga* species to be occurring in southern Africa. The distribution and possible hosts of *Striga* spp found in Africa are listed in Table 1.

Table 1. Distribution and possible hosts of African species of *Striga*.

Species	Distribution	Host	Remarks
S. hermonthica (Del.) Benth.	Western Africa (especially in Nigeria, Ghana, Burkina Faso, Niger, Chad, Mali, Senegal, and Mauritania); Eastern Africa (especially in Sudan, Ethiopia, Yemen, Kenya, and Uganda); and the SADCC (especially in Angola, Tanzania, and Mozambique)	Sorghum, pearl millet, maize, and wild grasses	
S. asiatica (L.) Kuntze	Eastern Africa (Ethiopia, Kenya), more countries in the SADCC region. Very limited occurrence in western Africa, Burkina Faso (yellow type)	Sorghum, pearl millet, finger millet, maize, upland rice, sugarcane, and wild grasses	Red-flowered types mainly, with occasional orange and yellow forms
S. gesnerioides (Willd Vatke = (*S. orobanchioides*)	Western Africa (Nigeria, Niger, Burkina Faso, Mali, and Senegal); SADCC region (Botswana, Malawi, Swaziland, Tanzania, Zimbabwe, and Angola?)	Cowpea, tobacco, convolvulaceae, and fabaceae	
S. aspera (Willd) Benth	Malawi, Tanzania, Mozambique, Burkina Faso, and Nigeria	Rice, wild grasses, and occasionally maize, sorghum, and sugarcane	
S. euphrasioides Benth = (*S. angustifolia* (DON) Saldanha)	Malawi, Tanzania, Zimbabwe, Zambia, and Mozambique	Sorghum, maize, sugarcane, upland rice, and wild grasses	
S. forbesii Benth	Tanzania, Botswana, Zimbabwe, Swaziland, and possibly in Angola, Zambia, and Mozambique	Sorghum, maize, rice, and a few wild grasses	

Continued

Table 1. *Continued*

Species	Distribution	Host	Remarks
S. bilabiata (Thunb.) Kuntze	In the eastern African lakes region and across western Africa in the Sahel and Sudanian zones; Zimbabwe	Mostly on wild grasses	
S. elegans Benth	Tanzania, Kenya, Botswana, South Africa, Zimbabwe	Wild grasses	
S. macrantha Benth	Sudan, Zimbabwe and Sudanian zone of western Africa	Wild grasses	
S. aequinoctialis Chev. ex Hutch. and Dalz	Burkina Faso, Mali, Nigeria, Niger, and Chad (western Africa)	Wild grasses	
S. klingii (Skann) Engler = (*S. Dalzielii* Hutch.)	Burkina Faso	Sorghum, millet, and wild grasses	With small bluish flowers
S. junodkii Schinz	Southern Africa, including Mozambique		
S. hallaei A. Raynal	Gabon, Zaire		
S. fulgens Hepper	Tanzania		
S. elegans Benth	Tanzania, Kenya, Botswana, and South Africa	Cereals and wild grasses	
S. chrysantha A. Raynal	Zaire and Central African Republic		
S. brachychalyx Skan	Sahelian and Sudanian zones of Africa	Cereals and wild grasses	
S. baumanii Engler	Zaire, Kenya, and western Africa		
S. latericea Vatke[3]	Kenya, Tanzania, Ethiopia, and Somalia	Sugarcane and cereals	With vegetative propagation
S. ledermannii Pilger	Cameroon		
S. linearifolia Hepper (Schumach et Thona.) = (*S. strictissima* Skan) = (*S. canescens* Engl)	Western Africa		
S. primuloides Chev	Côte d'Ivoire		
S. pubiflora Klotzsch (=*S. agnustoifolia*?) (=*S. zanzibarensis* Vatke?)	Ethiopia and Somalia		Status not confirmed

Information about this genus is not sufficient to provide comprehensive keys or descriptions, but Raynal-Roques (1987) has made an excellent attempt for the species found in western and central Africa. A *Striga* identification booklet, describing the species of agronomic importance, was earlier published (Ramaiah et al. 1983), followed by a recent pamphlet (Obilana et al. 1987). There is considerable confusion in distinguishing closely related species complexes such as *Striga hermonthica - S. aspera - S. curviflora*; and *Striga euphrasioides - S. angustifolia - S. pubiflor - S. zanzibarensis*. Fairly detailed work and discussions on taxonomic and biosystematic issues of *Striga* species in southern Africa have been presented by Ralston et al. (1986), Riches et al. (1987, pp. 358–372) and Obilana et al. (1988, pp. 342–364).

The chromosome number of *Striga* species is not well studied. Interspecific relationships and the origin of different species are yet to be understood. Electrophoretic studies, chromosome studies, and interspecific hybridization should provide useful information on these aspects (Musselman, personal communication). The preliminary results on crosses between *S. hermonthica* and *S. aspera* indicate that each is cross-compatible in either direction. The sympatric nature of *S. asiatica* (red-flowered) and *S. forbesii* observed in the southern African region needs study.

The floral biology of these *Striga* species needs more study. There are two distinct reproductive systems—autogamy and allogamy—in this genus (Musselman and Parker 1983). Autogamy appears to be widespread compared to allogamy. Autogamy characterized by a mass of pollen persisting on the stigma was observed in *S. asiatica, S. gesnerioides, S. bilabiata, S. angustifolia, S. forbesii, S. passargeii* and *S. densiflora* (Musselman 1987). Allogamy appears to be restricted to *S. hermonthica* and *S. aspera*, the two closely related species. In recent field surveys in Burkina Faso, we observed several pollen vectors—butterflies, bee flies, and moths—on these two species. They are now being identified at the Commonwealth Institute of Entomology. Though we (Obilana et al. 1988, pp. 342–364) confirmed autogamy in *S. forbesii* by field-cage studies, the exerted stigmas observed in few plants seemed to be persistent. This characteristic, in addition to *S. forbesii*'s bright long-lasting corollas, can be considered evidence of outcrossing (allogamy).

Physiology

Many aspects of *Striga* biology are not fully understood. They include the physiological processes of photosynthesis, respiration, transpiration, water relations, cause of heavy crop-yield reductions, morphology, and analoging of the haustorium in relation to its function. The overall physiology of resistance or tolerance to *Striga* in cereal species is included.

Some understanding of these aspects is provided by the excellent work of Stewart's group in University College, London (Press and Stewart, In Press; Shah et al. 1987; Press et al. 1987a,b). The rates of photosynthesis of *Striga* species were found to be very low in contrast to the respiration and transpiration rates of higher plants. Stomata of drought-stressed *Striga* plants were found to be open in darkness and their control is poor. This high rate of transpiration is interpreted as a means of maximizing solute transfer from host to parasite. Such reliance on high rates of transpiration has suggested use of antitranspirants in *Striga* control.

Field experiments in Sudan, using Synchemicals' antitranspirant wilt proof S600 spray, were very encouraging. Aerial shoots of treated *Striga* plants turned black after a few days and collapsed. On newly emerged shoots, the spray treatment caused blackening and collapse within 24 h. The straw and grain yields of sorghum increased by 25% and 18%, respectively. *Striga* yields were markedly reduced by 40% in 1985 and 20% in 1986. These results, if confirmed, will be very valuable; antitranspirants could effectively replace herbicides, and they are less hazardous to the environment.

Press and his colleagues demonstrated that sorghum varieties infected with *Striga* show massive reduction in photosynthesis and that *Striga* plants derive more than 35% of their carbon from the photosynthetic activity of the host plant. The loss of a substantial portion of the host plant's photosynthetic production and drought stress account for the reduction in grain and stover yield attributed to *Striga*. The reduction in photosynthetic activity of the host in the presence of *Striga* may be useful as a screening indicator of susceptibility.

Control Approaches

Cultural

Several cultural methods were reported to be efficient in controlling *Striga* species (Ramaiah 1987; Bebawi 1987; Ogborn 1987). Some include use of costly inputs such as fertilizers, some are labor-intensive, like hand weeding, and some involve minor adjustments in the cropping systems–such as crop rotations. Among these, hand weeding is most often recommended to Third World farmers and is also the one often rejected, mainly because of its low cost-benefit ratio. Extension workers emphasize the importance of hand weeding as soon as the *Striga* plants emerge, but most of the damage to the host has occurred while the *Striga* is still in the soil. Hand removal of *Striga* helps reduce population build-up, and thus is helpful in the long run, but farmers seldom think beyond 2 or 3 years ahead. In most nations of western and southern Africa, land ownership is nonexistent. Farmers seldom undertake long-term strategies to control *Striga* or improve the soil by anti-erosion measures, crop rotation, or trap cropping.

Recent experiments on farms in northern Burkina Faso have produced encouraging results on a single hand-weeding of *Striga* in millet fields just before harvest (Ramaiah 1985; ICRISAT 1986). The dramatic reduction of *Striga* populations in the following 2 years provided evidence that a majority of the seeds still on the *Striga* plants were destroyed when the plants were pulled and burned. Significant increases in millet yields occurred in the two cropping seasons that followed. Verification is expected in experiments now underway in Mali, Cameroon, and Ethiopia; the practice needs testing in Tanzania, Botswana, Malawi, Swaziland, and Zimbabwe. This method along with resistant varieties, fertilizers, and other cultural practices known to reduce *Striga* populations could be used very effectively as a supplementary control method. Hand weeding may then become more practical, and perhaps a common practice.

In Jordan, Syria, and Israel solar energy was used successfully for the control of *Orobanche*, a closely related parasitic species. The process is called soil solarization: it killed *Orobanche* seeds effectively when the soil was covered with polyethylene for about 40 days following summer irrigation. This was found to be very effective, like methyl bromide fumigation, but less toxic. It has been effective on other soilborne pathogens, including the pigeonpea fusarium wilt disease in India (ICRISAT 1987, pp. 177–178). Limitations include availability of water and of the large polyethylene sheets needed to cover the soil for about 40 days. It appears to have potential in countries like Sudan where *Striga* is a serious problem in the irrigated areas, and in the wetter savanna areas of Nigeria and Tanzania; rainfall here may be up to 1200 mm and the sorghum fields become carpeted with *Striga*.

Chemical

Excellent pre- and postemergence *Striga* herbicides are now available (Eplee and Norris 1987a). Goal® and 2,4-D are pre- and post-emergence applications, respectively. In addition, fumigants like methyl bromide and germination stimulants like ethylene gas effectively reduce *Striga* seed reserves in infested soils. Fumigants, however, are most expensive for use by the small-scale farmer.

Though herbicides are available to control *Striga* as it emerges or later to prevent seed production, using them does not prevent the damage inflicted by preemergent parasitic attachment. This limitation is now overcome by the discovery of systemic herbicides (Eplee and Norris 1987b). In USA, Dicamba® is used on sorghum and maize to kill *Striga* before crop yields are reduced. Dicamba® sprayed at 0.5 kg a.i. ha^{-1} on sorghum or maize no taller than 1 m gave excellent control of *Striga* without reduction in yield.

Host resistance

Resistant varieties offer an excellent approach to avoiding yield losses to *Striga* in subsistence farmers' fields. Breeding efforts at ICRISAT in Burkina Faso have developed resistant varieties acceptable to farmers. We now have identified a few varieties with very high levels of resistance to *S. hermonthica*. They include IS 6961, IS 7777, IS 7739, IS 14825, and IS 14928. Grain yields are very low in these varieties, and efforts are underway to transfer their resistance into high-yielding backgrounds. Screening for host resistance in sorghum at SADCC/ICRISAT Sorghum and Millets Improvement Program Bulawayo, found three cvs—SAR 19, SAR 29, and SAR 33—to be resistant to *S. forbesii*. These were earlier found resistant to *S. asiatica* (white-flowered) in India. SAR 19 and two others, SAR 16 and SAR 35, were resistant to *S. asiatica* (red-flowered) in Botswana.

Ramaiah (1987) presented a detailed review of host resistance to *Striga*. Except for finding these new sources of resistance, there is no major breakthrough in understanding resistance mechanisms, genetics of resistance, physiological variability within *Striga* species in relation to stability of host resistance, screening/infestation techniques, and evaluation for resistance. Comparison of root-distribution patterns of resistant and of susceptible sorghums (Housely et al. 1987) is of interest. Resistant varieties, like Framida and P 967083, have deeper root systems with significantly less total root length in the upper 30 cm of the soil core than the susceptible variety, Dabar. Babikar et al. (1987) observed that *Striga* seeds buried deeper than 24 cm undergo dormancy not easily broken by normal preconditioning. *Striga* seeds 5 to 10 cm below the soil surface are the most active. Thus host varieties with shallow root systems will be more exposed to active *Striga* seeds than those like P 96083 which have less of the root system in the *Striga*-active soil zone. Breeding for deep-rootedness in sorghum varieties may have an advantage for resistance to *Striga* and perhaps to drought stress as well.

More intensive efforts to develop high-yielding stable and broad-spectrum resistant varieties are desirable. This is possible only if we have reliable screening techniques to identify resistant plants and we know more about the biochemistry of the resistance mechanism, and about other factors conditioning resistance. Meanwhile, we have observed indications of multispecies resistance in some ICRISAT sorghum materials, especially SAR 19 with resistance to *S. asiatica* (white- and red-flowered) and *S. forbesii* (Obilana et al. 1988).

Control Strategies

It is now well recognized that no single method of control can effectively solve the *Striga* problem. *Striga* species are weeds and they are para-

sites. As weeds, they appear late in the season, and escape the normal early weeding operations. As parasites, they have the advantage of remaining out of sight (and frequently out of mind) while feeding on the host. The three major principles of *Striga* control are: (1) reduction of seed numbers in the soil; (2) prevention of new seed production; and (3) prevention of movement of seeds from infested to noninfested areas. Any control strategy should include an integration of at least one method from each of these three major principles (Table 2).

With minor adjustments in traditional farming

Mixed cropping and rotation. Mixed cropping of host and nonhost crops is a very common practice in western Africa; cereal crops (mostly sorghum) are grown mixed with trap crops such as cowpea, groundnut, cotton, soybeans, field peas, and sesame. Our recent laboratory experiments with these trap crops show that most of these do indeed act as trap crops. These crops are mixed and sown at random, instead of seeding and cultivating them as sole crops on areas of variable length and width in keeping with their importance in the mixture. However, sole cropping would permit application of *Striga* control methods to the cereal row area only, thus reducing the labor required and reducing the possibility of herbicide damage to the associated crops. It will also allow rotation of these strips, which is useful in two ways; trap crops, such as groundnut, will encourage abortive germination of *Striga* seeds in the soil. Secondly soil fertility is improved through fixation of atmospheric nitrogen. Other beneficial effects of trap crops in reducing *Striga* and improving the yields of cereals can be listed.

Subsistence farmers need their cereal staple each year. Crop rotation is an ideal practice; the cereal is grown on a different piece of land each year. This recommendation should meet with least resistance in central and western Africa.

A similar approach, together with crop rotation, could be used for *Striga* control in communal cereal farms in southern Africa where "monocropping" or "double cropping" is more common.

Table 2. *Striga* control methods described under three principles.

I Reducing number of *Striga* seeds in soil	II Preventing production of new seeds	III Preventing spread from infested to noninfested soils
1. **Cultural**	1. **Cultural**	1. **Cultural**
Trap crops Catch crops Deep plowing Soil solarization	Resistant varieties Hand weeding Irrigation/flooding Date of sowing Fertilizers/manures Density of sowing/shading	Antisoil erosion 2. **Phytosanitary practices** Clean seed
2. **Chemical**	2. **Chemical**	3. **Feeding only *Striga*-free fodder to livestock**

Methyl-bromide fumigation
Germination stimulants
  (Ethylene gas and strigol
  analogs)
	Herbicides	

	3. **Biological**	

Crop substitution, hand weeding, cattle manure. Sorghum and millet, the most important cereals in the semi-arid tropics, occupy different ecological zones in the sub-Saharan Africa. Sorghum and maize are dominant in the Sudanian and northern Guinean zones and the dry veld where rainfall is relatively high (700-1200 mm) and the soils are heavy and relatively more fertile. Millet is the most important crop on sandy soils of the Sahel and the dry veld. Accordingly, *Striga* has developed extreme crop-specificity in these zones of western Africa (Ramaiah 1984). Therefore, introduction of early-maturing and drought-resistant sorghum varieties adapted to the sandy soils of the Sahel or dry veld will greatly solve the *Striga* problem there. Conversely, cultivation of millets in the Sudanian zone will be advantageous. This practice need not be followed throughout the region, but will be very useful in *Striga*-endemic areas. It may also have limited use in very dry and hot environments, as in Botswana, where the crop-specificity of *Striga* has not been confirmed.

In the Sudan a similar system is followed by the farmers of southern Darfur (Ogborn 1987), where farmers have started to utilize their freedom from the *Striga* menace.

Farmers must be careful not to allow *Striga* on these introduced crops to set seed, as it is common to find that *Striga* populations build up to intolerable levels rapidly. Crop substitution should, therefore, be accompanied by hand weeding. As fewer *Striga* plants emerge, the task should not be extensive.

If supplementary control measures are ignored, the substituted crops can succumb to *Striga*. This has occurred in western Africa (Porters 1952) and in the Sudan (Bebawi 1987).

We recommend supplementary hand weeding to prevent *Striga* buildup and the parking of cattle, sheep, and goats in continuously cropped fields in dry season. The manure will help sustain productivity of the soil.

Varietal substitution, hand weeding, and cattle manure. Local varieties of sorghum and millet in Africa have some resistance to *Striga*, though they suffer greatly under heavy infestations. Resistant varieties of sorghum have been known since the 1930s (Saunders 1933), but their use by the farmers was limited because of certain undesirable agronomic traits.

Nevertheless, varieties like Dobbs, Framida, Radar, and Mugd are grown by farmers in eastern and southern Africa. Framida was introduced into western African countries and is doing extremely well in Togo, Ghana, Benin, and Burkina Faso. Under heavy infestations, this variety is known to produce several times more than the local susceptible varieties. In addition to its resistance to *Striga*, Framida is excellent in seedling establishment and resistance to drought, has higher and stable yields, possesses partial photoperiod sensitivity, resists bird damage (due to brown grains), and good food quality when properly processed. Its brown-grain characteristic limits its cultivation to red-sorghum growing areas, and to production for brewing opaque beer in southern Africa.

The use of recent resistant varieties and improved derivatives from older cvs, in place of susceptible cvs, plus hand weeding and cattle manure can reduce significantly the *Striga* menace and increase cereal productivity. Two of the most promising varieties are ICSV 1002 BF and ICSV 1007 BF. ICSV 1002 BF combines the additional good characteristics of white grains, lodging resistance (nonsensescence character), and resistance to important leaf diseases. This variety is in the prerelease stage in Burkina Faso and Gambia.

The second selection, ICSV 1007 BF, proved to be the most resistant variety throughout the *Striga* belts of all continents, and in on-farm tests in heavily infested irrigated areas of the Sudan (Nour, personal communication). In the southern African region, we have found the promising white-grained SARs 16, 19, and 33 to be resistant to species of *Striga*. Their productivity is being evaluated in variety trials in the nine countries of the SADCC.

Another local variety, IS 9830 from the Sudan, has good levels of resistance, is early, and yields well under a range of growing conditions.

Recently a derivative of Framida, SAR 1, was released in India for cultivation in *Striga*-endemic areas (ICRISAT 1986).

Varietal substitution should always be accompanied by hand weeding and farm animal manure to sustain long-term productivity of the soil.

Preventing buildup of new strains of *Striga* to levels that overcome the resistance of a new variety is an absolute requirement. Breakdowns in resistance to *Striga* were reported from the Sudan (Ogborn 1984) and southern Africa (Riches et al. 1987).

Date of sowing, land preparation, supplementary hand weeding. Farmers in Africa sow their local varieties following the first heavy rain very early in the season. Early sowing on land not well-tilled, before normal rainfalls begin, exposes seedlings to drought stress. This is known to favor early emergence of *Striga* in western Africa and Ethiopia (Ramaiah 1987) and in the Sudan (Andrews 1945).

In a delayed-sowing operation, it is advised to prepare the land with first rains and sow improved varieties. The late local varieties suffer significant yield losses if there is an early cessation of rainfall. Early varieties sown early should possess resistance to mold and grain weathering. Early varieties (such as the CK 60 B sorghum), support the emergence of many *Striga* plants, but because grain in these varieties fills and matures quite early, yields do not seem to be reduced significantly.

In the northern Guinean zone of western Africa, the rainfall is greater and the season is relatively longer; relay cropping is traditional. This system could be modified to suit our *Striga* control strategy. In northern Nigeria we have observed a relay-cropping system of three crops: photosensitive sorghum, millet, and groundnut. Millet is harvested at the end of September; sorghum not until the end of November. In September, hardly any *Striga* is seen on sorghum, but by the end of October, it is in full flower and has plenty of time to set seed before sorghum harvest in November. For this situation, we recommend sowing an early- to medium-maturing sorghum that can be harvested by mid-October. *Striga* by this time is only starting to flower, thus preventing to some extent enormous seed buildup. These early varieties have a three-fold advantage—escape damage, prevent *Striga* seed buildup, and act as catch crops. This practice has been recommended in Kenya where the sorghum cv "Dobbs" would be ready for harvest before most of the *Striga* matures.

One practical possibility is to move the photosensitive varieties (flowering in mid-September) from the Sudanian zone into the north Guinean zone, as replacements for the local photosensitive varieties that flower in mid-October. But somehow this did not happen in the traditional system, perhaps because of different local-adaptation requirements. A second possibility is to introduce improved high-yielding varieties for sowing late in the relay-cropping system.

is has produced excellent results, in terms of proved overall productivity in Mali (ICRISAT 86). We recommend improved resistant varieties, wherever available, for relay sowing on nd well-prepared and fertilized, supported by ate hand weeding.

ith major adjustments in traditional systems

here are very effective *Striga*-control methods r farmers who can afford them—costly soil fugations, herbicides, fertilizers, phytosanitary gulations, and antitranspirants. Some of the ntrol strategies using these methods are listed:

rtilizers and improved varieties, late-season nd weeding. We recommend the use of ferizers with input-responsive resistant varieties ch as Framida, ICSV 1002 BF, and SAR 19 cked with late hand weeding. In Gambia, Carn (1986) has recommended a similar package r *Striga*-endemic areas.

hylene gas, herbicides, phytosanitary regula- ns. This method is being successfully used eradicate *Striga* in USA. It represents the best m each of the three major principles of *Striga* ntrol described earlier: (1) reduction of seeds infested soils (ethylene gas), (2) preventing w seed production (herbicides), and (3) prention of spread of *Striga* from infested to nonfested fields (phytosanitary regulations). Each mands a lot of resources and infrastructure to ake it effective. A comprehensive review of ese methods was recently published by Eplee d Norris (1987b).

Recent research on *Striga* physiology has proded clues for *Striga* control through use of anranspirants [Press et al. (In press)]. *Striga* ants were reported to have higher transpiran rates, and any action that interferes with its nspiration function restricts cooling, raises ternal temperatures, and ultimately causes ath of the plant. Antitranspirants are less hazdous than herbicides to humans and the envinment, in general, and therefore should be an tractive alternative to herbicides.

It is our hope that extension services in riga-affected areas will demonstrate these conl methods, explaining their potential so that rmers become motivated to use them.

Research Needs

The foregoing shows that *Striga* is not a new problem on cereals and legumes in Africa; some early work was done and much is now being done to understand *Striga* and its control. However, new *Striga* problems are being found, and they need to be researched.

We recognize three levels of research leading to understanding and control of *Striga*: (1) adaptive, (2) applied, and (3) basic. Adaptive research includes verification of research recommendations through on-farm testing. These recommendations, resulting from applied research are practices that will be suggested to, and hopefully adopted, by farmers. Applied research consists of experiments to determine how the knowledge gained in basic research can be used in practical ways to help solve the *Striga* problem — in other words, to become useful in devising a new control practice. Basic research seeks new knowledge of the *Striga* plant and its hosts. Conducting research into more biological aspects using the latest research techniques are employed. In our view the research needs are as follows.

Host resistance

Adaptive research. (1) Demonstration of host resistance in sorghum to farmers and extension agents through multilocational on-farm testing.

Applied research. (1) Breeding stable highyielding acceptable (to farmers) resistant varieties of sorghum and millet; (2) screening of wild relatives of pearl millet to find resistance; (3) development of reliable infestation and screening techniques; (4) physiological variability in relation to stability of host resistance; (5) field designs and appropriate statistical analyses.

Basic research. (1) Understanding *Striga* resistance mechanisms active in sorghum and millets; (2) inducting resistance by changing host physiology, increasing lignification in root tissue, root exudation, defense chemicals, seed hardening, etc.; (3) rapid screening techniques, based on "chemical markers" produced in *Striga*-infected plants; (4) application of molecular genetics to isolate resistance genes useful as probes to screen germplasm and varieties.

Agronomic practices

Adaptive research. (1) Demonstrate value of late hand weeding; (2) demonstrate value of fertilizers, wherever available, particularly with maize; (3) demonstrate value of rotating modified mixed and host and nonhost crops.

Applied research. (1) Field experiments to determine optimal crop densities, dates of sowing, etc.; (2) identification of suitable trap crops for different *Striga* species; (3) study *Striga* chasers like coriander, etc.; (4) determine seasonal emergence patterns.

Basic research. (1) *Striga* seed behavior under different conditions in the soil; (2) determine factors such as soil, host, parasite, and climate that dictate narrow and broad spectrum adaptation of *Striga*; (3) develop growth models, etc.

Chemicals

Adaptive research. (1) Demonstrate value of 2, 4-D in sole crops; (2) demonstrate other herbicides in crop mixtures.

Applied research. (1) Conduct trials with antitranspirants and systemic herbicides.

Basic research. (1) Synthesis of stable stimulants based on strigol, ethylene, or other compounds; (2) study combinations of stimulant and herbicide compounds.

Biological control

Applied research. (1) Controlled exchanges of insects between India and Africa; (2) survey of organisms parasitic or pathologic on *Striga*; (3) use of the gall-forming weevil *Smicronyx umbrinus* in reducing *Striga* seed production.

Basic research. (1) Develop bioherbicides based on biological "enemies" of *Striga*; (2) study taxonomy of African *Smicronyx* spp; (3) study diseases and other pests of *Striga*.

Crop specificity

Adaptive research. (1) Crop substitution (millet in sorghum zones and sorghum in millet zones); (2) crop rotations using nonhost (trap) crops; (3) study socioeconomic implications of these in the respective ecologies.

Basic research. (1) Determine basis for immunity in nonhost (trap) crops.

Taxonomy

Applied research. (1) Sympatric existence of *Striga* species and its implications; (2) clarification of relationships among closely linked species.

Basic research. (1) Genetic diversity and genetic interchanges among and within the *Striga* species by inter- and intra-specific hybridization, chromosome studies, and seed morphology, floral morphology, and breeding systems.

Biology

Applied research (1) Field experimentation with antitranspirants.

Basic research (1) Relationship between *Striga* seed germination and host crop phenological development; (2) photosynthesis; (3) transpiration; (4) water relationships; (5) chemical compounds produced by the host that initiate parasitic shoot growth; (6) *Striga*-specific herbicides; (7) toxins of *Striga*, and antitoxins of host plants.

Outlook

Striga research in Africa is inadequate, and the volume of work required for rapid progress in *Striga* research is almost certainly beyond present resources. Even so, increased effort is essential if the *Striga* problem is to be managed.

Considerable manpower will be required to intensify work on the series of research and extension needs identified here. Intensive training of national research and extension staff, especially those in endemic areas for *Striga* is essential; involvement with international scientists in this effort for regional cooperation would be required.

Practical research results should be made available to the farmers as soon as practical through pilot projects demonstrating the value of simple recommended cultural practices.

There is need for continued cooperation and collaboration among all groups working on the *Striga* problem in Africa and elsewhere; these include IARCs (International Agricultural Research Centers), NARSs (National Agricultural Research Systems), funding agencies, and coordinating organizations.

A continental-action program in the form of a "*Striga* Research and Control Network" with strong regional components and strong national commitments to research, training, and extension would be of genuine value, with payoff at the farmer level.

References

Andrews, F.W. 1945. The parasitism of *Striga hermonthica* Benth. on sorghum spp. under irrigation. I. Preliminary results and the effect of heavy and light irrigation on *Striga* attack. Annals of Applied Biology 32(3):193–200.

Babikar, A.G. J., Hamdoun, A.M., and **Mansi, M.G.** 1987. Influence of some soil and environmental factors on response of *Striga hermonthica* (Del.) Benth. seeds to selected germination stimulants, Pages 53–56 *in* Parasitic flowering plants (Weber, H.C., and Forstreuter, W., eds.). Hesse, Germany: Marburg an der Lahn.

Bebawi, F.F. 1987. Cultural practices in witchweed management. Pages 159–172 *in* Parasitic weeds in agriculture, Vol I. *Striga* (Musselman, L.J. ed.) Boca Raton, FL, USA: CRC Press Inc.

Burtt-Davey, J. 1905. Botanical notes: Witchweed. Transvaal Agricultural Journal 3:130.

Carson, A. 1986. Improved weed management control for the small scale farmer with particular reference to *Striga* in Gambia. Paper presented at the FAO OAU All-African Government consultation on *Striga* control, 20–24 Oct 1986, Maroua, Cameroon.

Doggett, H. 1965. *Striga hermonthica* on sorghum in East Africa. Journal of Agricultural Science 65(2):183–194.

Eplee, R.B., and **Norris, P.S.** 1987a. Chemical control of *Striga*. Pages 173–182 *in* Parasitic Plants in Agriculture. Vol I. *Striga* (Musselman, L.J., ed.). Boca Raton, FL, USA: CRC Press Inc.

Eplee, R.B., and **Norris, P.S.** 1987b. Control of *Striga asiatica* F (L) Kuntze with systemic herbicides. Pages 183–186 *in* Parasitic Flowering Plants (Weber, H.C., and Forstreuter, W., eds.). Hesse, Germany: Marburg an der Lahn.

Housley, T.L., Ejeta, H., Sherif-ari, O., Netzley, D.H., and **Butler, L.J.** 1987. Progress towards an understanding of sorghum resistance to *Striga*. Pages 411–419. *in* Parasitic Flowering Plants (Weber, H. C. and Forstreuter, W., eds.). Hesse, Germany: Marburg an der Lahn.

ICRISAT (International Crops Research Institute for the Semi-Arid Tropics). 1986. ICRISAT/Burkina Faso Annual Report, 1985. Integrated *Striga* management, Burkina Faso. Ouagadougou, Burkina Faso: ICRISAT.

ICRISAT. (International Crops Research Institute for the Semi-Arid Tropics).1987. Annual Report, 1986. Patancheru, A. P. 502 324, India: ICRISAT.

Musselman, L.J. 1987. Taxonomy of witchweeds. Pages 3–12 *in* Parasitic Weeds in Agriculture. Vol I. *Striga* (Musselman, L.J., ed.). Boca Raton, FL, USA: CRC Press Inc.

Musselman, L.J., and **Parker, C.** 1983. Biosystematic Research in the genus *Striga* (Scrophulariaceae). Pages 19–24 *in* Proceedings of the Second International Workshop on *Striga*. (Ramaiah, K.V., and Vasudeva Rao, M.J., eds.), Patancheru, India; International Crops Research Institute for the Semi-Arid Tropics.

Obilana, A.T. 1983a. Socioeconomic impact of *Striga* in cereal crops in West Africa. Paper presented at the International Training Course on the Control of *Striga* in Cereal Crops. 8–26 Aug 1983. Raleigh, NC, USA: North Carolina State University. 31 pp.

Obilana, A.T. 1983b. *Striga* studies and control in Nigeria. Pages 87–98 *in* Proceedings of the Second International Workshop on *Striga* (Ramaiah, K.V., and Vasudeva Rao, M.J., eds.). Patancheru, India; International Crops Research Institute for the Semi-Arid Tropics.

Obilana, A.T. 1987. Breeding cowpeas for *Striga* resistance. Pages 243–253. *In* Parasitic Weeds in Agriculture (Vol. I) (Musselman, L.J., ed.). Boca Raton, FL, USA: CRC Press Inc.

Obilana, A.T., Knepper, D., and Musselman, L.J. 1988. *Striga* (witchweeds) in Zimbabwe. Paper presented at the Fourth Regional SADCC/ICRISAT Sorghum and Millet Improvement Program Workshop, 21–24 Sep 1987, Matapos, Zimbabwe.

Obgorn, J.E.A. 1984. Research priorities in agronomy. Pages 195–212 *in* Proceedings, Workshop on the Biology and Control of *Striga*, 14–17 Nov 1983, Dakar, Senegal, (Ayensu et al., eds.). ICSU Press/IDRC.

Ogborn, J.E.A. 1987. *Striga* control under peasant Farming conditions. Pages 145–158 *in* Parasitic weeds in agriculture Vol I. *Striga* (Musselman, L.J., ed.). Boca Raton, FL, USA: CRC Press Inc.

Parker, C. 1983. *Striga*: analysis of past research and summary of the problem. Pages 145–158 *in* Proceedings of the Second International Workshop on *Striga*, 5–8 Oct 1981, IDRC/ICRISAT, Ouagadougou, Burkina Faso. Patancheru, A.P. 502 324, India: International Crops Research Institute for the Semi-Arid Tropics.

Pearson, H.H.W. 1913. The problem of the witchweed. South African Department of Agriculture. Agricultural Journal 4:34.

Porters, R. 1952. Linear cultural sequences in primitive systems of agriculture in Africa and their significance. Sols Afr. 2:15.

Press, M.C., and Stewart, G.P. (In press). Growth and photosynthesis in *Sorghum bicolor* infected with *Striga hermonthica*. Annals of Botany.

Press, M.C., Tuohy, J.M., and Stewart, G.R. 1987a. Gas-exchange characteristics of the sorghum-*Striga* host-parasite association. Plant Physiology 84:814–819.

Press, M.C., Shah, N., Tuohy, J.M., and Stewart, G.R. (In press). Carbon isotope ratios demonstrate carbon flux from C4 host to C3 parasite. Plant Physiology.

Press, M.C., Mallaburn, P., Nour, J., and Stewart, G.R. 1987b. The characterization of resistance and tolerance mechanism against the parasitic angiosperm *Striga hermonthica*. Annual Report 1986–87. Research project R 4022. London WCIE 6BT: University College London.

Ralston, D.M., Riches, C.R., and Musselman, L.J. 1986. Morphology, floral syndrome, and host specificity of *Striga* spp in Botswana. OICD Report. P.B. 0033 Gaborone, Botswana: Dept. of Agricultural Research. 41 pp.

Ramaiah, K.V. 1984. Patterns of *Striga* resistance in sorghum and millets with special emphasis on Africa. Pages 71–92 *in Striga*: Biology and Control, Proceedings Workshop on the biology and control of *Striga*, 14–17 Nov 1983, Dakar, Senegal. ICSU Press/IDRC.

Ramaiah, K.V. 1985. Hand pulling of *Striga hermonthica* on Pearl Millet. Tropical Pest Management 31(44):326–327.

Ramaiah, K.V. 1987. Control of *Striga* and *Orobanche* species - a review. Pages 637–664 *in* Parasitic Flowering Plants (Weber, H.C. and Forstreuter, W., eds.). Hesse, Germany: Marburg an der Lahn.

Ramaiah, K.V., Parker, C., Vasudeva Rao, M.J., and Musselman, L.J. 1983. *Striga* Identification and Control Hand Book. Information Bulletin no. 15. Patancheru, A.P. 502 324, India: International Crops Research Institute for the Semi-Arid Tropics. 52 pp.

Raynal-Roques, A. 1987. The genus *Striga* (Scrophulariaceae) in western and central Africa - a survey. Pages 675–689 *in* Parasitic Flowering Plants (Weber, H.C. and Forstreuter, W., eds.). Hesse, Germany: Marburg an der Lahn.

Riches, C.R., de Milliano, W.A.J., Obilana, A.T., and House, L.R. 1987. Witchweeds (*Striga* spp) of sorghum and pearl millet in the SADCC Region—Distribution and Control. *In* Proceedings, Third Regional SADCC/ICRISAT Sorghum and Millet Improvement Project Workshop, Lusaka, Zambia, 20–25 Sep 1986.

Saunders, A.R. 1933. Studies on phanerogamic parasitism with particular reference to *Striga lutea*. Bulletin No. 128. Pretoria, South Africa: Department of Agricultural Science, South Africa. 56 pp.

Saunders, A.R. 1942. Field experiments at Potchefstroom. A summary of investigations conducted during the period 1903–1940. South African Department of Agriculture and Forestry Science Bulletin 214:19–31, 49–50.

Shah, N., Smirnoff, N., and Stewart, G.R. 1987. Photosynthesis and stomatal characteristics of *Striga hermonthica* in relation to its parasitic habit. Physiologia Plantarum 69:699–703.

Timson, S.D. 1931. Witchweed. Progress report and a warning. Rhodesia Agricultural Journal 28:1101–1111.

Timson, S.D. 1934. Witchweed. Rhodesia Agricultural Journal 31:792–801.

Vasudeva Rao, M.J., Chidley, V.L., and House, L.R. (In press). Estimates of grain yield losses caused in sorghum (*Sorghum bicolor* L. Moench) by *Striga asiatica* (L.) Kuntz obtained using the regression approach. ICRISAT Journal Article JA 718.

Visser, J.H., and Botha, P.J. 1974. Chromatographic Investigation of the *Striga* seed germination stimulant. Zeitschrift für Pflamzenphysiologie 72:352–358.

Anthracnose of Sorghum

M.E.K. Ali[1] and H.L. Warren[2]

Abstract

Anthracnose is one of the most destructive diseases of sorghum growing in warm, humid regions. The pathogen, Colletotrichum graminicola, infects leaves, stalks, peduncles, and panicles, including the grain. A brief review of anthracnose of sorghum, with emphasis on the biology of pathogen and the economic importance of the disease is presented. C. graminicola is a highly variable pathogen, and there is evidence of races among pathogen populations. Variability in virulence among pathogen populations may pose a threat to sorghum production in areas where the conditions are favorable for anthracnose development. Grain yield is influenced by the aggressiveness of the pathogen, sorghum genotype, environment, and the physiological status of the host. Grain losses of up to 50% have been reported.

Introduction

Sorghum anthracnose, caused by *Colletotrichum graminicola* (Cesati) Wilson, has been reported in most areas where sorghum is grown, but is more prevalent and severe in warm, humid regions (Bergquist 1973; Porter 1962; Powell et al. 1977; Sundaram et al. 1972; Tarr 1962). Anthracnose has been reported (Harris and Sowell 1970) as the most destructive disease of sorghum in Georgia, USA.

History and Distribution

Anthracnose of sorghum was first reported from Togo in 1902 (L.K. Mughogho, ICRISAT, personal communication, March 1988). The disease now occurs worldwide (Bergquist 1973; Edmunds et al. 1970; Frederiksen 1984; Heald and Wolf 1912; Hsi 1956; Miller 1956; Noble 1977; Pastor-Corrales and Frederiksen 1980; Porter 1962; Sundaram et al. 1972; Tarr 1962).

Causal organism

The anthracnose fungus are described by a number of reviewers (Mordue 1967; Pastor-Corrales and Frederiksen 1980; Tarr 1962; Warren 1986).

The imperfect fungus, *Colletotrichum graminicola* (Cesati) Wilson, causes anthracnose of other cereals and many grasses as well. *C. graminicola* of the Melanconiales was erected by Wilson (1914) to include most forms of *Colletotrichum* with falcate conidia. Synonyms encountered in the early literature include *Dicladium graminicola* Cesati and several species of *Vermicularia*. Based on morphological and cultural characteristics, Politis (1975) was able to differentiate between the perfect states of *C. graminicola* and *C. falcatum* by size of perithecia, asci, and ascospores, as well as on shape of ascospores. Accordingly, the perfect state of *C. graminicola* was named *Glomerella graminicola*.

1. Research Plant Pathologist, Agricultural Research Corporation, Hudeiba Research Station, PO Box 31, Ed-Damer, Sudan.
2. Research Plant Pathologist and Professor, USDA-ARS, Department of Botany and Plant Pathology. Purdue University, West Lafayette, IN 47907, USA.

Ali, M.E.K., and Warren, H.L. 1992. Anthracnose of sorghum. Pages 203–208 *in* Sorghum and millets diseases: a second world review. (de Milliano, W.A.J., Frederiksen, R.A., and Bengston, G.D., eds). Patancheru, A.P. 502 324, India: International Crops Research Institute for the Semi-Arid Tropics.

Symptoms

C. graminicola infects leaves, stalks, peduncles, panicle, and the grain, either separately or all together (Pastor-Corrales and Frederiksen 1980). These disease symptoms include a foliar phase, stalk rot, and panicle and grain anthracnose. These different phases of anthracnose often manifest as apparently different diseases.

The characteristic symptoms caused by *C. graminicola* in sorghum have been described elsewhere (Edmunds et al. 1970; Edmunds and Zummo 1975; Tarr 1962). Variability in symptoms may be due to variation in the pathogen, host resistance, or physiological status of the host following infection (Frederiksen 1984).

Survival, seasonal persistence, and spread of inoculum

C. graminicola survives from season to season on infected crop residues and infected weed hosts (Tarr 1962). Naylor and Leonard (1977) reported that *C. graminicola* survived through the winter and the following growing season in exposed infected maize stalks in North Carolina. *C. graminicola* has been reported to persist for 18 months as a saprophyte in plant tissue colonized during parasitism (Vizvary 1975), but conidia or mycelia in the absence of residue lysed within a few days (Vizvary and Warren 1982).

A few reports suggest that *C. graminicola* is seed transmitted (Tarr 1962). However, seed transmission was demonstrated by the water agar seedling symptom test and in a pot experiment (Basu Chaudhary and Mathur 1979). *C. graminicola* has been reported to survive on maize seeds for longer than 1 year (Warren and Nicholson 1975; Warren 1977). Warren and Nicholson (1975) reported that seedlings derived from infected maize seeds usually are infected.

Primary infection is commonly from the mycelium and conidia on the crop residue (Dickson 1956). In wet weather, pink masses of spores ooze from the acervuli and are spread by splashing rains (Edmunds et al. 1970). Nicholson and Moraes (1980) demonstrated that spores of *C. graminicola* are produced in association with water-soluble mucilaginous matrix (spore matrix) which protects them against desiccation and aids in their survival and dissemination.

They also suggested that spores in the field may survive and be dispersed in dry particulate matter. The role of infected seed as a possible source of primary infection and transmission was demonstrated (Basu Chaudhary and Mathur 1979). Infected weed hosts such as *Echinochloa colonum, Digitaria sanguinalis,* and *Dactyloctenium aegyptum* were reported to act as sources of primary infection (ICRISAT 1983).

Host range and physiological specialization

C. graminicola has a wide range among cultivated and wild species of cereals and grasses (Tarr 1962), including wheat, barley, oats, rye, maize, sorghum, sudangrass, and johnsongrass (Bruehl and Dickson 1950; Luke and Seecher 1963; Sanford 1935; Shurtleff 1973; Wiese 1977). However, isolates from one host do not necessarily infect other hosts. The capacity of an isolate of *C. graminicola* to infect several hosts is not a subject of general agreement. Some writers report that an isolate of *C. graminicola* is capable of infecting several hosts (Chowdhury 1936; LeBeau et al. 1951; Lohman and Stokes 1944; Wheeler et al. 1974), but many others report evidence of a high degree of host specificity among isolates of *C. graminicola* (Ali 1986; Bruehl and Dickson 1950; Dale 1963; Hooker 1977; LeBeau 1950; Pupipat and Mehta 1969; Shahnaz and Nicholson 1979; Tarr 1962; Williams and Willis 1963; Zwillenberg 1959). Because of this contradicting information, additional studies to identify reservoir hosts of *C. graminicola* isolates that attack sorghum become essential. Crop sequencing, cultural practices, and breeding for disease resistance are effective measures of disease control, but require accurate knowledge of host specificity.

Pathogen variability

C. graminicola is a highly variable species. Harris and Johnson (1967) have previously suggested the existence of races of *C. graminicola*; and some sorghum cultivars are reported to react differently in Texas, Mississippi, and Georgia (Frederiksen and Rosenow 1971). Harris and Cunfer (1976) reported shifts in plant reaction to anthracnose in some cultivars and suggested the existence of races of *C. graminicola*. Pastor-Corrales and Frederiksen (1979) reported that many of the resistant sorghum entries in Georgia were susceptible in Brazil; but Redlan, a traditionally susceptible entry in Georgia, was resistant in Brazil. Nakamura (1982) identified five races of *C. graminicola* using seven differential sorghum cultivars (Tx 2536, Martin, TAM 428, Tx 430, Brandes, SC 170-6-17, SC 175-14). Ferreira and Casela (1986) identified seven races of *C. graminicola* using 13 differential cultivars of sorghum (Tx 2536, Martin, TAM 428, Tx 430, Brandes, SC 170-6-17, SC 175-14, SC 112-14, Theis, Reis, Redlan, SC 326-6, and SC 283). The seven isolates used in their study were collected from different locations in Brazil. Ali and Warren (1987) identified three races of *C. graminicola* using six differential cultivars (IS 4225, IS 8361, 954130, 954062, Br 64, and 954206). The three races were identified from nine isolates of *C. graminicola* obtained from Florida, Georgia, Puerto Rico, and Indiana. They also reported the different races even among isolates from the same area. The existence of races among pathogen populations present challenging problems to breeders and pathologists trying to develop resistant cvs.

Casela and Ferreira (In press) have proposed a system of nomenclature of races of *C. graminicola* in Brazil. In this system, three sorghum cultivars (Redlan, SC 326-6, SC 283) are used to separate eight groups of races, then sorghum cultivars Tx 623, Brandes, SC 112-14, Martin, Tx 2536, and Theis are used to distinguish 32 races within each group.

Economic importance

Sorghum anthracnose may limit grain sorghum production in the humid southeastern USA (Harris et al. 1964; Harris and Sowell 1970), Latin America, Brazil, and Venezuela (Pastor-Corrales and Frederiksen 1979), and other humid tropical and subtropical areas (Bergquist 1973; Powell et al. 1977).

Losses in grain yield were estimated to exceed 50% on susceptible sorghum cultivars in a severe anthracnose epiphytotic in Georgia (Harris et al. 1964). They reported a negative (r = -0.632) and highly significant ($P = 0.01$) correlation between grain yield and leaf anthracnose rating. Highly significant ($P = 0.01$) negative correlations between grain yield and leaf, head, and

stalk anthracnose ratings were reported. In these studies, leaf anthracnose and stalk rot were independent of each other, but the head infection was associated with both the leaf and stalk phases of the disease. Powell et al. (1977) reported that grain yield was reduced by 70% and more than half the yield loss resulted from incomplete grain fill as verified by 42% decrease in 1000-seed mass and 17.2% decrease in seed density. Early seed abortion's role in yield reduction was also suggested to be important. Gorbet (1977) reported that grain production of susceptible sorghum cultivars is severely limited when the disease develops during heading or early grain filling. Correlation coefficients between anthracnose severity and grain yield in 42 sorghum hybrids were negative and highly significant for all disease rating dates; however, the r^2 values decreased as grain developed from the milk stage to harvest (Harris and Fisher 1974). The percentage loss in sorghum grain yield varied from 1.2 to 16.4, depending upon anthracnose severity (Mishra and Siradhana 1979). Luttrel (1950) reported that serious yield losses may not occur if leaf symptoms do not appear until after the plants mature. Ali et al. (1987) reported that highly significant ($P = 0.01$) positive correlations between percentage loss in grain yield and anthracnose leaf blight (ALB) severity index occurred in 1984 and 1985 and for the pooled data of both years, with correlation coefficients of 0.86, 0.84, and 0.85, respectively. Correlations between percentage loss in 1000-seed mass and ALB severity also were highly significant ($P = 0.01$) for the individual and pooled data of both years. The highly significant ($P = 0.01$) positive correlations between percentage loss in grain yield and percentage loss in 1000-seed mass indicate that ALB reduces grain yield of sorghum largely by decreasing seed mass. They also reported a loss in grain yield of about 30% in the most-susceptible sorghum cultivar. They suggested that the amount of loss in grain yield due to ALB was influenced by the aggressiveness of the pathogen, sorghum genotype, and environmental conditions favoring anthracnose development.

The extent of damage of loss due to anthracnose stalk rot has been reported as a reflection of the host susceptibility, environment, aggressiveness of the pathogen, and physiological status of the host (Frederiksen 1984).

References

Ali, M.E.K. 1986. *Colletotrichum graminicola*, physiologic specialization, characterization of races, and the effect of anthracnose leaf blight on grain yield of sorghum. Ph.D. thesis, Purdue University, West Lafeyette, IN 47907, USA. 110 pp.

Ali, M.E.K., and **Warren, H.L.** 1987. Physiological races of *Colletotrichum graminicola* on sorghum. Plant Disease 71:402–404.

Ali, M.E.K., Warren, H.L., and **Latin, R.X.** 1987. Relationship between anthracnose leaf blight and losses in grain yield of sorghum. Plant Disease 71:803–806.

Basu Chaudhary, K.C., and **Mathur, S.B.** 1979. Infection of sorghum seeds by *Colletotrichum graminicola* 1. Survey, location in seed, and transmission of the pathogen. Seed Science and Technology 7:87–92.

Bergquist, R.R. 1973. *Colletotrichum graminicola* in *Sorghum bicolor* in Hawaii. Plant Disease Reporter 57:272–275.

Bruehl, G.W., and **Dickson, J.G.** 1950. Anthracnose of cereals and grasses. USDA Technical Bulletin 1005. Washington, D.C., USA: U.S. Government Printing Office. 37 pp.

Casela, C.R., and **Ferreira, A.S.** (In press). Proposta de um sistema de classificacao de races de *Colletotrichum graminicola* (*Ces.*) Wils. (Sensu Arx, 1957), agente causal da antracnose em sorgo (*Sorghum bicolor* (L.) Moench). Fitopatologia Brasileira.

Chowdhury, S.C. 1936. A disease of *Zea mays* caused by *Colletotrichum graminicola* (*Ces.*) Wils. Indian Journal of Agricultural Science 6:833–843.

Dale, J.L. 1963. Corn anthracnose. Plant Disease Reporter 47:245–249.

Dickson, J.G. 1956. Diseases of Field Crops. New York, NY, USA: McGraw Hill Book Company, Inc. 517 pp.

Edmunds, L.K., Futrell, M.C., and **Frederiksen, R.A.** 1970. Sorghum diseases. Pages 200–234 *in* Sorghum production and utilization (Wall, J.S., and Ross, W. M., eds.). Westport, CT, USA: AVI Publishing Company, Inc.

Edmunds, L.K., and **Zummo, N.** 1975. Sorghum diseases in the United States and their control. USDA Handbook No. 468. Washington, D. C., USA: U.S. Government Publishing Office.

Ferreira, A.S., and **Casela, C.R.** 1986. Racas patogenicas de *Colletotrichum graminicola*, agente causal de antracnose em sorgo (*Sorghum bicolor* (L.) Moench). Fitopatologia Brasileira 11(1): 833–87.

Frederiksen, R.A. 1984. Anthracnose stalk rot. Pages 37–42 *in* Sorghum root and stalk rots, a critical review: proceedings of the Consultative Group Discussion on Research Needs and Strategies for Control of Sorghum Root and Stalk Rot Diseases, 27 Nov–2 Dec 1983, Bellagio, Italy. Patancheru, A.P. 502 324, India: International Crops Research Institute for the Semi-Arid Tropics.

Frederiksen, R.A., and **Rosenow, D.T.** 1971. Disease resistance in sorghum. Pages 71–82 *in* Proceedings of the 26th Annual Corn and Sorghum Research Conference. Washington, D. C., USA: American Seed Trade Association.

Gorbet, D.W., 1977. Anthracnose of grain sorghum in North Florida. Sorghum Newsletter 20:98–99.

Harris, H.B., and **Cunfer, B.M.** 1976. Observations on sorghum anthracnose in Georgia. Sorghum Newsletter 19:100–101.

Harris, H.B., and **Fisher, C.D.** 1974. Yield of grain sorghum in relation to anthracnose expression at different developmental stages of host. Plant Breeding Abstracts 44(4):2455.

Harris, H.B., and **Sowell, G., Jr.** 1970. Incidence of *Colletotrichum graminicola* on *Sorghum bicolor* introductions. Plant Disease Reporter 54:60–62.

Harris, H.B., and **Johnson, B.J.** 1967. Sorghum anthracnose symptoms, importance, and resistance. Pages 48–52 *in* Proceedings of the Fifth Biennial Grain Sorghum Research and Utilization Conference, Grain Sorghum Producers Association (GSPA) and Sorghum Improvement Conference of North America, Lubbock, Texas. Abernathy, TX, USA: Grain sorghum Producers Association.

Harris, H.B., Johnson, B.J., Dobson, J.W., and Luttrell, E.S. 1964. Evaluation of anthracnose on grain sorghum. Crop Science 4:460–462.

Heald, F.D., and Wolf, F.A. 1912. A plant-disease survey in the vicinity of San Antonio, TX. US Bureau of Plant Industry Bulletin No. 226. 129 pp.

Hooker, A.L. 1977. Corn anthracnose leaf blight and stalk rot. Pages 167–182 in Proceedings of the 31st Annual Corn and Sorghum Research Conference. Washington, D.C., USA: American Seed Trade Association.

Hsi, C.H. 1956. Stalk rots of sorghum in Eastern New Mexico. Plant Disease Reporter 40:369–371.

Huguenin, M., Lourd, M., and Geiger, J.P. 1982. Comparative morphological, physiological, and pathogenic studies of isolates of Colletotrichum falcatum and Colletotrichum graminicola. Phytopathologische Zeitschrift 105:293–304 (in French).

ICRISAT. 1983. Annual Report 1982. Patancheru, A.P. 502324, India: International Crops Research Institute for the Semi-Arid Tropics.

King, S.B., and Frederiksen R.A. 1976. Report on the International Sorghum Anthracnose Virulence Nursery. Sorghum Newsletter 19:105–106.

LeBeau, F.J. 1950. Pathogenicity studies with Colletotrichum from different hosts on sorghum and sugarcane. Phytopathology 40:430–438.

LeBeau, F.J., Stokes, I.E., and Coleman, O.H. 1951. Anthracnose and red rot of sorghum. USDA Bulletin 1035. Washington, D.C.: U.S. Government Printing Office. 21 pp.

Lohman, M.L., and Stokes, I.E. 1944. Stem anthracnose and red rot of sorghum in Mississippi. Plant Disease Reporter 28:76–80.

Luke, H.H., and Seecher, D.T. 1963. Rye Anthracnose. Plant Disease Reporter 47:936–937.

Luttrell, E.S. 1950. Grain sorghum diseases in Georgia—1949. Plant Disease Reporter 34:45–52.

Miller, P.R. 1956. Plant Disease situation in the United States. FAO Plant Production Bulletin 4:152–156.

Mishra, A., and Siradhana, B.S. 1979. Evaluation of losses due to anthracnose of sorghum. Indian Journal of Mycology and Plant Pathology 9(2):257.

Mordue, J.E.M. 1967. Colletotrichum graminicola. CMI Descriptions of Pathogenic Fungi and Bacteria. No. 132 Kew, UK: Commonwealth Mycological Institute. 2 pp.

Nakamura, K. 1982. Especializacao fisiologica em Colletotrichum graminicola (ces.) Wils. (Sensu Arx, 1957) agente causal da antracnose em sorgo. Ph.D. thesis, Universidade Estadual Paulista, Jaboticabal, Brazil. 143 pp.

Naylor, V.D., and Leonard, K.J. 1977. Survival of Colletotrichum graminicola in infected corn stalks in North Carolina. Plant Disease Reporter 61:382–383.

Nicholson, R.L., and Moraes, W.B.C. 1980. Survival of Colletotrichum graminicola: Importance of the spore matrix. Phytopathology 70:255–261.

Noble, R.J. 1937. Australia: Notes on plant diseases recorded in New South Wales for the year ending 30th June, 1937. International Bulletin of Plant Production 11:246–247.

Pastor-Corrales, M.A., and Frederiksen, R.A. 1979. Sorghum reaction to anthracnose in the United States, Guatemala, and Brazil. Sorghum Newsletter 22:127–128.

Pastor-Corrales, M.A., and Frederiksen, R.A. 1980. Sorghum anthracnose. Pages 289–294 in Sorghum diseases, a world review: proceedings of the International Workshop on Sorghum Diseases, 11–15 Dec 1978, Hyderabad, India. Patancheru, A.P. 502 324, India: International Crops Research Institute for the Semi-Arid Tropics.

Politis, D.J. 1975. The identity and perfect state of *Colletotrichum graminicola*. Mycologia 67:56–62.

Politis, D.J., and Wheeler, H. 1972.The perfect stage of *Colletotrichum graminicola*. Plant Disease Reporter 56:1026–1027.

Porter, R.H. 1962. A preliminary report of surveys of plant diseases in East Asia. Plant Disease Reporter 46:58–166. (Supplement).

Powell, P., Ellis, M., Alameda, M., and Sotomayor, A. 1977. Effect of natural anthracnose epiphytotic on yield, grain quality, seed health, and seed-borne fungi in *Sorghum bicolor*. Sorghum Newsletter 20:77–78.

Pupipat, U., and Mehta, Y.R. 1969. Stalk rot of maize caused by *Colletotrichum graminicola*. Indian Phytopathology 22:346–348.

Sanford, G.B. 1935. *Colletotrichum graminicola* (Ces.) Wils. as a parasite of the stem and root tissue of *Avena sativa*. Scientific Agriculture 15:370–376.

Shahnaz, F.F., and Nicholson, R.L. 1979. *Colletotrichum graminicola* host specificity and plant age. Phytopathology 69:542–543. (Abstract).

Shurtleff, M.C. 1973. A compendium of corn diseases. St. Paul, MN, USA: American Phytopathological Society. 64 pp.

Sundaram, M.V., Palmer, L.T., Nagarajan, K., and Prescott, J.M. 1972. Disease survey of sorghum and millets in India. Plant Disease Reporter 56:740–743.

Sutton, B.C. 1968. The appressoria of *Colletotrichum graminicola* and *C. falcatum*. Canadian Journal of Botany 46:873–876.

Tarr, S.A.J. 1962. Diseases of sorghum, Sudan grass, and broom corn. Kew, Surrey, UK: Commonwealth Mycological Institute. 380 pp.

Vizvary, M.A. 1975. Saphrophytism and survival of *Colletotrichum graminicola* in soil and corn stalk residues and pathogenic behavior in corn roots. M.S. thesis, Purdue University, West Lafeyette, IN 47907, USA. 115 pp.

Vizvary, M.A., and Warren, H.L. 1982. Survival of *Colletotrichum graminicola* in soil. Phytopathology 72:522–525.

Warren, H.L. 1977. Survival of *Colletotrichum graminicola* in corn kernels. Phytopathology 67:160–162.

Warren, H.L. 1986. Panicle and grain anthracnose. Page 40 *in* Compendium of sorghum diseases (Frederiksen, R.A., ed.). St. Paul, MN, USA: American Phytopathological Society. 82 pp.

Warren, H.L., and Nicholson R.L. 1975. Kernel infection, seedling blight, and wilt of maize caused by *Colletotrichum graminicola*. Phytopathology 65:620–623.

Wheeler, H., Politis, D.J., and Poneleit, C.G. 1974. Pathogenicity, host range, and distribution of *Colletotrichum graminicola* on corn. Phytopathology 64:293–296.

Wiese, M.V. 1977. Compendium of wheat diseases. St. Paul, MN, USA: American Phytopathological Society. 106 pp.

Williams, L.E., and Willis, G.M. 1963. Diseases of corn caused by *Colletotrichum graminicola*. Phytopathology 53:364–365.

Wilson, G.W. 1914. The identity of the anthracnose of grasses in the United States. Phytopathology 4:106–112.

Zwillenburg, H.H.L. 1959. *Colletotrichum graminicola* (Ces.) Wils. auf Mais und verschiedenen anderen Pflanzen. Phytopathologische Zeitschrift 14:417–425.

Physiological Races of
Colletotrichum graminicola in Brazil

C.R. Casela[1], A.S. Ferreira[2], and R.E. Schaffert[3]

Abstract

Annual surveys in Brazil since 1984-85 have indicated the existence of several physiological races, some high in virulence, of Colletotrichum graminicola. *The sorghum research program in Brazil has been considering the variability of this pathogen in order to select genotypes highly resistant to anthracnose.*

Introduction

Differential reactions of several cultivars in the National Uniform Sorghum Diseases Nursery in different years and in different regions of Brazil suggest existence of pathogenic races of *Colletotrichum graminicola* in that South American nation. The first positive evidence for the existence of races came from the work of Nakamura (1982), who identified five races of *C. graminicola*. The races were identified by reactions produced by several single-spore isolates, obtained from different parts of the country, on the cvs Tx 2536, Martin, TAM 428, Brandes, SC 170-6-17, and SC 175-14.

Ferreira and Casela (1986) studied single spore isolates of *C. graminicola* obtained from infected plants at seven locations in Brazil and inoculated on the cultivars Tx 2536, Martin, TAM 428, Tx 430, Brandes, SC 170-6-17, SC 175-14, SC 112-14, Theis, Tx 378, SC 326-6, and SC 283. Differential interactions between cultivars and isolates (Table 1) confirmed Nakamura's work.

Casela and Ferreira (In press) recently proposed a system of nomenclature of races of *C. graminicola* in Brazil. In this system, three sorghum cultivars, Tx 378 (Redlan), SC 326-6, and

Table 1. Reaction of 12 sorghum cultivars to seven single-spore isolates of *Colletotrichum graminicola*.[1]

Cultivar	Isolate[2]						
	I 1	I 2	I 3	I 4	I 5	I 6	I 7
Tx 2536	S	S	S	S	S	R	R
Tx 398	S	S	R	R	R	S	S
TAM 428	S	S	S	S	S	S	S
Tx 430	S	S	S	S	S	S	S
SC 170-6-17	R	R	R	R	S	R	R
SC 175-14	S	S	S	R	S	S	S
Brandes	S	S	R	R	S	R	R
SC 112-14	R	S	S	S	S	S	S
Theis	R	R	R	R	S	S	R
Redlan	S	R	S	S	R	R	R
SC 326-6	R	R	S	R	R	R	R
SC 283	R	R	R	R	R	R	R

1. S = susceptible, R = resistant.
2. Source of isolates: Sete Lagoas, Minas Gerais: I 1; Pelotas Rio Grande do sul: I 2; Jatai, Goias: I 3; Capinopolis, Minas Gerais: I 4; Goiania, Goias: I 5; Quixada, Cearea: I 6; and Anapolis, Goias: I 7.

1. Sorghum Breeder for Disease Resistance, National Maize and Sorghum Research Center - EMBRAPA (NMSRC), Sate Lagoas, Minas Gerais, Brazil.
2. Sorghum Pathologist at the above address.
3. Sorghum Pathologist at the above address.

Casela, C.R., Ferreira, A.S., and Schaffert, R.E. 1992. Physiological Races of *Colletotrichum graminicola* in Brazil. Pages 209–212 *in* Sorghum and millets diseases: a second world review. (de Milliano, W.A.J., Frederiksen, R.A., and Bengston, G.D., eds). Patancheru, A.P. 502 324, India: International Crops Research Institute for the Semi-Arid Tropics.

SC 283 are used to separate eight groups of races designated by the letters A to H (Table 2), and the genotypes Tx 623, Brandes, SC 112-14, Martin (Tx 398), Tx 2536, and Theis are used to distinguish 32 races within each group, according to Habgood (1970) (Table 2).

The surveys, conducted annually since 1984-85, indicate the existence of several races. The most important races of *C. graminicola* identified in the 1984-85 and 1985-86 surveys are listed in Tables 3 and 4, respectively. Races 13 H, 29 H, and 31 H were isolated at Sete Lagoas, Minas

Table 2. Nomenclature of races of *Colletotrichum graminicola*, based on differential reactions in selected cultivars.[1]

Differential cultivar	Race Group							
	A	B	C	D	E	F	G	H
1. Redlan	R	S	R	R	S	S	R	S
2. SC 326-6	S	R	R	S	R	S	R	S
3. SC 283	R	R	R	S	R	S	S	S

Differential cultivar	Binary value	Races within groups																															
		00	01	02	03	04	05	06	07	08	09	10	11	12	13	14	15	16	17	18	19	20	21	22	23	24	25	26	27	28	29	30	31
4. Tx 623	0	S	S	S	S	S	S	S	S	S	S	S	S	S	S	S	S	S	S	S	S	S	S	S	S	S	S	S	S	S	S	S	S
5. Brandes	2^0	R	S	R	S	R	S	R	S	R	S	R	S	R	S	R	S	R	S	R	S	R	S	R	S	R	S	R	S	R	S	R	S
6. SC 112-14	2^1	R	R	S	S	R	R	S	S	R	R	S	S	R	R	S	S	R	R	S	S	R	R	S	S	R	R	S	S	R	R	S	S
7. Tx 398	2^2	R	R	R	R	S	S	S	S	R	R	R	R	S	S	S	S	R	R	R	R	S	S	S	S	R	R	R	R	S	S	S	S
8. Tx 2536	2^3	R	R	R	R	R	R	R	R	S	S	S	S	S	S	S	S	R	R	R	R	R	R	R	R	S	S	S	S	S	S	S	S
9. Theis	2^4	R	R	R	R	R	R	R	R	R	R	R	R	R	R	R	R	S	S	S	S	S	S	S	S	S	S	S	S	S	S	S	S

1. R = resistant, S = susceptible.

Table 3. Races of *Colletotrichum graminicola* identified during 1984–85.

Cultivar[1]	Races[2]												
	0 A	06 A	12 A	15 A	29 A	31 A	01 B	02 B	05 B	13 B	09 C	30 C	31 C
Redlan	R	R	R	R	R	R	S	S	S	S	R	R	R
SC 326-6	R	R	R	R	R	R	R	R	R	R	S	S	S
SC 283	R	R	R	R	R	R	R	R	R	R	R	R	R
Tx 623	S	S	S	S	S	S	S	S	S	S	S	S	S
Brandes	R	R	R	S	S	S	S	R	S	S	R	S	S
SC 112-14	R	S	R	S	R	S	R	S	R	R	R	S	S
Tx 398	R	S	S	S	S	S	R	R	S	S	R	S	S
Tx 2536	R	R	S	S	S	S	R	R	S	S	S	S	S
Isolates found (no.)	3	1	2	2	1	3	1	1	1	1	1	1	3

1. R = resistant, S = susceptible.
2. Source: OA: Sete Lagoas, Minas Gerais; Goiania, Goias; and Pelotas, Rio Grande do Sul; 06A: Sete Lagoas, Minas Gerais; 12A: Sete Lagoas, Minas Gerais and Goiania, Goias; 15A: Goiania, Goias, and Jaboticabal, Sao Paulo; 29A and 31A: Goiania, Goias; 01B: Sete Lagoas, Minas Gerais; 02B: Goiania, Goias; 05B and 13B: Sete Lagoas, Minas Gerais; 09C, 30C, and 31C: Goiania, Goias.

Gerais, and at Jatai, Goias from the cultivar SC 283. SC 283 is used as a source of aluminum tolerance and, up to now, has been resistant to *C. graminicola* in Brazil.

Research workers seeking anthracnose-resistant sorghums in Brazil have been considering the existence of physiological races of *Col-letotrichum graminicola* to select highly resistant genotypes in greenhouse screening. Reactions of several experimental grain sorghum hybrids developed in the National Maize and Sorghum Research Center (NMSRC) Sorghum Breeding Program are presented in Table 5. The first two hybrids, CMSXS 350 and CMSXS 351, will be

Table 4. Races of *Colletotrichum graminicola*, by source[1], identified during 1985-86.

Race	Isolates, by source location						Race	Isolates, by source location					
	A	B	C	D	E	Total		A	B	C	D	E	Total
00 A	1	3	3	1	0	8	13 B	9	0	0	3	0	12
01 A	1	0	0	0	0	1	14 B	0	0	0	0	1	1
02 A	0	0	0	1	0	1	15 B	1	0	0	0	0	1
04 A	0	0	1	1	0	2	01 C	1	0	0	0	0	1
08 A	1	2	1	1	0	5	15 C	0	0	0	0	2	2
09 A	1	0	0	0	0	1	31 C	1	0	1	0	1	3
10 A	0	0	2	2	0	4	12 E	0	0	1	0	0	1
12 A	1	1	1	0	0	3	13 E	3	0	0	0	0	3
13 A	2	0	1	1	0	4	15 E	0	0	0	1	0	1
14 A	1	0	1	0	2	4	28 E	1	0	0	0	0	1
15 A	0	1	0	1	0	2	31 E	0	0	1	0	1	2
26 A	0	0	0	3	0	3	30 E	1	0	0	1	2	4
31 A	0	0	0	0	3	3	13 H	1	0	0	0	0	1
04 B	0	1	0	0	0	1	29 H	0	0	1	0	0	1
09 B	0	1	0	0	0	1	31 H	0	0	1	0	0	1

1. A: S.Lagoas, Minas Gerais; B: Capinopolis, Minas Gerais; C: Jatai, Goias; D: Pelotas, Rio Grande do Sul; E: Taquari, Rio Grande do sul.

Table 5. Reaction[1] of 11 sorghum hybrids and BTx 623 to five races of *Colletotrichum graminicola*.

Hybrid	Pedigree[2]	Race and reaction				
		13 A	31 A	13 B	14 B	15 C
CMSXS 350	BR 001 A × CMSXS 178	R	R	R	R	R
CMSXS 351	BR 007 A × CMSXS 178	R	R	R	R	R
CMSXS 352	BR 001 A × CMSXS 179	S	R	S	S	R
CMSXS 353	BR 007 A × CMSXS 179	R	R	S	R	S
CMSXS 354	BR 008 A × CMSXS 179	R	R	S	S	R
CMSXS 355	CMSXS 156 A × CMSXS 179	S	S	S	R	S
CMSXS 356	BR 001 A × CMSXS 180	S	S	S	S	S
CMSXS 357	BR 007A × CMSXS 180	R	S	R	R	R
CMSXS 358	BR 008 A × CMSXS 180	R	R	S	S	R
CMSXS 359	CMSXS 156 A × CMSXS 180	S	S	S	R	S
CMSXS 5360	BR 001 A × CMSXS 181	S	R	S	S	R
	BTx 623	S	S	S	S	S

1. R = resistant, S = susceptible.
2. BR 001 A : wheatland derivative
   BR 007 A : Redbine derivative
   BR 008 A : Redlan
   CMSXS 156 A : 1391 (Texas)
   CMSXS 178 : (SC 326-6 × SC 748-5) Selection
   CMSXS 179 : (BRP 3 R × SC 326-1) Selection.

released in 1988 as commercial hybrids for their agronomic characteristics and genetic resistance to the principal races of *C. graminicola*.

References

Casela, C.R., and **Ferreira, A.S.** (In press). Proposta de um sistema de classificacao de racas de *Colletotrichum graminicola* (Ces.) Wils. (Sensu Arx., 1957), agente causal da antracnose em sorgo (*Sorghum bicolor* (L.) Moench.) Fitopatologia Brasileira.

Ferreira, A.S., and **Casela, C.R.** 1986. Racas patogenicas de *Colletotrichum graminicola*, agente causal da antracnose em sorgo *Sorghum bicolor*. Fitopatologia Brasileira 11(1):83–87.

Habgood, R.M. 1970. Designation of physiologic races of plant pathogens. Nature 227:1268–1269.

Nakamura, K. 1982. Especializacao fisiologica em *Colletotrichum graminicola* (Ces.) Wils. (Sensu Arx., 1957) agente causal da antracnose em sorgo. Ph.D. Thesis, Universidade Estadual Paulista, Jaboticabal, Brazil. 143 pp.

Current Status of Sorghum Downy Mildew Control

J. Craig[1] and G.N. Odvody[2]

Abstract

Resistance to sorghum downy mildew (Peronosclerospora sorghi) *is vulnerable to the variability for virulence found in* P. sorghi. *Several pathotypes (races) of* P. sorghi *have been identified. Successive sorghum genotypes possessing monogenic, physiological resistance have become susceptible to new virulent pathotypes of* P. sorghi. *The sorghum cultivar QL 3 with oligogenic factors for resistance to downy mildew may provide a source of durable resistance to the pathogen. The fungicide metalaxyl provides excellent chemical control of sorghum downy mildew. However, the ability of fungi related to* P. sorghi *to produce metalaxyl-resistant strains raises the possibility that a metalaxyl-resistant strain of* P. sorghi *will occur. The practice most likely to provide durable control of downy mildew is use of metalaxyl only on DM-resistant sorghum genotypes.*

Introduction

Research on sorghum downy mildew (SDM) (*Peronosclerospora sorghi*) has successfully determined most of the etiological aspects of this disease. The pathogen's host range, environmental requirements, modes of infection, and other characteristics have been established (Frederiksen 1980; Safeeulla 1976; Williams 1984).

Disease-control procedures involving host-plant resistance or fungicides have been developed to reduce crop damage by SDM. These procedures have been used successfully, but our observations suggest that their continued effectiveness is in question. The current status and future possibilities of host-plant resistance and fungicides in the control of SDM are discussed.

Host-plant Resistance

Host-plant resistance is the most widely used method of reducing SDM losses. Downy mildew resistance as presently utilized in sorghum-improvement programs is expressed as a physi-ological incompatibility between host and pathogen that prevents infection. In most genotypes, this type of resistance is probably monogenic or infrequently oligogenic. Sorghum improvement programs in Texas have made extensive use of physiological resistance to SDM.

Sorghum downy mildew first appeared in Texas in 1961. Within 6 years it became economically important. Resistant sorghum hybrids were developed and widely grown. Six years after popular acceptance of those first resistant hybrids, a new SDM pathotype (physiological race), capable of attacking some of these hybrids, appeared (Craig and Frederiksen 1980). However, hybrids with the resistant sorghum inbred Tx 430 as a parental line were resistant to this new pathotype. Again, 2 years later, a third pathotype of *P. sorghi* appeared, and was capable of overcoming the resistance of Tx 430 (Craig and Frederiksen 1983).

Again, sources of resistance effective against the new race of *P. sorghi* were identified and utilized in breeding programs. These new hybrids are resistant to all the known Texas pathotypes of SDM, but it is probable that a new race of the

1. Agricultural Research Service, United States Department of Agriculture, PO Drawer DN, College Station, TX 77841, USA.
2. Texas Agricultural Experiment Station, Route 2, PO Box 589, Corpus Christi, TX 78410, USA.

Craig, J., and **Odvody, G.N.** 1992. Current status of sorghum downy mildew control. Pages 213–217 *in* Sorghum and millets diseases: a second world review. (de Milliano, W.A.J., Frederiksen, R.A., and Bengston, G.D., eds). Patancheru, A.P. 502 324, India: International Crops Research Institute for the Semi-Arid Tropics.

pathogen virulent to these hybrids will appear in the near future. The variability for virulence found in *P. sorghi* presents a problem for the control of SDM with host-plant resistance.

To date, five races of *P. sorghi* have been identified in the Americas, three in Texas (Frederiksen 1980), one in Brazil (Fernandes and Schaffert 1983), and one in the Honduras (C. D. H. Meckenstock, Soil Crops Department, Texas A&M University, College Station, Texas, personal communication). These five races represent only a small portion of the variability present in the pathogen. In a study of virulence in *P. sorghi*, Pawar tested 75 sorghum cultivars for reaction to a collection of isolates of the pathogen from Asia, Africa, Central America, and North America (Pawar 1986). Differential reactions of the test cultivars identified each of the 16 isolates as a different pathotype. The pathotypes from Africa and India had a much wider range of virulence among the 75 cultivars than did pathotypes from the Americas. Ten of the 75 sorghum cultivars were resistant to all of the pathotypes.

There is no shortage of sorghum cultivars resistant to one or more races of *P. sorghi*. In a recent test I screened 2000 accessions in an Ethiopian sorghum collection, and 17% of these were resistant to the three Texas pathotypes of *P. sorghi*. These tests identify resistant genotypes but provide no information about the range of genetic diversity for SDM resistance represented by these resistant cultivars. Diversity of resistance genes, rather than numbers of resistant cultivars, is the relevant factor in the use of host-plant resistance for control of SDM.

Recent research conducted in Texas on the modes of inheritance of SDM resistance in the sorghum lines SC 414-12 and QL 3 indicated that resistance in SC 414-12 was conferred by a dominant gene (Craig and Schertz 1986; Sifuentes 1985). In QL 3, resistance was conditioned by each of two independent dominant genes; neither at the same locus as the resistant factor in SC 414-12. An earlier study of SDM resistance in QL 3, conducted in India (Bhat 1981), found that the mode of inheritance was relatively complex and involved six genes. The discrepancy between results obtained in India and those reported in Texas could have been caused by differences in the methodology of testing, pathogen biotypes, or the QL 3 selections.

Our experience in Texas indicates that monogenic physiological resistance will not provide horizontal, or even durable, resistance to SDM. We seem to have embarked on an endless cycle of development, deployment, breakdown, and replacement of SDM resistance. The oligogenic resistance found in QL 3, and possibly present in some other cultivars, may provide the mechanism for durable resistance to SDM. Selections of QL 3 were resistant to *P. sorghi* at several geographically disparate test sites for many years (Mughogho 1983). In addition, QL 3 was resistant to an international collection of 16 pathotypes of *P. sorghi* in greenhouse trials (Pawar 1986). However, translation of these past performances of QL 3 into agronomically desirable sorghum hybrids is yet to be reported.

Another approach to durable host-plant resistance that should be given attention is the use of minor, nonspecific genes to reach an acceptable level of resistance to SDM. Presumably, these minor factors exist and account for differences in SDM incidence between sorghum cultivars when neither possesses major genes for physiological resistance to *P. sorghi*. The task of identifying these minor genetic factors and increasing their frequencies in sorghum selections will be difficult, but success in this endeavor would provide stable resistance to SDM.

Chemical Control

Until the advent of acylanilide fungicides in the 1970s, there was no satisfactory chemical control for systemic downy mildews of cereals. Other types of contact and systemic fungicides were partially effective against local lesion infection by downy mildew pathogens, but only as protectants against further infections (Cohen and Coffey 1986; Schwinn 1980). The acylanilides have a narrow spectrum of activity, restricted to the Peronosporales and the Hypochytridiomycetes. The acylanilides are taken up by the roots and leaves, and are primarily translocated acropetally to become systemic (Cohen and Coffey 1986). The acylalanines include metalaxyl, furalaxyl, and benalaxyl. Of this group, metalaxyl, marketed by Ciba-Geigy as Apron®, is the only fungicide used to control sorghum downy mildew. Metalaxyl's principal use is as a seed treatment to protect plants against infection by oospores in the soil (Odvody and Frederiksen 1984a; Odvody et al. 1984; Anahosur and Patil 1980). However, the fungicide is also effective when applied to the foliage to protect plants against infection by conidia or to eradicate the pathogen after infection has occurred (Odvody and Frederiksen 1984b; Anahosur and Patil 1980). Metalaxyl treatment at rates as low as 0.63 g a.i. kg^{-1} of seed provides excellent protection against SDM (Odvody et al. 1984).

Metalaxyl acts against the pathogen after penetration of the host has occurred. The chemical does not affect spore germination or host penetration, but inhibits further development of the pathogen after the initial invasion. Metalaxyl's mode of action is reported to be the inhibition of the synthesis of RNA in sensitive fungi (Davidse and de Waard 1984).

Metalaxyl was approved for control of sorghum downy mildew in Texas because most of the sorghum hybrids grown in Texas were susceptible to a new pathotype of *P. sorghi* (Odvody et al. 1984). The use of metalaxyl in Texas was recently expanded by an iatrogenic relationship between SDM and the herbicide antidote Concep II®. Chloroacetanilide herbicides are widely used to control grasses in sorghum fields in Texas; sorghum is sensitive to these herbicides, but can be protected from damage by seed treatment with a herbicide antidote. In 1984, it was discovered that the herbicide antidote CGA 92194, marketed as Concep II® by Ciba-Geigy, increased the incidence of SDM in sorghum (Craig et al. 1987). Concep II® did not affect sorghum genotypes that had physiological resistance to *P. sorghi*, but the disease incidence in plants of susceptible sorghum from seed treated with Concep II® was significantly greater than that in plants from nontreated seed of the same genotypes. The combination of metalaxyl with Concep II® in the treated seed provided adequate protection against downy mildew (Craig et al. 1987).

The narrow spectrum of fungicidal activity exhibited by metalaxyl made it highly probable that resistant biotypes of the target fungi would develop. This has occurred relatively quickly. Metalaxyl-resistant strains have appeared in the genera *Peronospora*, *Phytophthora*, *Plasmopara*, *Pseudoperonospora*, and *Pythium*. However, no resistant strains of the graminaceous downy mildew pathogens have been reported (Cohen and Samouche 1986; Davidse and de Waard 1984).

The disease systems in which pathogens have developed metalaxyl-resistant strains are those in which several cycles of spore production and host infection occur in one season and metalaxyl was applied repeatedly under conditions favorable to multiplication of the pathogen. Such systems apply extremely heavy selection pressure for the development and increase of resistant biotypes of the pathogen.

The disease cycle of SDM in grain sorghum differs significantly from these systems and is much less likely to produce metalaxyl-resistant biotypes of *P. sorghi*. Metalaxyl is usually applied only once as a seed treatment. Systemic infection of the plant is required to produce the oospores needed for survival over the fallow season. This systemic infection occurs only at the early stages of plant growth. As a result, conidia produced on systemically diseased plants by a metalaxyl-resistant biotype of the pathogen are unlikely to induce systemic infection and oospore production in surrounding plants. Consequently, the buildup of a metalaxyl-resistant population of *P. sorghi* would be appreciably slower than in the systems with several oospore-producing cycles per season.

These conditions reduce the probability of producing a metalaxyl-resistant biotype of *P. sorghi*, but the possibility remains (Odvody and Frederiksen 1984a). Practices that increase this possibility should be avoided. Metalaxyl seed treatment should not be applied at concentrations less than required for maximum control. Where infection is primarily mediated by oospores in young seedlings, the control goal should be to achieve as close to 100% control of SDM as possible to reduce the number of plants systemically infected with biotypes of *P. sorghi* possessing variable tolerance to metalaxyl (Odvody and Frederiksen 1984a).

The use of metalaxyl to control SDM in forage sorghum should be avoided. Conidial infection of the regrowth after each cutting provides an opportunity for large increases of oospore inoculum of any metalaxyl-resistant biotype present. Foliar application of metalaxyl to plants systemically infected with *P. sorghi* should be avoided, because it would select resistant strains of the pathogen.

A metalaxyl-resistant biotype of *P. sorghi* would complicate continued chemical control of SDM, because no equally effective fungicide is available. The pathogen is sensitive to other acylalanine fungicides (Odvody, unpublished). However, the metalaxyl-resistant strains that have occurred in other Peronosporales were also resistant to other acylalanines (Cohen and Samouche 1984). The same cross protection would be probable in metalaxyl-resistant biotypes of *P. sorghi*.

The use of fungicide mixtures in some disease systems effectively prevents or slows the selection and increase of pathogen strains tolerant to specific fungicides in the mixture (Cohen and Coffey 1986). This is currently impractical for graminaceous mildews, because all effective fungicides are in the acylanilide group. An exception is the phosphonate fosetyl-A1, but even at high rates fosetyl-A1 is sometimes inconsistent in control (Odvody, unpublished). In addition, recent research demonstrated that metalaxyl-resistant strains of *Phytophthora infestans* and *Pseudoperonospora cubensis* were also resistant to fosetyl-A1 (Cohen and Samouche 1984).

Conclusion

The combined use of the best available host-plant resistance with metalaxyl should simultaneously provide the best protection against fungicide-tolerant strains and conserve resistant germplasm. Virulent, metalaxyl-sensitive pathotypes, and avirulent metalaxyl-resistant pathotypes would not survive. The probability of a pathotype occurring which combines virulence to the resistant host and resistance to metalaxyl is comparatively small. However, this option may soon be lost in regions where protection of susceptible sorghum genotypes with metalaxyl has become routine.

References

Anahosur, K.H., and Patil, S. 1980. Chemical control of sorghum downy mildew in India. Plant Disease 64: 1004–1006.

Bhat, M.G. 1981. Studies on inheritance of resistance to downy mildew *Peronosclerospora sorghi* (Weston and Uppal) Shaw in sorghum. M.Sc. thesis. University of Agricultural Science, Bangalore, India. 156 pp.

Cohen, Y., and Coffey, M.D. 1986. Systemic fungicides and the control of oomycetes. Annual Review of Phytopathology 24: 311–338.

Cohen, Y., and Samouche, Y. 1984. Cross resistance to four systemic fungicides in metalaxyl resistant strains of *Phytophthora infestans* and *Pseudoperonospora cubensis*. Plant Disease 68: 137–139.

Craig, J., and Frederiksen, R.A. 1980. Pathotypes of *Peronosclerospora sorghi*. Plant Disease 64: 778–779.

Craig, J., and Frederiksen, R.A. 1983. Differential sporulation of pathotypes of *Peronosclerospora sorghi* on inoculated sorghum. Plant Disease 67: 278–279.

Craig, J., Frederiksen, R.A., Odvody, G.N., and Szerszen, J. 1987. Effects of herbicide antidotes on sorghum downy mildew. Phytopathology 77: 1530–1532.

Craig, J., and Schertz, K.F. 1985. Inheritance of resistance in sorghum to three pathotypes of *Peronosclerospora sorghi*. Phytopathology 75: 1077–1078.

Davidse, L.C., and de Waard, M.A. 1984. Systemic Fungicides. Pages 191–257 in Advances in Plant Pathology, Vol. 2. (Ingram, D.S., and Williams, P.H., eds.) London, UK, and New York, NY, USA: Academic Press.

Fernandes, F.T., and Schaffert, R.E. 1983. The reaction of several sorghum cultivars to a new race of sorghum downy mildew (*Peronosclerospora sorghi*) in southern Brazil in 1982–1983. 1983 Agronomy Abstracts: 63.

Frederiksen, R.A. 1980. Sorghum downy mildew in the United States: Overview and outlook. Plant Disease 64: 903–908.

Mughogho, L.K. 1983. The occurrence and control of sorghum downy mildew in India and Africa. Newsletter, International Working Group on Graminaceous Downy Mildews 5(2):5.

Odvody, G.N., and Frederiksen, R.A. 1984a. Use of systemic fungicides metalaxyl and fosetyl-A1 for control of sorghum downy mildew in corn and sorghum in south Texas. I: Seed Treatment. Plant Disease 68: 604–607.

Odvody, G.N., and Frederiksen, R.A. 1984b. Use of systemic fungicides metalaxyl and fosetyl-A1 for control of sorghum downy mildew in corn and sorghum in south Texas. II: Foliar Application. Plant Disease 68: 608–609.

Odvody, G.N., Frederiksen, R.A., and Craig, J. 1984. The integrated control of downy mildew. Proceedings 38th Annual Corn and Sorghum Research Conference 38: 28–36.

Pawar, M.N. 1986. Pathogenic variability and sexuality in *Peronosclerospora sorghi* (Weston and Uppal) Shaw, and comparative nuclear cytology of *Peronosclerospora* species. Texas A&M University. Ph.D. thesis. 92 pp.

Safeeulla, K.M. 1976. Biology and control of the downy mildews. Mysore, India: Wesley Press. 304 pp.

Schwinn, F.J. 1980. Prospects for chemical control of the cereal downy mildews. Pages 220–222 *in* Sorghum diseases, a world review: proceedings of the International Workshop on Sorghum Diseases, 11–15 Dec 1978, ICRISAT, Hyderabad, India. Patancheru, A.P. 502 324, India: International Crops Research Institute for the Semi-Arid Tropics.

Sifuentes, J.A. 1985. Inheritance of resistance to pathotypes 1, 2, and 3 of *Peronosclerospora sorghi* in *Sorghum bicolor*. M.Sc. thesis, Texas A&M University. 48 pp.

Williams, R.J. 1984. Downy mildews of tropical cereals. Pages 1–103 *in* Advances in Plant Pathology, Vol. 2. (Ingram, D.S., and Williams, P.H., eds.) London, UK, and New York, NY, USA: Academic Press.

Stalk Rots

S. Pande[1] and R.I. Karunakar[2]

Abstract

Stalk rots of sorghum are diseases of great destructive potential. Rots caused by the fungi Macrophomina phaseolina *and* Fusarium moniliforme *appear to be widely distributed stalk diseases of sorghum. Recently a vascular pathogen,* Acremonium strictum *that causes leaf and stalk death, has become important on sorghum.*

Improved high-yielding varieties tend to be highly susceptible to these diseases. Losses vary from season to season and region to region. Grain losses exceeding 15% are not uncommon; as much as 60% can occur. Several fungi and bacteria are often associated in diseased roots and stalks, suggesting that stalk rot diseases are of complex etiology. Etiology and host resistance to charcoal rot, fusarium root and stalk rot, and acremonium wilt are discussed.

Introduction

Stalk rots are universally important and among the most destructive diseases of sorghum throughout the world. Development of stalk rots is favored by early grain-filling and a late-season stress. Postflowering stresses include various leaf diseases, excessive cloudiness, high plant densities, drought, hail damage, low K coupled with high N, root rots, and injury by root and stalk boring insects. Almost any factor that reduces photosynthate production by the sorghum plant favors stalk rot.

In most cases stalk rots follow root rots, and are caused by several fungi and bacteria that attack plants approaching maturity. A number of organisms often can be isolated from diseased roots and stalks; the well known causal agents are the fungi *Macrophomina phaseolina* (charcoal rot), *Fusarium moniliforme* (fusarium root and stalk rot complex), *Periconia circinata* (milo disease), *Pythium* spp (root and seedling rots), and *Colletotrichum graminicola* (anthracnose stalk rot).

M. phaseolina and *F. moniliforme* appear to be widely distributed in sorghum-growing areas.

P. circinata, once thought to be restricted to USA, has been reported from Australia (Mayers 1976).

Recently *Acremonium strictum*, a vascular pathogen causing leaf and stalk death, has been recognized as an important disease in the Americas.

Root and stalk diseases reported by Tarr (1962) include pink root rot (*Pyrenochaeta terrestris*), southern sclerotial rot (*Sclerotium rolfsii*), and rhizoctonia stalk rot (*R. solani*). Bacteria, particularly *Erwinia* spp, have also been implicated as sporadically important causal agents of stalk rots in the Philippines (Karganilla and Exconde 1972), in India (Anahosur 1979), Nigeria (King 1973), and USA (Zummo 1969). Little is known about the etiology of these diseases and the topic is an obvious area that demands research.

Root and stalk rots are reported to cause crop losses, but data from experiments are limited. Losses vary from season to season and from region to region worldwide. Yield losses from 15 to 60% may occur on susceptible cultivars. Losses may be direct (due to poor grain filling) or indirect (due to stalk breakage or lodging).

1. Plant Pathologist, Cereals Program, ICRISAT Center, Patancheru, Andhra Pradesh 502 324, India.
2. Research Associate, Cereals Program, ICRISAT Center, Patancheru, Andhra Pradesh 502 324, India.

Pande, S., and **Karunakar, R.I.** 1992. Stalk rots. Pages 219–234 *in* Sorghum and millets diseases: a second world review. (de Milliano, W.A.J., Frederiksen, R.A., and Bengston, G.D., eds). Patancheru, A.P. 502 324, India: International Crops Research Institute for the Semi-Arid Tropics. (CP 742).

Eighteen papers in the proceedings (Mughogho 1984) of the 1983 Consultative Group Discussion on Research Needs and Strategies for Control of Sorghum Root and Stalk Rot Diseases, Bellagio, Italy, discussed in detail the physiological and environmental factors that influence these diseases and their control by fungicides, cultural practices, and host resistance. This review attempts to summarize the present state of knowledge and progress on basic aspects of charcoal rot and fusarium root and stalk rot, the two major root and stalk rot diseases of complex etiology.

Acremonium wilt has become an important disease of sorghum. Anthracnose stalk rot, another damaging disease of sorghum particularly in warm humid sorghum-growing areas, was reviewed by Ali (this publicaction). He discussed two other phases of the disease: foliar anthracnose and panicle and grain anthracnose phase.

Charcoal Rot

Charcoal stalk rot of sorghum [*Sorghum bicolor* (L.) Moench] caused by the fungus *Macrophomina phaseolina* (Tassi Goid.), is a disease of great destructive potential when vigorously growing sorghums fill grain under drought stress (Edmunds 1964; Odvody and Dunkle 1979). Charcoal rot has been reported from all the ecologically diverse areas of sorghum culture in the tropics, subtropics, and temperate regions (Tarr 1962; ICRISAT 1980). In general, the worldwide distribution of the disease indicates its occurrence on many different soil types.

In diseased roots and stalks, *M. phaseolina* is often associated with other fungi, suggesting that the disease is of complex etiology (Mughogho and Pande 1984).

Economic importance

The literature contains several reports on the destruction of sorghum crops by charcoal root and stalk rot, but sound and reliable quantitative data on yield losses are not available. Reports by Uppal et al. (1936) from Maharashtra state, India, and Harris (1962) from Kano, Nigeria, stated that charcoal rot caused "sufficient to considerable loss in yield." S. B. King and D. Barry (Major cereals *in* African Project, Samaru, Nigeria, 1970; unpublished report of a trip to Cameroon and Chad) saw severe symptoms of charcoal rot and estimated yield losses up to 50%. Similarly "serious losses" in several states in USA were reported but quantitative data were not given (Leukel et al. 1951).

In spite of the lack of such data, the destructive potential of charcoal rot in susceptible cultivars may be recognized in four ways: (1) loss in grain yield and quality due to plants smaller than normal and premature ripening; (2) poor crop stand due to seedling blight; (3) complete loss of yield of lodged plants where mechanical harvesting is practiced; destruction of lodged plants by termites or other animal pests before the grain or fodder is manually collected; and (4) loss in quality and quantity of fodder due to infection and destruction of the stalk.

Mughogho and Pande (1984) estimated grain yield losses of 23–64% in research sowings of the CSH 6 hybrid. Similarly Anahosur and Patil (1983) noted 15–55% loss in grain mass in their experiments at Dharwad, India. In these experiments, grain yield from plots subjected to drought was measured. Although drought alone must have contributed to some of the yield reduction, the combined effects of drought and charcoal rot that caused plants to lodge must have greatly increased the level of yield loss.

To separate the effect of drought from the combined effect of drought and charcoal rot on sorghum yield, we conducted an experiment during 1985 postrainy season at ICRISAT Center and at Dharwad. In this experiment the soil fungus flora was eliminated by methyl bromide gas fumigation of the test plots (500 g a.i. m⁻²) before sowing. Control plots were not fumigated.

Natural charcoal rot infection was induced by continuously reducing soil moisture from postflowering to grain maturity (Mughogho and Pande 1984). We obtained 95–100% lodging and grain yield losses of 20% in nonfumigated plots of CSH 6 hybrid (Table 1). These data on grain yield losses clearly show the potential economic importance of the disease. There is still need for data, particularly from surveys in farmers' fields, on the various types of losses.

Table 1. Lodging and yield of root and stalk rot-infected CSH 6 sorghum under continuously receding soil moisture and fumigated treatments at ICRISAT Center and at Dharwad, India, postrainy season 1985/86.

	ICRISAT Center			Dharwad		
Treatment	Lodging (%)	Yield [kg (9m²)⁻¹]	1000-grain mass (g)	Lodging (%)	Yield [kg(9m²)⁻¹]	1000-grain mass (g)
Fumigation[1]	6.5	2.0	14.1	23.3	3.6	18.5
No fumigation	99.7	1.6	12.5	95.2	2.9	15.2
SE	±2.1	±0.1	±1.0	±6.8	±0.1	±0.7
Yield loss (%)[2]		20	11.34		19.44	17.83

1. Plots were fumigated with methyl bromide @ 500 g a.i. $(5 m^2)^{-1}$
2. [(Fumigated − Nonfumigated)/fumigated] × 100.

Symptoms

Symptoms of the disease include root discoloration, root rots, soft stalks and lodging of plants, premature drying of stalks, and poorly developed panicles with small, inferior-quality grains. Vascular bundles of infected roots and stalks are profusely covered with tiny charcoal-colored sclerotia (Uppal et al. 1936; Tarr 1962).

The most striking and usually the first indication of the disease is lodging of plants approaching maturity.

Causal organism

The causal organism of charcoal rot is a common soilborne fungus known in its imperfect state as *Macrophomina phaseolina* (Tassi) Goid. (Domsch et al. 1980). In its perfect state it is called *Sclerotium bataticola* Taub. Eight synonyms that may be encountered in the literature are: *Macrophomina phaseoli* (Maubi.) Ashby, *Macrophomina philippines* Petr., *Macrophomina crochori* Sawada, *Macrophomina cajani* syd. & Butl., *Macrophomina sesami* Sawada, *Rhizoctonia bataticola* (Raub.) Butl., *Rhizoctonia lamellifera* Small, and *Dothiorella cajani* syd. & Butl. (Holliday and Punithalingam 1970).

Fungal colonization

Sorghum roots may be infected by various fungi from the seedling stage until maturity. In dis-
eased roots and stalks with conspicuous signs of charcoal rot, fungal isolations usually reveal the association of *M. phaseolina* with other fungi. In Argentina, where *F. moniliforme* was the predominant fungus isolated from diseased plants, 40% of the isolations were *M. phaseolina*. Others isolated included unidentified *Fusarium* spp, *Rhizoctonia solani*, *Helminthosporium sativum*, and *Nigrospora sphaerica* (Frezzi and Teyssandier 1980). Similarly, in New South Wales, Australia, systematic surveys to assess the relative importance of fungi associated with root and stalk rots revealed that, although *F. moniliforme* was predominate, *M. phaseolina* and *N. sphaerica* were also regularly isolated from these diseased roots and stalks (Trimboli and Burges 1983).

Patridge et al. (1984) reported early colonization of sorghum roots by species of *Fusarium*, *Alternaria*, and *Epicoccum*, all common root inhabitants. Under certain external stress situations, the plant becomes quasidefenseless, and some of these fungi become pathogenic, causing stalk rot.

Recently we monitored fungal colonization of roots and stalks of susceptible hybrid CSH 6 at seven growth stages, from seedling to grain maturity (black-layer formation), in non-drought-stressed plots and in drought-stressed plots from onset of flowering to grain maturity. This was done by planting surface-sterilized pieces of roots, crown, and first internode on potato dextrose agar, Czapeck dox agar, and Meyer's medium. Six *Fusarium* species (*F. moniliforme*, *F. moniliforme* var *subglutinans*, *F. mon-*

iliforme var intermedium, F. solani, F. semitectum, and F. chlamydosporum), Macrophomina phaseolina, Rhizoctonia solani, Phoma sorghina, Exserohilum rostratum, and Trichoderma harzianum were found to colonize sorghum roots and stems. Among these fungi we found F. moniliforme var. subglutinans and F. oxysporum to be early colonizers, increasing in abundance after the induction of drought stress from onset of flowering to matu-

rity. M. phaseolina was not isolated until after the hard-dough growth stage and then only from drought-stressed plants (Fig. 1).

Data cited above show that in most cases of charcoal rot, M. phaseolina is not the sole cause of the disease under natural field conditions, but acts in combination with other pathogens to produce it. In other words, what is visually identified as charcoal rot is a sign of one fungus

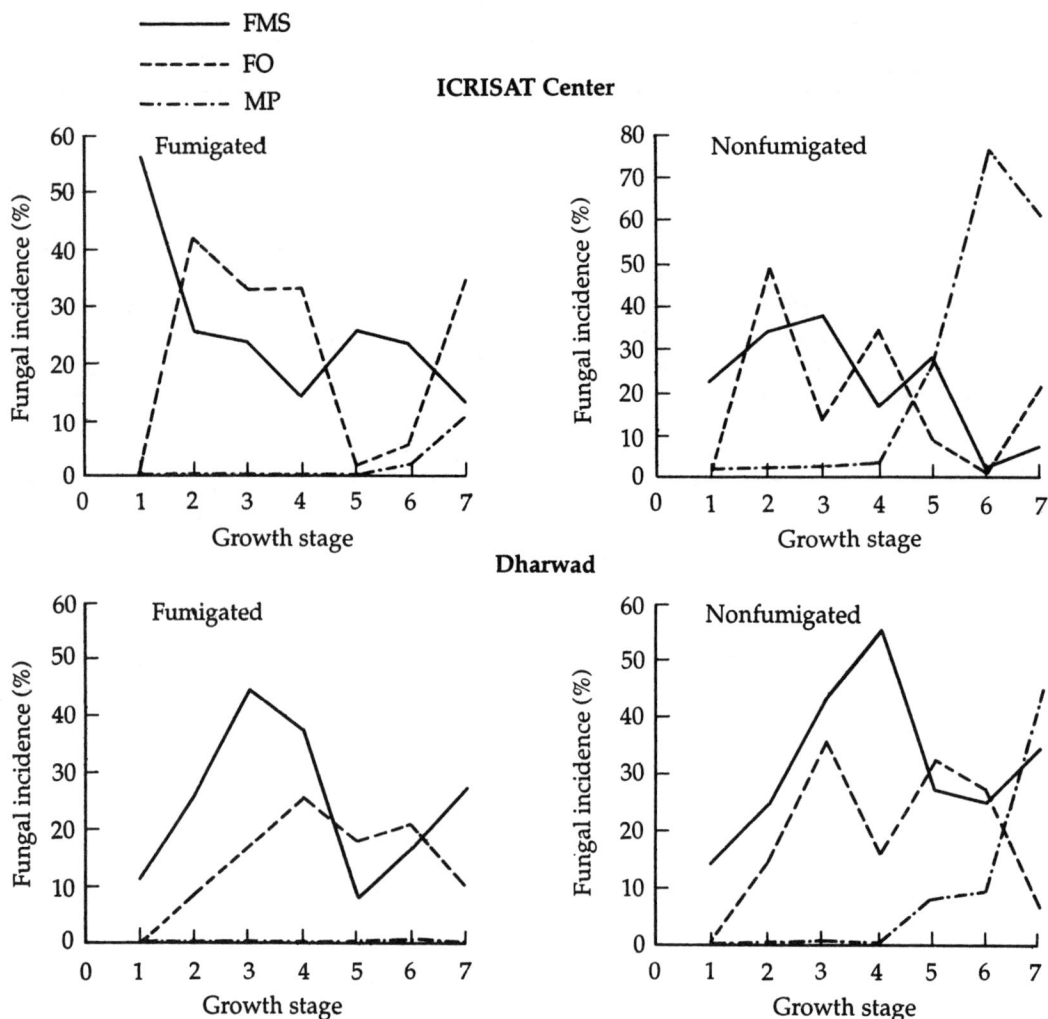

FMS = *Fusarium moniliforme* var. *subglutinans*, FO = *Fusarium oxysporum*, MP = *Macrophomina phaseolina*

1 = 5-6 leaf stage, 2 = Boot, 3 = Anthesis, 4 = Milk, 5 = Soft dough, 6 = Hard dough, 7 = Physiological maturity.

Figure 1. Fungal incidence (%) at different growth stages of CSH 6 under fumigated and nonfumigated conditions.

among several in a disease of complex etiology. Wadsworth and Sieglinger (1950) suggested that the several fungi associated with stalk rots attack in some orderly sequence, with *M. phaseolina* being the last and most conspicuous. The pathological significance of this involvement of several fungi in root and stalk rot infection remains unknown, and calls for prompt detailed investigation.

Variability and host range

M. phaseolina is highly variable in pathogenicity and mycological characteristics. Some isolates of the fungus are host specific (Hildebrand et al. 1945); others can attack a wide range of hosts (Holliday and Punithalingam 1970). Physiological races have been reported for isolates of some crops, such as jute (Ahmed and Ahmed 1969). Variability in cultural characteristics and pathogenicity of isolates from different parts of the same plant have been reported in soybean (Dhingra and Sinclair 1973). The fungus, a plurivorous pathogen, can affect 75 different plant families and about 400 plant species (Dhingra and Sinclair 1977).

Pathogen variation and physiological specialization of *M. phaseolina* are not known in charcoal rot of sorghum. It would be useful to know if sorghum is susceptible to isolates of the pathogen from other plant species and if physiological races exist among sorghum isolates of the pathogen.

Biology and epidemiology

Most of our knowledge of the biology of *M. phaseolina* is derived from the results of research with isolates from crops other than sorghum. It is assumed that the general biology of sorghum isolates is similar to that of isolates from other crops, although the pathosystem may be different. In this review only those aspects of the biology that influence the pathosystem will be discussed.

Inoculum source and survival

M. phaseolina is a root-inhibiting fungus with little or no saprophytic growth in soil or in dead host cells of infected plants (Norton 1953; Edmunds 1964). It survives over seasons predominantly as small black sclerotia in diseased root and stem debris or, upon decay of the plant material in which they were formed, in soil. Inoculum density has a direct implication in disease management strategies.

Drought stress and host colonization

Drought causes harmful physiological or metabolic changes in the plant. It reduces plant vigor; affected plants are predisposed to attack by non-aggressive pathogens such as *M. phaseolina* (Schoeneweiss 1978). From a review of stalk rot problems in maize and sorghum and the associated environmental factors, Dodd (1977, 1980) developed a photosynthetic stress-translocation balance concept to explain the predisposition of sorghum to charcoal rot. His hypothesis implies that the interaction of drought stress and pathogens causes stalk rots and lodging. Direct evidence of this has been provided by Henzell et al. (1984).

Recent research (1984–86) in sorghum pathology at ICRISAT Center has further clarified the role of soilborne fungi in root and stalk rot disease. The role of pathogens was examined in nontreated plots and plots fumigated with either granular fumigant tetrahydro-3, 5-dimethyl-2H-1,3, 5-thiadiazine-2-thione or with methyl bromide gas. The importance of soil moisture was studied in plots receiving adequate irrigation up to physiological maturity and in plots where drought stress was created by withdrawing irrigation at the boot-leaf stage. Plant lodging in the plots irrigated until plants matured was low (3–18%), as compared to the drought-stressed treatments (27–100%). However, in the drought-stressed treatments, lodging (100%) was significantly higher in the nonfumigated than in the fumigated plots (Table 2).

Results of plant senescence and visible stem infection by *M. phaseolina* and *Fusarium* spp (especially *F. moniliforme* var *subglutinans*) followed the same pattern as that of plant lodging and soft stalks. Our results show that soil fumigation significantly reduced plant senescence and lodging in the drought-stressed plots, supporting the findings of Henzell et al. (1984) that soilborne root-infecting fungi (especially *F. moniliforme*),

Table 2. Lodging, soft stalk, and yield of sorghum hybrid CSH 6 and sorghum variety E 36-1 under different treatment combinations of fumigation and drought stress at ICRISAT Center, India, postrainy season 1986/87.

		CSH 6			
Drought stress	Fumigation[2]	Lodging (%)	Soft stalk (%)	Plot yield [(kg(27m^2)$^{-1}$]	Grain mass (g)
Stress[1]	MB	27.4	11.3	11.6	19.1
	BAS	50.1	19.1	13.1	21.0
	NFC	99.5	96.6	8.8	17.0
	NF	100.0	99.6	9.8	17.0
Yield loss (%)[3]				20.0	15.0
No stress[4]	MB	2.9	0.8	14.2	23.0
	BAS	12.2	3.8	15.6	33.0
	NFC	16.4	9.3	3.7	24.0
	NF	17.6	3.7	14.1	22.0
SE Fum × stress × cv		±1.79	±1.97	±0.86	±0.14
		E 36-1			
Drought stress	Fumigation	Lodging (%)	Soft stalk (%)	Plot yield [(kg(27m^2)$^{-1}$]	Grain mass (g)
Stress	MB	0	0	12.9	27.0
	BAS	9.6	3.4	15.6	28.0
	NFC	8.9	6.5	13.2	26.0
	NF	10.4	5.2	12.9	26.0
No stress	MB	0	0	15.7	32.0
	BAS	0	0	17.0	32.0
	NFC	0	0	15.4	31.0
	NF	0	0	15.2	30.0
SE Fum × stress × cv		± 1.7	±2.0	±0.86	±0.14

1. Stress = Irrigation stopped at final leaf stage.
2. MB = methyl bromide, BAS = basamid granules, NFC = no fumigation, but covered, NF = no fumigation (control).
3. [(Average yield, fumigated plots – yield, nonfumigated plots)/average yield, fumigated plots] × 100.
4. No stress = irrigation continued until physiological maturity.

along with drought stress, play an important role in the root and stalk rot complex.

Henzell and Gullieron (1973) and Chamberlin (1978), on the other hand, hold the view that plant lodging under drought stress is purely physiological. Drought stress reduced assimilate supply to the lower part of the stalk for maintenance respiration. This results in senescence and disintegration of pith cells, and hence lodging.

Drought stress alone can cause lodging without assistance from pathogens where inoculum is absent (Henzell et al. 1984). However, where pathogens are present, drought-stressed plants are invariably invaded by them, and this leads to increased damage of plants. It is possible that low or intermediate levels of drought stress, in the absence of the pathogen, may be tolerated by the plant.

Crop management

Crop management can influence soil moisture and in turn incidence and severity of root and stalk rots. Sorghums grown in close spacings show more charcoal rot than those in wider spacings. Differences were reported in the effect of plant densities on charcoal rot incidence, attributing increased disease incidence in higher plant populations to increased drought stress. In India, nitrogen fertilization adequate to maximize the yield potential of improved cultivars increased the severity of charcoal rot (Avadhani et al. 1979; Mote and Ramshe 1980).

A factorial experiment using line-source irrigation provided highly significant positive correlations between drought stress and plant density, but not N fertilization (Fig. 2). The effect of nitrogen in increasing charcoal rot may be due to its effect on shoot growth. Nitrogen promotes shoot growth, and partially restricts root development. A restricted root system reduces the ability of a plant to obtain moisture. At the same time its water needs increase, because of the increased foliage growth (Ayers 1978).

Practices that reduce pathogen inoculum and conserve soil water decrease the incidence of charcoal rot. Sorghum grown under wheat-sorghum-fallow rotation had 11% stalk rots, compared to 39% in conventional tillage (Doupnik and Boosalis 1975).

Sorghum growing in a mixed-cropping situation has been reported to suffer less charcoal rot damage than when growing as a sole crop (Khune et al. 1980).

Control

Incidence of charcoal rot of sorghum can be minimized by maintaining soil moisture during the postflowering stage. This can be done by conservation tillage, seeding lower plant populations, selecting resistant sorghums, assuring good plant nutrition, and using fungicides and plant-growth regulators (such as antitranspirants). Host resistance is the most practical long-term solution for the control of charcoal rot.

Host resistance

Four essential requirements for the identification and utilization of host resistance to charcoal

rot have been discussed by Mughogho (1982). My objective is to review briefly techniques used to identify resistance, resistance sources, and factors associated with resistance. Rosenow (1980, 1984), Maunder (1984), and Henzell et al. (1984) have comprehensively reviewed the breeding for host resistance to sorghum root and stalk rot complex.

Resistance-screening technique

The procedure followed by most investigators to screen for charcoal rot resistance is essentially that reported by Edmunds et al. (1964). Sorghum is grown under irrigation in an environment favoring charcoal rot development. Drought stress is induced by withdrawing irrigation at selected stages of plant growth, and the stalks are inoculated by inserting mycelium- and sclerotia-infested toothpicks into holes punctured into the stalk just above the first node. Lodging, soft stalks, and distance the fungus travels upstem from the point of inoculation are measured to assess reaction of genotypes to the disease.

Toothpick inoculation and other methods where inoculum is introduced into the plant through the stalk are unsatisfactory, primarily because natural infection begins in the roots and only later goes up to the stem. The level of disease development with toothpick inoculation is usually less than that occurring naturally; thus stalk inoculation is thought unsatisfactory for meaningful resistance assessment (Edmunds et al. 1964).

At ICRISAT Center we have successfully induced, without artificial inoculation, charcoal rot in field-grown susceptible sorghums. One method is to sow the crop just before the end of the rainy season so that it grows and matures under continuously receding soil moisture. This timing is similar to that of the postrainy season (rabi) crop in India; postrainy season sorghum suffers most from charcoal rot. A second method is to grow the crop, under irrigation, during the dry season and to withdraw irrigation at 50% flowering. In either method, charcoal incidence and severity varies according to location—probably due to soil type, level of drought stress, and the pathogen inoculum potential in the soil. Nevertheless, disease development in susceptible genotypes is sufficiently high for useful evaluation of test genotypes, but a reliable, effi-

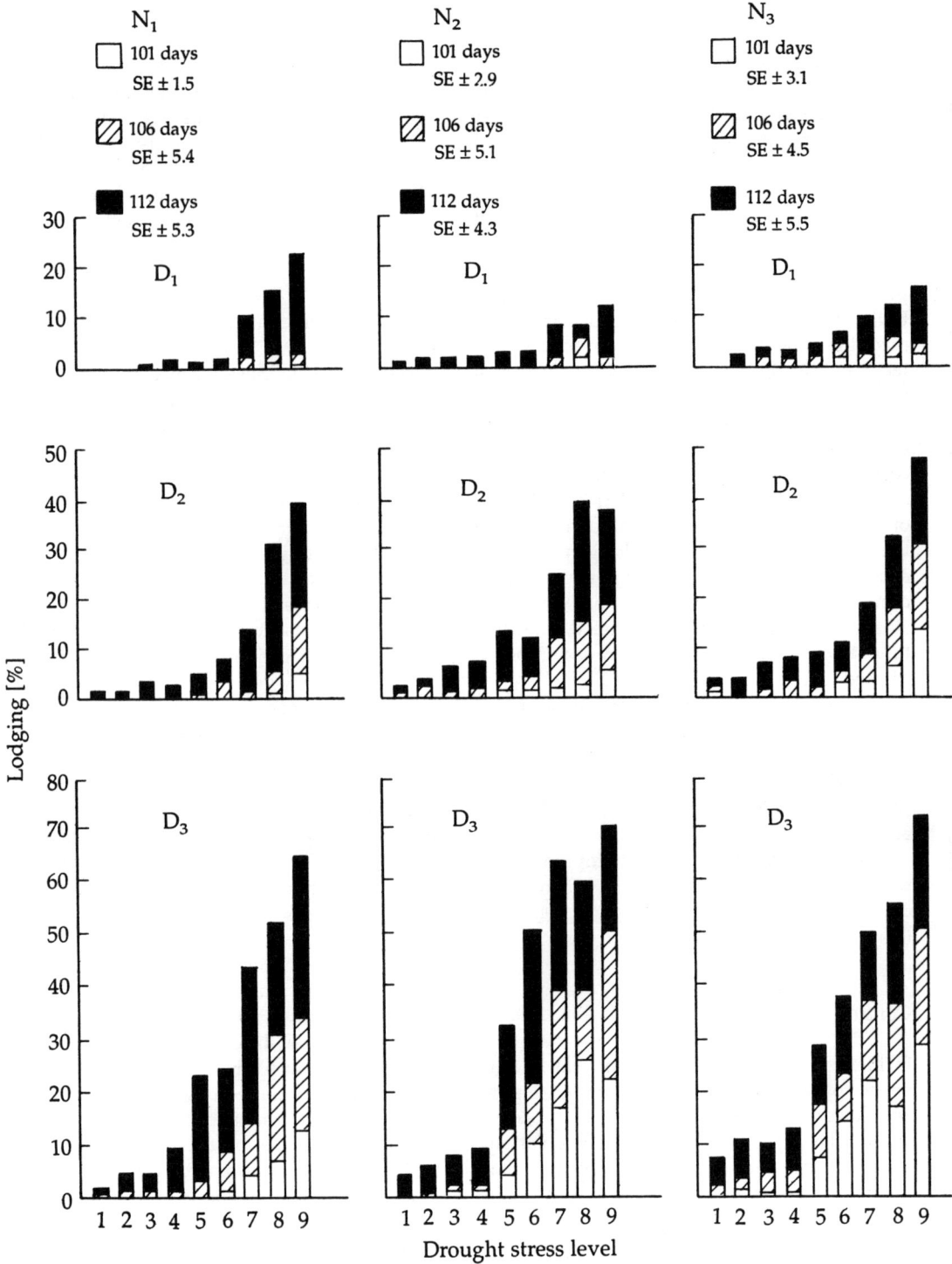

Figure 2. Periodical lodging (%) in three nitrogen levels (N_1 = 20, N_2 = 60, and N_3 = 120 kg N ha^{-1}), three plant densities (D_1 = 66 675, D_2 = 133 350 and D_3 = 266 700 plants ha^{-1}) and nine drought stress levels [1-9, 1(S_1) = nearest to LS and 9(S_9) farthest from LS] created by LS at ICRISAT Center, Patancheru, during the 1981/82 postrainy season.

cient, and epidemiologically sound resistance-screening technique for charcoal rot is yet to be developed.

Plant characters associated with resistance

The plant character most promising to be positively correlated with charcoal rot resistance is nonsenescence, increasingly being used as a selection criterion. Rosenow (1980) reported significant correlations between nonsenescence, lodging resistance, and charcoal rot resistance in Texas, USA. Similar results were obtained by Mughogho and Pande (1984) at ICRISAT, India. However, they could not find a consistent and stable nonsenescent line among the genotypes tested. Recently we have tested a set of 47 lines (reported nonsenescent in trials in Australia, India, and Mali) at four locations in India (Patancheru, Dharwad, Nandyal, and Bijapur). Four lines from Australia (Q 101, Q 102, Q 103, and Q 104) retained 9–63% green leaf area and were free from root and stalk rot (Tables 3, 4). Stability of nonsenescence would most probably depend on the level of drought stress. Up to a

Table 3. Time to 50% flowering, green leaf area, plot yield, and 1000-grain mass of four sorghum genotypes (rated as nonsenescents) at four locations in India, postrainy seasons 1985/86 and 1986/87.

Geno-type	Location	Time to 50% flowering (days)		Green leaf area (%)		Plot yield [kg(3m²)⁻¹]		1000-grain mass (g)	
		1985/86	1986/87	1985/86	1986/87	1985/86	1986/87	1985/86	1986/87
Q 101	Patancheru	69	59	17.5	45.5	1.4	2.3	26	28
	Dharwad	58	51	14.3	50.0	1.2	1.6	24	29
	Nandyal	54	59	40.8	30.0	0.7	1.5	17	22
	Bijapur	65	-[1]	10.7	-	0.6	-	19	-
Q 102	Patancheru	61	53	14.1	34.8	0.8	1.5	22	20
	Dharwad	56	51	14.3	29.2	1.1	1.7	21	25
	Nandyal	57	59	50.0	28.2	0.9	1.1	21	18
	Bijapur	63	-	8.7	-	0.7	-	18	-
Q 104	Patancheru	67	60	25.0	37.5	1.2	2.0	21	20
	Dharwad	59	55	17.9	38.9	1.3	1.7	20	22
	Nandyal	61	69	52.3	13.3	0.7	1.3	15	15
	Bijapur	65	-	9.9	-	0.6	-	20	-
Q 104	Patancheru	61	52	54.2	54.9	0.8	2.5	35	27
	Dharwad	57	52	32.2	63.3	0.9	2.2	35	36
	Nandyal[2]	58	57	63.3	15.0	1.1	1.2	21.3	19
	Bijapur	57	-	19.6	-	0.5	-	32.5	-
Control	Patancheru	62	51	0	0	1.4	3.9	23	27
(CSH 6)	Dharwad	53	46	4	6.3	1.7	2.0	27	22
	Nandyal	52	56	0	0	0.8	1.4	16	18
	Bijapur	63	-	0.2	-	0.9	-	17	-
SE	Patancheru	±2.1	±0.96	±4.2	±0.3	±1.63	±1.3	±1.3	±1.4
	Dharwad	±1.5	±0.61	±5.2	±4.53	±0.2	±0.17	±1.5	±0.8
	Nandyal	±2.4	±1.31	±5.9	±3.40	±0.1	±0.18	±1.2	±0.14
	Bijapur	±0.3	-	±2.4	-	-	-	±1.3	-

1. Poor emergence, experiment abandoned.
2. At Nandyal, Q 104 was harvested 5 weeks after physiological maturity in 1986/87.

Table 4. Lodging, soft stalk, nodes crossed, and root infection scores of four sorghum genotypes (rated as nonsenescents) grown at four locations in India, postrainy seasons 1985/86 and 1986/87.

Geno-type	Location	Lodging (%)		Soft stalk (%)		No. of nodes crossed		Root infection score	
		1985/86	1986/87	1985/86	1986/87	1985/86	1986/87	1985/86	1986/87
Q 101	Patancheru	0	0	0	0	0	0	3.6	2.9
	Dharwad	0	0	0	0	0	0	3.0	2.1
	Nandyal	0	0	6.3	1.5	0.3	0.05	4.3	3.75
	Bijapur	0	-[1]	0	-	0	-	2.0	-
Q 102	Patancheru	0	0	0	0	0	0	3.3	1.8
	Dharwad	0	0	0	0	0	0	2.4	2.0
	Nandyal	2.5	0	3.8	1.5	0.2	0.05	4.3	2.95
	Bijapur	0	-	0	-	0	-	1.8	-
Q 103	Patancheru	0	0	0	0	0	0	3.1	2.3
	Dharwad	0	0	0	0	0	0	2.4	2.0
	Nandyal	0	2.5	0	9.35	0	0.25	3.7	3.35
	Bijapur	0	-	0	-	0	-	2.3	-
Q 104	Patancheru	0	0	0	0	0	0	2.72	2.3
	Dharwad	0	0	0	0	0	0	2.5	2.0
	Nandyal	1.8	6.85	1.3	60.7[2]	0.1	2.9	4.7	4.5
	Bijapur	0	-	0	-	0	0	1.3	-
Control	Patancheru	71	100	71	100	2.7	4.6	5.0	4.6
(CSH 6)	Dharwad	73	100	73	100	3.1	5.0	4.9	5.0
	Nandyal	100	100	100	100	5.1	5.7	5.0	5.0
	Bijapur	100	-	0	-	0	-	5.0	-
SE	Patancheru	±7.5	± 6.92	±7.4	±3.42	±0.3	±2.06	±0.4	±0.41
	Dharwad	±10.8	± 5.59	±11.4	±7.47	±0.7	±0.20	±0.4	±0.33
	Nandyal	±4.9	± 9.56	±5.6	±8.63	±0.4	±0.29	±0.3	±5.14
	Bijapur	±4.3	-	±5.4	-	±0.3	-	±0.4	-

1. Poor emergence, experiment abandoned.
2. At Nandyal, Q 104 was harvested 5 weeks after physiological maturity in 1986/87.

specific stress level, a genotype would show stability in nonsenescence at several locations, but beyond that it may not. Further research is needed to explain this.

Resistance sources

Hoffmaster and Tullis (1944), in a most comprehensive testing program, screened 232 sorghum lines of diverse genetic background for 4 years at four locations. Although they found differences in stability to charcoal rot resistance in these lines, data showed no stability in performance of the lines from season to season. They concluded, "it is impossible, from the data available, to recommend certain varieties for lo-

calities in which macrophomina dry rot is a limiting factor."

In ICRISAT's charcoal rot research project we have also found inconsistencies in reaction to the disease by a large number of germplasm lines. This lack of stability is due to different levels of predisposition to the disease. However one line, E 36-1, has consistently shown resistance to lodging at several locations in 3 to 5 years of testing. Fungal isolations from roots and stalks showed presence of charcoal rot, but the infection was not severe enough to cause lodging (ICRISAT 1983). Rosenow (1980) identified 13 nonsenescent lines as good sources of resistance to charcoal rot. The stability of these lines in other countries where charcoal rot is a problem needs further evaluation. The need for stable and better sources of resistance is obvious. Most of ICRISAT's sorghum germplasm collection (more than 20 000 lines) has not been screened, and it is conceivable that these (especially among lines from drought-prone areas) include lines resistant to charcoal rot. However, the priority should be to develop a reliable screening technique that can be used to distinguish resistance from susceptible lines under graded levels of drought stress.

Fusarium Root and Stalk Rot

Fusarium root and stalk rot caused by the fungus *Fusarium moniliforme* has become an increasingly common stalk rot disease of sorghum in many areas of western Africa (Saccas 1954; Tarr 1962; Zummo 1980). In USA, the disease is generally found in the areas where charcoal rot occurs, particularly on the High Plains from Texas to Kansas (Edmunds and Zummo 1975). *Fusarium moniliforme* incites seed rots, seedling blights, and root and stalk rots of numerous crops, including maize, rice, millet, sudangrass, sugarcane, and sorghum (Bolle 1927, 1928; Dickson 1956; Bourne 1961; Sheldon 1904; Ullstrup 1936; and Voorhies 1933). *F. moniliforme* affects sorghum plants at all growth stages. It causes diseases such as pokkah boeng.

Etiology and symptoms

The fungus persists on plant residues that remain in the soil or rest on its surface. Mycelia, conidia, and (in its perfect state, *Gibberella fujikuroi*), ascospores may be produced on or in plants or the soil at any time during the growing season, and secondary infections of host tissue may occur when environmental conditions favor disease development.

Infection starts in the roots, typically involving the cortical and then the vascular tissues. Newly formed roots may exhibit distinct lesions. Older roots are often destroyed, leaving little plant anchorage. If root rot is extensive, the sorghum plants are often easily uprooted.

Fusarium stalk rot is usually accompanied by root damage. Under irrigation and heavy nitrogen fertilization, root damage may not change above-ground appearances before the stalk begins to rot. Stalk rot may reduce seed fill; grain

weight losses may be as high as 60%. Fusarium stalk rot apparently requires some predisposing conditions for disease development as plants approach maturity, and is usually most damaging during cool wet weather following hot, dry weather.

Trimboli and Burges (1983), in greenhouse trials, reproduced basal stalk rot and root rot on grain sorghum plants growing in *Fusarium moniliforme*-infected soil at optimal soil moisture, then at flowering subjected the plants to a gradual development of severe drought stress between the flowering and the middough stages, followed by rewetting. Stalk rot did not develop, nor was root rot severe, in plants grown to maturity at optimal soil moisture, although many of these plants were infected by *F. moniliforme*. Stalk and root rot developed in the majority of stressed plants grown in soil initially noninfested but contaminated by *F. moniliforme* after sowing.

Fusarium stalk rot can usually be distinguished from charcoal rot by the less-pronounced pigmentation and disintegration of pith tissues and the slower rate of tissue damage by *fusarium*. Charcoal rot may destroy a field of sorghum in 2 or 3 days; Fusarium stalk rot may take 2 or 3 weeks. The appearance of sclerotia in the later stages of charcoal rot is confirmation of its identity.

Acremonium Wilt

Acremonium wilt, caused by *Acremonium strictum* W. Gams (Syn. *Cephalosporium acremonium* Cord. Gams), has recently appeared in many sorghum-growing regions. *A. strictum* has a worldwide distribution in soil and the atmosphere (Domsch et al. 1980), but as an incitant of disease in sorghum, it has been reported from Argentina (Forbes and Crespo 1982), the Honduras (Wall et al. 1985), and USA (Frederiksen et al. 1980). A *Cephalosporium* sp reported as the cause of sorghum wilt in Egypt (El-Shafey et al. 1979) was probably *A. strictum*, and the disease probably occurs in additional sorghum-growing countries (Frederiksen 1984). Recently the disease was reported in India (Bandyopadhyay et al. 1987). Similar disease symptoms on a few isolated sorghum plants were observed in the African countries of Lesotho, Zimbabwe, Malawi, Tanzania, Burkina Faso, Mali, and Zambia (L.K.

Mughogho, R. Bandyopadhyay, and S. Pande, personal observations), but does not appear to be a serious disease on varieties currently grown by farmers. In the Honduras, however, the disease is important on local landraces (Wall et al. 1985). In maize, *A. strictum* (= *Cephalosporium acremonium* Corda) causes wilting of mature plants, infects seedlings (Christensen and Wilcoxson 1966), and internally colonizes grain (Hesseltine and Bothast 1977).

Symptoms

Symptoms of acremonium wilt include foliar desiccation and vascular discoloration of the lateral leaf veins. Initially only a portion of a leaf is affected, but as the disease progresses, large areas of wilted tissue develop on one axis of the leaf on either side of its midrib. Vascular plugging continues through the leaf sheath and into the vascular bundles of the stalk. In severely affected susceptible plants, the upper leaves and the shoot die. Vascular browning in the stalk associated with foliar wilting is the most distinguishing symptom of the disease.

Etiology

Natural infection probably begins in the leaf blades or leaf sheath and spreads through the vascular system. Infection and colonization through roots appear to be the exception (Frederiksen 1984). However, reports from Egypt suggest that the pathogen is soilborne and colonizes the roots prior to invading vascular tissue (El-Shafey and Refaat 1978). Frederiksen et al. (1981) reproduced disease symptoms experimentally through root dippings, soil amending, and hypodermic infections of conidia of *A. strictum* in sorghum whorls. Sorghum cultivars with field resistance (Redlan, Martin, and Wheatland) developed severe wilting when inoculated by these techniques. According to Salama (1979), wilting occurs commonly in regions of Egypt, similarly Frederiksen (1984) observed foliar infections in the farmer's field in the Nile delta near Numberia, Egypt. These authors hold the view that acremonium wilt is not a stalk rot, because *A. strictum* acts like a true vascular parasite. However, stalk-rotting fungi often develop

in wilted plants; in this respect *A. strictum* acts as a predisposing agent.

Greenhouse studies of several inoculation methods for the establishment of infection and the disease's systemic colonization of the host plants and transmission through seed were reported by Bandyopadhyay et al. (1987). In pots containing the susceptible cultivar IS 18442, the soil was drenched on the 10th day post-emergence with a conidial suspension of *A. strictum*; more than 70% of the plants became diseased. Disease symptoms in the greenhouse were similar to those in the field, and suggested that root injury was not essential for entry of *A. strictum*. In these experiments, it was possible to isolate *A. strictum* from any part of the infected plants, from roots to grains, confirming systemic colonization of the host. Infected seeds produced diseased plants when sown in autoclaved soil, but the actual potential of seed transmission under field conditions remains to be studied.

Host resistance

Most sorghum varieties appear to be resistant to acremonium wilt, or the pathogen develops so slowly that it does not cause serious losses. However, tropically adapted sorghums derived from IS 12610 are super susceptible to acremonium wilt. Elite inbreds such as ATx 623 and ATx 625 and many hybrids produced with these inbreds are unusually susceptible in the Honduras (Frederiksen 1984).

Research Recommendations

1. **Crop losses.** Quantitative crop-loss data of root and stalk rot diseases on sorghum in farmers' fields are limited or nonexistent. It is recommended that systematic crop-loss surveys be conducted in farmers' fields in areas where these diseases are economically important.
2. **Pathogenesis.** Knowledge of host-pathogen environment interactions is scanty. There is need to investigate pathogenesis, with emphasis on temperature, moisture, and nutritional stress as predisposing factors in charcoal rot, fusarium root and stalk rot, and acremonium wilt diseases of sorghum.

3. **Associated pathogens.** Interactions among the different pathogens involved in charcoal rot, different *Fusarium* species in fusarium root and stalk rot complex, and relations between *Acremonium strictum* and stalk rotting and lodging needs detailed study.

4. **Reliable field screening.** Development of a reliable field-screening technique for these diseases is essential for success in breeding for resistance.

5. **Plant-character and disease-resistance relationships.** Determine physical and physiological plant characteristics associated with resistance to the pathogen and to lodging. Study the physiological basis of nonsenescence, its stability under different environmental conditions, and its relationship to charcoal rot and fusarium root and stalk rot resistance.

References

Ahmed, N., and Ahmed, Q.A. 1969. Physiological specialization in *Macrophomina phaseoli* (Maubi) Ashby causing root rot of jute, *Corchorus* species. Mycopathologia et Mycologia Applicata 39:129–138.

Anahosur, K.H. 1979. Bacterial stalk rot of sorghum in regional research, Dharwar. Sorghum Newsletter 22:121.

Anahosur, K.H., and Patil, S.H. 1983. Assessment of losses in sorghum seed weight due to charcoal rot. Indian Phytopathology 36:85–88.

Ayers, P.G. 1978. Water relations of diseased plants. Pages 1–60 *in* Water deficits and plant growth (Kozlowski, T.T., ed.). New York, USA: Academic Press. 323 pp.

Avadhani, K.K., Patil, S.S., Mallanagoude, B., and Parvaticar, S.R. 1979. Nitrogen fertilization and its influence on charcoal rot. Sorghum Newsletter 22:119–120.

Bandyopadhyay, R., Mughogho, L.K., and Satyanarayana, M.V. 1987. Systemic infection of sorghum by *Acremonium satrictum* and its transmission through seed. Plant Disease 71:647–650.

Bolle, P.C. 1927. Een onderzoek near de oorzaak van pokkah boeng en toprot (An investigation into the cause of pokkah boeng and toprot). Archief Sulkerindustrie Nederlands-Indie III, 35: 589–609.

Bolle, P.C. 1928. Verdere onderzoek ingen over pokkah boeng en toprot (Further investigations in pokkah boeng and toprot). Archief Sulkerindustrie Nederlands-Indie I., 36:116–129.

Bourne, B.A. 1961. Fusarium soft or stem rot. Pages 182–202 *in* Sugarcane Diseases of the World (Vol. 1). (Martin, J.P., Abbor, E.U., and Hughes, C.G., eds.). New York, NY, USA: Elsevier Publishing Company.

Chamberlin, R. J. 1978. The physiology of lodging of grain sorghum (*Sorghum bicolor* L. Moench). Ph.D. thesis, University of Queensland, Australia.

Christensen, J.J., and Wilcoxson, R.D. 1966. Stalk rot of corn. Monograph, No. 3. St. Paul MN, USA: American Phytopathological Society. 59 pp.

Dhingra, O.D., and Sinclair, J.B. 1973. Location of *Macrophomina phaseoli* on soybean plants related to culture characteristics and virulence. Phytopathology 63:934–936.

Dhingra, O.D., and Sinclair, J.B. 1977. An annotated bibliography of *Macrophomina phaseolina* 1905–1975. Vicosa, Brazil: Imprensia Universitaria, Universidade Federal de Vicosa. 244 pp.

Dickson, J.G. 1956. Diseases of field crops. New York, NY, USA: McGraw Hill Book Co. 517 pp.

Dodd, J.L. 1977. A photosynthetic stress-translocation balance concept of corn stalk rot. Pages 122–130 in Proceedings, 32 Annual Corn and Sorghum Research Conference (Loden, H.D., and Wilkinson, D., eds.). Washington, DC, USA: American Seed Trade Association.

Dodd, J.L. 1980. The photosynthetic stress-translocation balance concept of sorghum stalk rots. Pages 300–305 *in* Sorghum diseases, a world review: proceedings of the International Workshop on Sorghum Diseases, 11–15 Dec 1978, ICRISAT, Hyderabad, India. Patancheru,

Andhra Pradesh 502 324, India: International Crops Research Institute for the Semi-Arid Tropics.

Domsch, K.H., Gams, W., and Anderson, T.H. 1980. Compendium of soil fungi (2 volumes) London, UK: Academic Press. 859 and 405 pp.

Doupnik, B., and Boosalis, M.G. 1975. Eco-fallow reduced stalk rot in grain sorghum. Phytopathology 65:1021–1022.

Edmunds, L.K. 1964. Combined relation of plant maturity, temperature, and soil moisture to charcoal stalk rot development in grain sorghum. Phytopathology 54:514–517.

Edmunds, L.K., Voight, R.L., and Carasso, F.M. 1964. Use of Arizona climate to induce charcoal rot in grain sorghum. Plant Disease Reporter 48:300–302.

Edmunds, L.K., and Zummo, N. 1975. Sorghum diseases in the United States and their control. United States. Department of Agriculture Handbook No. 469. Washington, DC, USA: US Government Printing Office. 47 pp.

El-Shafey, H.A., and Refaat, M.M. 1978. Studies on the stalk rot diseases of grain sorghum in Egypt. Agricultural Research Review 56:71–79. (Ministry of Agriculture, Cairo, Egypt)

El-Shafey, H.A., Abid-El-Rahim, M.F., and Refaat, M.M. 1979. A new Cephalosporium wilt of grain sorghum in Egypt. Pages 514–532 in Proceedings, Third Egyptian Phytopathological Congress.

Forbes, G.A. and Crespo, L.B. 1982. Marchitamiento en sorgo causado por Acremonium strictum Gams. Information Tecnica 89, Argentina: Estacion Experimental Agripecuaria Manfredi, INTA.

Frederiksen, R.A. 1984. Acremonium wilt. Pages 49–51 in Sorghum root and stalk rots, a critical review: proceedings of the Consultative Group Discussion on Research Needs and Strategies for Control of Sorghum Root and Stalk Rot Diseases, 27 Nov–2 Dec 1983, Bellagio, Italy. Patancheru, A.P. 502 324, India: International Crops Research Institute for the Semi-Arid Tropics.

Frederiksen, R.A., Natural, M., Rosenow, D.T., Morton, J.B., and Odvody, G.N. 1980. Acremonium wilt of sorghum. Sorghum Newsletter 23:134.

Frederiksen, R.A., Rosenow, D.T., and Natural, M. 1981. Acremonium wilt of sorghum, Pages 77–79 in Proceedings, 12th Biennial Grain Sorghum Research and Utilization Conference. Abernathy, TX, USA: Grain Sorghum Producers' Association.

Frezzi, M., and Teyssandier, E.E. 1980. Summary and historical review of sorghum diseases in Argentina. Pages 11–15 in Sorghum diseases, a world review: proceedings of the International Workshop on Sorghum Diseases, 11–15 Dec 1978, ICRISAT, Hyberabad, India. Patancheru, Andhra Pradesh 502 324, India: International Crops Research Institute for the Semi-Arid Tropics. 469 pp.

Harris, E. 1962. Diseases of guinea corn. Samaru Technical Notes 21–13.

Henzell, R.G., Dodman, R.L., Done, A.A., Brengman, R.L., and Mayers, O.E. 1984. Lodging, stalk rot, and root rot in sorghum in Australia. Pages 225–236 in Sorghum root and stalk rots, a critical review: proceedings of the Consultative Group Discussion on Research Needs and Strategies for Control of Sorghum Root and Stalk Rot Diseases, 27 Nov–2 Dec 1983, Bellagio, Italy. Patancheru, Andhra Pradesh 502 324, India: International Crops Research Institute for the Semi-Arid Tropics.

Henzell, R.G., and Gillieron, W. 1973. Effect of partial and complete panicle removal on the rate of death of some Sorghum bicolor genotypes under moisture stress. Queensland Journal of Agricultural and Animal Sciences 30:291–299.

Hesseltine, C.W., and Bothast, R.J. 1977. Mold development in ears of corn from tasselling to harvest. Mycologia 69:328–340.

Hilderbrand, A.A., Miller, J.J., and Koch, L.W. 1945. Some studies on Macrophomina phaseoli

(Maubi) Ashby in Ontario. Scientific Agriculture 25:690–706.

Hoffmaster, D.E., and **Tullis, E.C.** 1944. Susceptibility of sorghum varieties to Macrophomina dry rot (charcoal rot). Plant Disease Reporter 28:1175–1184.

Holliday, P., and **Punithalingam, E.** 1970. *Macrophomina phaseolina.* CMI Descriptions of pathogenic fungi and bacteria, No. 275. Kew, Surrey, UK: Commonwealth Mycological Institute.

ICRISAT. 1980. Sorghum diseases, a world review. proceedings of the International Workshop on Sorghum Diseases, 11–15 Dec 1978, ICRISAT, Hyberabad, India. Patancheru, Andhra Pradesh 502 324, India: International Crops Research Institute for the Semi-Arid Tropics. 478 pp.

ICRISAT (International Crops Research Institute for the Semi-Arid Tropics). 1983. Annual Report. 1982. Patancheru, Andhra Pradesh 502 324, India: ICRISAT.

Karganilla, A.D., and **Exconde. O.R.** 1972. Bacterial stalk rot of corn and sorghum. Philippine Phytopathologie 8–4 (abstract).

King, S.B. 1973. Plant pathology annual report (sorghum): Major Cereals in Africa Project. Samaru Nigeria: Institute of Agricultural Research.

Khune, N.N. Shiwankar. S.K., and **Wangikar, P.D.** 1980. Effect of mixed cropping on the incidence of charcoal rot of sorghum. Food Farming and Agriculture 12:292–293.

Leukel, R.W., Martin, J.H., and **Lefebvre. C.L.** 1951. Sorghum diseases and their control. Farmers' Bulletin No. 1959. Washington, D.C., USA: U.S. Department of Agriculture. 50 pp.

Maunder, A.B. 1984. Breeding for stalk rot resistance as a component of acceptable agronomic performance. Pages 219–224 *in* Sorghum root and stalk rots, a critical review: proceedings of the Consultative Group Discussion on Research Needs and Strategies for Control of Sorghum Root and Stalk Rot Diseases, 27 Nov–2 Dec 1983, Bellagio, Italy. Patancheru, A.P. 502 324, India: International Crops Research Institute for the Semi-Arid Tropics.

Mayers, P.E. 1976. The first recording of milo disease and *Periconia circinata* on sorghums in Australia, Australian Plant Pathology Society Newsletter 5:59–60.

Mote, U.N., and **Ramshe, D.G.** 1980. Nitrogen application increases the incidence of charcoal rot in rabi sorghum cultivars. Sorghum Newsletter 23:129.

Mughogho, L.K. 1982. Strategies for sorghum disease control. Pages 273–282 *in* Sorghum in the Eighties: proceedings of the International Symposium of Sorghum, 2–7 Nov 1981, ICRISAT Center, India. Pantancheru, A.P. 502 324, India: International Crops Research Institute for the Semi-Arid Tropics.

Mughogho, L.K., and **Pande, S.** 1984. Charcoal rot of sorghum. Pages 11–24 *in* Sorghum root and stalk rots, a critical review: proceedings of the Consultative Group Discussion on Research Needs and Strategies for Control of Sorghum Root and Stalk Rot Diseases, 27 Nov–2 Dec 1983, Bellagio, Italy. Patancheru, A.P. 502 324, India: International Crops Research Institute for the Semi-Arid Tropics.

Norton, D.C. 1953. Linear growth of *Sclerotium bataticola* through soil. Phytopathology 32:633–636.

Norton, D.C. 1958. The association of *Pratylenchus hexincisus* with charcoal rot of sorghum. Phytopathology 48:355–358.

Odvody, G.N., and **Dunkle, L.D.** 1979. Charcoal stalk rot of sorghum: effect of environment in host-parasite relations. Phytopathology 69:250–254.

Partridge, J.E., Reed, J.E., Jensen, S.G. and **Sidhu, G.S.** 1984. Spatial and temporal succession of fungual species in sorghum stalks as affected by environment. Pages 59–78 *in* Sorghum root and stalk rots, a critical review: proceedings of the Consultative Group Discussion on Research Needs and Strategies for Control of Sorghum Root and Stalk Rot Diseases, 27 Nov–2 Dec 1983, Bellagio, Italy. Patancheru, A.P. 502 324, India: International Crops Research Institute for the Semi-Arid Tropics.

Rosenow, D.T. 1980. Stalk rot resistance breeding in Texas. Pages 306–314 *in* Sorghum diseases, a world review: proceedings of the International Workshop on Sorghum Diseases, 11–15 Dec 1978, ICRISAT, Hyderabad, India. Patancheru, A.P. 502 324, India: International Crops Research Institute for the Semi-Arid Tropics.

Rosenow, D.T. 1984. Breeding for resistance to root and stalk rots in Texas. Pages 209–218 *in* Sorghum root and stalk rots, a critical review: proceedings of the Consultative Group Discussion on Research Needs and Strategies for Control of Sorghum Root and Stalk Rot Diseases, 27 Nov–2 Dec 1983, Bellagio, Italy. Patancheru, A.P. 502 324, India: International Crops Research Institute for the Semi-Arid Tropics.

Saccas, A.M. 1954, Les Champignons parasites des sorghos (*Sorghum vulgare*) et des penicillaries (*Pennisetum typhoidum*) en Afrique Equatoriale Francaise, Agronomie Tropicale Nogent 9:135–173, 263–301, 647–686.

Salama, I.S. 1979. Investigations of the major stalk, foliar, and grain diseases of sorghum (*Sorghum bicolor*) including studies on the general nature of resistance. Fourth Annual Report, Field Crops Research Institute, United States Agricultural Research Program, PL 480, Project No. E6-ARS-29, Giza, Egypt.

Schoeneweiss, D.F. 1978. Water stress as a predisposing factor in plant disease, Pages 61–99 *in* Vol. 5, Water deficits and plant growth (Kozlowski, T.T., ed) New York, NY, USA: Academic Press. 323 pp.

Sheldon, J.L. 1904. A corn mold (*Fusarium moniliforme* n. sp.). Nebraska Agricultural Experiment Station Annual Report (1903) 17:23–32.

Tarr, S.A.J. 1962. Diseases of sorghum, sudangrass and broom corn, Kew, Surrey, UK: Commonwealth Mycological Institute. 380 pp.

Trimboli, D.S., and Burges, L.W. 1983, Reproduction of *Fusarium moniliforme* basal stalk rot and root rot of grain sorghum in the greenhouse. Plant Disease 67:891–894.

Tullis, E.C. 1951. *Fusarium moniliforme*, the cause of a stalk rot of sorghum. Texas Phytopathology 41:529–535.

Ullstrup, A.J. 1936. The occurrence of *Gibberela fujikuroi* var. *subglutinans* in the United States. Phytopathology 26:685–693.

Uppal, B.N., Kolhatkar, K.G., and Patel, M.K. 1936. Blight and hollow-stem of sorghum. Indian Journal of Agricultural Science 6:1323–1334.

Voorhies, R.K. 1933. *Gibberella moniliforme* on corn. Phytopathology 23:368–378.

Wadsworth, D.F., and Sieglinger, J.B. 1950. Charcoal rot of sorghum, Oklahoma Agricultural Experiment Station Bulletin No. B-355. Stillwater, Oklahoma, USA: Oklahoma A&M College and U.S. Department of Agriculture. 7 pp.

Wall, G.C., Meckenstock, D.H., Nolasco, R., and Frederiksen, R.A. 1985, Effect of Acremonium wilt on sorghum in Honduras. (Abstr.) Phytopathology 75:1341.

Zummo, N. 1969. Bacterial soft rot, a new disease of sweet sorghum. Phytopathology 59:119 (abstract).

Zummo, N. 1980. Fusarium disease complex of sorghum in West Africa. Pages 297–299 *in* Sorghum diseases, a world review: proceedings of the International Workshop on Sorghum Diseases, 11–15 Dec 1978, ICRISAT, Hyderabad, India. Patancheru, A.P. 502 324, India: International Crops Research Institute for the Semi-Arid Tropics.

Sorghum Ergot

R. Bandyopadhyay[1]

Abstract

Ergot, caused by Sphacelia Sorghi *is an increasingly important disease of sorghum in eastern and southern Africa as well as Asia. The disease reaches economically significant levels particularly in sterile sorghums. Currently there are no distinct sources of host resistance, and other methods for control in areas where the pathogen is endemic, have not been established. Management of ergot in sorghum remains a major concern for the improvement of sorghum in much of the sorghum-growing region of Africa and Asia.*

Ergot is known to occur in Botswana, Ethiopia, Ghana, Kenya, Mozambique, Nigeria, Rwanda, Senegal, South Africa, Sudan, Tanzania, Uganda, Zambia, and Zimbabwe in Africa; and India, Japan, Myanmar, the Philippines, Sri Lanka, Thailand, and Yemen Arabic Republic in Asia. It has not been reported in the Americas, Australia, or Europe.

Information of the disease is reviewed, gaps in knowledge indicated, and areas for future research suggested.

Introduction

Ergot causes grain yield reduction because infected spikelets do not produce grain. Honeydew exuded from infected florets can smear healthy grains; saprophytes grow and reduce the quality of grains. Published data on crop loss are not available.

Ergot is a serious limiting factor in production of hybrid seed, particularly if seed-set in male-sterile lines is delayed due to lack of viable pollen caused by nonsynchronous flowering of male steriles and restorers. Moreover, environmental conditions favorable for disease development are not congenial for rapid seed-set, thus making spikelets more vulnerable to ergot attack. In India, where hybrid seeds are in great demand but in short supply, the damage is twofold: directly through loss in seed yield, and indirectly through rejection of ergot-sclerotia contaminated seed lots due to poor quality. If the disease strikes in new areas where farmers are totally dependent on hybrids, the effect would be devastating. Ergot can also cause widespread damage of male-fertile cultivars in farmers' fields when environmental conditions favorable to the pathogen occur at flowering (Molefe 1975; Kukadia et al. 1982).

Dihydroergosine is the major alkaloid in sorghum ergot sclerotia and honeydew, but four clavine alkaloids are also produced (Mantle 1968). The alkaloid spectrum in sclerotia collected from diverse sources may be uniform, but their concentrations differ (Molefe 1975). From rat-feeding trials, Mantle (1968) concluded that sorghum ergot sclerotia was the least toxic, if toxic at all, of all alkaloid-containing ergots.

Symptoms

The earliest symptom of the disease may be seen on the ovary if the glumes are opened 3–5 days after infection. The infected ovary appears dull

---

1. Plant Pathologist, Cereals Program, ICRISAT Center, Patancheru, Andhra Pradesh 502 324, India.

Bandyopadhyay, R. 1992. Sorghum ergot. Pages 235–244 *in* Sorghum and millets diseases: a second world review. (de Milliano, W.A.J., Frederiksen, R.A., and Bengston, G.D., eds). Patancheru, A.P. 502 324, India: International Crops Research Institute for the Semi-Arid Tropics. (CP 740).

green and slightly shriveled, in contrast to dark green and round appearance of a healthy, fertilized ovary. Within 2 days, superficially visible white mycelial stromata appear in the base of the ovary and gradually extend upwards. The ovary is converted into fungal stromata with shallow folds. The first external symptoms, clear to pinkish drops exuding from infected ovaries, appear 5–10 days after inoculation. The name, 'sugary' disease, for sorghum ergot originates from this sticky sweet fluid. It is also called honeydew and contains numerous conidia. Under humid conditions, a saprophyte *Cerebella volkensii* (Syn. *C. sorghi-vulgaris*) grows on honeydew and converts it into a matted, black mass. However, warm and dry conditions after the formation of honeydew will dry it, forming an easily removable, hard, white crust on the panicle. Finally, the fungal stromata are transformed into the hard, resting structure (sclerotia) that may or may not be concealed by the glumes.

Causal organism

The fungus is best known by its imperfect stage, *Sphacelia sorghi* McRae, but the perfect stage, *Claviceps sorghi*, has been described by Kulkarni et al. (1976). The imperfect stage is associated with the honeydew phase of symptoms. Closely packed pallisade-cell-like conidiophores are produced, either in the interior of folded stromata that replace the ovary or on the surface of the stromata. Numerous apical, hyaline, unicellular conidia are borne one at a time on the short conidiophores, possibly by a constriction mechanism. Conidia are oval or elliptical or oblong, and often have distinct vacuole-like bodies at the ends. Conidia are 5–8 × 12–20 µm in size (Mantle 1968; Kulkarni et al. 1976) and are released in the honeydew exuding from infected ovaries. More complete description of the asexual stage is required, considering the existence of two distinct conidial types (macroconidia and secondary conidia) of the pathogen (see Frederickson and Mantle this publication).

Abundant honeydew with conidia have been produced in culture on modified Kirchoff's medium (Nagarajan and Saraswathi 1975). Trace elements affect growth and sporulation of the sphacelial stage (Chinnadurai 1972). Conidial shape and size was reported to vary with nitro-

gen content of the culture substrate (Chinnadurai and Govindaswamy 1971b).

The fungal stromata transform into mature sclerotia within 4 weeks (Futrell and Webster 1965) to 2 months (Sangitrao and Bade 1979b) following inoculation. The shape and size of mature sclerotia depends on host genotype, environment, and nutritional factors (Sangitrao and Bade 1979b). Sclerotial development is often hampered by fungal contaminants that grow on honeydew and developing sclerotia. Velvety olive to black growth of *Cerebella* spp is the most common contaminant. *Fusarium moniliforme*, *F. roseum* f. sp *cerealis*, and *Cladosporium* spp may also be found (Futrell and Webster 1966). Most reports (Futrell and Webster 1966; Sangitrao and Bade 1979b) consider *Cerebella* as a parasite, but proof of parasitism is lacking.

Langdon reported in 1942 that *Cerebella* inhabiting sclerotial formation is a saprophyte nearly always associated with honeydew.

The perfect stage of *Claviceps* is initiated when the sclerotium germinates to produce two or three stipes bearing stromatal heads containing embedded flask-shaped perithecia with slight protrusion at the ostiolar region. Perithecia measure 66.4–124.5 × 132.8–232.4 µm and contain hyaline cylindrical asci 2.4–3.2 × 56–112 µm in size, bearing hyaline caps at their apices. Each ascus contains eight aseptate, filiform ascospores measuring 0.4–0.8 × 40–85 µm. This description of the perfect stage of *Claviceps* is based on the single report to date (Kulkarni et al. 1976). Additional descriptions of the *Claviceps* stage are required of isolates from different regions to determine if morphological and pathogenic variations occur in this pathogen.

Several researchers have reported varying degrees of success in attempts to germinate sclerotia in the laboratory. In some instances sclerotia did not germinate at all (Sundaram 1970; Nagarajan and Saraswathy 1975). In others, grayish bulges appeared from the cortex that broke through the rind, but instead of producing a clear stipe and capitulum, the bulge continued to proliferate to form a callus-like, spherical growth with mycelial tuft (Mantle 1968; R. Bandyopadhyay, unpublished). Mantle suggested the possibility of lack of infection by heterothallic strains as the reason for the lack of production of sclerotia that can germinate to produce ascospores.

Singh (1964) was the first to report that sorghum ergot sclerotia germinate to produce clear stipe and capitulum, but he did not describe the perfect stage. Sexual germination of sclerotia was confirmed by Kulkarni et al. (1976), and Sangitrao and Bade (1979b) in India, and by W.A.J. de Milliano and coworkers (W.A.J. de Milliano, SADCC/ICRISAT Sorghum and Millets Improvement Program, Bulawayo, personal communication 1987) in Zimbabwe. In India, sclerotia appeared to have a 3- to 4-month dormancy period. Ascospores obtained from germinated sclerotia, when inoculated repeatedly on panicles of male-sterile plants, produced honeydew and sclerotia (Singh 1964). Most reports indicated that very few sclerotia germinate, but Kulkarni et al. (1976) obtained 80% sclerotial germination. Details of the methodology used in these investigations were not reported.

Mode of infection and colonization

Information in this section is based on unpublished data. Conidia germinate on stigma to produce short germ tubes that penetrate stigmatic papillae. Sundaram (1970) reported that conidia germinate to produce secondary conidia, which then produce long germ tubes that penetrate stigma. This has not been confirmed. Occasionally, conidia can also germinate and penetrate style and ovary wall, but further route of invasion inside the host tissue is not known. Infection hyphae passes intercellularly through the style and ovary wall, and attaches itself to the vascular bundles at the base of the ovary 3 days after inoculation. Bundles of hyphal strand then grow rapidly along the inner layer of the ovary wall and envelop the ovule, without colonizing the ovule tissues. At the same time mycelial strands grow along the outer epidermis of the ovary wall, and then within the tissues of the ovary wall. Ovule tissues are the last to be colonized. The ovary is gradually digested by the fungal stromata, and in 5 to 7 days the ovary has been displaced by deeply involuted fungal stroma. Internal stromatic tissues possess locules in which conidia are produced on short conidiophores. They are also produced on the surface of the stromata. Conidia are released from the honeydew that oozes from the infected spikelet. Histological changes associated with the transformation of stroma into sclerotia have not been studied.

A more detailed histopathological study, such as those on rye (Luttrell 1980) and on wheat (Shaw and Mantle 1980a,b), is required to understand the host-pathogen relationship in sorghum ergot.

Flowering Biology and Environment in Disease Development

Flowering biology

Since *C. sorghi* infects and colonizes only the gynecia, knowledge of the inflorescence is essential to understand the disease. Flowering of the spikelets has not been studied in detail, but is thought to be a factor in susceptibility of genotype.

Flowering behavior of sorghum was described by Stephens and Quinby (1934) and again by Quinby (1958). It is known to vary in different environments and genotypes. For example, almost all spikelets flower during the night and pollen loses viability after 5 h when sorghum flowers at 26–39°C and 25 to 60% RH (Stephens and Quinby 1934); but at 10–30°C and 30 to 95% RH, anthesis occurs only after sunrise and pollen remain viable for longer periods (Sanchez and Smeltzer 1965).

The spikelet parts most critical for infection and colonization are its stigma and anthers. Normally, a stigma is pollinated soon after it is exposed, pollen germinates within 30 min, and fertilization occurs within another 2–12 hours (Stephens and Quinby 1934; Artschwager and McGuire 1949). On the other hand, conidia require 8–12 hours for germination on the stigma, and 36–48 hours to reach the base of the ovary (R. Bandyopadhyay, unpublished data). Therefore, under normal conditions, pollen germinates earlier than conidia and reaches the embryo sac faster than the colonizing infection hyphae. However, the times for these events vary from floret to floret.

It is generally believed that ovaries resist infection and colonization after fertilization (Fuentes et al. 1964; Futrell and Webster 1965). However, Puranik et al. (1973) reported that ovaries could be infected 5 days after pollination. It is not clear if infected ovaries were fertilized at all, because not all ovaries are fertilized by hand pollination (Patil and Goud 1980). Clearly, histological studies upon simultaneous inoculation and pollination, to trace the progress of pollen tube and infection hyphae to determine if fertilized ovaries can be colonized, would be valuable.

The susceptible period of spikelets, or what has been termed as the 'window of infection' in pearl millet ergot (Willingdale et al. 1986), begins when stigma become receptive and ends (most likely) when the ovary is fertilized. Any factor that prolongs the susceptible period predisposes sorghum to ergot infection. Conversely, factors that ensure rapid fertilization would serve to reduce ergot attack. For example, protogynous spikelets of male-fertile plants would be more susceptible than spikelets in which anthers and stigma emerge together, because stigma would remain unpollinated but receptive to conidia longer.

Male sterility promotes ergot infection because viable pollen is not produced (Futrell and Webster 1965). The utility of readily available fertile pollen in reducing ergot severity has been shown in pearl millet ergot (Thakur and Williams 1980; Thakur et al. 1983). Conidia germination is reduced on sorghum stigma pollinated by viable pollen (Chinnadurai et al. 1970a).

Environmental factors

Favorable environmental conditions for infection and disease development should coincide with anthesis. Futrell and Webster (1966) reported that near 100% relative humidity for 24 h during anthesis of a male-sterile line was optimal for infection. Molefe (1975) reported that inoculated panicles held for 20 hours at high humidity and at 20°C and 25°C, but not at 40°C, showed honeydew within 7 days. From field experiments, Anahosur and Patil (1982) concluded that minimum temperatures ranging from 19° to 21°C and relative humidity from 67 to 84% during the 10-day period following panicle emergence from boot leaf is sufficient to favor serious ergot infection in a male-sterile. Temperature and relative humidity are also reported to affect the incubation period (Sangitrao and Bade 1979a).

In Botswana, Molefe (1975) observed 25–90% ergot incidence, presumably on male-fertile lines under natural conditions, only in a year characterized by high rainfall (750 mm for the epidemic growing season). Infected panicles flowered when relative humidity was 90%, minimum temperature 14°C, and maximum temperature 27°C. Cloudiness during anthesis aids disease development (Anahosur and Patil 1982). Sangitrao and Bade (1979a) felt that in conditions of high humidity, rainfall is not essential for ergot development. Their conclusions are from experiments in which abundant inoculum was available in other plots sown nearby.

Much of the present knowledge on how environmental factors affect ergot development is mostly hypothetical and lacking experimental evidence. Environmental conditions such as cloudy weather and wetness (rainfall) affect anther dehiscence, pollen viability, pollen deposition, and pollen activity (Quinby 1958). Ovule abortion and pollen sterility can occur due to abnormal mega-sporogenesis and microsporogenesis at 5–21°C (Brooking 1979; Dhopte 1984). These conditions are likely to enhance ergot severity because pollination and fertilization are delayed or reduced.

Environmental factors, mainly through their effects on fungal saprophytes, affect sclerotial development. Sclerotia develop poorly in the rainy season due to the growth of *Cerebella* (Sangitrao and Bade 1979b), but wet conditions congenial for infection followed by sclerotial development in dry weather allow formation of less contaminated mature sclerotia (Futrell and Webster 1966).

Research to explain the relationship of sorghum flowering biology to ergot development is needed. Precise information on epidemiology is required, not only for male-sterile lines on which ergot is usually more severe, but on male-fertile hybrids and varieties as well. Future epidemiological studies should consider altered floral behavior of the host and its effect on pathogen development resulting from changes in environment.

Disease cycle

Primary infection in the field is probably established by conidia from collateral hosts and infected panicle debris in soil (Futrell and Webster 1966). Ascospores may also serve as primary inoculum (Singh, 1964), but proof is lacking. Conidia in dried honeydew in infected panicles that fall on the ground remain viable for 7 months (Futrell and Webster 1966). Singh (1964) and Futrell and Webster (1966) hypothesize that ascospores and conidia may infect collateral hosts that flower prior to sorghum, and conidia in the honeydew of collateral hosts provide fresh inoculum to initiate primary infection in sorghum. This has not been documented.

Within 5–12 days after infection in sorghum, the pathogen produces millions of conidia, in honeydew, to infect spikelets that flower subsequently in the same panicle or in different panicles. The pathogen spreads rapidly, probably carried by flies, bees, and other insects (Futrell and Webster 1966) and rain splashes. However, insects may facilitate pollination, and that in turn reduces infection.

In wet conditions *Cerebella* overgrows honeydew and sclerotial production is arrested (Futrell and Webster 1966), but in dry weather well-differentiated sclerotia are produced and the honeydew forms a hard white crust. In the next season, ascospores are produced from sclerotia (Singh 1964); conidia are produced from dried honeydew (Futrell and Webster 1966) or germinating sclerotia (C.S. Sangitarao and B.N. Ghoderao, Punjabrao Krishi Vidyapeeth, Akola, India, personal communication 1981). The sexual and asexual spores may infect collateral hosts or they may infect sorghum to complete the disease cycle, but what actually occurs is still unknown. Demonstration, in the absence of sorghum, of inoculum production in collateral hosts through several seasons would help clear up this matter. The role of wild sorghums in the disease cycle needs investigation.

Host range

Several workers have examined the host range of *Sphacelia sorghi*; their reports are listed in Table 1. Except for that of Ramakrishnan (1947), others are based on artificial inoculation, but in none of the experiments was an adequate non-inoculated control maintained. Cross inoculation tests to prove pathogenicity of isolates were performed only by Reddy et al. (1969) on *Pennisetum typhoides*, by Sundaram et al. (1970) on *P. orientale* × *P. typhoides*, and by Sundaram and Singh (1975) on *Ischaemum pilosum*. Some workers

reported *Cenchrus ciliaris, Panicum maximum,* and *P. typhoides* as hosts of *S. sorghi,* whereas other workers failed to confirm these. Futrell and Webster (1966) reported maize to be a collateral host and subsequently Chinnadurai and Govindaswamy (1971a) found maize ovaries replaced by fungal stromata containing conidia that did not ooze out. Futrell and Webster (1966) also reported that pearl millet ovaries are killed by *S. sorghi* and do not support the fungal stromata. Clearly, more studies are required—not only to determine the collateral hosts but also to elucidate their roles in the disease cycle.

Control

Several strategies for the control of sorghum ergot have been advocated. Quarantine has been effective in excluding the pathogen from countries where ergot is not found. Australia and USA, for example, are countries with strict quarantine regulations against the entry of the pathogen on seeds.

Several researchers in India have shown, under experimental conditions, the utility of early sowing to avoid ergot (Singh 1964, Sangitrao et al. 1979; Anahosur and Patil 1982). Early sowing was invariably required to avoid the disease. Singh (1964) thought that an advantage similar to that of early sowing might be achieved by using early-maturing varieties and hybrids, but early-maturing cultivars sown early are likely to suffer from grain molds.

Sundaram (1968) suggested sowing pathogen-free seed by steeping seeds in 5% salt solution to remove sclerotia (C.S. Sangitrao and B.N. Ghoderao, Punjabrao Krishi Vidyapeeth, Akola, India, personal communication 1981). Removal of collateral hosts in and around fields and rouging of infected plants are other cultural methods of control (Sundaram 1968). Ensuring availability of pollen during flowering reduces pearl millet ergot (Thakur et al. 1983), but evidence of its practicability for large-scale use in sorghum is lacking.

Fungicides were reported to control ergot (Gangadharan et al. 1976; Anahosur 1979), but is considered impractical, ineffective, and non-economical except in small plots of valuable breeding lines.

Table 1. Reported hosts and nonhosts of *Sphacelia sorghi*.

Name	Reference
Host	
Cenchrus ciliaris	Chinnadurai and Govindaswamy (1971a)[1,2]
C. setigerus	Chinnadurai and Govindaswamy (1971)[1,2]
Ischaemum pilosum	Singh (1964)[1,2]; Sundaram and Singh (1975)[2]
Pennisetum orientale	Sundaram et al. (1970)[2]
P. typhoides	
P. typhoides	Reddy et al. (1969)[2], Sundaram (1974)[2]
Sorghum arundinaceum	Ramakrishnan (1947)[1,2]
S. caffrorum	Ramakrishnan (1947)[1,2]
S. halepense	Ramakrishnan (1947)[1,2]
S. membraceum	Ramakrishnan (1947)[1,2]
S. nitens	Ramakrishnan (1947)[1,2]
S. Verticelliflorum	Ramakrishnan (1947)[1,2]
Zea mays	Futrell and Webster (1966)[1,2], Chinnadurai and and Govindaswamy (1971a)[1,2]
Nonhost	
Avena sativa	Singh (1964)
Brancharia ramosa	Sundaram et al. (1970)
Cenchrus ciliaris	Sundaram et al. (1970)
Cyandpm dactylon	Sundaram et al. (1970; Chinnadurai and Govindaswamy (1971a)
Dicanthium annulathum	Chinnadurai and Govindaswamy (1971a)
Echinocloa colonum var. *frmuantacea*	Chinnadurai and Govindaswamy (1971a)
Elusine coracana	Singh (1964), Chinnadurai and Govindaswamy (1971a)
Heteropogon contortus	Sundaram et al. (1970); Chinnadurai and Govindaswamy (1971a)
Iseilem laxum	Chinnadurai and Govindaswamy (1971a)
Panicum antidotale	Chinnadurai and Govindaswamy (1971a)
P. maximum	Sundaram et al. (1970); Chinnadurai and Govindaswamy (1970a)
P. miliaceum	Chinnadurai and Govindaswamy (1971a)
P. milliare	Chinnadurai and Govindaswamy (1971a)
P. scrobiculatum	Chinnadurai and Govindaswamy (1971a)
Pennisetum ciliaris	Chinnadurai and Govindaswamy (1971a)
P. hohenackeri	Chinnadurai and Govindaswamy (1971a)
P. massicum	Chinnadurai and Govindaswamy (1971a)
P. orientale	Sundaram et al. (1970)
P. polystachyon	Sundaram et al. (1970)
P. purpureum	Chinnadurai and Govindaswamy (1971a)
P. ruppelli	Chinnadurai and Govindaswamy (1971a)
P. squamulatum	Chinnadurai and Govindaswamy (1971a)
P. typhoides	Singh (1964); Futrell and Webster (1966); Chinnadurai and Govindaswamy (1971a)
Rhynchylertrium roseum	Sundaram et al. (1970)
Secale cereale	Singh (1964)
Setaria italica	Singh (1964)
Triticum vulgare	Singh (1964)

1. Cross inoculation tests (sorghum-host-sorghum-host) were not conducted.
2. Noninoculated controls not maintained.

Host resistance

Sowing of resistant cultivars is probably the most practical, economical, and effective method to control ergot. However, resistant lines are not available.

Identification of sources of resistance requires a long-term systematic approach, and a major prerequisite is an effective screening technique to identify resistant lines in a large population.

Screening technique

Artificial inoculation methods must take into account the importance of flowering biology in disease development. It is improper to pollinate spikelets prior to inoculation, because pollination interferes with infection. Evaluation of pollinated and then inoculated spikelets would measure escape, rather than resistance. Therefore, it is important to remove all spikelets flowering prior to inoculation. This was not taken into account in the inoculation methods followed earlier (Singh 1964; Sundaram 1971; Khadke et al. 1978). At ICRISAT Center, we have devised a screening methodology after carefully considering these factors. It was effectively used for resistance screening by the national programs of Ethiopia and Rwanda, in cooperation with ICRISAT. In brief, the technique is as follows:

All spikelets flowering before inoculation were cut and removed, retaining all spikelets in primary branches borne on two or three nodes below the cut portion. Only spikelets in these primary branches were inoculated and evaluated. Trimmed panicles were inoculated with a suspension of fresh honeydew (10^6 spores mL^{-1}) twice at 2-day intervals and bagged for 15 to 20 days. Overhead sprinkler irrigation was used for 1 h each morning for about 20 days after flowering.

For evaluation, 100 spikelets were observed in the inoculated primary branches. The number of spikelets with fungal stromata were counted, and ergot severity was recorded as the percentage of ergot-infected florets. Qualitative visual rating (Sundaram 1980), although simpler, is less appropriate than the quantitative ergot-severity measurement employed at ICRISAT Center; this is because all of the infected spikelets may not be visible (sorghum grains, unlike those of pearl millet, are not on one plane in the panicle), and honeydew from an infected spikelet may smear a neighboring healthy grain, causing it to appear to be infected.

Sources of resistance

Reliable information on sources of resistance is meager. Several thousand germplasm accessions have been screened under artificial inoculation to identify resistant sources (Singh 1964; Chinnadurai et al. 1970b; Sundaram 1970, 1980; and Khadke et al. 1978). However, these reports were based on results of 1 year's nonreplicated trials, and lines thought resistant proved to be otherwise in subsequent trials.

A major resistance screening activity, in collaboration with ICRISAT, is presently underway in the east African countries of Ethiopia and Rwanda. It is being done without, overhead sprinkler irrigation at 'hot spots', using the screening technique described above. In 1987, the sorghum collection in Rwanda was screened at the Rubona and Karama research stations of Rwanda. Not one accession showed a high level of resistance. In Ethiopia, 248 genotypes with intermediate and highland adaptation were tested at Arsi Negele Research Station. We selected 27 accessions with less than 5% infected florets, and among these ETS 2454, ETS 3135, ETS 3147, ETS 2148, ETS 3286, ETS 3753, ETS 4762, and ETS 4485 were least infected (1–2% infected florets). These selections will be screened again in Ethiopia and Rwanda to see if resistance can be confirmed.

The lack of confirmed sources of resistance to sorghum ergot underscores the urgency to identify ergot-resistant sources. The greatest need is in male-sterile lines, although chances of identifying or breeding ergot-resistant male steriles appear slim in view of the role of pollen in the disease. Control of ergot must integrate all methods known to reduce the disease, but more research is needed in this direction.

Research Needs

This review shows that research on sorghum ergot has been largely superficial, and that wide gaps exist in knowledge of the biology of the pathogen and epidemiology of the disease. The process of pathogenesis and how it is influenced by flowering biology and environment is not well understood.

Following are some of the areas that need research attention in the future:

1. **Crop loss.** Data showing loss in grain yield and quality in hybrid seed production and crop production are needed. Much of the present seed certification standard for ergot contamination appears to be arbitrary. Research is also required to determine practical levels of sclerotia in seed-lots.
2. **Biology.** Studies are needed to determine why sclerotia germinate poorly (e.g., the role of heterothallism) and to identify conditions that will improve germinability of sclerotia. Variability in the pathogen should be studied.
3. **Pathogenesis.** Histopathology of host-pathogen interaction should include the routes of infection (e.g., can the pathogen penetrate the ovary wall?), and the progress of colonization until the formation of the fungal stromata and its differentiation into sclerotia.
4. **Flowering biology.** The biology of sorghum flowering is poorly understood. How flowering biology is associated with opportunities for infection and how the host-pathogen interaction is modified by variable flowering behavior due to altered environmental conditions is largely unknown. Can the pathogen infect and colonize the fertilized ovary? Can infection occur prior to flower opening? Fluorescence microscopy is a powerful tool for the study of competition between pollen and conidia.
5. **Epidemiology.** Environmental conditions favorable for disease development are largely known. However, it is necessary to define cool and humid conditions and develop useful and nonambiguous terminology for describing a particular climatic condition. We also need to know the role of preflowering environmental conditions on production of primary inocula by collateral hosts and sclerotia.
6. **Disease cycle.** Further research should seek sources of primary inoculum, and determine how these are maintained and disseminated. Information on the role of sclerotia in disease cycle is required.
7. **Host resistance.** Development of simple, reliable, large-scale field screening techniques requires major attention; identification of resistant

sources and efficient breeding for resistance is dependent upon such tools.

Acknowledgment. I wish to thank L.K. Mughogho for critically reviewing this manuscript.

References

Anahosur, K.H. 1979. Chemical control of ergot of sorghum. Indian Phytopathology 32:487–489.

Anahosur, K.H., and **Patil, S.H.** 1982. Effect of date of sowing on the incidence of ergot of sorghum. Indian Phytopathology 35:507–509.

Artschwager, E., and **McGuire, R.C.** 1949. Cytology of reproduction in *Sorghum vulgare.* Journal of Agricultural Research 78:659–673.

Brooking, I.R. 1979. Male sterility in *Sorghum bicolor* (L.) Moench. Induced by low night temperature. Part 2. Genotypic differences in sensitivity. Australian Journal of Plant Physiology 6:143–147.

CABI. (Commonwealth Agricultural Bureau International).1987. Commonwealth Mycological Institute Distribution Maps of Plant Diseases. Map No. 582, edn 1, issued 1 Oct 1987. Kew, Surrey, UK:CABI. 2 pp.

Chinnadurai, G. 1972. Effect of certain trace elements on the growth and sporulation of *Sphacelia sorghi.* Indian Phytopathology 25:599–600.

Chinnadurai, G., and **Govindaswamy, C.V.** 1971a. Host range of sorghum sugary disease pathogen. Madras Agricultural Journal 58:600–603.

Chinnadurai, G., and **Govindaswamy, C.V.** 1971b. Influence of nitrogen nutrition on the spore size of *Sphacelia sorghi.* Indian Phytopathology 24:177–178.

Chinnadurai, G., Govindaswamy, C.V., and **Ramakrishnan, K.** 1970a. Studies on the effect of stigmatic exudates of sorghum on the parasitism of *Sphacelia sorghi* McRae. Phytopathologische Zeitschrift 69:56–63.

Chinnadurai, G., Lalithakumari, D., and Govindaswamy, C.V. 1970b. Reaction of various species and varieties of sorghum to sugary disease. Madras Agricultural Journal 57:735–736.

Dhopte, A.M. 1984. Influence of night temperature on microsporogenesis and megasporogenesis in *Sorghum bicolor* (L.) Moench. Ph.D. thesis, University of Nebraska, Lincoln, NE, USA. 283 pp.

Fuentes, S.F., Lourdes, De La Isla, M., Ullstrup, A.J., and Rodriquez, A.E. 1964. *Claviceps gigantea*, a new pathogen of maize in Mexico. Phytopathology 54:379–381.

Futrell, M.C., and Webster, O.J. 1965. Ergot infection and sterility in grain sorghum. Plant Disease Reporter 49:680–683.

Futrell, M.C., and Webster, O.J. 1966. Host range and epidemiology of the sorghum ergot organism. Plant Disease Reporter 50:828–831.

Gangadharan, K., Subramanian, N., Kandaswamy, T.K., and Sundaram, N.V. 1976. Control of sugary disease of sorghum. Madras Agricultural Journal 63:411–413.

Khadke, V.D., More, B.B., and Kone, B.K. 1978. Note on screening of sorghum varieties and selections against sugary disease. Indian Journal of Agricultural Research 12:257–258.

Kukadia, M.U., Desai, K.B., Desai, M.S., Patel, R.H., and Raja, K.R.V. 1982. Natural screening of advanced sorghum varieties to sugary disease. Sorghum Newsletter 25:117.

Kulkarni, B.G.P., Seshadri, V.S., and Hegde, R.K. 1976. The perfect stage of *Sphacelia sorghi* McRae. Mysore Journal of Agricultural Science 10:286–289.

Langdon, R.G. 1942. The genus *Cerebella cesati* - its biologic status and use. Phytopathology 32:613–617.

Luttrell, E.S. 1980. Host-parasite relationships and development of the ergot sclerotium in *Claviceps purpurea*. Canadian Journal of Botany 58:942–958.

Mantle, P.G. 1968. Studies on *Sphacelia sorghi* McRae, and ergot of *Sorghum vulgare* Pers. Annals of Applied Biology 62:443–449.

Molefe, T.L. 1975. Occurrence of ergot on sorghum in Botswana. Plant Disease Reporter 59:751–753.

Nagarajan, K., and Saraswathi, V. 1975. Production of 'honey-dew' like secretions in the culture of *Sphacelia sorghi*. Indian Phytopathology 28:110.

Patil, R.C., and Goud, J.V. 1980. Viability of pollen and receptivity of stigma in sorghum. Indian Journal of Agricultural Research 50:522–526.

Puranik, S.B., Padaganur, G.M., and Hiremath, R.V. 1973. Susceptibility period of sorghum ovaries to *Sphacelia sorghi*. Indian Phytopathology 26:586–587.

Quinby, J.R. 1958. Grain sorghum production in Texas. Texas Agricultural Experiment Station Bulletin No. 912. College Station, TX, USA: Texas A&M University. 10 pp.

Ramakrishnan, T.S. 1947. The natural occurrence of ergot in South India–III. Proceedings of the Indian Academy of Sciences (B) 26:136–141.

Reddy, K.D., Govindaswamy, C.V., and Vidhyasekaran, P. 1969. Studies on ergot disease of Cumbu (*Pennisetum typhoides*). Madras Agricultural Journal 56:367–377.

Sanchez, R.L., and Smeltzer, D.G. 1965. Sorghum pollen viability. Crop Sciences 5:111–113.

Sangitrao, C.S., and Bade, G.H. 1979a. Meteorological factors associated with honey dew development and sclerotial stage in sorghum ergot. Sorghum Newsletter 22:107–108.

Sangitrao, C.S., and Bade, G.H. 1979b. Morphological studies of sclerotia of sorghum ergot. Sorghum Newsletter 22:108.

Sangitrao, C.S., Taley, Y.M., and Moghe, P.G. 1979. Effect of planting date on the appearance of sugary disease. Sorghum Newsletter 22:111.

Shaw, B.I., and Mantle, P.G. 1980a. Host infection by *Claviceps purpurea*. Transactions of the British Mycological Society 75:77–90.

Shaw, B.I., and Mantle, P.G. 1980b. Host infection by *Claviceps purpurea*. Transactions of the British Mycological Society 75:117–121.

Singh, P. 1964. Sugary disease of sorghum. Pages 29–33 *in* Progress report of the Accelerated Hybrid Sorghum Project and the Millet Improvement Programme. Indian Council of Agricultural Research and cooperating agencies, New Delhi, India: Indian Council of Agricultural Research. 81 pp.

Stephens, J.C., and Quinby, J.R. 1934. Anthesis, pollination, and fertilization of sorghum. Journal of Agricultural Research 49:123–136.

Sundaram, N.V. 1968. Sugary disease of jowar - how to recognize and control it. Indian Farming, April 1968 issue: 21–22.

Sundaram, N.V. 1970. Sugary disease of sorghum. Pages 435–439 *in* Plant disease problems: Proceedings of the First International Symposium on Plant Pathology, 27 Dec 1966 – 1 Jan 1967, New Delhi, India. Indian Agricultural Research Institute. New Delhi. India: Indian Phytopathological Society.

Sundaram, N.V. 1971. Possible resistance to sugary disease in sorghum. Indian Journal of Genetics and Plant Breeding 31:383–387.

Sundaram, N.V. 1974. Natural Occurrence of *Sphacelia sorghi* on *Pennisetum typhoides*. Indian Phytopathology 27:124–125.

Sundaram, N.V. 1980. Sorghum ergot. Pages 377–379 *in* Sorghum diseases, a world review: Proceedings of the International Workshop on Sorghum Diseases, 11–15 Dec 1978, ICRISAT, Hyderabad, India. Patancheru, Andhra Pradesh 502 324, India: International Crops Research Institute for the Semi-arid Tropics. 478 pp.

Sundaram, N.V., and Singh, S.D. 1975. A new host for sugary disease of sorghum caused by *Sphacelia sorghi*. Science and Culture 41:528.

Sundaram, N.V., Bhowmik, T.P., and Khan, I.D. 1970. A new host for *Sphacelia sorghi*. Indian Phytopatholgy 23:128–130.

Thakur, R.P., and Williams, R.J. 1980. Pollination effects on pearl millet ergot. Phytopathology 70:80–84.

Thakur, R.P., Williams, R.J., and Rao, V.P. 1983. Control of ergot in pearl millet through pollen management. Annals of Applied Biology 103:31–36.

Willingdale, J., Mantle, P.G., and Thakur, R.P. 1986. Postpollination stigmatic constriction, the basis of ergot resistance in selected lines of pearl millet. Phytopathology 76:536–539.

Sorghum Smuts

El Hilu Omer[1] and R.A. Frederiksen[2]

Abstract

The four smuts of sorghum are reviewed in respect to their economic importance, biology, and control. The two kernel smuts (Sporisorium sorghi and Sphacelotheca cruenta) have been reduced to minor importance because of seed treatment with fungicides, whereas head smut (Sporisorium reilianum) and long smut (Tolyposporium ehrenbergii) continue to be significant. Resistance to long smut has been developed and work on identifying sources of resistance is underway. Programs to reduce head smut in endemic areas have changed little over the past decade.

Introduction

Four fungi cause four different smuts in sorghum. The pathogens *Sporisorium reilianum* (Kühn) Langdon and Fullerton, *Sporisorium sorghi* (Ehrenberg) Link, *Sphacelotheca cruenta* (Kühn) Potter and *Tolyposporium ehrenbergii* (Kühn) Patouillard cause head, covered kernel, loose kernel, and long smut, respectively (Table 1).

Head Smut

The life cycle of head smut (Fig. 1) has been recognized for some time. The fungus is seedling-infecting from soilborne teliospores. Teliospores survive for many years in the field. In Texas, these spores are estimated to survive up to a decade in soil. Spores germinate by the production of a promycelium (epibasidium). On artificial media these structures produce sporidia. The function or role of sporidia under natural conditions remains unknown. Similarly the nuclear condition of the vegetative and sexual stages of the fungus has not been elucidated, although the parasitic mycelium is dikaryotic (Natural 1983). Colonization of the host following initial penetration and formation of the sorus has been studied in sorghum by Wilson and Frederiksen (1970a, 1970b), and Natural (1983). The process of colonization appears to be similar in sorghum and maize. Mycelium of the pathogen grows toward differentiating meristem, where it grows systemically with the host. When differentiation begins in the host, the smut fungus also organizes its sorus within the host. The fungus differentiates into sporogenous hyphae, cells that ultimately produce spores and spore balls; partitioning hyphae, mycelium that grows within the sorus among the sporogenous hyphae and nonsporogenous hyphae for all other hyphae in the sorus. Some of the nonsporogenous hyphae, along with two or three cell layers of host cells form the peridium. As the sorus matures and dries, the peridium ruptures and releases spores that fall to the soil for the next cycle. As the sorus continues to release spores, the vascular strands of the host remain. There are a wide range of symptoms for sorghum head smut, based in part on the time and extent of colonization and the length of time after maturity that the diseased plant survives (Wilson and Frederiksen 1970a).

1. Plant Pathologist, Agricultrural Research Corporation, Gezira Agricultural Research Station, Wed Medani, Sudan.
2. Professor, Plant Pathology and Microbiology, Texas A&M University, College Station, TX 77843, USA.

El Hilu Omer and Frederiksen, R.A. 1992. Sorghum smuts. Pages 245–252 *in* Sorghum and millets diseases: a second world review. (de Milliano, W.A.J., Frederiksen, R.A., and Bengston, G.D., eds). Patancheru, A.P. 502 324, India: International Crops Research Institute for the Semi-Arid Tropics.

Table 1. Comparison of sorghum smuts.

Disease	Pathogen	Method of infection	Control method
Head smut	*Sporisorium reilianum*	Seedling, from soilborne teliospores	Host resistance
Covered kernel smut	*Sporisorium sorghi*	Seedling, from seedborne teliospore	Fungicide, seed treatment
Loose kernel smut	*Sphacelotheca cruenta*	Seedling, from seedborne teliospore or shoot infection from air-borne Teliospores	Fungicide, seed treatment
Long smut	*Tolyposporium ehrenbergii*	Local, floral infecting	Host resistance (?)

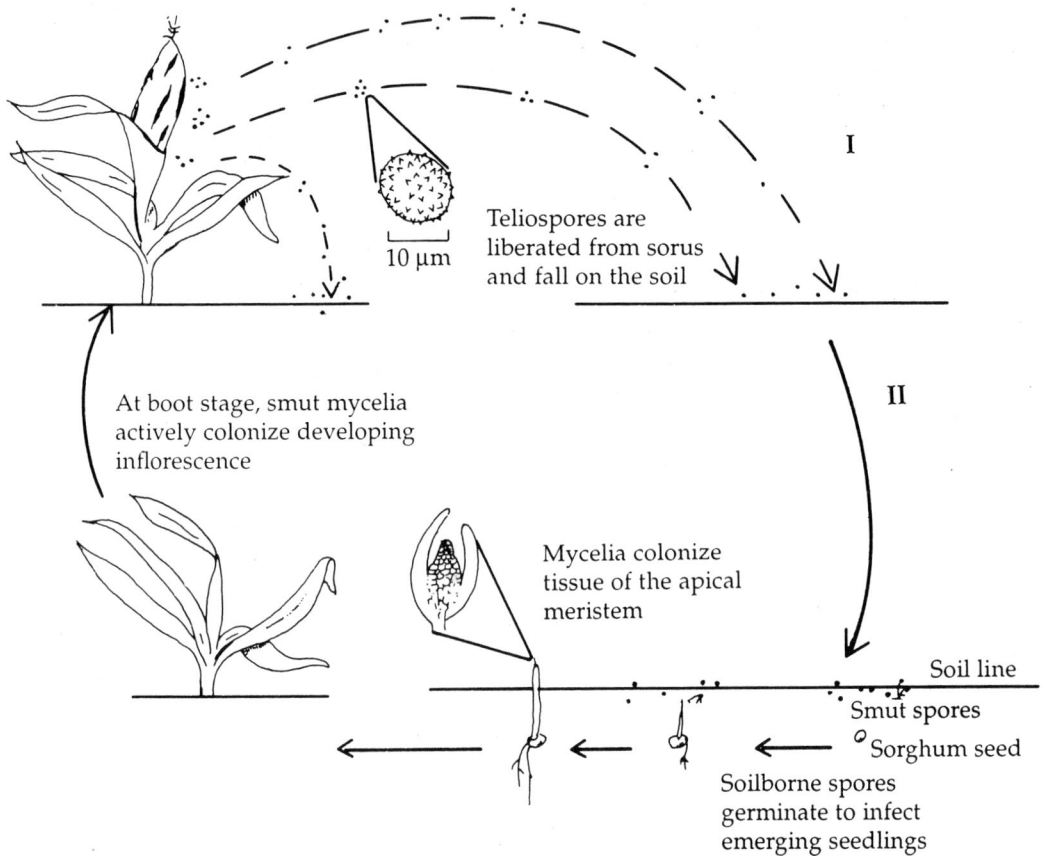

At boot stage, smut mycelia actively colonize developing inflorescence

Teliospores are liberated from sorus and fall on the soil

10 μm

I

II

Mycelia colonize tissue of the apical meristem

Soil line

Smut spores

Sorghum seed

Soilborne spores germinate to infect emerging seedlings

Figure 1. Life cycle of *Sporisorium reilianum*, causal agent of head smut.

Variation in pathogen populations

Sporisorium reilianum is composed of physiological races. The importance and status of our understanding of the populations of these fungi were reviewed by Frederiksen and Reyes (1980). For the most part, the populations of the pathogens described then (Frowd 1980; Frederiksen and Reyes 1980) have changed little. The same differential cultivars are used in Texas and the widely grown uniform head smut nursery (UHSN) contains national program differential cultivars, along with cultivars with resistance from other areas.

Sources of resistance to head smut in USA

Many of the lines identified as resistant by Frederiksen and Reyes (1980) remain useful in 1988. Changes in the pathogen population have not caused an unusual erosion of host resistance. Currently, populations and cultivars of sorghum are being bred for higher levels of smut resistance in Texas.

Long Smut

The disease has wide distribution in middle eastern countries, the Indian subcontinent, and in the far eastern countries. (Briton-Jones 1922; Tarr 1962; Mahdi 1962; Ragab and Mahdi 1966; and Frowd 1980).

Long smut has gained prominence as a serious problem in certain countries in eastern and western Africa. In Sudan, on hybrids especially, annual crop loss was estimated to be 2–5%, but a much higher value (40–60%) had been reported from individual fields in Pakistan (Hafiz 1958).

Symptoms of long smut

Long smut appears as conspicuous sori affecting individual grains. The sorus is cylindrical and elongate, usually slightly curved with a thick membrane. The spore sac, about 4 cm long and 7 mm wide, splits when mature, exposing a black mass of spore balls. Within the spore mass are eight to ten filamentous structures arising from the base. These are the remnants of the vascular tissues of the ovary. The spore balls are variable in size, globose to oblong, black to dark brown (Fig. 2). Within the spore ball, spores are cemented together and their number varies, depending on the degree of breakage. Thus the size of the spore ball has no significance since its permanency is greatly affected by physical factors. Spores in the middle of a spore ball are somewhat angular, light colored, and thin walled, whereas those on the periphery are dark, echinulate, and globose. Teliospores are about 12 µm in diameter.

Seasonal persistence and spread

Since the first International Workshop in Hyderabad, in 1978, our knowledge of long smut has expanded. Mahdi (1962) reported that the thermal death point of teliospores was 67°C. However, recent investigations indicate that the spores remain viable after 1 week at 85°C (Hago, personal communication 1988). The highest recorded soil-surface temperature in sorghum-growing areas of the Sudan is 65°C. The longevity of spore balls may extend to 2 years or longer. Sporidia in the laboratory retained their viability for at least 255 days (Mahdi 1962) whereas dry sporidia, presumably short-lived, remained viable at room temperature for longer than 1 month (Hago, personal communication 1988).

Kamat (1933) claimed that most spores in a spore ball readily germinate on different media and over a wide range of temperature (19–39°C). In sterile distilled water only a few teliospores in a spore ball germinate, forming three- to six-celled stout promycelium bearing sporidia terminally and near the septa. Sporidia are 8–12 mm, spindle shaped, single-celled, and rarely with a transverse septum. In solid medium, branched germ tubes with clusters or chains of sporidia develop. The optimal temperature for growth is 25–30°C. The fungus grows well in a variety of media and does not lose pathogenicity after a series of transfers.

Evidence from various sources refutes the reports (Mahdi 1962; Ragab and Mahdi 1966) that the long smut pathogen is internally seedborne. The work of Vasudeva et al. (1950), Manzo (1976), and Omer et al. (1985) demonstrates that the pathogen is airborne and infects single florets. Attempts to simulate the disease using arti-

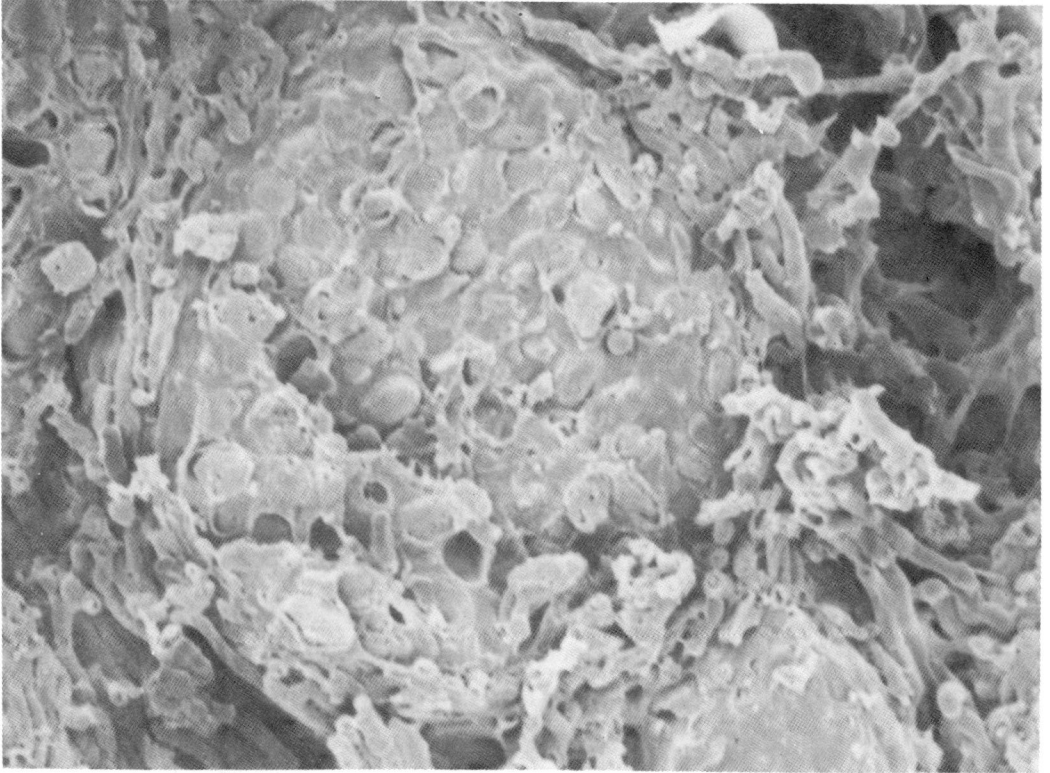

Figure 2. Scanning electron micrograph of a spore ball of *Tolyposporium ehrenbergii*. Most of the peripheral spores have fallen away.

ficially contaminated seed have failed to reproduce symptoms. Similarly, inoculations of plants at juvenile stages of growth did not develop smut. The only method that permits infection is when plants are inoculated at the boot stage before the ear emerges. Infection occurs when the inflorescence is still enclosed in the boot or about the time it emerges when the flowers are very young. Manzo (1976) wrote that floral infection by *T. ehrenbergii* takes place when inoculation is done at boot stage, with decreasing infection from preanthesis to early anthesis, and infection does not occur during early seed filling. In this respect the mode of infection of *T. ehrenbergii* is identical to that of *T. penicillariae*, the long smut of millet (Bhatt 1946).

We propose that the annual recurrence of the disease is from spore balls picked up from the soil surface by winds during a dry period, deposited in the flag leaf, then they are washed down in the flag leaf where conditions are conducive for sporidia production and penetration (Fig. 3). This accounts for the occurrence of infected heads under bags. Frequently, perhaps because of the higher humidity, infection is higher under a bagged head than an uncovered head. Bagged sterile heads also develop long smut. Freshly produced spores are again wind blown and could serve as inoculum within the same location. Consequently, experiment station nurseries with several dates of sowing or flowering are more likely, in some seasons, to have high levels of long smut.

The fungus penetrates the developing ovary, mostly through the style (rarely through the ovary wall), and rapidly colonizes the internal tissues. Ramification of the invading hyphae is soon followed by the aggregation of spore balls that progress from the upper part of the young sorus. In about 15 days, following inoculation, symptoms of the disease should appear.

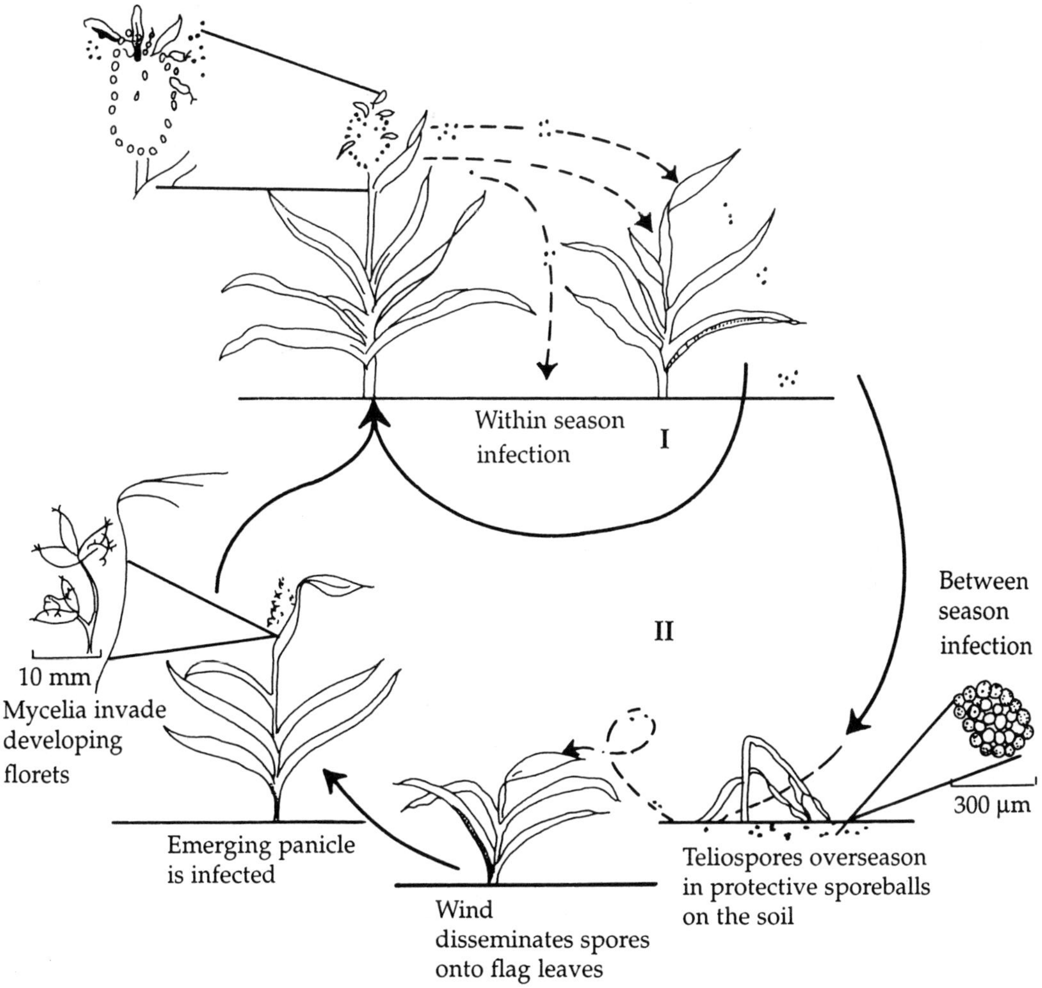

Figure 3. Life cycle of *Tolyposporium ehrenbergii*, causal agent of long smut.

Screening techniques

Screening techniques were standardized by using a single petri dish of a 72-h sporidial culture harvested in one liter of water. The teliospore inoculum consisted of two smutted seeds L^{-1} of water. Three mL inoculum were injected per head. The results (Table 2) indicate the superiority of teliospore inoculum over sporidia; 98% of the teliospore inoculum injections resulted in infection, compared to 62% for inoculations made with sporidia.

Disease intensity was highest with the fresh teliospores. The method is reliable, and can be easily adapted into a breeding program. It is particularly efficient when using an automatically refilling syringe. Agitation of the container is necessary for uniform delivery.

In areas where annual recurrence of the disease is certain, the screening can be done through inoculating early susceptible border rows, which are left for natural spread. The method hopefully helps in exploring all possible means of plant resistance.

Table 2. Comparison of disease intensities between sorghum heads inoculated with a 72 h sporidal inoculum and with fresh teliospores of *Tolyposporium ehrenbergii*.

Disease intensity no. of sori head[-1]	With sporidial inoculum (no.)	With fresh teliospores (no.)
(0)	33	1
(1-20)	26	20
(21-50)	13	11
(51-100)	9	10
(10-200)	3	10
(>200)	3	3
Total heads inoculated	87	55
Total heads infected	54	54
Infection (%)	62%	98%
Mean disease intensity	2.2	3.3

Cross inoculation with long smut isolates

Tarr (1955) reported pearl millet as a host of *T. ehrenbergii*. The causal organism of long smut on the wild sorghum, *Sorghum purpureosericeum*, in India was classified as *T. ehrenbergii* var *grandiglobum* (Uppal and Patel 1943). Other wild sorghum species are predominant in most areas where cultivated sorghum is grown. Cross inoculation tests carried out by the authors indicated no difference between isolates infecting sorghum and the wild sorghum, *S. purpureosericeum*, but millet in the Sudan was not susceptible to *T. ehrenbergii*; neither did long smut isolated from millet infect sorghum (Table 3).

Control of long smut

At this time, the control of long smut by host resistance is under consideration. Clearly, differences in reaction exist. Some old sorghum culti-

vars, such as BTx 378, are resistant whereas BTx 623 is not. Currently, the sorghum improvement program in Sudan inoculates sorghums for reaction to long smut and a similar program is being considered for Niger and Mali. Evidence of good resistance among the cultivars tested in the Sudan is reported in Table 4. QL 3 (India), SC 326-6, and SC 630-11 E were among the best performers in two seasons' testing.

Uniform flowering of sorghum should reduce plant-to-plant spread of the pathogen and lessen the impact of the disease. Study of the effects of the level of smut on disease development has not been carried out. However, the relationship between the level of infection and damage appears to be linear.

The Kernel Smuts

Covered kernel and loose kernel smut are seed-borne seedling-infecting smut fungi. While early workers exploited host resistance and detected

Table 3. Cross inoculation with long smut isolates.

Long smut isolate	*S. purpureosericium*	*Sorghum feterita*	Pearl millet Baladi	Pearl millet Ugandi
S. purpureosericium	+	+	–	–
S. bicolor	+	+	–	–
Pearl millet	–	–	+	+

Table 4. Reaction of selected sorghum cultivars to long smut inoculation.

Cultivar	1983		1985	
	Number of infected/ Number of inoculated	Number of sori infected heads	Number of infected/ Number of inoculated	Number of sori infected heads
SC 326-6	8/21	3	0/11	0
SC 630-11 E	4/30	1	8/56	5
QL 3 (India)	2/26	2	2/22	2
CR 54:36	48/48	101	14/52	28
Dabar	–	–	40/41	89
G. Elhaman	–	–	31/39	11
HD 1	–	–	22/34	12

physiological races of these pathogens, virtually nothing outside of the evaluation of fungicides for control of these fungi has been done over the past decade (Frowd 1980). If seed-dressing fungicides are not used, high levels of loose kernel smut develops in ratooned sorghum.

Since loose smut may under some conditions be shoot-infecting (Dean 1966), an occasional endemic is possible.

References

Bhatt, R.S. 1946. Studies in the ustilaginales. I. The mode of infection of the Bajra plant (*Pennisetum typhoides* Stapf) by the smut *Tolyposporium penicillariae* Brefeld. Journal of Indian Botanical Society 25:163–186.

Briton-Jones, H.R. 1922. The smuts of millets. Egypt Ministry Agricultural Technology and Science Service Bulletin. No. 49. 8 pp.

Dasgupta, S.N., and Narain, A. 1960. A note on physiologic races of *Sphacelotheca sorghi*. Current Science 26:226–227.

Dean, J.L. 1966. Local infection of sorghum by Johnson grass loose kernel smut fungus. Phytopathology 56:1342–1344.

Frederiksen, R.A. (ed.). 1986. Compendium of sorghum diseases. St. Paul, MN, USA: American Phytopathological Society Press. 82 pp.

Frederiksen, R.A., and Reyes, L. 1980. The head smut program at Texas A&M University. Pages 367–372 *in* Sorghum diseases, a world review: proceedings of the International Workshop on Sorghum Diseases, 11–15 Dec 1978, ICRISAT, Hyderabad, India. Patancheru, Andhra Pradesh 502 324, India: International Crops Research Institute for the Semi-Arid Tropics.

Frowd, J.A. 1980. A world review of sorghum smuts. Pages 331–338 *in* Sorghum diseases, a world review: proceedings of the International Workshop on Sorghum Diseases, 11–15 Dec 1978, ICRISAT, Hyderabad India. Patancheru, Andhra Pradesh 502 324, India: International Crops Research Institute for the Semi-Arid Tropics.

Hafiz, A. 1958. Some studies on long smut of sorghum. Pakistan Journal of Scientific Research 10:83–87.

Kamat, M.N. 1933. Observations on *Tolyposporium filiferum*, cause of 'long smut' of sorghum. Phytopathology 23:985–92.

Mahdi, M.T. 1962. Studies on *Tolyposporium ehrenbergii*, the cause of long smut of sorghum in Egypt. Proceedings of the International Seed Association. 27:870–877.

Manzo, S.K. 1976. Studies on the mode of infection of sorghum by *Tolyposporium ehrenbergii*, the causal organism of long smut. Plant Disease Reporter 60:948–952.

Natural, M. 1983. Ultrastructural and histopathological studies of the host-parasite relations of sorghum head smut. Ph.D. thesis, Texas A&M University, College Station, TX, USA. 93 pp.

Omer, M.E.H., Frederiksen, R.A., and Ejeta Gebisa. 1985. A method for inoculating sorghum with *Tolyposporium ehrenbergii* and other observations on long smut in Sudan. Sorghum Newsletter 28:95–97.

Ragab, M.A., and Mahdi, M.T. 1966. Studies on *Tolyposporium ehrenbergii*, the cause of long smut of sorghum in Egypt (UAR). Mycologia 58:184–191.

Tarr, S.A.J. 1955. The fungi and plant diseases of the Sudan. Kew, Surrey, UK: The Commonwealth Mycological Institute.

Vaheeduddin, S. 1950. Two new physiologic races of *Sphacelotheca sorghi*. Indian Phytopathology 3:162–4.

Vasudeva, R.S., Seshadri, M.R., and Iyengar. 1950. Mode of transmission of the long smut of Jowar (Sorghum). Current Science 19:123–124.

Uppal, B.N. and Patel, M.K. 1943. Long smut of *Sorghum purpureosericeum*. Indian Journal of Agricultural Science 13:520–521.

Wilson, J.M., and Frederiksen, R.A. 1970a. Histopathology of the interaction of *Sorghum bicolor* and *Sphacelotheca reiliana*. Phytopathology 60: 828–832.

Wilson, J.M., and Frederiksen, R.A. 1970b. Histopathology of the resistance in *Sorghum bicolor–Sphacelotheca reiliana* interaction. Phytopathology 60:1365–1367.

A Review of Sorghum Grain Mold

G.A. Forbes[1], R. Bandyopadhyay[2], and G. Garcia[3]

Abstract

Terminology and definitions, symptoms, causal agents, importance, and control of fungal-related grain deterioration of sorghum are reviewed. The term grain mold (GM) has gained general acceptance as the most satisfactory descriptor of this condition.

Two concepts of fungal-related grain deterioration may be found in literature. In one, GM is a condition caused by parasitic and/or saprophytic interactions of numerous fungal spp and the plant at anytime between anthesis and harvest. In the other, only a few fungi infecting and colonizing spikelet tissues prior to grain maturity are involved. Fungi involved in postharvest deterioration (weathering) are not considered part of the GM complex. Numerous forms of GM damage have been described, but little work is reported on quantification of losses. A potential mycotoxin contamination in molded grain samples has been demonstrated.

New techniques, including serial dilutions and ergosterol concentration, have been useful in evaluating GM severity. Screening of more than 7000 accessions has identified more than 150 GM-resistant lines.

Introduction

The purpose of this review is to summarize research done on fungal-related deterioration of sorghum grain, frequently referred to as grain mold (GM). Grain mold, in its broadest sense, is certainly one of the major biotic constraints of sorghum for feed and food production. The historical development of GM and its perceived importance were reviewed by Williams and Rao (1981).

GM is usually the result of a complex of fungus-host interactions. Because of this complexity, it is difficult to synthesize a coherent review of the related literature. This review discusses GM from these viewpoints: (1) description, (2) importance, and (3) control.

This information is intended to complement the review of Williams and Rao (1981). Readers are advised to refer to that review for further discussion and references on various aspects of sorghum GM.

What is Grain Mold?

Terminology and definition

Williams and Rao (1981) reported that numerous and diverse terms have been used to describe fungal infection and colonization of sorghum spikelet tissues. Since publication of their review, consensus has developed among several major institutions for the exclusive use of the term "grain mold" (GM) to describe the condition resulting from fungal deterioration of sorghum grain (Canez and King 1981; Castor 1981; Frederiksen et al. 1982; Forbes 1986; ICRISAT 1986). However, other terms still appear in the recent literature: seedborne fungi (Bhale and

1. Plant Pathologist, Centro Internacional de la Papa, International Potato Center, PO Box 5969, Lima, Peru.
2. Plant Pathologist, Cereals Program, ICRISAT Center, Patancheru, Andhra Pradesh 502 324, India.
3. Plant Pathologist, CIBA-GEIGY, Arias 1851 (1429) Buenos Aires, Argentina.

Forbes, G.A., Bandyopadhyay, R., and Garcia, G. 1992. A review of sorghum grain mold. Pages 265–272 *in* Sorghum and millets diseases: a second world review. (de Milliano, W.A.J., Frederiksen, R.A., and Bengston, G.D., eds). Patancheru, A.P. 502 324, India: International Crops Research Institute for the Semi-Arid Tropics.(CP 738).

Khare 1982; El Shafie and Webster 1981; Hepperly et al. 1982; Kissim 1985); seed mycoflora (Kabore and Couture 1983; Khairnar and Gambhir 1985; Shree 1984), fungus associated with sorghum seed (Munghate and Faut 1982), head mold (Dayan 1980; Mathur and Naik 1981; Naik et al. 1981), seed-rotting fungi (Anahosur et al. 1984), and weathering (Ibrahim et al. 1985).

It would be difficult to demonstrate the intrinsic value of any one term over the others, but the advantage of researchers agreeing on the use of a single term seems obvious. The relatively significant level of acceptance for the term "GM" at present should be predictor of greater degree of uniformity in future publications as well. Some divergence in terminology may continue, however, as a reflection of an even-more fundamental level of dispute among researchers, that of definition.

Definitions of GM are only rarely given in explicit terms (Castor 1981; Williams and Rao 1981). Therefore the following discussion is based on implicit definitions inferred from GM-related literature and is subject to interpretive bias. Nonetheless, most definitions of GM found in recent literature appear to fit into one of two general concepts of fungal-related grain deterioration.

The first concept (A, Fig. 1) describes a condition resulting from fungal infection and colonization of grains occurring any time between anthesis and harvest. Here GM can be broadly defined as a fungal component of preharvest grain deterioration, involving numerous fungal species interacting in different ways with the plant (i.e., parasitically and/or saprophytically).

The second concept (B, Fig. 1) restricts the definition of GM to a condition caused by infection and colonization of spikelet tissues prior to grain maturity. In this limited definition, only a few fungi are thought to be involved. The multitude of field fungi that colonize grain after physiological maturity are not part of GM per se, but rather constitute a component of weathering, or general postharvest grain deterioration.

On a practical level, the two concepts are similar. For example, early and late infections in concept A can be seen as analogous to the GM and weathering of concept B. Fungal-related grain deterioration, whether occurring before or after grain maturity, can cause important losses. The objective of plant-improvement programs, therefore, is sorghum cultivars resistant to all aspects of fungal-related grain deteriorations.

These concepts differ mainly in the way that infections occurring before grain maturity are related to fungal colonization of the mature grain. In concept A, the difference is quantitative. The earlier the infection occurs, the greater the potential for damage and the fewer the fungal species involved.

In concept B, infections occurring prior to grain maturity could be considered qualitatively different from postmaturity colonization. The early infections involve relatively few fungi acting as true parasites on living tissue. Postmaturity grain colonization involves many genera of field fungi that colonize primarily nonliving tissue.

For practical purposes, the more generalized concept A sufficiently explains what one sees in the field. The qualitative distinction between GM and weathering (concept B), however, helps explain many aspects of fungal deterioration of sorghum grain, including resistance, symptom expression, infection process, and etiology.

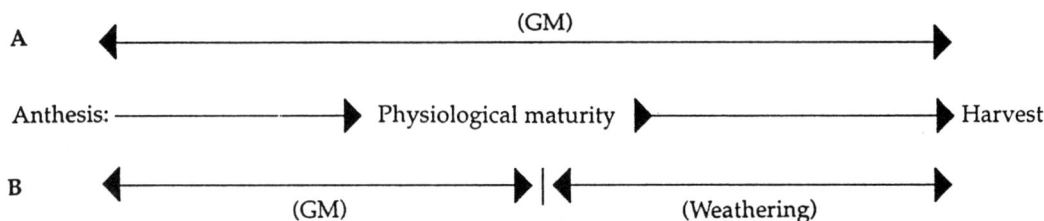

Figure 1. Two concepts A, and B, of fungal-related grain deterioration in sorghum.

For purposes of this review, GM refers to a condition resulting from all fungal associations with sorghum spikelet tissues occurring from anthesis to harvest. However, the qualitative distinction between early infection and post-maturity colonization will be employed when needed to facilitate discussion of certain aspects of the disease.

Symptoms

In discussing symptoms, one cannot help re-turning to the qualitative difference between early infections and postmaturity colonization. Symptoms of the two conditions can be very different.

Early infection by a GM pathogen probably occurs on the apical portions of spikelet tissues: glumes, lemma, palea, etc. Colonization then proceeds toward the base of the spikelet, either in the spikelet tissues or in voids between these tissues. A more-detailed discussion of this infection pattern will follow later.

Infection of the grain itself occurs at the base, near the pedicel, and can interfere with grain filling (Frederiksen et al. 1982) and/or cause a premature formation of the black layer (Castor 1981). Either condition causes a reduction in grain size, a symptom often associated with GM.

Visible superficial growth (the first signs of the fungus) occurs at the hilar end of the grain, and subsequently extends acropetally on the pericarp surface. Climatic conditions determine whether this growth will eventually spread to that part of the grain not covered by the glumes.

Infection induced by inoculation in green-house plants growing under low humidity pro-duces very small grains without visible signs of the fungus on the exposed stylar end of the grain (Forbes 1986). That part of the grain hid-den by the glumes is covered by a dense fungal mat. In contrast, the result of severe infection in the field usually is grains that are pink, white, or black (depending on the pathogen). This is be-cause of coverage of the grain by fungal my-celium (Williams and Rao 1981).

Early infections also involve spikelet tissues other than the grain. One of the first visible symptoms following inoculation is pigmenta-tion of the lemma, palea, glumes, and lodicules. This factor is highly cultivar dependent, and

may be linked with mechanisms of resistance (discussed later).

Fungal colonization of sorghum grain pro-duces a different set of symptoms. Colonization occurs primarily on the exposed part of the grain and may be limited to that area. Removal of the glumes will often show a sharp line of demarcation between protected and exposed areas (authors' observations). Postmaturity colo-nization is generally what produces the "moldy appearance" of grain maturing in humid envi-ronments. The color of the moldiness depends on the fungi involved.

Differences between early infections and postmaturity colonization can be difficult to sub-stantiate in the field. Both conditions occur to-gether, and late-season colonization can mask symptoms of infection occurring during grain development.

Causal fungi

It is thought that only a few fungi infect sor-ghum spikelet tissues during early stages of grain development. These are (in approximate order of importance) *Fusarium moniliforme* Sheld., *Curvularia lunata* (Wakker) Boedijn, *F. semitectum* Berk., & Rav., and *Phoma sorghina* (Sacc.). *F. moniliforme* and *C. lunata* are of signifi-cance worldwide (Castor 1981; Frederiksen et al. 1982; Williams and Rao 1981; Bandyopadhyay 1986). The pathogenicity of these fungi has been established by inoculation of plants in the field and in the greenhouse.

If sorghum grains of harvest maturity are in-cubated on nonselective agar, the above fungi may be isolated in low frequencies relative to many other fungi. This is because the pericarp of sorghum routinely supports a rich and varied mycoflora that is not eradicated with conven-tional techniques of surface sterilization.

Williams and Rao (1981) list the species most frequently isolated in studies of mycoflora asso-ciated with sorghum grain. Subsequent studies list much the same spectra of fungal species. Re-cent papers in this area of research include El Shafie and Webster 1981, Granja and Zambolim 1984, Kabore and Couture 1983, Kissim 1985, Khairnar and Gambhir 1985, Novo and Menezes 1985, Pachkhede et al. 1985, and Shree 1984.

The importance of this mycoflora is not well known. These fungi are generally thought to be restricted to the pericarp, but penetration into the endosperm can occur if the mature grain is exposed to high relative humidity or moisture for an extended period. Under severe climatic conditions, the endosperm can be completely colonized and partially degraded by field fungi (Glueck and Rooney 1980).

Fungal colonization of pericarp tissues of many cereal grains is common. Depending upon the timing and degree of penetration, these fungi are considered to be saprophytes or apathogenic weak parasites (Neergaard 1977).

Researchers in Australia have recently described *F. nygami* in association with sorghum grain (Burgess and Trimboli 1986). This new species resembles *F. moniliforme*, but produces chlamydospores. Its role in the etiology of grain mold is unknown.

Head Blight

Williams and Rao (1981) described head blight as "an invasion of tissues of the inflorescence by *F. moniliforme* Sheld. which results in the florets being killed to various degrees, up to complete destruction of the head." Symptoms include discoloration and necrosis of the panicle, extending into inflorescence branches, and reddening of the pith in affected areas. Severe head blight results in open panicles with drooping rachis branches (Frederiksen et al. 1982).

Many researchers feel that head blight is distinct from GM (Williams and Rao 1981), but there appears to be no differentiation at the pathogen level (Frederiksen et al. 1982). Grain mold symptoms are routinely induced by inoculation with *F. moniliforme* Sheld., but head blight does not always occur. This would seem to indicate that certain causal or predisposing factors for head blight and for GM may differ.

Researchers in Argentina report that resistance to *F. moniliforme* may be tissue dependent. A resistance reaction for head blight is not always indicative that a cultivar will be resistant to *F. moniliforme* in spikelet tissues (Forbes et al., unpublished data).

The actual losses to head blight are not known, but its potential for economic loss has been demonstrated. In 1979, losses of between U.S. \$3.2 million and U.S. \$7.2 million were attributed to head blight in Texas (Castor and Frederiksen 1981). In general, head blight appears to be more important in Mexico and the humid southeastern USA than in Texas (Frederiksen et al. 1982) In southern France, panicle discoloration and necrosis is common in some genotypes (author's observations), but the etiology of this condition has not been studied.

Importance of Grain Mold

There is little doubt that GM in its broadest sense constitutes one of the most important biotic constraints to sorghum improvement and production. Sorghum workers worldwide, queried in 1977, indicated GM as one of the most important diseases of sorghum (Williams and Rao 1981). More recently, the real and potential importance of GM has been emphasized for Africa (Louvel and Arnoud 1984), the Americas (Frederiksen et al. 1982), and India (ICRISAT 1987).

Damages caused by grain mold

Williams and McDonald (1983) pointed out that in spite of general agreement that GM is important, there have been few attempts to quantify losses resulting from the disease. This problem does not arise from a lack of evidence that GM causes damage. Certain GM pathogens have repeatedly been associated with losses in seed mass (Castor and Frederiksen 1980; Hepperly et al. 1982; Singh and Makne 1985); grain density (Castor 1981; Ibrahim et al. 1984), and percentage germination (Castor 1981, Maiti et al. 1985). Other types of damage relating to storage quality, food and feed processing quality, and market value that may result from GM have been discussed by Williams and Rao (1981).

Mycotoxin Research

One consequence of GM that has received much attention in the last decade is contamination. There is growing concern for the deleterious nature of subacute doses on animals. Mycotoxins in feed slow the growth rate, predispose animals to other infections and are teratogenic and carcinogenic (Lacey 1985). Mycotoxin content of grains contaminated during preharvest stages usually increases when the grains are stored.

Since the 1980s, several instances of sorghum contamination by mycotoxins have been reported from USA, Australia, Africa, and India.

McMillian et al. (1981, 1983a, 1983b, 1985) collected preharvest grain samples from several sorghum fields in Georgia and Mississippi from 1980 to 1982 and in 1984, and reported variable mycotoxin contamination with respect to the nature of mycotoxin, region, and species (e.g., maize shows more aflatoxin than sorghum). In 64 fields sampled on Georgia's coastal plain, 56% showed 1–90 ppb aflatoxin and 31% had 2–1468 ppb zearalenone. Grain harvested in Mississippi had neither of the mycotoxins.

Mold damage was severe in 1982, and mycotoxicosis was suspected in grain-fed swine. Of the 25 Georgian fields sampled, 84% showed aflatoxin [7–148 ppb (median 16 ppb)], and 8% contained zearalenone- [1515–10 420 ppb (median 6120 ppb)]. None of the 1984 samples showed aflatoxin. but one sample contained 80 ppb zearalenone. Shotwell et al. (1980) reported more than 1000 ppb zearalenone in 18% of the samples; 1000 ppb is the threshold value of physiological significance (Mirocha and Christensen 1974).

Australian reports of mycotoxin contamination of pre- and postharvest sorghum have been reviewed by Blaney (1985). He cites cases of suspected mycotoxicosis in four commercial swine operations — two due to aflatoxin, another due to aflatoxin and ochratoxin A, and the fourth due to zearalenone. Very high µg g^{-1} concentrations of these mycotoxins (aflatoxin <9.6, ochratoxin <0.1, and zearalenone <8) were detected in grain harvested and improperly stored.

In Nigeria, Salifu (1981) studied mold invasion and mycotoxin contamination in developing grains of short- and long-duration genotypes. The short-duration cultivars filled grains in unusually wet weather; no rains occurred from milk stage onward until harvest of the long-duration cultivars. All mature samples of the four short-duration cultivars had aflatoxin (10–80 µg g^{-1}). Aflatoxin and zearalenone were first detected at the hard-dough stage. None of the long-duration genotypes in this study produced mycotoxin, but the author cites another instance of aflatoxin contamination (100 µg g^{-1}) in a long-duration cultivar grown in a wetter region in northern Nigeria.

Bhradraiah and Ramarao (1982) reported the occurrence of aflatoxin B$_1$, B$_2$, and G$_1$ from pre-harvest and mature grain samples of some widely grown cultivars in India. They reported more aflatoxin in the early-maturing hybrids CSH 5 than on medium- and long-duration cultivars; additional studies are needed. Aflatoxin B$_1$ content in their study was 25–180 ppb.

Grains are also contaminated with toxic metabolites produced by species of *Alternaria*, particularly *A. alternata*. Although alternariol and its monomethyl ether, altenuene and altertoxin I were found in moldy grain, no sign of toxicity was noticed in rats or chicks fed with these mycotoxins (Seitz 1984).

Tenuazonic acid is a potent mycotoxin, anti-neoplastic and protein-inhibiting, primarily produced by some species of *Alternaria*, but it has not been detected in sorghum grain. However, *Phoma sorghina*, a widely distributed GM fungus, is known to produce tenuazonic acid (Steyn and Rabie 1976) and may be responsible for onyalai, a human disorder prevalent in Africa. Onyalai is diagnosed by haemorrhagic vesicles in the mouth that appear when *Phoma*-infected grain is ingested.

Most of the mycotoxin research has been carried out in countries that use sorghum grain as feed. It is important to analyze the situation in countries where sorghum is consumed by human beings. Many questions remain concerning mycotoxin contamination. How prevalent is mycotoxin in food prepared from contaminated grain? What is the epidemiology of mycotoxin production in the field? Is it possible to breed for reduced mycotoxins, as has been done in groundnut (Mehan et al. 1986). Several toxigenic fusaria are known to occur on sorghum, so how widespread is the occurrence of trichothecins in field-grown sorghum grains? Intensification of research on mycotoxins as it relates to GM was advocated in the last review on the subject (Williams and Rao 1981).

Measuring grain mold

The above discussion illustrates the potential damage resulting from GM. To accurately assess the importance of GM, however, it becomes necessary to correlate the level of damage with the corresponding level of disease. Assessment of GM importance, therefore, is effective only to the degree of accuracy in measuring GM. Measurement of GM severity is also important for other areas of research, including epidemiology and host resistance.

Visual appraisal has been the most common means of quantifying GM to date. Visual appraisal involves a complex of factors and can estimate severity (degree of colonization per grain indicated by signs or discoloration), incidence (proportion of grain affected), or damage (reduction is grain size), depending upon the method of assessment.

Visual appraisal, obviously the quickest and easiest method of disease assessment, is used for screening large numbers of samples (Bandyopadhyay and Mughogho 1988a). Advances in the search for resistance to grain mold achieved to date can be attributed to screening techniques based primarily on visual appraisal.

This form of estimation often has a surprisingly close association with other measures of severity. In several independent studies, a significant correlation has been established between visual appraisal and ergosterol concentration (discussed below) (Bandyopadhyay and Mughogho 1988b; Forbes 1986; ICRISAT 1986; Seitz et al. 1983).

Several factors can bias visual appraisal. For example, light-colored grains show more grain mold than dark-colored grains with equal severity. To avoid this problem, and be more accurate in general, workers at ICRISAT compare grain samples with light-grained and dark-grained standards of known severity levels (Bandyopadhyay and Mughogho 1988a). Comparing threshed grain is the most accurate method of visual assessment of GM (Frederiksen et al. 1982).

If visual assessments of GM severity are to be useful elsewhere, a common scale is required. Scales using well-defined units, such as percentage of grain surface affected (Forbes 1986; Bandyopadhyay and Mughogho 1988a) would seem to standardize comparison methods.

Because visual appraisal is a global evaluation of the condition of sorghum grain, it can provide only limited information about severity of GM per se. Extraneous factors, perhaps cultivar dependent, may mask the effects of GM. To get more accurate measurement of GM, researchers have used several techniques that have the commonality of estimating the quantity or incidence of the pathogen (fungal tissue or propagules) in a given amount of host tissue.

Most attempts to quantify GM pathogens in grain tissue have involved measures of incidence, and are based on the proportion of grains infected with certain pathogens (Hepperly et al. 1982; Gopinath and Shetty 1985; Granja and Zambolim 1984). Infection frequencies are measured by plating and incubating the entire kernel on blotting paper, or more often, agar.

Whole-grain plating can be biased by the competitive nature of the fungi making up the mycoflora (Neergaard 1977). Some scientists have attempted to compensate for this bias by using selective agar (Castor 1981) or chemical treatment of grain (Gopinath and Shetty 1985). The importance of competitive nature in a fungal sp is demonstrated by the fact that the incidence of *F. moniliforme* often increases when a Fusarium-specific agar is used (Castor 1981).

The relationship between GM severity and incidence is poorly understood. One can assume, however, that incidence would not reflect the important effects of infection timing on severity, since a grain infected late would count the same as one with early infection. Incidence-severity relationship studies for other diseases have proved to be complex, and have been impossible to determine for certain diseases (Seem 1984). It is doubtful that incidence studies will give much information about the severity of GM.

Some researchers have tried to quantify the degree of fungal colonization of sorghum grain. Forbes (1986) spread suspensions of ground seed tissues on a *Fusarium*-specific agar to quantify colonization by *F. moniliforme*. This technique, proposed as an indicator of disease severity, estimates the amount of viable fungal tissue (propagules g^{-1} of seed tissue).

Fungal biomass in a sample of sorghum grain is also estimated by measuring the concentration of ergosterol, a sterol produced by fungi but not by plants (Seitz et al. 1977). Ergosterol measurements are routine at ICRISAT (ICRISAT 1986). The procedure is sensitive and has the attractive attribute of estimating total (viable and nonviable) fungal biomass. Differences in ergosterol concentrations are often found among grain samples with similar degrees of superficial mold growth (Seitz et al. 1983).

Relationship of disease severity and damage

The potential for GM to damage grain is often demonstrated, but the relationship between severity and damage has seldom been quantified. New methods of assessing severity, such as measuring ergosterol, may make this an active area of future research. Even now, certain patterns are emerging from the few studies reported.

Severity appears to be more closely associated with viability than with yield. In a recent study, two measures of severity (ergosterol concentration, and propagules of *F. moniliforme* g^{-1} seed tissue) were more highly correlated with percentage germination than with seed mass or grain density (Forbes 1986).

The sensitivity of percentage germination as an indicator of GM was also demonstrated by Castor and Frederiksen (1980). They suggested its use as a means of evaluating resistance. Automatic measuring of seed leachates and correlation with germination could become an efficient technique for studying the effects of GM severity on viability (Forbes 1986).

Control of Grain Mold

Avoidance

Avoidance of GM has often been described as one of the most important traditional control strategies (Castor 1981; Williams and Rao 1981). In areas where photosensitive cultivars have grown, GM is avoided because flowering and grain fill occur in the dry season. Avoidance is one of the most important control strategies still in commercial seed production. Most seed is produced with irrigation in arid regions to avoid GM and other problems.

Chemical control

Chemical control appears to provide some protection against GM. In another experiment, fungicide sprays at milk stage and 10 days later were shown to reduce GM infection (Naik et al. 1981).

Most studies involving fungicides and GM-related fungi, however, deal with the efficacy of seed dressings for improving seedling emergence and vigor (Patil et al. 1986; Munghate and Faut 1982; Vidhyasekaran 1983). Certain fungi have also been eradicated from sorghum grains with hot water treatment (Bhale and Khare 1982).

Resistance

In most cases, avoidance or chemical control in farmers' sowings is impractical. For this reason, major research efforts have focused on development of resistant cultivars. Improvement of screening techniques is a major effort in this research.

Screening at one of the major research institutes is currently done with natural inocula (Bandyopadhyay and Mughogho 1988a). High moisture levels are assured by sprinkling on rain-free days. Sprinklers are used as necessary throughout the period of grain development, as well as after physiological maturity. Materials to be screened are compared with grain samples with known levels of GM severity. More than 7000 accessions have been screened, and 156 lines selected as resistant (Bandyopadhyay and Mughogho 1988b).

Worldwide, many screening techniques are used. In Argentina a seed company has developed a method of separately screening for resistance to GM fungi occurring before grain maturity, those colonizing seed tissues after grain maturity, and those causing head blight. The different types of resistance are identified by screening at different stages of plant development and in the case of head blight, on the basis of symptom development in the peduncle (G. Garcia, unpublished data).

Screening with this technique revealed that resistance to early-season GM pathogens is not always associated with resistance to fungi causing damage late in the season. The independence of these two types of resistance was demonstrated earlier (Castor 1981).

Another approach to the identification of resistance is used by some researchers in northern Africa. Multivariate statistical techniques are used to determine cultivar reaction based on incidence of important GM pathogens, grain quality, germination and seedling viability, and visual assessment of moldiness (Louvel and Arnoud 1984).

Resistance mechanisms

In the last 10 years there has been a great deal of research directed toward the elucidation of resistance mechanisms. Much of this research has involved biochemical analyses of infected and noninfected tissues. Waniska reviews this work elsewhere in this volume.

Histological study has produced some insight of the process of early infection by GM pathogen on susceptible and resistant cultivars.

Several independent histological studies indicate similar patterns of initial infection and subsequent colonization of sorghum spikelet tissues (Castor 1981; Forbes 1986; Bandyopadhyay 1986). These studies were designed to determine the infection pattern following inoculation with *F. moniliforme* and with *C. lunata*. Resistant and susceptible cultivars inoculated with *F. moniliforme* were also compared (Castor 1981; Forbes 1986).

On a susceptible cultivar, initial infection by *F. moniliforme* occurs on the apical ends on the spikelet tissues: lemma, palea, glumes, filaments, and senescing styles. Fungal mycelium advances basipitally, either by colonizing spikelet tissues or by growing in voids between these tissues. Early colonization of glumes (3–4 days following inoculation) was found to be very heavy and caused little cellular disruption or pigmentation in the host (Forbes 1986).

Within 5 days of inoculation, mycelium can be seen in all parts of the spikelet, with the denser growth around the ovary base. Lodicules appear to serve as an important energy source, and are always surrounded by dense fungal growth, but extensive colonization of lodicule tissue per se has been questioned (Forbes 1986). It is apparently from this energy source, near the point of attachment to the pedicel, that infection of the ovary wall occurs.

In the next stages of invasion, a dense mycelial mat progresses acropetally, between the aleurone layer and the pericarp. Subsequent invasion of the endosperm, embryonic tissues, and pericarp originates from this peripheral mat. Halloin (1983) has pointed out that peripheral growth on the inner layers of the true seed coat precedes embryonic colonization in many seed species.

When environmental conditions are appropriate, mycelial growth pushes through the pericarp, producing a white or pink fungal mass which can completely cover the grain.

Early invasion of a resistant spikelet appears to be as follows. As in the susceptible cultivar, mycelial growth can be seen in all parts of the spikelet at 5 days after inoculation. However, much of this growth is found in the voids between spikelet structures (Forbes 1986).

Pigmentation occurs rapidly in localized areas where host and fungal tissue were in close association. Fungal growth can involve cell disruption and cell wall depositions, inducing localized necrosis (Forbes 1986). Using another resistant cultivar, Castor (1981) likewise noticed heavy pigmentation associated with restricted fungal growth in and near the lodicules.

Castor proposed that localized pigmentation associated with resistance could be caused by luteolinidin that reddens sorghum stalks in response to pathogenic and nonpathogenic fungi. Pigmentation can also occur as a result of inoculation, suggesting that the mere presence of pigments does not confer resistance. However, pigmentation in susceptible cultivars appears to differ, from that in resistant cultivars, in coloration, intensity, location, and timing (Castor 1981; Forbes 1986).

After these early events, fungal invasion of the resistant spikelet is either arrested (Forbes 1986), or proceeds at a much slower pace than in a susceptible cultivar (Castor 1981), delaying infection of the ovary, and protecting it somewhat from damage.

Infection by *C. lunata* differs from that of *F. moniliforme* in the following way. During the initial period of spikelet invasion, *C. lunata* can infect the apical part of the ovary wall from the colonized lemma, palea, lodicules, filaments, pollen grain, and decaying style (Bandyopadhyay 1986). Within 5 to 10 days mycelium penetrates the pericarp and ramifies throughout the cross and tube cells. Colonization does not usually continue directly into the endosperm, but rather through the placental sac, which can also lead to invasion of the embryo.

Differences between the infection patterns of *F. moniliforme* and *C. lunata* may partially explain the fact that resistance to the two occasionally differ (Castor 1981; Louvel and Arnoud 1984).

For both fungi, infection pattern and the degree of damage caused undoubtedly is affected by the maturity of the spikelet at the time of infection. Either fungus can interfere with grain filling and cause premature formation of the black layer, reducing kernel size. If infection occurs early enough, invasion of the ovary base will cause the caryopsis to be aborted (Castor 1981).

In summary, colonization of a susceptible cultivar proceeds rapidly in all spikelet tissues without observable immediate host reaction. Colonization patterns are different for *F. moniliforme* and *C. lunata*. In resistant cultivars examined in these studies, the presence of *F. moniliforme* induced pigmentation and localized necrosis, involving cellular disruption and cell wall depositions. Infection of the embryo did not occur, or was retarded. Thus resistance mechanisms may involve spikelet tissues other than the ovary.

Epidemiology

Epidemiological studies may provide information that can be used to improve control strategies. Unfortunately, little is known about the epidemiology of GM. At the time of the review by Williams and Rao (1981), knowledge at the time was probably well-stated by their comment, "generally it seems that wet weather following flowering is necessary for GM development and the longer wet period, the greater the mold development." Even now, there have been but few studies on GM epidemiology, and little added to our knowledge of the subject.

In what appears to be one of the few epidemiological studies of GM, ICRISAT workers successfully monitored diurnal and seasonal trends of aerial spore densities of *Curvularia lunata* throughout the growing season (ICRISAT 1986). This type of study has been done for *F. moniliforme* spores in and above the canopy of maize (Ooka and Kommedahl 1977), indicating that techniques exist which could easily be applied to sorghum.

Some epidemiological insight may be gained indirectly from a study done in Texas (Forbes 1986). A conidial suspension of *F. moniliforme* was applied to panicles at anthesis, either by spraying or submerging. Plants were then incubated for 24 h and later moved, with nonincubated controls, to a greenhouse where conditions were not favorable for further infections. Severe GM developed on all incubated plants,

but not on the noninoculated plants, indicating that moisture is needed for initial infection but not for disease progression at the grain level. The severity of GM within a field is probably greatly influenced by the effect of moisture on repeated infections through time. Little is known, however, about the apparently critical relationship between moisture, inoculum availability, and host maturity.

Future Needs

Etiology and the role of host maturity

There are few published accounts of controlled inoculation studies with suspected GM pathogens. Institutions with appropriate facilities (growth chambers, inoculation chambers, and greenhouses with controlled environments) could do closely monitored and replicated studies using known and suspected grain mold pathogens at different stages of plant development. Such research might clear up a major area of confusion in GM-related literature—which organisms are capable of infecting which tissues at which stages of host development. Immunofluorescence techniques would add sensitivity and selectivity to histological methods.

Resistance mechanisms

There is a need to continue studies on the resistance mechanisms of sorghum without a testa or with low tannin content. Preliminary studies on two cultivars have indicated potential mechanisms of resistance, including localized necrotic reaction and inhibition of fungal growth associated with pigmentation. Confirmation of these characteristics would be most useful in other cultivars (Bechtel et al. 1985) Determination of the nature of the physiological changes associated with resistance likewise would be a valuable contribution.

Epidemiology

As mentioned, few quantitative studies on epidemiological aspects of grain mold are in the record. Research designed to determine the importance of environmental variables and inoculum dynamics in disease development should be of first priority. Knowledge of disease spread in time and space may facilitate many other areas of GM research.

References

Anahosur, K.H., Kulkarni, K.A., and Patil, S.H. 1984. Role of fungicide-cum-insecticide seed treatment in the control of seed rotting fungi and shootfly of sorghum. Mysore Journal of Agricultural Sciences 16:418–421.

Bandyopadhyay, R. 1986. Grain mold. Pages 36–38 *in* Compendium of sorghum diseases. (Frederiksen, R.A., ed.) St. Paul, MN, USA: The American Phytopathological Society.

Bandyopadhyay, R., and Mughogho, L.K. 1988a. Evaluation of field screening techniques for resistance to sorghum grain molds. Plant Disease 72:500–503.

Bandyopadhyay, R., and Mughogho, L.K. 1988b. Sources of resistance to sorghum grain molds. Plant Disease 72:504–508.

Bechtel, D.B. Kaleikau, L.A. Gains, R.L., and Seitz, L.M. 1985. The effects of *Fusarium graminicola* infection on wheat kernels. Cereal Chemistry 62:191–192.

Bhadraiah, B., and Ramarao, P. 1982. Isolates of *Aspergillus flavus* from sorghum seeds and aflatoxin production. Current Science 51: 1116–1117.

Bhale, M.S., and Khare, M.N. 1982. Seed-borne fungi of sorghum in Madhya Pradesh and their significance. Indian Phytopathology 35:676–678.

Blaney, B.J. 1985. Mycotoxins in crops grown in different climatic regions of Queensland. Pages 97–108 *in* Trichotechenes and other mycotoxins: proceedings of the International Mycotoxin Symposium, Aug 1984, Sydney, Australia (Lacey, J., ed.). Chichester, UK: John Willey & Sons.

Burgess, L.W., and Trimboli, D. 1986. Characterization and distribution of *Fusarium nygamai* sp nov. Mycologia 78:223–229.

Canez, V.M. Jr., and King, S.B. 1981. A comparison of methods for assessing grain mold in sorghum. Phytopathology 71:865. (Abstract).

Castor, L.L. 1981. Grain mold histopathology, damage assessment and resistance screening within *Sorghum bicolor* (L.) Moench lines. Ph.D. thesis, Texas A&M University, College Station, TX, USA. 192 pp.

Castor, L.L., and Frederiksen, R.A. 1980. *Fusarium* and *Curvularia* grain mold in Texas. Pages 93–192 *in* Sorghum diseases, a world review: proceedings of the international workshop on sorghum diseases, 11–15 Dec 1978, ICRISAT, Hyderabad, India. Patancheru, Andhra Pradesh 502 324 India: International Crops Research Institute for the Semi-Arid Tropics.

Castor, L.L., and Frederiksen, R.A. 1981. *Fusarium* head blight occurrence and effects on sorghum yield and grain characteristics in Texas. Plant Disease 64:1017–1019.

Dayan, M.P. 1980. Studies on sorghum head mold organisms in the Philippines. D.Sc. thesis, University of the Philippines, Los Banos, Luzon, Philippine Island.

El Shafie, A.E., and Webster, J. 1981. A survey of seed-borne fungi of *Sorghum bicolor* from the Sudan. Transactions of the British Mycological Society 77:339–342.

Forbes, G.A. 1986. Characterization of grain mold resistance in sorghum (*Sorghum bicolor* L. Moench). Ph.D. thesis, Texas A&M University, College Station, TX, USA. 75 pp.

Frederiksen, R.A., Castor, L.L., and Rosenow, D.T. 1982. Grain mold, small seed and head blight: the *Fusarium* connection in sorghum. Proceedings of the Thirty-seventh Annual Corn and Sorghum Industry Research Conference 37:26–36.

Glueck, J.A., and Rooney, L.W. 1980. Chemistry and structure of grain in relation to mold resistance. Pages 119–140 *in* Sorghum diseases, a world review: proceedings of the international workshop on sorghum diseases, 11–15 Dec 1978, ICRISAT, Hyderabad, India. Patancheru, Andhra

Pradesh 502 324, India: International Crops Research Institute for the Semi-Arid Tropics.

Gopinath, A., and Shetty, H.S. 1985. Occurrence and location of *Fusarium* species in Indian sorghum seed. Seed Science and Technology 13: 521–528.

Granja, E.M., and Zambolim, L. 1984. The incidence of phytopathogenic fungi on sorghum seeds. Revista Ceres 31:115–119.

Halloin, J.M. 1983. Deterioration resistance mechanisms in seeds. Phytopathology 73:335–339.

Hepperly, P.R., Feliciano, C., and Sotomayor, A. 1982. Chemical control of seedborne fungi of sorghum and their association with seed quality and germination in Puerto Rico. Plant Disease 66:902–904.

ICRISAT (International Crops Research Institute for the Semi-Arid Tropics). 1986. Annual Report 1985. Patancheru, Andhra Pradesh 502 324, India: ICRISAT.

ICRISAT (International Crops Research Institute for the Semi-Arid Tropics). 1987. Annual Report 1986. Patancheru, Andhra Pradesh 502 324, India: ICRISAT.

Ibrahim, O.E., Nyquist, W.E., and Axtell, J.D. 1985. Quantitative inheritance and correlations of agronomic and grain quality traits of sorghum. Crop Science 25:649–654.

Kabore, K.B., and Couture, L. 1983. Mycoflore des semences du sorgo cultivé en Haute-Volta. Naturaliste Canadien 110:453–457.

Khairnar, D.N., and Gambhir, S.P. 1985. Studies on seed mycoflora of some jowar varieties from Nalik area (Maharashtra). Indian Botanical Reporter 4:68–70.

Kissim, M.Y. 1985. Seed-borne fungi of sorghum in Saudi Arabia. Pakistan Journal of Botany 17:263–266.

Lacey, J. (ed.) 1985. Trichothecenes and other mycotoxins: proceedings of the International Mycotoxin Symposium, Aug 1984, Sydney, Australia. Chichester, UK: John Wiley & Sons.

Louvel, D., and Arnoud, M. 1984. Moisissures des grains chez le sorgho: définition d'une résistance. Etat des recherches sur le sorgho au Sénégal, Actes de la reunion de travail du 22 Avril 1983.

Maiti, R.K., Raju, P.S., and Bidinger, F.R. 1985. Studies on germinability and some aspects of preharvest physiology of sorghum grain. Seed Science and Technology 13:27–35.

Mathur, K., and Naik, S.M. 1981. Fungi associated with moldy sorghum grain and reaction of sorghum cultivars to head molds. Bulletin of Grain Technology 19:143–146.

McMillian, W.W., Widstrom, N.W. and Wilson, D.M. 1981. Aflatoxin contamination in 1984 preharvest sorghum. Sorghum Newsletter 24:123.

McMillian, W.W., Wilson, D.M., and Widstrom, N.W. 1985. Aflatoxin contamination in 1984 preharvest sorghum. Sorghum Newsletter 28:25.

McMillian, W.W., Wilson, D.M., Mirocha, C.J., and Widstrom, N.W. 1983a. Mycotoxin contamination in grain sorghum from fields in Georgia and Mississippi. Cereal Chemistry 60:226–227.

McMillian, W.W., Wilson, D.M., and Widstrom, N.W. 1983b. Aflatoxin and zearalenone contamination in grain sorghum samples collected in Georgia fields in 1982. Sorghum Newsletter 26:122.

Mehan, V.K., McDonald, D., and Ramakrishna, N. 1986. Varietal resistance in peanut to aflatoxin production. Peanut Science 13:7–10.

Mirocha, C.J., and Christensen, C.M. 1974. Oesterogenic mycotoxins synthesized by *Fusarium*. Pages 129–148 in Mycotoxins. Amsterdam, The Netherlands: Elsevier Science Publishers.

Munghate, A.G., and Faut, J.G. 1982. Efficacy of nine different fungicides against fungi frequently associated with sorghum seed. Pesticides 16:16–18.

Naik, S.M., Singh, S.D., and **Mathur, K.** 1981. Efficacy of fungicides in controlling head molds of sorghum. Pesticides 15:25–39.

Neergaard, P. 1977. Seed Pathology. London, UK: McMillian Press Ltd. 1187 pp.

Novo, R.J., and **Menezes, M.** 1984. Effectiveness of fungicides in seed treatment of grain sorghum, *Sorghum bicolor*. Fitopatologia Brasileira 9:543–549.

Novo, R.J., and **Menezes, M.** 1985. Fungi isolated from sorghum (*Sorghum bicolor*) seeds from the municipalities of Caruaru and Serra Talhada, Pe. Fitopatologia Brasileira 10:57–64.

Ooka, J.J., and **Kommedahl, T.** 1977. Wind and rain dispersal of *Fusarium moniliforme* in corn fields. Phytopathology 67:1023–1026.

Pachkhede, A.U., Vyawajare, S.V., and **Shreemali, J.L.** 1985. Two new species of *Phoma* from India. Current Science (India) 54:485–486.

Patil, M.R., Sakpal, P.N., and **Patil, V.N.** 1986. Efficacy of different fungicides against *Fusarium moniliforme* and *Curvularia lunata* (important fungi) associated with sorghum seed. Pesticides 20:27–30.

Salifu, A. 1981. Mycotoxins in short season sorghums in northern Nigeria. Samaru Journal of Agricultural Research 1:83–88.

Seem, R.C. 1984. Disease incidence and severity relationships. Annual Review of Phytopathology 22:133–150.

Seitz, L.M. 1984. Alternaria metabolites. Pages 443–455 *in* Mycotoxins—production, isolation, separation and purification (Betina, V., ed.). Amsterdam, The Netherlands: Elsevier Science Publishers.

Seitz, L.M., Mohr, H.E., Burroughs, R., and **Sauer, D.B.** 1977. Ergosterol as an indicator of fungal invasion in grains. Cereal Chemistry 54:1207–1217.

Seitz, L.M., Mohr, H.E., Burroughs, R., and **Glueck, J.A.** 1983. Preharvest fungal invasion of sorghum grain. Cereal Chemistry. 60:127–130.

Shree, M.P. 1984. Seed mycoflora of sorghum varieties with particular references to *Exserohilum turcicum* (Pass) Leo at Sug. and related genera, causing seed rot, seedling and leaf blight diseases. Seed Research 12:24–39.

Shotwell, O.L., Bennett, G.A., Goulden, M.L., Plattner, R.D., and **Hesseltine, C.W.** 1980. Survey for zearlenone, aflatoxin and ochratoxin in U.S. grain sorghum from 1975 and 1976 crops. Journal of the Association of Official Analytical Chemists 63:922–926.

Singh, A.R., and **Makne, V.G.** 1985. Correlation studies on seed viability and seedling vigor in relation to seed size in sorghum (*Sorghum bicolor*) Seed Science and Technology 13:139–142.

Steyn, P.S., and **Rabie, C.J.** 1976. Characterization of magnesium and calcium tenuazonate from *Phoma sorghina*. Phytochemistry 15:1977–1979.

Vidhyasekaran, P. 1983. Control of *Fusarium moniliforme* infection in sorghum seed. Seed Science and Technology 11:435–439.

Williams, R.J., and **McDonald, D.** 1983. Grain molds in the tropics: problems and importance. Annual Review of Phytopathology 21:153–178.

Williams, R.J., and **Rao, K.N.** 1981. A review of sorghum grain molds. Tropical Pest Management 27:200–211.

Cereal Chemistry and Grain Mold Resistance

R.D. Waniska[1], G.A. Forbes[2], R. Bandyopadhyay[3],
R.A. Frederiksen[4], and L.W. Rooney[5]

Abstract

Grain of sorghum at immature and mature stages of development is deteriorated by fungi. Sorghum cultivars that exhibit some resistance to deterioration have a more corneous endosperm, contain specific phenolic compounds, are thin in pericarp thickness, and show decreased fissuring or cracking in the field. Resistant cultivars respond more quickly than susceptible cultivars to fungal invasion via increased levels of phenolic compounds in glume tissues. Extensive deterioration of the grain does not occur before physiological maturity because the developing grain contains 3 to 10 times the level of specific phenolic compounds of mature grain. In sorghum grown under wet conditions, grain of resistant cultivars contains lower levels of free phenolic compounds at maturity compared with grain of susceptible cultivars. Hence, phenolic compound content in grain is a predictor of the sorghum cultivar's level of resistance to grain molding.

Introduction

Grain of *Sorghum bicolor* (L.) Moench is exposed to sunshine, rain, diseases, and pests during its development. Grain molding or weathering refers to deterioration of the grain by several mechanisms (Glueck and Rooney 1980; Williams and Rao 1981), i.e., chemical, enzymatic, bacterial, fungal, and due to insects. Regardless of the cause, changes in the physical properties of grain include (1) moldy and discolored pericarp, (2) a soft and chalky endosperm, (3) decreased grain filling and size, (4) sprouting (reduced germination of seed), (5) mycotoxins, (6) decreased dry matter, density and test weight, and (7) altered composition of phenolic compounds (Glueck and Rooney 1980; Williams and Rao 1981; Jambunathan et al. 1986.

The changes that occur in the seed when sorghum deteriorates in the field as a result of at-

tack by molds are reviewed. The terms 'molding' and 'weathering' will herein be used interchangeably.

Invasion by Fungi

Fungi are responsible for much of the damage during early and later stages of deterioration (Glueck and Rooney 1980; Williams and Rao 1981). Bacteria and fungi invade and colonize spikelet tissues at anthesis (Fig. 1) (Castor and Frederiksen 1980). After the glumes open at anthesis, reabsorption of the lodicules and related tissues by the plant commences (Castor and Frederiksen 1980). Many saprophytes, including species of *Curvularia*, *Fusarium*, and *Colletotrichum*, also utilize these tissues as nutrients (Castor and Frederiksen 1980; Glueck and Rooney 1980). These fungi colonize glume tis-

1. Associate Professor, Department of Soil and Crop Sciences, Texas A&M University, College Station, TX 77843-2474, USA.
2. Plant Pathologist, Centre Internacional de la Papa, International Potato Center, PO Box 5969, Lima, Peru.
3. Plant Pathologist, Cereals Program, ICRISAT Center, Patancheru, Andhra Pradesh 502 324, India.
4. Professor of Plant Pathology and Microbiology, Texas A&M University, College Station, TX 77843-2474, USA.
5. Professor, Department of Soil and Crop Sciences, Texas A&M University, College Station, TX 77843-2474, USA.

Waniska, R.D., Forbes, G.A., Bandyopadhyay, R., Frederiksen, R.A., and **Rooney, L.W.** 1992. Cereal chemistry and grain mold resistance. Pages 265–272 *in* Sorghum and millets diseases: a second world review. (de Milliano, W.A.J., Frederiksen, R.A., and Bengston, G.D., eds). Patancheru, A.P. 502 324, India: International Crops Research Institute for the Semi-Arid Tropics. (CP 735)

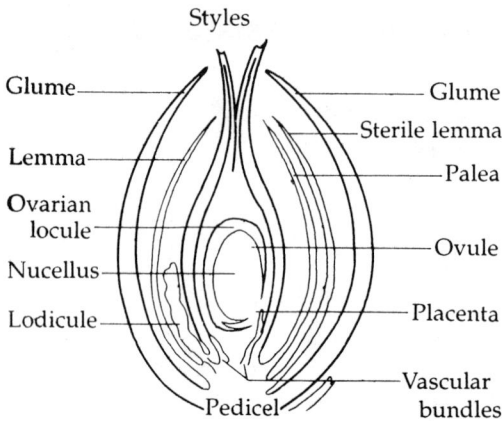

Figure 1. Diagram of sorghum spikelet at anthesis showing the ovule, styles, lemma, palea, lodicules, and glumes.

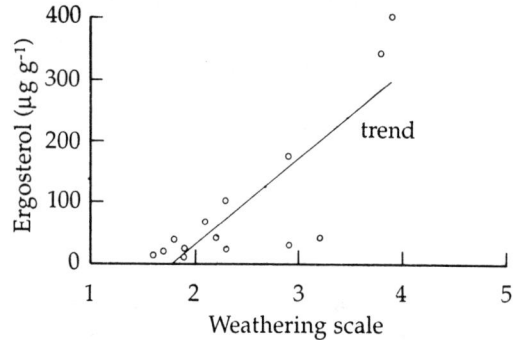

Figure 2. The relationship of ergosterol content in sorghum grain to a visual rating of weathering on a 1 to 5 scale where 1 = less mold and 5 = severe mold. Source: Data from Seitz et al. (1983).

sues and the pericarp of the developing caryopses during the next 4 weeks (Castor and Frederiksen 1980). At physiological maturity, colonies of fungi were observed in the starchy mesocarp and the cross and tube cells of the pericarp in all cultivars (Glueck and Rooney 1980).

Rain and warm temperatures following anthesis increase fungal colonization of developing caryopses (Castor and Frederiksen 1980; Williams and Rao 1981). In mature grains, wet weather and early fungal colonization increase the rate and extent of deterioration (Glueck and Rooney 1980; Williams and Rao 1981). One measure of fungal metabolism is the presence of ergosterol (Seitz et al. 1983). The ergosterol content of grain and the rate of deterioration of mature grain increased geometrically with degree of rainy weather (Fig. 2). Hence, visual ratings may not be a good predictor of fungal colonization of the grain.

Inhibition by Phenolic Compounds

Resistant cultivars contain less fungi than susceptible cultivars in the caryopses and glumes (Castor and Frederiksen 1980; Forbes et al. 1989). Resistant cultivars also respond more quickly to fungal invasion via increased levels of phenolic compounds and pigmentation of spikelet tissues than do susceptible cultivars (Forbes et al. 1989).

It seems that specific phenolic compounds in spikelet tissues limit the growth of fungi.

Most cultivars with pigmented testa containing polymers of flavan-3-ols (tannins) resist weathering (Glueck and Rooney 1980; Hahn et al. 1983; Bandyopadhyay et al. 1988). These cultivars normally contain higher levels of phenolic compounds, phenolic acids, and tannins than do cultivars without pigmented testae (Hahn et al. 1984). Apparently, tannins or precursors of tannins impart antifungal and antibacterial activity in the developing and mature caryopses.

Cultivars with a red pericarp (with or without a pigmented testa) containing significant levels of flavan-4-ols exhibit resistance to weathering [Jambunathan et al. 1986; Mukuru (this publication)]. Cultivars with a white pericarp without a pigmented testae do not contain significant levels of flavan-4-ols; yet some of these cultivars exhibit resistance to weathering. Apparently, flavan-4-ols and related compounds are involved in some way with resistance to grain weathering.

Free phenolic compound and tannin contents in caryopses increase significantly during development (Doherty et al. 1987; Forbes 1986), reaching maximum levels 7–18 days after anthesis (Figs. 3, 4). At maximum, the levels of free phenolic compounds and tannins were two to eight times higher than those observed in the mature grain. Phenolic compounds and tannins are apparently being bound to cellular tissues, and therefore are not extractable for analysis. The

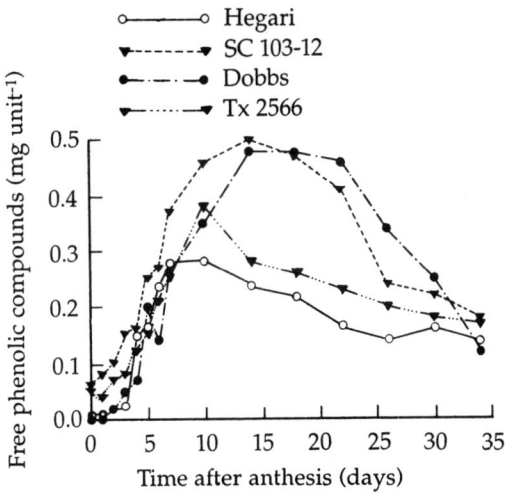

Figure 3. Effect of pericarp color and spreader gene on free phenoloic compounds in caryopses of group II and group III sorghums. Source: Doherty et al. (1987).

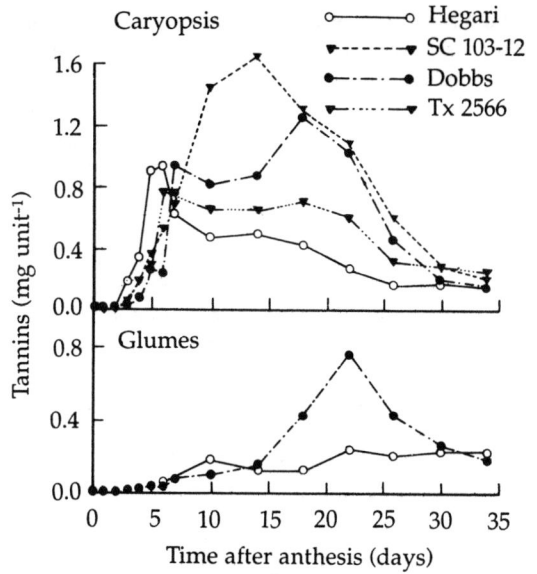

Figure 4. Effect of pericarp color and spreader gene on tannis in caryopses and glume tissues of group II and group III sorghums. Source: Doherty et al. (1987).

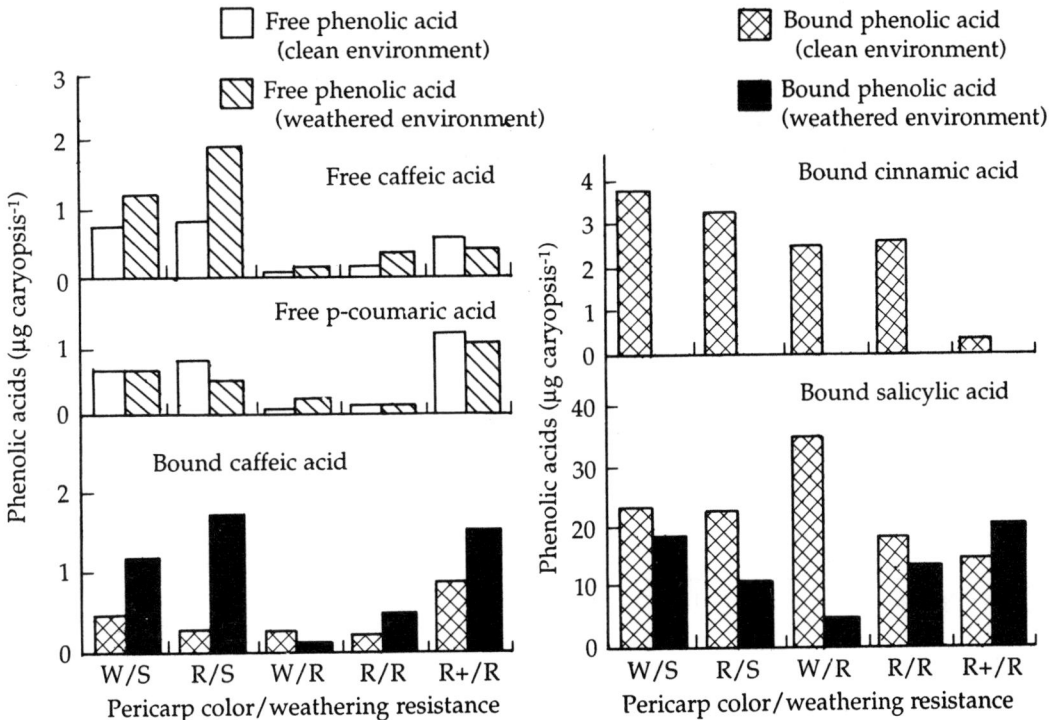

White pericarp (W); red pericarp (R); presence of a pigmented testa (+); weathering resistant (R); and susceptible (S).

Figure 5. Effect of weathering on free caffeic acid, bound caffeic acid, cinnamic acid, p-coumaric acid, and salicylic acid on caryopses of sorghum. Source: Waniska et al. (1989).

high levels of free phenolic compounds and tannins occurred during the period of early invasion and colonization of fungi of the caryopses. Hence, it is likely that these compounds or specific components of these compounds are involved in the resistance mechanisms of grain weathering.

Phenolic compounds and acids were quantified in sorghum grown under conditions that yielded clean and weathered grain (Waniska et al. 1989). Higher levels of free phenolic acids, especially p-coumaric and caffeic acids, were observed in susceptible cultivars grown in the wet environment (Fig. 5). Bound phenolic acids were present at higher levels than free phenolic acids in all cultivars; however, bound phenolic acids did not significantly correlate with weathering resistance. Resistant and susceptible cultivars without a pigmented testae were properly grouped using a statistical procedure called the principal component analysis (Fig. 6). When the cultivars were grouped by pericarp color, weathering resistance, and environmental conditions, a scatter plot of free p-coumaric acid vs. free phenolic compounds separated the resistant (left) and susceptible (right) groups of the scatter plot. The higher levels of phenolic acids in weathered, susceptible cultivars probably resulted from fungal metabolites. Hence, analysis of phenols in the grain can predict the level of

resistance to weathering. Verification is the objective of current projects at Texas A&M University and ICRISAT.

A developmental study of phenols in sorghum caryopses and glumes after inoculation with *Fusarium moniliforme* revealed differences in tissues, cultivars, and phenolic components (Figs. 7-12) (Forbes 1986). Glumes (including other spikelet tissues) contained relatively high

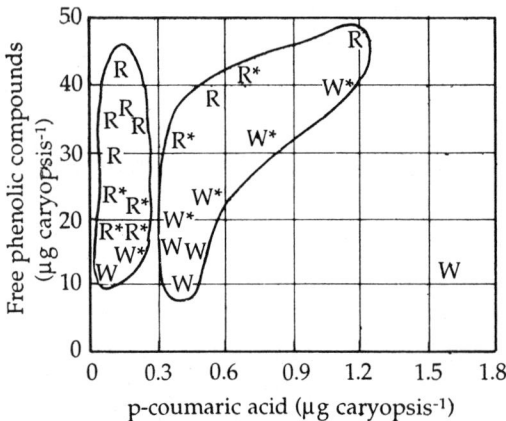

Figure 6. Classification of sorghum cultivars into resistant (*left*) and susceptible (*right*) groups according to the phenolic acid composition.

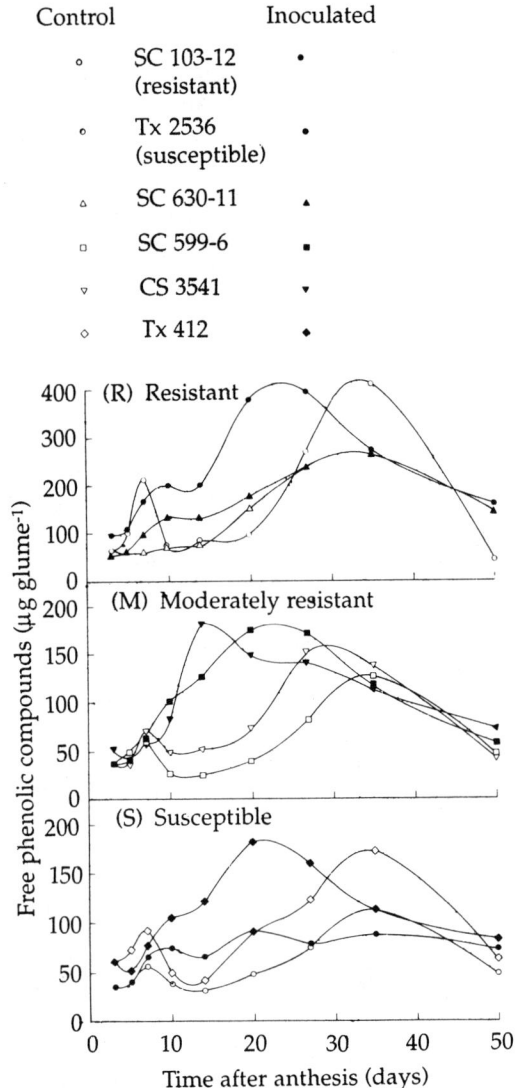

Figure 7. Free phenolic compounds in sorghum glumes that were inoculated with *Fusarium moniliforme* at anthesis and evaluated during caryopsis development.

levels of free phenolic compounds (1–2%) at anthesis, and these levels increase to 2–10% during normal development. Inoculation with fungi caused an increase in phenolic compounds at 10, 14, and 20 days after anthesis (Fig. 7), especially in resistant and moderately resistant cultivars. Levels of free phenolic compounds in sorghum caryopses were similar for inoculated and control treatments during development, i.e., concentration and time effects were not apparent in caryopses.

Free p-hydroxybenzoic acid significantly contributed to the higher levels of phenolic compounds at 7, 10, and 14 days after anthesis for both glumes and caryopses (Figs. 8, 9). However, higher levels of free p-hydroxybenzoic acid did not correspond to resistance to grain molding. Free p-coumaric acid also contributed to the higher levels of phenolic compounds in glumes and caryopses; and the resistant cultivars and one of the moderately resistant cultivars had

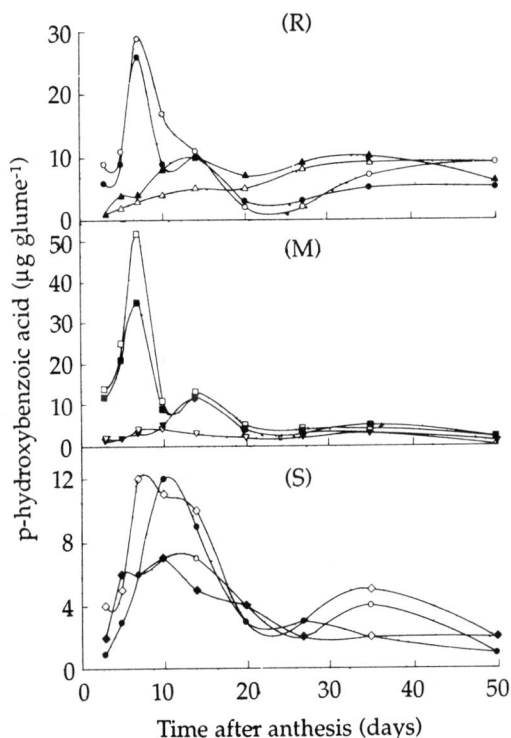

Figure 9. Free p-hydroxybenzoic acid in sorghum caryopses that were inoculated with *F. moniliforme* at anthesis and evaluated during caryopsis development. Legend: (see Fig.7).

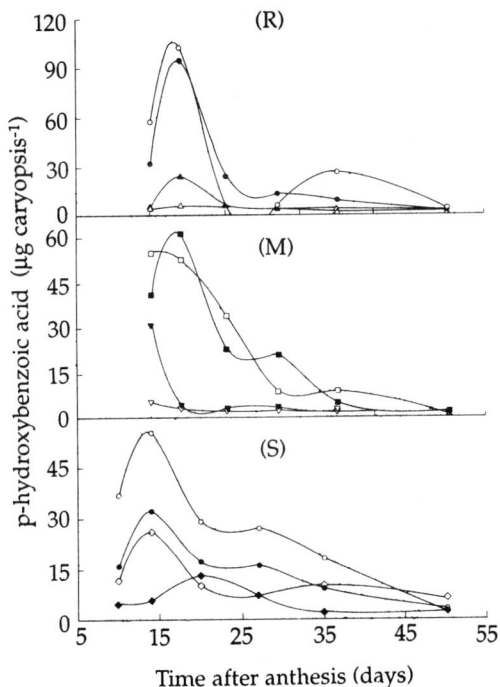

Figure 8. Free p-hydroxybenzoic acid in sorghum glumes that were inoculated with *F. moniliforme* at anthesis and evaluated during caryopsis development. Legend: (see Fig. 7).

earlier or higher levels than their control treatments (Figs. 10, 11). Free caffeic acid in the caryopsis was higher in concentration for all cultivars, except one of the susceptible cultivars (Fig. 12). Hence p-coumaric and caffeic acids appear to be involved in the biochemical response to fungal invasion of sorghum glumes and caryopses.

Significant changes in concentrations of these (and other) phenolic acids during development is noteworthy, as it implies that significant biochemical activity occurs in glumes and caryopses during development. Glumes appear to be the plant's first defense against fungal invasion and colonization; later it is the chemical and physical properties of the caryopsis that appear to be more important in the resistance of grain to molding.

Other Resistance Mechanisms

Several structural characteristics of the sorghum grain correspond to improved resistance to

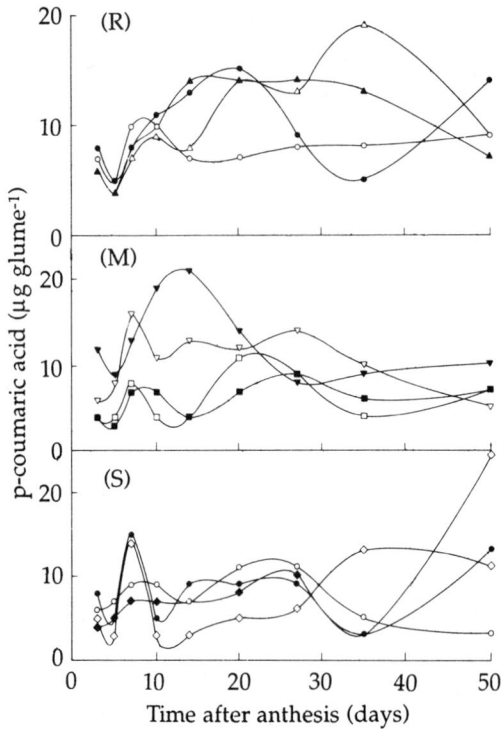

Figure 10. Free p-coumaric acid in sorghum glumes that were inoculated with *F. moniliforme* at anthesis and evaluated during caryopsis development. Legend: (see Fig. 7).

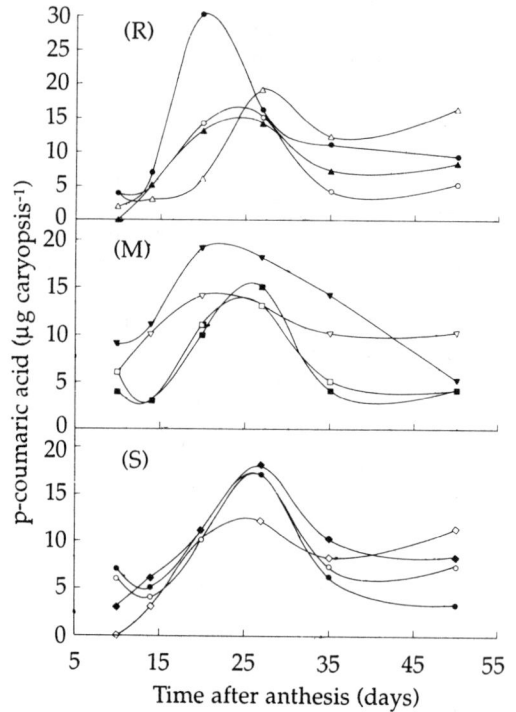

Figure 11. Free p-coumaric acid in sorghum caryopses that were inoculated with *F. moniliforme* at anthesis and evaluate during caryopsis development. Legend: (see Fig. 7).

weathering: endosperm texture, pericarp thickness, surface wax, and grain integrity. A more corneous endosperm texture, e.g., grain with a larger proportion of corneous to floury endosperm, usually exhibit less weathering (Glueck and Rooney 1980). Mycelia of fungal colonies do not easily penetrate the dense, tightly packed aleurone and peripheral endosperm layers of mature caryopses.

A thin paricarp on sorghum caryopsis normally corresponds to less weathering (Glueck and Rooney 1980). Since fungal colonies were observed inside the pericarp in all sorghum cultivars, nutrients in the pericarp are readily hydrolyzed by saprophytic fungi. Hence, a thick mesocarp that contains starch and protein supports more fungal colonies than sorghums with a thin mesocarp. Also, most of the free phenolic compounds are located in the pericarp and the adjoining testae layer. These apparently bioactive compounds would be diluted with starch, etc., in a thick pericarp.

Glueck and Rooney (1980) reported that the surface wax of the grain is thicker and continuous, and thus more protective, in grain of weathering resistant cultivars. Fungi were not able to penetrate the wax-covered, epicarp surface. Nor could water penetrate the epicarp. Hence, water and fungi probably enter the grain through holes on the surface of the epicarp, i.e., through the hylar and stylar areas.

The rate of absorption of water by the endosperm affects its rate of deterioration (Glueck and Rooney 1980) (Fig. 13). Water enters the cross and tube cells of the pericarp, and rapidly moves around the grain. Concurrently, water moves through the hilum (black layer) into the germ. After 30 min, water had moved into the area where the endosperm, germ, and pericarp meet. Some water also enters the grain through the style, and moves around the grain in the cross and tube cells. Water then moves readily through the less organized floury endosperm.

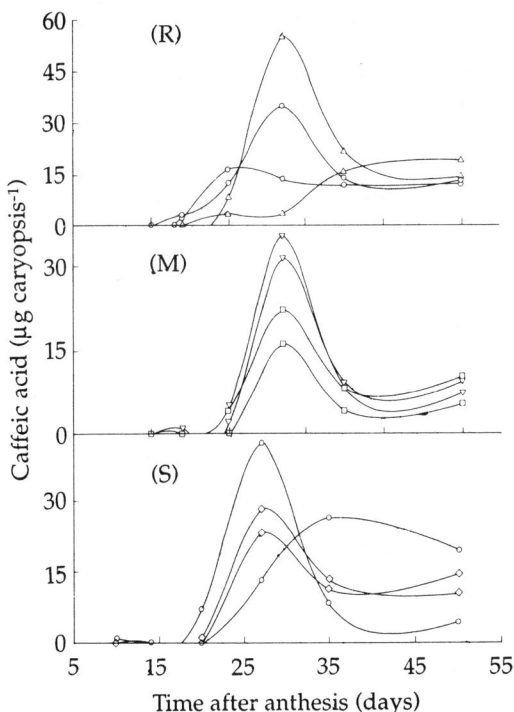

Figure 12. Free caffeic acid in sorghum caryopses that were inoculated with *F. moniliforme* at anthesis and evaluated during caryopsis development. Legend: (See Fig. 7).

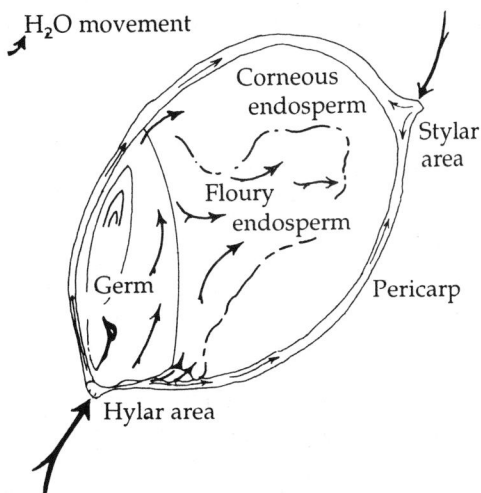

Figure 13. Water uptake and movement in a sorghum seed. Arrows indicate primary water-entry areas and subsequent movement in the seed. Source: Glueck and Rooney (1980).

Water begins to move into the corneous endosperm after 3 h.

Mature grains of some cultivars crack or break apart when exposed to rain after a dry period. Breaks in the grain surface increased the rate and altered the pattern of water uptake. Cultivars with more breakage or with a more rapid rate of water uptake (intact grains) exhibited less resistance to weathering. Composition of leachate from these cultivars was richer in nutrients. A thicker mesocarp and a softer endosperm texture usually corresponded to increased water absorption and richer leachate.

Panicle shape, glume character, wet-season avoidance, seed size, and other factors contribute to increased resistance to grain molding (Glueck and Rooney 1980). A panicle that is more open than compact does not provide as much shelter to insects and disease pathogens. Drooping panicles shelter the grain (instead of catching water) and decrease the exposure of the grain to moisture required for fungal growth. Longer glumes are considered protective to the grain, but only if the glumes do not trap water. Grain maturing during the nonrainy season avoids the environmental conditions required for maximum fungal growth. Grain size does not seem to be a factor, except very large grains tend to have a softer endosperm texture and are more susceptible to grain molding.

References

Bandyopadhyay, R., Mughogho, L.K., and **Prasada Rao, K.E.** 1988. Sources of resistance to sorghum grain molds. Plant Disease 72:504–508.

Castor, L.L., and **Frederiksen, R.A.** 1980. *Fusarium* and *Curvularia* grain molds in Texas. Pages 93–102 *in* Sorghum diseases: a world review: proceedings of the International Workshop on Sorghum Diseases, 11–15 Dec 1978, ICRISAT, Hyderabad, India. Patancheru, Andhra Pradesh 502 324, India: International Crops Research Institute for the Semi-Arid Tropics.

Doherty, C.A., Waniska, R.D., Rooney, L.W., Earp, C.F., and **Poe, J.H.** 1987. Free phenolic compounds and tannins in sorghum caryopsis and glumes during development. Cereal Chemistry 64:42.

Forbes, G.A., Frederiksen, R.A., and Seitz, L.M. 1989. Assessment of sorghum grain mold: disease intensity and disease loss. Seed Science and Technology 17:297–307.

Forbes, G.A. 1986. Characteristics of grain mold resistance in sorghum (*Sorghum bicolor* (L.) Moench). Ph.D. thesis, Texas A&M University, College Station, TX, USA.

Glueck, J.A., and Rooney, L.W. 1980. Chemistry and structure of grain in relation to mold resistance. Pages 119–140 *in* Sorghum diseases: a world review: proceedings of the International Workshop on Sorghum Diseases, 11–15 Dec 1978, ICRISAT, Hyderabad, India. Patancheru, Andhra Pradesh 502 324, India: International Crops Research Institute for the Semi-Arid Tropics.

Hahn, D.H., Rooney, L.W., and Faubion, J.M. 1983. Sorghum phenolic acids, their HPLC separation, and their relation to fungal resistance. Cereal Chemistry. 60:255–259.

Hahn, D.H., Rooney, L.W., and Earp, C.F. 1984. Tannins and phenols of sorghum. Cereal Foods World 29:776–779.

Jambunathan, R., Butler, L.G., Bandyopadhyay, R., and Mughogho, L.K. 1986. Polyphenol concentrations in grain, leaf, and callus tissues of mold-susceptible and mold resistant sorghum cultivars. Journal of Agriculture and Food Chemistry. 34:425–430.

Seitz, L.M., Mohr, H.E., Burroughs, R., and Glueck, J.A. 1983. Preharvest fungal invasion of sorghum grain. Cereal Chemistry 60:127–130.

Waniska, R.D., Poe, J.H., and Bandyopadhyay, R. 1989. Effects of weathering on phenolic acids in sorghum caryopses. Journal of Cereal Science 10:217–225.

Williams, R.J., and Rao, K.N. 1981. A review of sorghum grain moulds. Tropical Pest Management 27:200–211.

Breeding for Grain Mold Resistance

S.Z. Mukuru[1]

Abstract

Colored-grained germplasm sorghum collection lines with high levels of mold resistance were used for developing white-grained genotypes with good levels of resistance. Segregating progenies were intensively screened for white-grained genotypes, and five white-grained advanced selections with mold resistance similar to their colored-grain parental lines are now in the breeding program. Mold resistance in the white-grained types was associated with grain hardness, while that of brown-grained types was associated with either high-tannin or flavan-4-ol or grain hardness. Cultivars with a combination of these factors were highly resistant. It appears that flavan-4-ol is not produced in white-grained types, either mold-resistant or mold-susceptible. At present the only factor known to be responsible for mold resistance in white-grained cultivars is grain hardness. Intensification of efforts to breed mold resistance into high-yielding cultivars is in order so that farmers can efficiently grow mold-free sorghums.

Introduction

Grain mold (GM), caused by several unspecified fungi, is a major and widespread problem in the semi-arid tropics (SAT) because of significant reductions in yield and quality of the harvested sorghum grain. The problem has become particularly widespread where early-maturing improved cultivars are cultivated and mature under warm and wet conditions. Grain deteriorates because of breakdown of grain structure and eventual loss of viability from mold-induced physical, physiological, and chemical changes. Endosperm of molded grain appears chalky because of partial hydrolysis of starch and protein. Molded grains may be contaminated with mycotoxins and present health hazards to consumers (Castor and Frederiksen 1980).

The only practical and economical methods for control of GM in sorghums are avoidance and/or sowing GM-resistant cultivars. Chemical control of GM in SAT nations is usually thought to be impractical and too costly. Seed dressings may be an exception, but many SAT farmers who grow their own seed will be slow to adopt this practice.

Most peasant farmers of limited means in Africa and India have always used the avoidance method of grain mold control—i.e., growing traditional photoperiod-sensitive landraces. These avoid, or escape, the mold problem because the grains develop and ripen during dry weather—after the rains. Pest and pathogen pressures are much reduced at this time. These landraces, however, are bulky plants with poor harvest indices, and often are affected by terminal drought-stress.

Another method of GM control used in some areas is to grow high-tannin, colored-grain landraces; these resist grain mold attack, and bird damage as well.

Sources with low levels of mold resistance, identified at ICRISAT Center in 1975, were used to develop improved cultivars with mold resistance. The high-yielding, white-grained cultivars that were developed, however, do not have sufficient levels of resistance to control grain

1. Principal Plant Breeder, Cereals Program, ICRISAT Center, Patancheru, Andhra Pradesh 502 324, India.

Mukuru, S.Z. 1992. Breeding for Grain Mold Resistance. Pages 273–285 *in* Sorghum and millets diseases: a second world review. (de Milliano, W.A.J., Frederiksen, R.A., and Bengston, G.D., eds). Patancheru, A.P. 502 324, India: International Crops Research Institute for the Semi-Arid Tropics. (CP 739).

molds. In the early 1980s, sources with high and stable levels of mold resistance were identified and are being used in developing white-grained breeding lines with good levels of mold resistance. Results and progress made in breeding for mold-resistant, white-grained sorghum genotypes using these high and stable sources of resistance are presented here.

Sources of Resistance

The sorghum germplasm collection was screened for resistance to grain molds at ICRISAT Center in mid-1970s and low levels of resistance were identified. Not one of the germplasm collection lines was immune to grain mold, and none showed a consistently high degree of resistance at all locations. However, in some lines mold development was consistently low at many locations and across seasons (Rao and Williams 1980). Search for higher levels of mold resistance in the sorghum germplasm collection continued and, by 1985, 7132 germplasm lines that flowered and matured during the rainy season had been screened; of these 156 colored-grain lines exhibited high levels of mold resistance (Bandyopadhyay et al. 1988). Threshed grain-mold rating (TGMR) of the selected colored-grain germplasm lines was 3 or lower on a 1 to 5 scale, (1 = 0%; 5 = mold on more than 50% of the grain surface). Most significant in these observations was that these lines maintained resistance for 2 to 3 weeks after physiological maturity. Pericarps of the highly resistant lines were colored; a testa layer was present in all but 14 lines. The 14 lines without the testa layer were low in tannin content, and diverse in geographical origin and taxonomic race. No white-grained germplasm line, however, has showed such a high level of grain-mold resistance.

Mechanisms of Resistance

Certain plant and grain characteristics are reported to be associated with grain-mold resistance (Glueck and Rooney 1980). Grain hardness and density and the structure of the outer layers of the seeds and chemical composition are reported to be partially responsible for improved weathering resistance. Brown seeds with high tannin contents and the presence of a pigmented testa layer have been associated with GM resistance (Harris and Burns 1970, 1973). However, some lines highly resistant to molds have been found to be low in tannin. These have been found to contain significant amounts of flavan-4-ol; the susceptible cultivars did not (Jambunathan et al. 1986). Hahn and Rooney (1986) reported that GM-resistant sorghum cvs contain a greater variety and quantities of phenolic acids than do susceptible cultivars.

It is apparent that several factors, independently or in combination, contribute to GM resistance. The most important are tannin, flavan-4-ol, and phenolic acids content, and grain hardness. A sorghum cv combining grain hardness with all or some of these polyphenolic compounds should possess a high level of mold resistance. Unfortunately, high-tannin content is nutritionally undesirable.

Genetics of Resistance

There is little information available on the genetics of mold resistance, but it is known that GM resistance is complex and may be the result of additive effects of many genes affecting several plant characteristics (Williams and Rao 1981). The high level of grain-mold resistance in colored-grain sorghums is associated with high tannin or flavan-4-ol content, or both.

Inheritance of the testa layer is controlled by the complementary B_1 and B_2 testa genes (Glueck and Rooney 1980). Presence of the testa layer in the presence of a dominant spreader gene "S" produces grain with high tannin content.

Information on the inheritance of grain-hardness is likewise scanty. That available indicates that grain hardness is governed by several additive genes.

The inheritance of flavan-4-ol is not yet understood.

Breeding for Resistance

The use of host-plant resistance to control grain mold is recognized as the logical approach to this problem. Breeding for grain-mold resistance therefore receives considerable emphasis in many national and international sorghum-improvement programs. At ICRISAT Center, breeding for grain-mold resistance began in the mid-1970s. Murty et al. (1980) reviewed the progress and achievements of ICRISAT's GM-breeding research. ICRISAT scientists crossed a large number of low mold-susceptible, white-grained sources and adapted high-yielding lines, then intensively screened and selected for mold resistance in early-segregating populations. They identified lines with moderately high grain-yield potentials and resistance to grain molds under low disease pressure. However, under moderate to high mold disease pressure, these lines became susceptible. We still do not have improved white-grained cultivars that farmers can use without a potential for GM loss.

I believe that the lack of progress can be attributed to the low levels of resistance in the material available for use in breeding programs. Various sources of low susceptibility were intermated to generate variability and concentrate the scattered resistance genes, but this did not significantly increase the levels of GM resistance.

The identification, in 1981, of high levels of mold resistance in colored-grain sorghums without the tannin-containing testa layer was a significant achievement, as it indicates that mold resistance in colored-grain sorghums isn't always associated (as previously believed) with high tannin levels. Thus, it may be possible to transfer their genes for mold resistance to white-grained sorghums. In 1982, we started using mold resistant, colored-grain sorghums to investigate the possible transfer of the high levels of resistance in colored-grain to white-grained sorghums through breeding.

Screening and Selection

We made crosses between mold-resistant, colored-grain sorghums (IS 14384, IS 14385, and IS 14388) and mold-susceptible, white-grained, high-yielding cultivars (ICSV 1 and SPV 104). IS 14384 does not have the tannin-containing testa layers, but IS 14385 and IS 14388 do.

Using the GM field-screening techniques developed at ICRISAT (Bandyopadhyay and Mughogho 1988), we screened F_1 and F_2 segregating progenies for resistance. Five hundred or more plants were screened from each cross and for each individual plant, time to 50% flowering was recorded. Harvesting of each panicle was carried out 54 days after flowering (i.e., 2 to 3 weeks after reaching physiological maturity), threshed grains of each panicle were rated for GM resistance on the 1 to 5 scale. Recording the time to 50% flowering for each individual plant is tedious and time consuming, but essential to ensure that developing grains of each plant were exposed to wet and humid environmental conditions (ideal for mold development) for an equal number of days. This avoids mistaking late-maturing genotypes as GM resistant, and improves the probability of identifying mold resistance in early-maturing genotypes.

Table 1. Frequency of mold-resistant, white-grained genotypes in F_2 segregating populations derived from crosses between mold-resistant, colored-grain sorghums, and mold-susceptible, white-grained sorghums at ICRISAT Center, rainy season 1983.

Pedigree	Total F_2 populations	Total white-grained progenies in F_2 Population		Selected white-grained F_2 progeny	
		(no.)	(%)	(no.)	(%)
IS 14384[1] × KSV 1	550	48	9	12	2.0
IS 14384[1] × SPV 104	340	59	17	12	3.5
KSV1 × IS 14385[1]	554	152	27	18	3.2
SPV 104 × IS 14385[1]	561	68	12	15	2.7
IS 14388[1] × KSV1	627	77	12	2	0.3
IS 14388[1] × SPV 104	521	63	12	1	0.2

1. Mold-resistant, colored-grain sorghum parent.

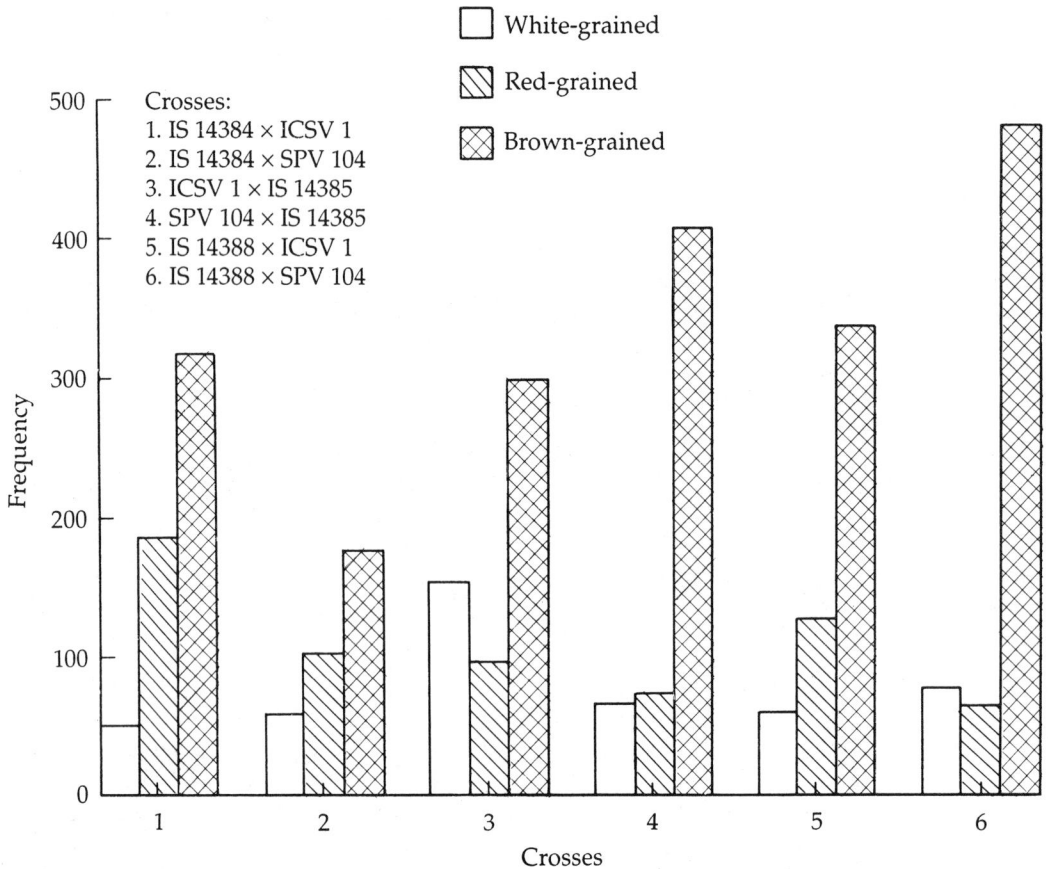

Figure 1. Frequency distribution of white, red, and brown grained F_2 segregates derived from each of six crosses between colored grain and white-grain sorghums.

Grains of all the F_1s were brown and possessed the tannin-containing testa layers; all were resistant to molds. All the F_2s segregated into white-, red- (without the testa layer), and brown-grained (with the testa layer) (Fig. 1). Few segregates for each F_2 population were white-grained; only 12–27% of the total F_2 population was in this category (Table 1). White-grained F_2 segregates resistant to grain mold (TGMR = 3 or below) were also very few compared to red- and brown-grained segregates (Figs. 2, 3).

We selected all white-grained F_2 segregates with Threshed Grain Mold Rating (TGMR) of 3 or less but selected only a few highly mold-resistant red-, and brown-grained segregates with TGMR of 2. The selected mold-resistant F_2 segregates were advanced during the postrainy sea-

sons and screened and reselected for mold resistance during rainy seasons 1983 and 1984. Selection and advancement of red- or brown-grained F_2 segregates proved very useful as a number of them were heterozygous for grain color and therefore segregated for white-grained types; these were screened and selected for GM-resistance in later generations.

Evaluation in Advanced Selections

Advanced mold-resistant selections were evaluated for resistance in a randomized-block design with three replications at ICRISAT Center during rainy season 1985. We included in the trial (as controls) the three colored parental lines (IS 14384, IS 14385, and IS 14388), two white-grained

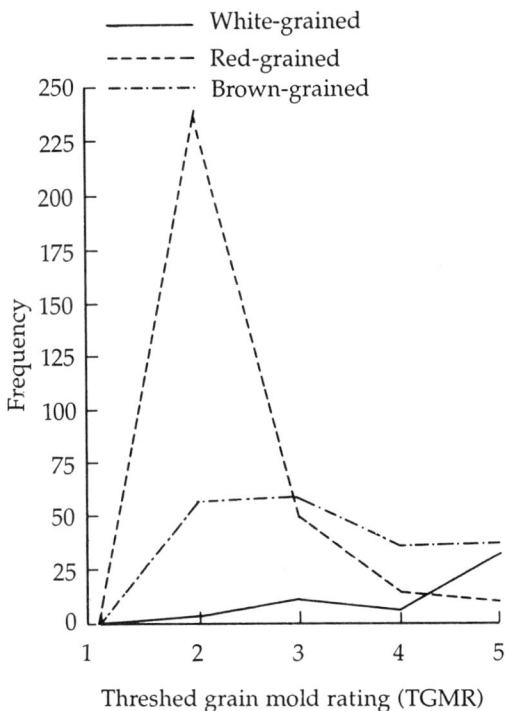

Figure 2. Frequency distribution of Threshed Grain Mold Rating (TGMR) of F_2 segregates derived from a cross between IS 14384 (red-grained and mold-resistant) and ICSV 1 (white-grained and mold-susceptible) sorghums.

grains high in flavan-4-ol but low in tannin content. The white-grained, mold-resistant lines are tall with long open panicles and are very low in grain yield, just like their red-grained parental lines. We are now using these lines as sources of resistance in the breeding program to transfer resistance into high-yielding, white-grained cultivars with desirable agronomic attributes. We screened for mold resistance in segregating F_2 progenies derived from crosses involving these sources during rainy season 1987 at ICRISAT Center and identified a few mold-resistant, white-grained segregates with acceptable agronomic traits.

Associated Traits

Detailed records of plant and grain characteristics of individual F_2 segregates include time to

parental lines (ICSV 1 and SPV 104), and five mold-susceptible white-grained cultivars. Five white-grained advanced selections were highly resistant, with average mold ratings of 2.2, compared to 2.3 for red- and brown-grained selections, and 2.4 for the mold-resistant parental lines (Table 2). The white-grained susceptible selections and white-grained parental lines were susceptible with a maximum mold rating of 5.0. White-grained mold-resistant selections, unlike white-grained mold-susceptible selections and parental lines, were significantly lower in floaters (%) and higher in dehulling recovery (%), indicating that their grains were hard with vitreous endosperm; white-grained susceptible lines were soft with flowery endosperm. Brown-grained selections have soft grains high in tannin but moderate in flavan-4-ol content, while red-grained selections have hard and vitreous

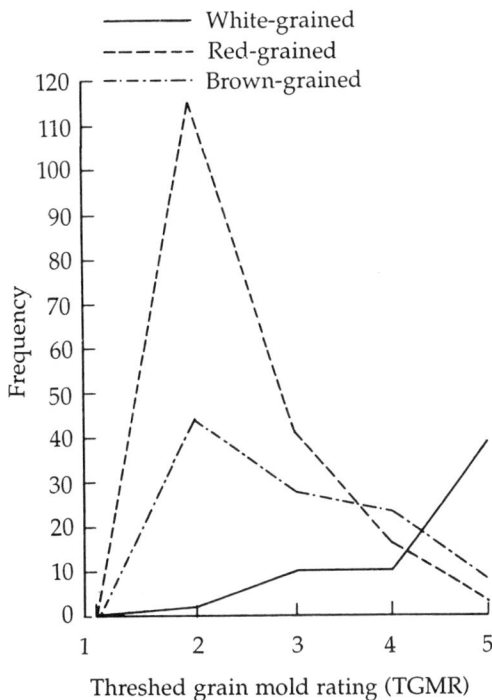

Figure 3. Frequency distribution of Threshed Grain Mold Rating (TGMR) of F_2 segregates derived from a cross between IS 14384 (red-grained and mold-resistant) and SPV 104 (white-grained and mold-susceptible) sorghums.

Table 2. Mean threshed grain mold rating (TGMR)[1], and physical characteristics of grains of four groups of advanced sorghum breeding lines, parental lines, and improved controls evaluated at ICRISAT Center, rainy season 1985.

Type of selection	Lines (no.)	Pericarp color	Testa[2]	TGMR	Floaters (%)[3]		Dehulling recovery (%)[4]	Tannin (cat.eq)[5]	Flavan- 4-ol
					A	B			
GM-resistant	2	Brown	+	2.3	52	56	75	1.53	3.19
GM-resistant	7	Red	–	2.3	23	29	84	0.12	4.57
GM-resistant	5	White	–	2.2	21	27	87	0.11	0.16
GM-susceptible	7	White	–	5.0	63	76	64	0.25	0.01
GM-susceptible	2	White	–	5.0	78	87	56	0.10	0.00
GM-resistant (parental lines)	3	Brown/ Red	+/–	2.4	52	49	70	2.27	5.44
GM-susceptible (controls)	5	White	–	5.0	73	86	60	0.11	0.01
SE	-	-	–	±0.15	±5.7	±3.8	±2.3	±0.10	±0.16

1. TGMR = Threshed grain mold rating on a 1 to 5 scale, where 1 = no mold, and 5 = 50% grain surface area molded.
2. Testa present (+); testa absent (-).
3. Floater A = grains of a sample (%) that floated when submersed in a solution of $NaNO_3$ (1.3 specific gravity). Floater B = grains in a sample (%) that floated in a 1.327 specific gravity solution of tetrachloro-ethylene and odorless kerosene.
4. Percentage of whole grains recovered after dehulling.
5. Cat.eq = Catechin equivalent.

50% flowering, plant height, plant color, panicle type, absence or presence of the testa layer, and grain mold resistance. We also determined grain density (% floaters of each F_2 segregate derived from three crosses (IS 14384 × ICSV 1, IS 14384 × SPV 104, and IS SPV 104 × IS 14385) and the tannin content of each F_2 segregate derived from one cross (IS 14384 × SPV 104). Frequency distribution plots of mold resistance of F_2 segregates derived from IS 14384 × ICSV 1 and IS 14384 × SPV 104 crosses are shown in Figures 2 and 3, respectively; that of floaters (%) of similar crosses in Figures 4 and 5, respectively.

Correlation coefficients between mold resistance and plant and grain traits were determined to identify plant or grain traits significantly associated with mold resistance. In a few crosses, mold resistance and time to 50% flowering were significantly correlated, indicating that late-flowering F_2 segregates in these crosses tended to escape mold damage. For the majority of crosses, however, correlation coefficients between mold resistance and time to 50% flowering

were small and not significant. Also correlation coefficients between panicle type and mold resistance were low and insignificant except in one cross (IS 14384 × SPV 104), indicating that open-panicle types are not always associated with mold resistance. We observed several F_2 segregates with open-panicle types severely molded.

Correlation coefficients between mold resistance and floaters (%) and tannin are presented in Table 3. Mold resistance was highly correlated with floaters in all crosses except among brown-grained F_2 segregates derived from IS 14384 × ICSV 1. This confirms a previous observation (Glueck and Rooney 1980) that grain hardness is associated with mold resistance. Tannin content was highly correlated with mold resistance among brown-grained F_2 segregates, confirming the observations of Harris and Burns (1970) that sorghum genotypes with high-tannin content resist grain molding. Correlation coefficients between TGMR and floaters (%), dehulling recovery (%), tannin and flavan-4-ol levels among white- and red-grained mold-resistant ad-

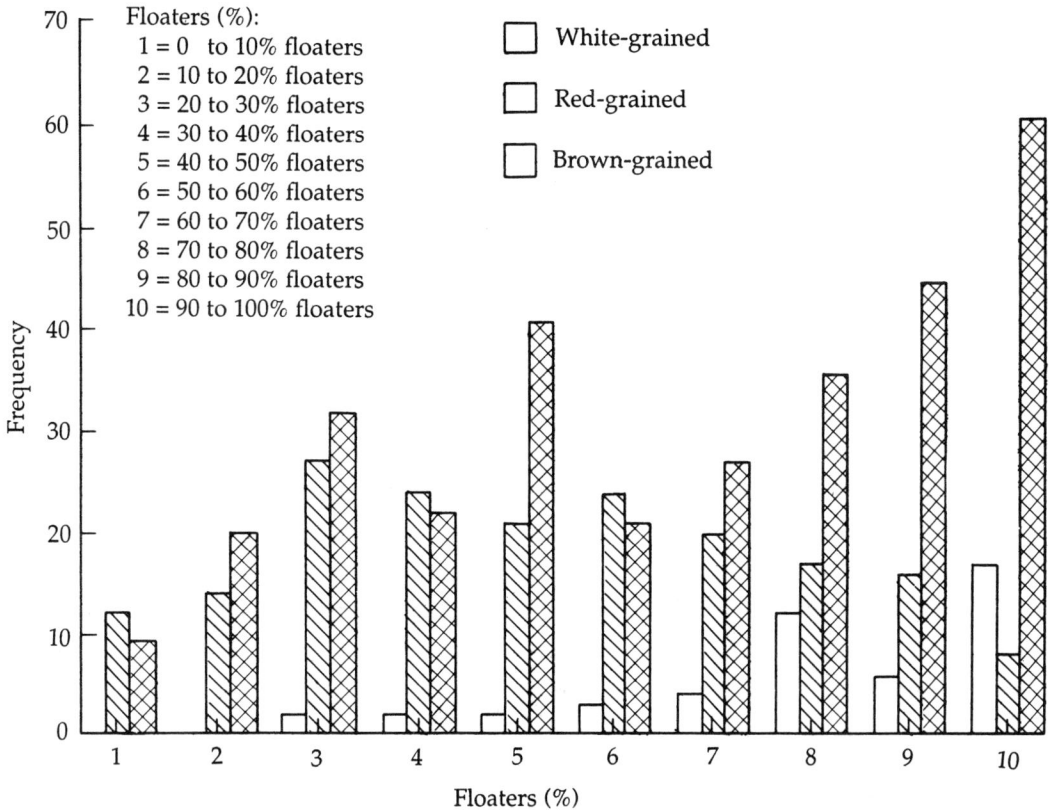

Figure 4. Frequency distribution of floaters (%) among white, red, and brown-grain F_2 segregates derived from a cross between IS 14384 (red-grained) and ICSV 1 (white-grained) sorghums.

vanced selections were calculated. Here again floater and dehulling recovery (0.8087 and − 0.8381, respectively, $P = 0.01$) were highly correlated with mold resistance. Tannin was not correlated with mold resistance (0.1801), since the advanced selections did not have the testa layer and were low in tannin content. Flavan-4-ol was highly correlated (-0.6612, $P = 0.01$) with mold resistance, confirming reports (Jambunathan et al. 1986) that high flavan-4-ol was responsible for resistance in low-tannin colored-grain sorghums. All the white-grained, mold-resistant advanced selections had zero or small amounts of flavan-4-ol.

We evaluated the stability of mold resistance and grain hardness in six low-tannin lines at ICRISAT Center during rainy season 1987. We scored mold resistance and determined floater percentage of the grains of each of these lines at physiological maturity, and at 1, 2, and 3 weeks

after physiological maturity (Table 4). In general, severity of mold increased and grain hardness decreased from physiological maturity to 3 weeks after physiological maturity. The increase was very small for the colored-grain mold-resistant IS 14384. Grain mold scores and floater percentage values of ICSV 1 and IS 14332 were very low and at physiological maturity not significantly different from IS 14384, but increased significantly during the first week of physiological maturity. This suggests that for genotypes to be mold-resistant their grains must remain hard and vitreous in the field under wet and warm conditions—the ideal environment for grain mold development. Sorghum grains with a hard and vitreous endosperm are also known to be least susceptible to storage weevils. However, it was reported that grains with a harder, more vitreous endosperm were less digestible than those having soft endosperm. Studies on grain

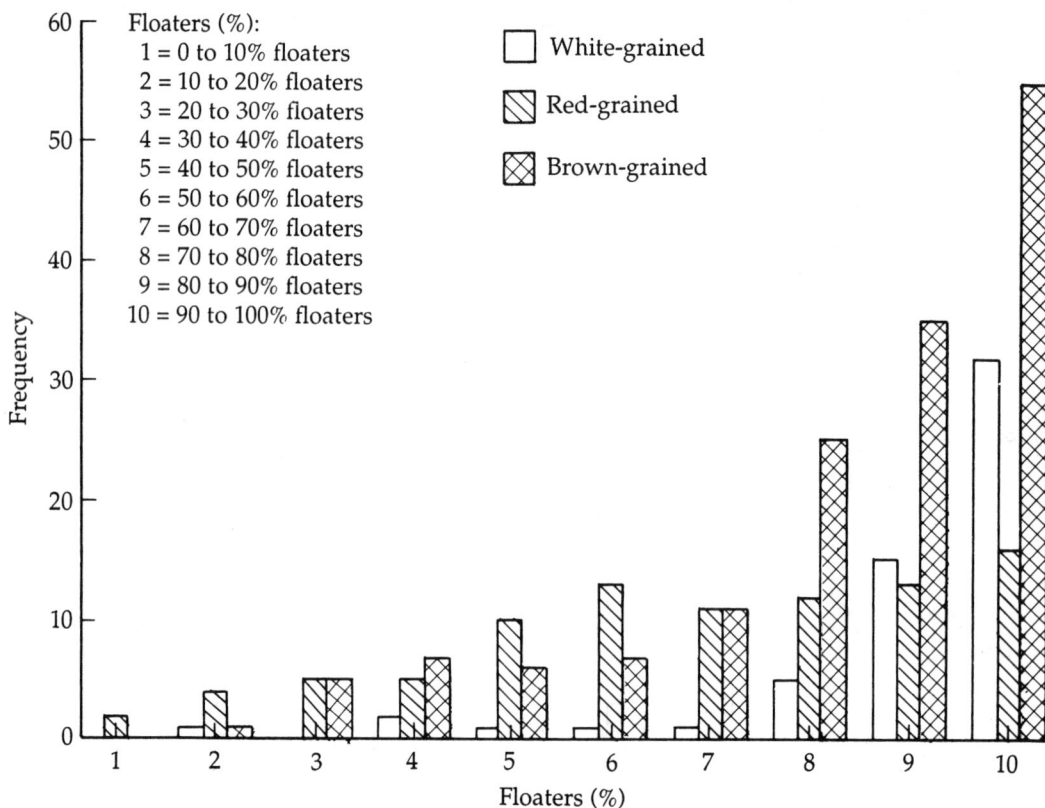

Figure 5. Frequency distribution of floaters (%) among white, red, and brown-grain F_2 segregates derived from a cross between IS 14384 (red-grained) and SPV 104 (white-grained) sorghums.

hardness and its relationship to digestibility and mold resistance deserve high priority.

Effect of Testa Layer and Spreader Genes

Sorghum lines high in tannin content and having the testa layer were reported to be more resistant to grain mold fungi by Glueck and Rooney (1980). The presence or absence of the testa layer is controlled by the B_1 and B_2 genes. The testa layer is present when the B_1 and B_2 genes are both dominant (B_1–B_2). A dominant spreader "S" gene in the presence of a testa layer results in brown-grained with high tannin. We investigated the effect of various combinations of the testa layer and spreader genes on mold resistance in 2 female and 11 male lines and their hybrids.

We crossed the 11 male lines with SPL 117 A and SPL 79 A male-steriles with $b_1b_1B_2B_2SS$ and $B_1B_1b_2b_2ss$ testa and spreader genotypes respectively (Bhola Nath et al. 1985) and produced 22 hybrids with various combinations of testa and spreader genotype. The male lines were grouped into white- and red-grained mold-susceptible; red-grained mold-resistant; and brown-grained mold-susceptible. These hybrids, together with their female and male parents, were evaluated for mold resistance in a randomized-block design with three replications during rainy season 1985 at ICRISAT Center. The testa and spreader genotypes, grain color, and mold resistance ratings of the 2 females, 11 males, and their hybrids are presented in Table 5.

Grains of the hybrids and parental lines were scored for mold resistance 54 days after 50% flowering and the floaters (%), dehulling recov-

Table 3. Phenotypic correlation between threshed grain mold rating (TGMR), floaters(%), and tannin content in F_2 segregating populations, ICRISAT Center, rainy season 1985.

Trait correlated with TGMR	Type of material	White grain	Red grain (testa absent)	Brown grain (testa present)
Floaters (%)	IS 14384 × ICSV 1 F_2	0.4423**	0.5239**	0.0830
Floaters (%)	IS 14384 × SPV 104 F_2	0.5256**	0.4689**	0.261**
Floaters (%)	SPV 104 × IS 14385 F_2	0.2734**	0.2452**	0.2138**
Tannin content	IS 14384 × SPV 104 F_2	–	–0.2374	–0.5653**

** significant at $P = 0.01$.

Table 4. Threshed grain mold rating (TGMR) and floaters (%) of six sorghum genotypes harvested at physiological maturity (pm) and 1, 2, and 3 weeks after physiological maturity, ICRISAT Center, rainy season 1987.

Genotype	Grain color	Trait	pm	Time of harvest[1]		
				1	2	3
SPV 104	White	TGMR	3.2	4.9	5.0	5.0
		Floaters (%)	99	97	100	100
ICSV 1	White	TGMR	2.1	4.7	5.0	5.0
		Floaters (%)	23	39	60	97
ICSH 153	White	TGMR	2.5	4.3	5.0	5.0
		Floaters (%)	75	61	77	85
IS 14332	White	TGMR	1.2	3.0	4.0	4.8
		Floaters (%)	11	44	60	65
B 42237	Red	TGMR	3.7	5.0	5.0	5.0
		Floaters (%)	90	96	99	96
IS 14384	Red	TGMR	1.6	2.0	2.4	2.9
		Floaters (%)	26	27	26	33
SE		TGMR	±0.1	±0.3	±0.2	±0.2
		Floaters (%)	±7.0	±8.7	±8.7	±4.5

1. Time of harvest = weeks after reaching physiological maturity (pm).

ery (%), and tannin and flavan-4-ol contents determined. Average TGMR, floaters (%), dehulling recovery (%), and tannin and flavan-4-ol contents of groups of females and males are presented in Table 6. All mold-susceptible male parents, irrespective of grain color, were low in tannin and flavan-4-ol content, high in percentage of floaters, and low in dehulling recovery (%). The red-grained mold-resistant male lines were different in that their grains contained sub-stantially greater amounts of flavan-4-ol and their grains were hard and vitreous.

Average TGMR, floaters (%), dehulling recovery (%), tannin and flavan-4-ol contents of the grains of all the hybrids are presented in Table 7. The white- or red-grained mold-susceptible male parents produced white- or red-grained mold-susceptible hybrids on SPL 117 A but brown-grained hybrids on SPL 79 A with variable mold resistance. The brown-grained

Table 5. Testa (B_1-B_2-) and spreader (S) genotype, grain color, and threshed grain mold rating (TGMR)[1] of 2 female sorghum parents, of 11 male parents, and their hybrid. Screened for GM-resistance, ICRISAT Center, rainy season 1985.

Male parent	White-seeded mold-susceptible female parent	
	SPL 117 A $b_1b_2B_2B_2SS$(5.0)	SPL 79 A $B_1B_2b_2b_2ss$ (5.0)
GM-susceptible		
SPV 104, $b_1b_1B_2B_2ss$, white (5.0)[1]	$b_1b_1B_2B_2Ss$, white (5.0)	$B_1b_1B_2b_2ss$, gray (5.0)
B 47229, $b_1b_1B_2B_2ss$, white (5.0)	$b_1b_1B_2B_2Ss$, white (5.0)	$B_1b_1B_2b_2ss$, brown (4.7)
B 47230, $b_1b_1B_2B_2ss$, White (5.0)	$b_1b_1B_2B_2Ss$, white (5.0)	$B_1b_1B_2b_2ss$, brown (4.0)
B 47234, $b_1b_1B_2B_2SS$, white (5.0)	$b_1b_1B_2B_2SS$, white (5.0)	$B_1b_1B_2b_2Ss$, brown (3.7)
GM-susceptible		
B 47233, $b_1b_1B_2B_2SS$, red (5.0)	$b_1b_1B_2B_2SS$, red (5.0)	$B_1b_1B_2b_2Ss$, brown (4.0)
B 47237, $b_1b_1B_2B_2SS$, red (5.0)	$b_1b_1B_2B_2SS$, red (5.0)	$B_1b_1B_2b_2Ss$, brown (3.7)
B 47083, $b_1b_1B_2B_2SS$, red (4.3)	$b_1b_1B_2B_2SS$, red (3.7)	$B_1b_1B_2b_2Ss$, brown (2.0)
GM-resistant		
IS 14390, $B_1B_1b_2b_2SS$, red (2.0)	$B_1b_1B_2b_2SS$, brown (2.0)	$B_1B_1b_2b_2Ss$, red (4.0)
IS 14384, $B_1B_1b_2b_2SS$, red (2.0)	$b_1b_1B_2b_2SS$, brown (2.0)	$B_1B_1b_2b_2Ss$, red (3.0)
GM-susceptible		
B 47238, $B_1B_1B_2B_2Ss$, brown (5.0)	$B_1B_1B_2B_2Ss$, brown (2.3)	$B_1B_1B_2B_2ss$, brown (5.0)
B 47239, $B_1B_1B_2B_2ss$, gray (5.0)	$B_1B_1B_2B_2Ss$, brown (3.0)	$B_1B_1B_2B_2ss$, gray (5.0)

1. TGMR: Threshed grain mold resistance score on a scale of 1 to 5, where 1 = no molds, and 5 = more than 50% grain surface molded.

Table 6. Mean threshed grain mold ratings (TGMR)[1] and physical and chemical characteristics of grains of one group of female sorghum lines and four groups of male lines evaluated at ICRISAT Center, rainy season 1985.

Type of material	Testa and spreader genotypes	Lines (no.)	Testa[2]	TGMR	Floaters (%)[3] A	B	Dehulling recovery (%)[4]	(Cat. eq)[5]	4-ol
White-grained female lines	$b_1b_1B_2B_2ss$/ $B_1B_1b_2b_2ss$	2	-	5.0	87	93	60	0.04	0.10
White-grained male lines	b_1b_1/B_2B_2ss/ $b_1b_1B_2B_2SS$	4	-	5.0	97	97	35	0.02	0.09
Red-grained male lines	$b_1b_1B_2B_2ss$	3	-	4.8	83	88	54	0.02	2.44
Red-grained male lines	$B_1B_1b_2b_2ss$	2	-	2.0	15	17	84	0.05	9.86
Brown-grained male lines	$B_1B_1B_2B_2ss$	2	+	5.0	88	93	44	0.56	1.13
SE				±0.10	±2.5	±2.36	±3.2	±0.07	±0.25

1. TGMR = Threshed grain mold rating on a 1 to 5 scale, where 1 = no mold, and 5 = more than 50% grain surface area molded.
2. Testa present (+); testa absent (−).
3. Floater A = % grains in a sample that floated in a solution $NaNO_3$ (1.3 specific gravity).
   Floater B = % grains in a sample that floated in a 1.327 specific gravity solution of tetrachloro-ethylene and odorless kerosene.
4. Percentage of whole grains recovered after dehulling.
5. Cat. eq = Catechin equivalent.

Table 7. Mean threshed grain mold rating (TGMR)[1], and physical and chemical characterstics of grain of six groups of sorghum hybrids, ICRISAT Center, rainy season 1985.

Testa and spreader genotypes	Pericarp color	Lines (no.)	Testa[2]	TGMR	Floaters (%)[3]		Dehulling recovery (%)[4]	Tannin (Cat. eq)[5]	Flavan 4-ol
					A	B			
$b_1b_1B_2B_2ss/$ $b_1b_1B_2B_2Ss$	White	4	-	5.0	76	84	64	0.01	0.07
$B_1B_1b_2b_2SS$	Red	3	-	4.6	61	74	65	0.03	1.10
$B_1B_1b_2b_2Ss$	Red	2	-	3.5	25	27	86	0.03	3.08
$B-B_2-ss$	Brown	5	+	4.7	89	94	53	0.32	0.35
B_1-B_2-S-	Brown	6	+	3.1	85	91	62	1.74	1.84
B_1-B_2-S-	Brown	2	+	2.0	15	25	86	3.86	3.96
SE				±0.16	±1.93	±3.53	±1.86	±0.07	±0.19

1. TGMR = Threshed grain mold rating on a 1 to 5 scale, where 1 = no mold, and 5 = more than 50% grain surface area molded.
2. Testa present (+); testa absent (–).
3. Floater A = % grains in a sample that floated in a solution of $NaNO_3$ (1.3 specific gravity).
   Floater B = % grains in a sample that floated in a 1.327 specific gravity solution of tetrachlro-ethylene and odorless kerosene.
4. Percentage of whole grains recovered after dehulling.
5. Cat. eq = Catechin equivalent.

hybrids with the spreader gene (B_1-B_2-S-) were high in tannin content and were mold-resistant, while the hybrids with a recessive spreader gene (B_1-B_2-ss) were low in tannin content and were mold-susceptible. The grains of these hybrids, irrespective of color of their male parents, were low in flavan-4-ol and their grains were soft with floury endosperm.

The mold-susceptible male parents having testa layers and testing low in tannin content (B_1-B_2-ss) produced brown-grained mold-resistant hybrids on SPL 117 A, but brown-grained mold-susceptible hybrids on SPL 79 A. The difference between the two groups of hybrids lies with the spreader gene. Hybrids on SPL 117 A have dominant spreader gene (B_1-B_2-S-) and are therefore high in tannin content, while those on SPL 79A have the recessive spreader gene (B_1-B_2-ss) and are therefore low in tannin content.

The red-grained mold-resistant male parents (IS 14384 and IS 14390) produced brown-grained mold-resistant hybrids on SPL 117 A and red-grained low mold-susceptible hybrids on SPL 79 A. Hybrids on SPL 117 A were highly resistant because they have a dominant spreader gene (B_1-B_2-S-), contained high tannin levels and moderate flavan-4-ol levels, and possessed hard

and vitreous grains. On the other hand, hybrids on SPL 79 A were low mold-susceptible because they did not have the testa layer and were therefore low in tannin, but differed from the red susceptible hybrids in that they contained moderate levels of flavan-4-ol and their grains were hard and vitreous. The red-grained male parents, however, had higher levels of mold resistance, probably because they contained significantly higher levels of flavan-4-ol and their grains were harder and more vitreous than those of the progeny.

Except for tannin, phenotypic correlations among hybrids and parental lines (white- or red-seeded without testa, and brown-seeded with testa) were significant at the $P=0.01$ level; (so only in the brown-seeded lines) indicating that high tannin level (testa with spreader gene), high flavan-4-ol, and grain hardness influence mold resistance, independently or in combination. White-grained sorghums do not contain tannin and so far none has been found to contain appreciable amounts of flavan-4-ol. Mold resistance in white-grained sorghum lines probably is due to grain hardness or unidentified polyphenolic compounds. Hahn et al. (1984) reported both a greater variety and amount of phenolic acids in mold-resistant sorghum varieties.

Mold Resistant Population

Population improvement using recurrent selection is regarded to be a powerful breeding technique for traits controlled by several additive genes, and we plan to use this breeding technique to combine mold resistance with agronomically desirable traits. We have started to build a mold resistant composite population that will be improved by recurrent selection. We introduced the male-sterile gene (ms_3) into colored-grain mold resistance sources as well as into white-grained, mold-resistant selections to form a broad-based mold resistance population. We also introgressed breeding lines into the population to increase the frequency for high-yield potential and agronomically desirable traits. We plan to stabilize the population by allowing it to random mate for three generations before imposing recurrent selection. We will apply minimum selection pressure during the random mating cycles, discarding only those genotypes that are very late and tall.

Conclusion

In the crossing program, colored-grain germplasm sorghum collection lines with high levels of mold resistance were used for white-grained genotypes with good levels of resistance. The segregating progenies were intensively screened for white-grained genotypes with good levels of resistance. Five white-grained advanced selections with similar mold resistance as their colored-grain parental lines were identified, and are now in the breeding program as sources of mold resistance into developing improved cultivars with good agronomic traits.

Mold resistance in the white-grained types was associated with grain hardness, while that of brown-grained types was associated with either high tannin or flavan-4-ol or grain hardness. Cultivars with a combination of these factors were highly resistant. It appears that flavan-4-ol is not produced in white-grained types, either those mold-resistant or those mold-susceptible. At present the only factor known to be

responsible for mold resistance in white-grained cultivars is grain hardness.

I believe that the grain mold problem is a major factor restricting adoption of improved cultivars in Africa. We should intensify our efforts to breed mold resistance into high-yielding cultivars so that farmers can grow mold-free sorghums.

References

Bandyopadhyay, R., and **Mughogho, L.K.** 1988. Evaluation of field screening techniques for resistance to sorghum grain molds. Plant Disease 72(6): 500–503.

Bandyopadhyay, R., Mughogho, L.K., and **Prasada Rao, K.E.** 1988. Sources of resistance to sorghum grain molds. Plant Disease 72(6): 504–508.

Bhola Nath, Omran, A.O., and **House, L.R.** 1985. Identification of a double recessive genotype for "B" genes controlling presence and absence of pigmented testa in sorghum. Cereal Research Communications 13:2–3.

Castor, L.L., and **Frederiksen, R.A.** 1980. *Fusarium* and *Curvularia* grain molds in Texas. Pages 93–102 *in* Sorghum diseases, a world review: proceedings of the International Workshop on Sorghum Diseases, 11–15 Dec 1978, ICRISAT, Hyderabad, India. Patancheru, Andhra Pradesh 502 324, India: International Crops Research Institute for the Semi-Arid Tropics.

Glueck, J.A., and **Rooney, L.W.** 1980. Chemistry and structure of grain in relation to mold resistance. Pages 119–140 *in* Sorghum diseases, a world review: proceedings of the International Workshop on Sorghum Diseases, 11–15 Dec 1978, Hyderabad, India. Patancheru, Andhra Pradesh 502 324, India: International Crops Research Institute for the Semi-Arid Tropics.

Hahn, D.H., and **Rooney, L.W.** 1986. Effect of genotype on tannins and phenols of sorghum. Cereal Chemistry 63:4–8.

Hahn, D.H., Rooney, L.W., and **Earp, C.F.** 1984. Sorghum phenolic acids, their high performance liquid chromatography separation and their relation to fungal resistance. Cereal Chemistry 60:255–259.

Harris, H.B., and Burns, R.E. 1970. Influence of tannin content on preharvest seed germination in sorghum. Agronomy Journal 62:957–959.

Harris, H.B., and Burns, R.E. 1973. Relationship between tannin content of sorghum grain and preharvest seed molding. Agronomy Journal 65:957–959.

Jambunathan, R., Bulter, L.G., Bandyopadhyay, R., and Mughogho, L.K. 1986. Polyphenol concentration in the grain, leaf and callus tissue of mold susceptible and mold resistant sorghum cultivars. Journal of Agricultural and Food Chemistry 34:425.

Murty, D.S., Rao, K.N., and House, L.R. 1980. Breeding for grain mold-resistant sorghums at ICRISAT. Pages 154–162 *in* Sorghum diseases, a world review: proceedings of the International Workshop on Sorghum Diseases, 11–15 Dec. 1978, ICRISAT, Hyderabad, India. Patancheru, Andhra Pradesh 502 324, India: International Crops Research Institute for the Semi-Arid Tropics.

Rao, K.N., and Williams, R.J. 1980. Screening for sorghum grain mold resistance at ICRISAT. Pages 103–108 *in* Sorghum diseases, a world review: proceedings of the International Workshop on Sorghum Diseases, 11–15 Dec 1978, Hyderabad, India. Patancheru, Andhra Pradesh 502 324, India: International Crops Research Institute for the Semi-Arid Tropics.

Williams, R.J., and Rao, K.N. 1981. A review of sorghum grain molds. Tropical Pest Management 27(2):200–211.

Part 4

Short Communications

Short Communications

Abstract:

*Sweet sorghum varieties resistant to anthrac-
nose in USA are highly susceptible to the diseases in Brazil*

N. Zummo, Plant Pathologist, Agricultural Research Service, United States Department of Agriculture, and Professor, Department of Plant
Pathology and Weed Science, Mississippi State
University, Mississippi State, MS 39762, USA.

All sweet sorghum varieties (except Tracey)
produced at the USDA Sugar Crops Field Station, Meridian, MS, have been highly resistant to
anthracnose and red rot incited by *Colletotrichum
graminicola* (Ces.) G. W. Wilson. These varieties
have been grown throughout the syrup areas of
USA and have shown no evidence of anthracnose infection. When the sweet sorghum varieties Brandes, Theis, Rio, Wray, and Keller were
introduced into Brazil prior to 1987, all were resistant to *C. graminicola*. Since 1987, one or more
new races of the anthracnose fungus have appeared in Brazil and these sweet sorghum varieties are now highly susceptible. The varieties
retain resistance when grown in USA.

Abstract:

*Diseases of sorghum and millet in the breeding
nurseries of Tanzania*

H.M. Saadan, Sorghum and Millet Breeder in
Tanzania and graduate student, Department of
Agronomy and Crop Science, Mississippi State
University, and **L. Gourley,** Agronomist and
Professor, Department of Agronomy and Crop
Science, Mississippi State University, Box 5248,
Mississippi State, MS 39762, USA.

The sorghum and millet disease spectra in
Tanzania are markedly broad, as these crops are
cultivated under wide ranges of environments.
Most of the diseases recorded in breeding nurseries are commonly found in farmers' fields. Foliar diseases are sometimes difficult to diagnose
because of overlapping of symptoms of different
diseases on the same leaf.

Diseases observed in areas where sorghum is
grown include stalk rots, charcoal rot, bacterial
soft rot, anthracnose, fusarium stalk rot, leaf
blight, pokkah boeng (twist top) zonate leaf
spot, rust, sooty stripe, oval leaf spot, downy
mildew, virus, bacterial leaf stripe, bacterial leaf
streak, four smuts, ergot and other grain molds.
Diseases commonly found in millet fields include grain mold smut, ergot, downy mildew,
rust, and leaf spots.

Striga hermonthica and *S. asiatica* are common,
and *S. forbesii* occurs in localized areas.

Abstract:

Diseases of finger millet in Uganda

E. Adipala, Department of Crop Science, Make-
rere University, Box 7062, Kampala, Uganda

During 1976–84, a study of finger millet diseases was undertaken at four locations in
Uganda. The most important diseases were leaf
and head blast, *Cylindrosporium* leaf spot, tar
spot, and bacterial blight. Blast damaged seedlings, leaves, nodes, necks, and heads, but neck
and head infections were the most damaging.
Cylindrosporium sp caused heavy leaf spotting
leading to a blight. The newly recorded bacterium, *Xanthomonas coracanae* Desai et al., killed a
number of lines, but some lines showed no field
symptoms of this bacterium.

Other diseases found on finger millet include
sclerotium wilt (*Sclerotium rolfsii*), damping-off
(*Pythium* spp), foot rot (*Fusarium* sp), *Pyrenophora* seedling blight, leaf streak, a stunt disease resembling the rice stunt virus, and other
'virus-like' infections.

At high altitude (Kalengyene), even the relatively head blast-resistant lines, like P 304, developed the disease. The same was noticed in
lines resistant to *Cylindrosporium* leaf spot, tar
spot, and *Helminthosporium* leaf spots.

Abstract:

Sorghum diseases in Rwanda

C. Sehene, Breeder and Program Leader, Institute of Agricultural Sciences in Rwanda, PO Box 138, Rwanda

Sorghum is a traditional crop grown throughout Rwanda from lowland to highland (<2500 m altitude), rainfall is 800–1200 mm annually. Several diseases, varying in importance, are reported on sorghum according to ecological zone. The main diseases are downy mildew, ergot, leaf blight, smuts, and sooty stripe. *Striga* is a serious problem. Chemical control of diseases on a wide scale is not practiced. Research in sorghum pathology began in Rwanda only 3 years ago. We have collaborative projects with ICRISAT on ergot and downy mildew. It is desirable that sorghum pathology work in Rwanda be strengthened.

Abstract:

Predominant sorghum diseases in the Bay Region of Somalia

Ahmed Sheikh Hassan, Agronomist and Research Director, ARC BRADP, Mogadishu, Somalia

Somalia's sorghum-growing areas total 400 000 ha; the major portion is in its Bay Region. Bonka Research Station, funded under the Bay Region Agricultural Development Project (BRADP), emphasizes the improvement of sorghum and elimination of its production constraints. The 1984 surveys, made under efforts of this Station, are summarized. In the Bay Region, 95% of the cropping is to sorghum. Sorghum diseases predominant here include covered kernel smut, head smut, charcoal spot, sooty stripe, and anthracnose. Sorghum downy mildew, head molds, leaf blight, and leaf spot also are present in the region, but at lesser frequencies. Incidence of most of these diseases differed during the *du* and the *ger* seasons, but covered smut was the most serious in either season. Fernasan D® treatment of seed was found to be effective against covered smut in research plots; efforts to make this chemical available to farmers are underway.

Abstract:

Pyricularia spp causing head blight of finger millet (Eleusine coracana), and other fungi associated with finger millet in Tanzania

C. A. Kuwite, Plant Pathologist, Kibaha Agricultural Research Institute, Box 30031, Kibaha, Tanzania, and **F.M. Shao,** Plant Pathologist, Ministry of Agriculture and Livestock Development, PO Box 9192, Dar-es-Salaam, Tanzania.

A head blast disease of finger millet, *Eleusine coracana* (L) Goertn., caused by *Pyricularia* spp, was found in Tanzania in July 1985. Attempts to isolate *Pyricularia* spp on a nearby rice crop were not successful. There was no evidence of seed-borne-infection by the pathogen. Other fungi— *Helminthosporium* spp, *Curvularia* spp, *Aspergillus* spp, *Fusarium* spp, *Rhizopus* spp, *Alternaria* spp, and *Penicillium* spp—were associated with the source seed.

Abstract:

A review of sorghum and pearl millet diseases of Malawi

E.M. Chintu and **L.M. Gourley,** Graduate Student and Professor, respectively, Department of Agronomy, Mississippi State University, Mississippi state, MS 39762, USA.

Some of the most important diseases of sorghum (leaf blights, sooty stripe, oval leaf spot, anthracnose, head mold, and covered kernel smut) and pearl millet (downy mildew, ergot, and smut) in Malawi are discussed with reference to environment, locality, crop variety, and other factors. These diseases occur in the Lower Shire (southern region) and other dry sorghum-growing areas of Malawi.

Abstract:

Summary of pearl millet pathology in Zimbabwe from a breeder's standpoint

F.R. Muza, Millet Breeder (Zimbabwe) and graduate student, Mississippi State University, Mississippi State 39762, MS, USA; **L.M. Gourley,** Sorghum Breeder (INTSORMIL) and Professor of Agronomy, Mississippi State University, Mississippi State 39762, MS, USA; and **David J. An-**

drews, Sorghum and Millet Breeder, INTSOR-MIL, and Professor of Agronomy, University of Nebraska Lincoln, Lincoln NE 69503–0723, USA.

The major diseases of pearl millet in Zimbabwe are ergot, smut, rust, and downy mildew. Yields of traditional varieties growing in farmers' fields are not affected by current disease levels, but agriculture in many areas is becoming more intensive, and disease pressure is expected to increase, and resistance becomes a factor in selecting for improved varieties and hybrids. Top-cross hybrids and improved open-pollinated varieties appear best suited for most areas.

Abstract:

Maize dwarf mosaic on sorghum in Kenya: some preliminary observations

J.M. Theuri and J.G.M. Njuguna, Pathologist/Virologist and Plant Pathologist, respectively, Kenya National Agricultural Research Center, Muguga, PO Box 30148, Nairobi, Kenya; N.W. Ochanda, Alupe Agricultural Research Station, PO Box 278, Busia, Kenya; and Vartan Guiragossian, JP 31 SAFGRAD, PO Box 30786, Nairobi, Kenya.

In 1987 maize dwarf mosaic was observed (24–58% incidence) on sorghum in the Central Province highlands and in the Rift Valley and in the Western Provinces (0–100% incidence). Infection was not observed in the Eastern Province. At the Maguga Center (Central Province highlands), infected plants showed, on average, height reduction of 16% and head size reduction of 13%. At Alupe (Western Province), 10 of the 25 ICRISAT lines in sorghum trials were heavily attacked (disease index 2.5). Ten of the 11 local varieties in the trial showed little infection. Of the ICRISAT lines, ICSH 361 was most affected by the disease, showing the maximum infection score of 5. ICSH 193 was free of MDM virus symptoms.

Abstract:

Utilization of genetic resistance to diseases

D.J. Andrews, INTSORMIL Sorghum and Millet Breeder, University of Nebraska, Lincoln, NE 68583, USA.

The importance of incorporating sufficient durable disease resistance when breeding new crop varieties is usually of equal importance as improving grain yield and quality. However, as with any breeding situation, the greater the number of necessary objectives, the slower will be the overall progress.

To be successful, before embarking on breeding for disease resistance, the breeder needs basic information on the nature of the problem:

1. The existing and potential damage to yield.
2. The threshold incidence when yield loss begins.
3. Other factors that help control the disease.
4. The genetic basis for resistance—is it monogenic, multigenic, recessive or dominant, etc.?
5. Existing races or biotypes? Is the pathogen capable of rapidly evolving new biotypes?
6. Available sources of resistance? How good is their resistance and how agronomically good are their backgrounds?
7. Are there environments where resistant plants can be consistently identified?

In fact, complete information is never available and always some assumptions have to be made. Each breeder should be aware, however, that the more sparse the information, the more limited may be the success of the effort. In any case, elements 6 and 7 above are crucial to success.

Pathologists should know that the overall agronomic value of a source of resistance is very important to a breeder. The best degree of resistance is usually carried in some unadapted germplasm accession, but a source with moderate, though sufficient resistance (how much resistance is needed to lower incidence to the threshold value?) in an adapted background may be preferable because of ease of utilization. For 7, some method of statistically quantifying the degree and consistency of the disease pressure in the selection environment is necessary (e.g, from data on replicated control plots of known resistant and susceptible stocks).

Abstract:

Studies on the biology of long smut of sorghum caused by **Tolyposporium ehrenbergii**

J.G.M. Njuguna, Plant Pathologist, Kenya National Agricultural Research Center, Muguga, PO Box 30148, Nairobi, Kenya; **L.K. Mughogho,** ICRISAT Center, Patancheru, Andhra Pradesh 502 324, India; **V. Guiragossian,** JP 31 SAFGRAD, PO Box 30786, Nairobi, Kenya; and **R.O. Odhiambo,** National Agricultural Research Center, Muguga, PO Box 30148, Nairobi, Kenya.

Heads of sorghum plants growing in the greenhouse were inoculated with either chlamydospores or sporidia. A comparison of sporidia with chlamydospores as inoculum showed sporidia to be more efficient in causing infection. Plants inoculated with sporidia just before the heads emerged from the boot became infected, as did plants inoculated when heads were partially emerged. Infection did not occur on plants with fully emerged heads. Of the nine sorghum genotypes screened for resistance against long smut in the greenhouse experiments, ICSV 212 and QL 3 showed the least long smut sori.

Abstract:

Inheritance of resistance to systemic infection by downy mildew in sorghum

M.F. Beninati, ICRISAT Mali Program, BP 34, Bamako (via Paris), Mali.

Parents, F_1, F_2, and BCF generations from 60 resistant × susceptible or resistant × resistant crosses, involving 12 resistant parents, were screened (uncontrolled dosages) for resistance to systemic infection of downy mildew in 1986 and 1987. Field screening data from crosses of resistant parents IS 2266, IS 7215, and IS 18757(QL 3) with susceptible parents IS 643 and IS 18433 are reported. IS 2266 appears to possess a single dominant gene for resistance, while F_2 and BCF_1 data for crosses with IS 7215 suggest that resistance is controlled by the interaction between one dominant and one recessive locus. Resistance in IS 18757 appears to be controlled by at least two loci, and F_1 data indicate incomplete dominance. Further studies may be necessary to verify these findings and to determine the number of different loci involved in resistance in the lines under study.

Abstract:

Sorghum yellow banding virus (SYBV), a new polyhedral virus on **Sorghum spp**

L. M. Giorda, and **R. W. Toler,** Department of Plant Pathology and Microbiology, Texas A&M University, College Station, TX 77843, USA, and **G.N. Odvody,** Associate Professor Plant Pathology, Texas A&M University, Agricultural Experiment Station, Rt. 2 Box 589, Corpus Christi, TX 78410, USA.

Symptoms of virus activity include yellow streaks that coalesce into bands, followed by chlorosis, stunting, and eventual death of the plant. The host range of SYBV includes *Setaria italica, Sorghum bicolor, S. sudanense, S. halepense, Zea mays,* and sorghum × sudangrass hybrids. The sorghum cultivar, QL 3 India, immune to strains of sugarcane mosaic virus and maize dwarf mosaic virus, was susceptible to SYBV. QL 3 developed chlorotic streaking, bands, and severe necrosis. The size of virus particles from purified preparations ranged from 25 to 32 μm (mode: 27 μm). Antisera prepared against purified virus reached a titer of 1/2048 using the Ouchterlony gel double immunodiffusion test. Infected sap from sorghum and maize were tested against panicum mosaic virus (PMV), PMV-SAD (the St. Augustine strain of PMV), brome mosaic virus, cucumber mosaic virus, maize chlorotic mottle virus, and maize chlorotic dwarf virus antisera with negative results. Purified particles of SYBV formed a single band on sucrose gradient and in cesium chloride continuous-density gradient. Viral fractions had a 260/280 ratio of 1.53. Virions of SYBV were buoyant with a density of 1.386 g cm^{-3}, compared to 1.316 of tobacco mosaic virus. A single protein band with molecular weight of 29 kd was identified, using SDS-PAGE analysis. The RNA from purified SYBV, under nondenaturing conditions, migrate as five bands in 21.3% agarose gel with 10 mM sodium phosphate buffer pH 7.0 and 0.02% SDS. The major band was approximately 1.8 kb and the others ranged between 0.38 and 1.3 kb when compared to RNA ladder markers and lambda DNA/Hind III fragments.

Abstract:

Survival of **Colletotrichum graminicola** *in the soil at Burkina Faso*

B. Traore and **K.B. Kabore,** Laboratoire de Recherches du Service National de la Protection des Vegetaux, B.P. 403, Bobo-Dioulasso, Burkina Faso.

Survival of sorghum anthracnose and stalk red rot agent *Colletotrichum graminicola* on sandy-loamy soil near Bobo-Dioulasso, Burkina Faso, was studied in 1987. Spores of the fungus were counted in a sorghum field harvested and left fallow in November 1986, and on another field in 1987. The counts reveal that *Colletotrichum graminicola* can survive in the soil for 9 months following harvest, with 11 spores g^{-1} of soil. This residue of *Colletotrichum graminicola* retained pathogenicity and provoked stalk red rot of sorghum upon artificial inoculation. Soil samples from the cultivated field of 1987 shows that accumulation of *C. graminicola* spores begins near the last of July, at the end of tillering, and survive until the end of the rainy season.

These results show the insufficient effect of seed treatment and justify study of integrated pest management in search of a method for complete control of *C. graminicola* in the sorghum production.

Abstract:

Identification of **Striga** *germination stimulants from sorghum-root exudate and their potential significance in control of* **Striga.**

L. Butler, Professor of Biochemistry, Purdue University, West Lafayette, IN 47907, USA.

Sorghum root hairs exude a yellow oil containing four structurally similar compounds that are highly active as germination stimulants for *Striga hermonthica* and *Striga asiatica*. The structure of these compounds has been determined and their biological activities are being characterized. We have named these compounds sorgoleones. They are hydroquinones, readily oxidized to p- and o-benzoquinones with loss of their capacity to stimulate the germination of *Striga*. Their labile nature and their insolubility in water ensure that only those *Striga* seeds located very close to the host root will be stimulated to germinate. Sorgoleones have other biological activities which may benefit the sorghum plant and thus prevent selection against their protection under *Striga* stress. *Striga* apparently utilizes sorgoleones as specific host-recognition signals. Strategies for using analogs of sorgoleones, in the absence of the host, as *Striga* seed in infested fields, and as germination inhibitors of *Striga* are proposed.

Abstract

Secondary sporulation of **Sphacelia sorghi** *on sorghum*

D.E. Frederickson and **P.G. Mantle,** Department of Biochemistry, Imperial College, London SW 2AZ, UK.

Etiology of the sorghum ergot disease pathogen, *Sphacelia sorghi* McRae, becoming increasingly important in hybrid seed production in southern Africa, is not well understood. Collateral hosts have not been identified. Evidence of a role for sclerotia in the disease cycle, as with *Claviceps purpurea* in temperate cereals and grasses, is lacking. Current study of the disease cycle focuses on sphacelial fructification and its associated honeydew-containing conidia.

Part 5

Other Topics

Principles and Criteria of Seed Transmission of Sorghum Pathogens

D.C. McGee[1]

Abstract

More than 40 microorganisms are listed as seedborne on sorghum. This review categorizes them according to the criteria of transmissibility to the new crop and/or effects on seed quality. The main inoculum source is seed-transmitted in only three diseases. Several other important diseases are seed-transmitted, but with these crop residues are usually a more important inoculum source. Grain molds occurring in the seed-production field cause serious losses in seed viability. Fusarium spp and Curvularia spp are the major causal agents. Many seedborne microorganisms have been identified that do not cause sorghum disease just because of their presence on the seeds. Nonseedborne pathogens are of concern in sorghum-seed production in respect to adverse effects on yield and stand establishment.

Introduction

More than 40 microorganisms are listed as seedborne on sorghum. This review categorizes them according to the criteria of transmission to the new crop and/or effects on seed growth. The main inoculum source is seed-transmitted in only three sorghum diseases. Several other important diseases are seed-transmitted, but with these pathogens, residues are usually the more important inoculum source. Grain molds occurring in seed-production fields cause serious losses in seed viability.

Criteria for Determining Importance of Seedborne Pathogens

The annotated list of seedborne diseases records more than 1500 microorganisms as being seedborne on 600 genera of plants (Richardson 1979). To obtain a perspective of the large numbers of seedborne microorganisms, McGee (1981) described four classes into which they might be assigned. The first comprises seed-transmitted pathogens and the infected seeds become major inoculum sources. The second class contains pathogens that are seed-transmitted, but infected seeds are relatively insignificant as sources of inoculum. The third and largest class lists those microorganisms never before known to cause disease as a result of their presence on seeds. In the fourth class are those pathogens that attack new seeds during formation or in storage, thus affecting the seed's appearance, germination, and/or vigor.

Another consideration in assessing the importance of seedborne microorganisms is whether the seed is grown for sowing in the country where it's produced, or for export to another country. When grown for domestic use, control measures are needed only for important pathogens in the first or fourth classes described above. On the other hand, seed grown for export requires control measures so that it can meet the seed-health testing dictated by the plant quarantine regulations of the country of destination. Plant quarantine requirements are often unre-

---

1. Seed Pathologist, Department of Plant Pathology, Seed Science Center, Iowa State University, Ames, IA 50011, USA.

McGee, D.C. 1992. Principles and criteria of seed transmission of sorghum pathogens. Pages 297–302 *in* Sorghum and millets diseases: a second world review. (de Milliano, W.A.J., Frederiksen, R.A., and Bengston, G.D., eds). Patancheru, A.P. 502 324, India: International Crops Research Institute for the Semi-Arid Tropics.

lated to seed-health requirements for seeds grown for domestic use.

Epidemiological data available for the majority of seedborne microorganisms is insufficient to assign the microorganism to any of the above classes.

Seedborne Diseases of Sorghum

Approximately 40 microorganisms are listed as seedborne on sorghum (*Sorghum bicolor* L.) (Richardson 1979). These represent more than 75% of all sorghum diseases. Existing information on the seedborne phase of economically important sorghum diseases with respect to the above considerations is reviewed.

Infected seeds as the main inoculum source

Ergot, caused by *Sphacelia sorghi*, occurs in Africa and Asia. The disease affects individual florets, causing poor grain development. It is associated with sterility, and can cause problem in hybrid seed production where male-sterile lines are used as female parents (Futtrell and Webster 1965). Sclerotia mixed with the seeds are considered as an important source of primary inoculum for the new crop. Ergot has been added to this group only on the basis of morphology and not as the basis of scientific evidence. It may be important to sow seed free of sclerotia. Sclerotia can be separated from seed by immersing seeds in a salt solution.

Covered kernel smut (*Sporisorium sorghi*) affects floral parts, and individual seeds are replaced by sori. Seeds contaminated by teliospores released from infected plant parts are the main inoculum source. The disease is effectively controlled by seed treatment. If seed treatment is not applied, major losses can result.

Loose kernel smut (*Sphacelotheca cruenta*) also attacks the panicle. Spores become attached to seeds during maturation and harvest, and then become the primary inoculum source. Control can be achieved with seed treatment (Hansing 1957).

Infected seeds as a minor inoculum source

Head smut (*Sporisorium reilianum*) is of great economic importance, particularly in USA. All or part of the panicle is destroyed. Disease incidence is directly related to the number of spores of *S. reilianum* present in the soil; these infect young seedlings, grow systemically throughout the plant, then produce typical symptoms at flowering time. Spores released from infected floral parts fall to the soil; they survive for as long as 2 years (Saf'yanov et al. 1980). The pathogen also is seedborne, but as such is not an important inoculum source (Frowd 1980). Dalassus (1964) found that inoculation of seeds had no effect on disease incidence. Some degree of control can be achieved with cultivar resistance. Chemical seed treatment does not seem to be particularly effective, although one report indicated success with benomyl (Popov and Silaev 1978).

Long smut (*Tolyposporium ehrenbergii*) occurs in African and Asian nations. It is typified by long, curvular sacs scattered through the heads. Spore balls, consisting of masses of teliospores on the soil surface, are the main inoculum source. Contaminated seeds also can be a source. Hafiz (1958) showed that diseased heads developed on 10–20% of the plants grown from seeds coated with spores. Disease incidence was higher, however, when healthy seeds were grown in infested soil.

Sorghum downy mildew (*Peronosclerospora sorghi*) is a serious disease of maize and sorghum in many parts of the world, causing stunting and chlorosis. Thick-walled oospores are produced in systemically infected plants. These infest soil and become a major inoculum source. Bain and Alford (1969) detected fungal mycelium in the embryo of seeds. It has been consistently shown, however, that the pathogen does not survive in seed dried below 20% moisture, and there is little evidence that mycelium in dry seeds is transmitted to plants (Frederiksen 1980). Nonattached oospores could, however, be an inoculum source in seedlots. Kaveriappa and Safeeulla (1978) showed that transmission can occur when seeds are contaminated by glumes carrying oospores. In Texas, phytosanitary certificates are issued, after field inspection for the disease, to ensure that oospore-infested seeds do not occur (Frederiksen 1980). Seed treatment with metalaxyl has proved to be an effective control (Frederiksen 1980). Integrated control practices, combining chemical treatment with cultural practices, also have been investigated (Odvody et al. 1983). Resistant hybrids controlled the disease well in Texas in the 1970s, but new races appeared in 1979-80 (Odvody et al. 1983). The propensity for the development of new races has increased concern about transmission by infected seed.

Sooty stripe (*Ramulispora sorghi*) is common under warm, humid conditions in some countries. The pathogen survives as sclerotia. Seed infection has been detected (Lele et al. 1966), but proof of transmission is lacking.

Target leaf spot (*Bipolaris sorghicola*) survives on debris or on alternate hosts. It has been isolated from seeds (Purss 1953).

Charcoal rot (*Macrophomina phaseolina*) is a major disease in drier regions. It is reported as seedborne, but soilborne sclerotia are the major inoculum sources. One report (Chumaevskaya 1962) recommended seed treatment.

Milo disease (*Periconia circinata*), occurs in USA, causing a root and crown rot in some genotypes. Crop residues and soil are well recognized as the major inoculum sources, and it has been isolated from seeds (Hansing and Hartley 1962).

Zonate leaf spot (*Gloeocercospora sorghi*) commonly occurs during wet periods. It is seedborne (Bain 1950). Ciccarone (1949) suggested that zonate leaf spot was introduced into Venezuela by infected seeds. Sclerotia overwintering in dead leaf tissues provide inoculum for the following season's crop (Girard 1978).

Fusarium root and stalk rot (*Fusarium moniliforme*) is a common disease in tropical and temperate regions. Locally, losses may approach 100%. Infected debris is the main inoculum source. Seed can be a minor source (Mathur et al. 1967). Control can best be achieved by sowing hybrids with good stalk strength and then managing so as to avoid or reduce stress.

Bacterial leaf spot (*Pseudomonas syringae* pv *syringae*) has been reported from most countries. The pathogen survives on crop residues, and can survive for as long as 3 years in dry sorghum seeds. Sundaram (1980) indicated that sowings with seed from diseased plants, growing in sterile soil, developed the disease.

Other bacterial pathogens of sorghum recorded as being seedborne are *Pseudomonas andropogonis*, *Xanthomonas holcicola*, and *X. rubrisorghi* (Easwaran 1967; Sundaram 1980).

Nonpathogenic seedborne microorganisms

Many studies of the seed microflora of sorghum have been reported, usually as lists of microorganisms detected in seed-health tests. Many of these microorganisms are pathogens, such as *Fusarium* spp and *Curvularia* spp, but pathogenic activity for the majority has not been established. Papers by Mathur et al. (1975), Hansing and Hartley (1962), Pinheiro and Neto (1979), Rani et al. (1978), Tarr (1962), and Williams and Rao (1981) reflect the range of genera found in different parts of the world.

Pathogens affecting seed quality

The exposed nature of sorghum seeds render them susceptible to invasion by parasitic microorganisms and to effects of weathering in the production field (Frederiksen et al. 1982). Moldy seed shows as pink or blackened seed scattered through the head. Dead embryos are characteristic of the condition (Williams and Rao 1981). Wet weather during the 1966 harvest caused extensive losses from grain mold and weathering in Texas (Matocha et al. 1977). Three diseases, grain mold, small seed, and head blight, are all associated with reduced seed and grain quality. The fungus *Fusarium moniliforme* is implicated in each of these.

Grain mold is the result of infection of spikelets, perhaps as early as anthesis, by *F. moniliforme* and *Curvularia lunata*. The pathway of infection of the seed by *F. moniliforme* has been elucidated by Castor (1981). This work distinguished primary pathogenic effects from secondary effects associated with field weathering. In the latter case, fungi colonize mature grain (Castor 1981). There is value in these distinctions; they are important in screening sorghum lines for resistance to infection by *F. moniliforme*. Lines that consistently develop less grain mold have been identified (Frederiksen et al. 1982).

"Small seed" occurs, to some extent, in sorghum each year. Castor (1981) showed that infection at anthesis by *F. moniliforme* or *C. lunata* causes formation of a false black layer, resulting in light grain. Inoculation with *F. moniliforme* induced "small seed."

Head blight, caused by *F. moniliforme*, differs from grain mold in that it is the result of infection of the panicle, rachis branches, or peduncle. Grain mold, on the other hand, is a result of spikelet infection. Major losses were caused by head blight in southern Texas in 1979 (Castor and Frederiksen 1980). Sorghum genotypes differ in their response to this disease.

Colletotrichum graminicola, the causal pathogen of anthracnose, will cause a panicle and grain blight under cloudy, warm, and humid growing conditions. Infected seeds are discolored, germination is reduced, and seedling blights may occur. Fungicide seed treatment can reduce seedling blight. This disease also causes a stalk and leaf blight, but these are of lesser economic importance than the grain-blight phase (Kramer 1958).

Sorghum in storage, like any crop, is subject to invasion by *Aspergillus* and *Penicillium* spp. Lopez and Christensen (1963) reported extensive deterioration occurring at 15–16% moisture.

Pathogens not Seed Transmitted

Seed producers are concerned about yield as well as seed-quality factors. Therefore, they must contend with pathogens that do not affect seeds directly. Maize dwarf mosaic virus (MDMV) is one example. This virus disease has reached epidemic status in USA, Europe, South America, and Australia. Losses in susceptible cultivars infected during early stages of growth can approach 100%. Control measures include destruction of johnsongrass and other susceptible annual grasses that supply early-season inoculum. Use of tolerant or resistant cultivars is another approach. In addition to causing yield loss, MDMV has been associated with small-seed production when infection is followed by periods of low temperature during grain filling (Edmunds and Niblett 1973).

Rust, caused by *Puccinia purpurea*, can cause serious losses in cool, humid areas of Latin America, southeast Asia, and southern India. Control is best achieved by resistant cultivars.

Other significant diseases of sorghum that are not seedborne include gray leaf spot (*Cercospora sorghi*), tar spot (*Phyllachora sacchari*), yellow sorghum stunt, incited by a mycoplasma, and crazy top (*Sclerospora macrospora*).

Stand establishment can be a major problem in sorghum production, particularly if sown in cold wet soils (Tarr 1962). The problem is often caused by seedborne fungi, although soilborne *Pythium* spp also causes this. Fungicides have been effective against common seedborne fungi, but not against *Pythium* spp.

Conclusion

More than 40 microorganisms are listed as seedborne on sorghum. This review categorizes them according to the criteria, of transmission to the new crop and/or effects on seed growth. The main inoculum source is seed-transmitted in only three sorghum diseases. Several other important diseases are seed-transmitted, but with these pathogens, residues are usually the more important inoculum source. Grain molds occurring in seed-production fields cause serious losses in seed viability. *Fusarium* and *Curvularia* spp are the major causal agents. There are many seedborne microorganisms identified in sorghum that have not been shown to cause a disease as a result of their presence on seeds. Nonseedborne pathogens are also a concern in seed production, because of adverse effects on seed yields or on seed quality. Soilborne microorganisms can cause major problems in stand establishment.

References

Bain, D.C. 1950. Fungi recovered from seed of *Sorghum vulgare* Pers. Phytopathology 40: 521–522.

Bain, D.C., and **Alford, W.W.** 1969. Evidence that downy mildew (*Sclerospora sorghi*) of sorghum is seedborne. Plant Disease Reporter 53:802–803.

Baker, K.F. 1972. Seed Pathology. Pages 317–415 *in* Seed Biology, Vol. II. (Kowslowski, T.T., ed.) New York, NY, USA: Academic Press. 447 pp.

Castor, L.L. 1981. Grain mold histopathology, damage assessment, and resistance screening within *Sorghum bicolor* (L) Moench lines. Ph.D. thesis, Texas A&M University, College Station, TX, USA.

Castor, L.L., and **Frederiksen, R.A.** 1980. Fusarium blight occurrence and effects on sorghum yield and grain characteristics in Texas. Plant Disease 64:1017–1019.

Chumaevskaya, M.A. 1962. Charcoal rot of sorghum and maize. Zasch. Rast. Moskva 7:56.

Ciccarone, A. 1949. Zonate leaf spot of sorghum in Venezuela. Phytopathology 39:760–761.

Dalassus, M. 1964. The principal diseases of millet and sorghum in the Upper Volta in 1963. Agronomie Tropicale 19:489–498.

Easwaran, K.S.S. 1967. Seed transmission of bacterial leaf blotch disease (*Xanthomonas rubrisorghi*) of sorghum in south India. Plant Disease Reporter 51:63.

Edmunds, L.K., and Niblett, C.L. 1973. Occurrence of panicle necrosis and small seed as manifestation of maize dwarf mosaic virus infection in otherwise symptomless grain-sorghum plants. Phytopathology 63:388–392.

Frederiksen, R.A. 1980. Sorghum downy mildew. Plant Disease. 64:903–908.

Frederiksen, R.A., Castor L.L., and Rosenow, D.T. 1982. Grain mold, small seed, and head blight: the *Fusarium* connection. Pages 26–36 *in* Proceedings of the 37th Annual Corn & Sorghum Research Conference, Chicago, IL, USA. Washington D.C., USA: American Seed Trade Association.

Frowd, J.A. 1980. A world review of sorghum smuts. Pages 331–348 *in* Sorghum diseases, a world review: proceedings of the International Workshop on Sorghum Diseases, 11–15 Dec 1978, ICRISAT, Hyderabad, India. Patancheru, Andhra Pradesh 502 324, India: International Crops Research Institute for the Semi-Arid Tropics.

Futtrell, M.C., and Webster, O.C. 1965. Ergot infection and sterility in grain sorghum. Plant Disease Reporter 49:680–683.

Girard, J.C. 1978. A review of sooty stripe and rough, zonate, and oval spots. Pages 228–239 *in* Sorghum diseases, a world review: proceedings of the International Workshop on 11–15 Dec 1978, ICRISAT, Hyberabad, India. Patancheru, Andhra Pradesh 502 324, India: International Crops Research Institute for the Semi-Arid Tropics.

Hafiz, A. 1958. Some studies of long smut of sorghum. Pakistan Journal of Scientific Research 10:83–87.

Hansing, E.D. 1957. Effect of seed treatment with fungicides and with combinations of fungicides and insecticides on emergence and control of covered kernel smut of sorghum in 1956. Phytopathology 47:523.

Hansing, E.D., and Hartley, A. 1962. Sorghum seed fungi and their control. Proceedings of the Association of Official Seeds Analysts 52: 143–149.

Kaveriappa, K.M., and Safeeulla, K.M. 1978. Seed-borne nature of *Sclerospora sorghi* on sorghum. Proceedings of the Indian Academy of Science 87:303–308.

Kramer, N.W. 1958. The economic significance of disease in sorghum. Pages 123–128 *in* Proceedings of the 24th Annual Corn & Sorghum Research Conference, Chicago, IL, USA. Washington, D.C., USA: American Seed Trade Association.

Lele, V.C., Dhanraj, K.S., and Nath, R. 1966. Seed-borne infection of Sudan Grass: a new record for India. Indian Phytopathology 19: 357–358.

Lopez, F.L.C., and Christensen, C.M. 1963. Factors influencing invasion of sorghum seeds by storage fungi. Plant Disease Reporter 47:597–601.

Mathur, S.K., Mathur, S.B., and Neergard, P. 1975. Detection of seed-borne fungi in sorghum and location of *Fusarium moniliforme* in the seed. Seed Science and Technology 3:683–690.

Mathur, S.B., Sharma, R., and Joshi, L.M. 1967. Moldy head disease of sorghum: isolation of associated fungi on blotter and agar. Proceedings of the International Seed Testing Association 32:639–645.

Matocha, J.E., Reyes, L., and McCartor, M. 1977. Weathering effects on quality of grain sorghum in Coastal Bend. Texas Agricultural Experiment Station Progress Report. College Station, TX, USA: Texas A&M University.

McGee, D.C. 1981. Seed pathology: its place in modern seed production. Plant Disease 65:638–642.

Odvody, G.N., Frederiksen, R.A., and Craig, J. 1983. The integrated control of downy mildew. Pages 28–36 in Proceedings of the 38th Annual Corn & Sorghum Research Conference, Chicago, IL, USA. Washington D.C., USA. American Seed Trade Association.

Pinheiro, J.M., and Neto, J.P. da C. 1979. Identification of fungi on Sorghum bicolor (L) seeds in Rio Grande do Sul. Agronimia Sulriograndense 15:127–131.

Popov, V.I., and Silaev, A.I. 1978. Effectiveness of seed dressing of sorghum against two species of smut disease. Nauch. Tr. Leningrad S-KH Inst 351:85–87.

Purss, G.S. 1953. Mold growth on sorghum seed. Queensland Journal of Agricultural Science 10:125–126.

Rani, K., Mohan, M., and Mukerjit, K.G. 1978. Studies of seedborne fungi I: Occurrence of three pathogenic fungi on sorghum seeds. Seed Research 6:38–42.

Richardson, M.J. 1979. An annotated list of seed-borne diseases. 3rd edn. Kew, UK: Commonwealth Agricultural Bureau.

Saf'yanov, S.P., Bystrova, Z.F., and Silaev, A.I. 1980. Head smut of sorghum. Zaschita rastenii 8:43–44.

Sundaram, N.V. 1980. Bacterial diseases. Pages 385–390 in Sorghum diseases, a world review: proceedings of the International Workshop on Sorghum Diseases, 11–15 Dec 1978, ICRISAT, Hyberabad, India. Patancheru, Andhra Pradesh 502 324, India: International Crops Research Institute for the Semi-Arid Tropics.

Tarr, S.A.J. 1962. Diseases of sorghum, sudan grass, and broom corn. Kew, UK: Commonwealth Agricultural Bureau. 380 pp.

Williams, R.J., and Rao, K.N. 1981. A review of sorghum molds. Tropical Pest Management 27:200–211.

Epidemiology of Sorghum Diseases in Central America: a Case Study

G.C. Wall[1], R.A. Frederiksen[2], J. Craig[3], and M.J. Jeger[4]

Abstract

Field surveys were carried out in 1983, 1984, and 1985 in the Honduras, Central America. Twenty-one sorghum diseases were identified. Eight of these were studied with regard to their effect on yield. Near-isogenic populations that were resistant, intermediate, or susceptible to sorghum downy mildew (SDM) were sown in a disease-free area and in a SDM endemic area; a 43% disease incidence reduced yield by 44%.

Paired comparisons of diseased and healthy plants showed that maize dwarf mosaic (MDM) caused a 52% yield reduction on 'maicillos,' the traditional landraces. Acremonium wilt caused a 36% yield reduction on a susceptible cv (BTx 623) and a 33% loss on maicillos. A holistic, multivariate study, using data from subsistence farmers' fields, produced a multiple regression model for estimating yields on the basis of plant height, panicle length, and disease severity. The yield reduction estimate for gray leaf spot was 16%; for oval leaf spot, 6%; and for rust, 4%.

In fungicide-protected and nonprotected experiments, zonate leaf spot was estimated to cause 15% yield loss, and gray leaf spot 14%.

Comparing cropping systems, sorghum-maize intercropping was found to have lower oval leaf spot severity than sorghum growing alone, but MDM incidence was higher. Early sowing of maicillos resulted in higher disease severities than did late sowing. August sowings of improved cultivars had fewer diseases, but higher severities than did August-sown maicillos. Sorghum-bean intercropping did not affect sorghum disease severities. Various sources of disease resistance were identified by screening international and local nurseries.

Introduction

The area of highest sorghum production in Central America lies along its Pacific coast in parts of Guatemala, the Honduras, Nicaragua, and El Salvador. Epidemiological studies on sorghum diseases carried out in the Honduras from 1983 to 1985 are reported here.

As is typical of this area in Central America, sorghum production in the Honduras is mainly traditional landraces sown by subsistence farmers; 93% of all production in the Honduras is in the hands of these resource-poor farmers (SRRNN 1984). The crop is used for feeding poultry, swine, cattle, and horses, and as human food. One third of the total production is used for human consumption (De Walt 1982).

Sorghum is grown as a risk-reducing crop. In the semi-arid areas where growing maize is risky, sorghum offers a more reliable food source. Even though annual rainfall is ca. 1600 mm, the rainy season normally has an interrupted period called 'canicula'. This canicula often limits maize yields, but not so with sorghum.

1. Assistant Professor, College of Agriculture and Life Sciences, University of Guam, Mangilao, GU 96923, USA.
2. Professor, Plant Pathology and Microbiology, Texas A&M University, College Station, TX 77843, USA.
3. Research Plant Pathologist, ARS/USDA, PO Drawer DN, College Station, TX 77841, USA.
4. Head, Overseas Pest Research, Tropical Development and Research Institute, 56-62 Gray's Inn Road, London WC 1X 8LU, UK.

Wall, G.C., Frederiksen, R.A., Craig, J., and Jege, M.J. 1992. Epidemiology of sorghum diseases in Central America: a case study. Pages 303–317 in Sorghum and millets diseases: a second world review. (de Milliano, W.A.J., Frederiksen, R.A., and Bengston, G.D., eds). Patancheru, A.P. 502 324, India: International Crops Research Institute for the Semi-Arid Tropics.

In recent years, the government has attempted to increase sorghum production through the creation of a sorghum research program directed toward the development of improved cultivars and cultivation practices.

In a socioeconomic study of sorghum production in the southern parts of the Honduras, De Walt (1982) found that the constraints to sorghum production were many, but among these was no mention of diseases. This reflects an ignorance on the part of the subsistence farmer, as he blames only visible pests (mainly insects) for damage seen on his plants. If these farmers are asked to name the major constraints on their sorghum crop, they will talk about insect pests, which to them include what we know as diseases.

The vast majority of subsistence farmers cultivate marginal lands. Quite often, they sow on hilly slopes. Soil fertility is generally low. Little attention is given to sorghum fields through the growing season. Consequently, and in addition to the limitations dictated by the restricted gene pool within traditional landraces, yields are low and well below the world average (FAO 1985; Navarro 1979; SRRNN 1984).

Sorghum is sown according to several different cropping schemes, although the majority of farmers sow sorghum in association with maize. This is done in different ways and several spatial arrangements can be found. There are several pathogens, such as *Peronosclerospora sorghi*, the causal agent of sorghum downy mildew that have the ability to cause disease both in maize and in sorghum. Sorghum is also sown without intercropping with maize (De Walt and De Walt 1984; SRRNN 1984).

If sorghum production is to increase in the Honduras, it is essential to document the importance of sorghum diseases. The role of these diseases in relation to low yields must be determined. Means by which these losses can be reduced is a major priority. Whether landrace cultivars are going to be improved or replaced by new higher-yielding introductions (Guiragossian 1983) is of paramount concern, because little is gained by the use of otherwise well-adapted high-yielding sorghums if disease-management strategies are unavailable or not practiced.

Since the vast majority of sorghum growers sow on marginal land and experiment stations operated by the Ministry of Agriculture are located in fertility valleys where fields are flat, information generated by such experiment stations often does not reflect the realities of growing sorghum on hillsides and other poor soils (Chambers and Ghildyial 1985). For this reason, the experiments reported here were conducted, when possible, in farmers' fields.

The areas where sorghum is grown, as described by Nolasco (1983), are in Choluteca, Vella, El Paraiso, Limpira, Francisco Morazan, and Olancho. Most of it is grown in Choluteca and Valle (Fig. 1).

Throughout the 3-year period 1983–85, 21 sorghum diseases were identified (Table 1). Diseases found in all areas were acremonium wilt, anthracnose, gray leaf spot, grain molds, ladder spot, leaf blight, loose smut, maize dwarf mosaic, rust, sugarcane mosaic, and zonate leaf spot. Some diseases did not occur in all areas. Downy mildew was not observed in Valle and Choluteca. Maize chlorotic dwarf was seen only in one location in El Paraiso. Oval leaf spot was not found in Valle or Francisco Morazan. Other diseases, namely bacterial stripe, charcoal rot, covered smut, head blight, and sheath blight, occurred with less frequency and were observed in Choluteca, where the majority of the work was carried out. Head blight was also observed in El Paraiso. Charcoal rot was only seen on improved cultivars.

In the Comayagua area, downy mildew, ladder spot, gray leaf spot, and rust were highest in severity and prevalence in 1985.

In Francisco Morazan, rust, gray leaf spot, and ladder spot diseases were highest in disease index values (severity × prevalence). In El Paraiso, gray leaf spot and rust disease were highest. For Valle and Choluteca, the diseases with highest index values in 1985 were gray leaf spot and oval leaf spot.

Maize Dwarf Mosaic and Acremonium Wilt

Acremonium wilt (AW) is caused by the fungus *Acremonium strictum* W. Gams. It was only recently reported and described in the Americas (Frederiksen 1986). This vascular disease was found wherever sorghum was observed for diseases. The same is true for maize dwarf mosaic (MDM), a systemic disease caused by the maize dwarf mosaic virus (MDMV). The host range for

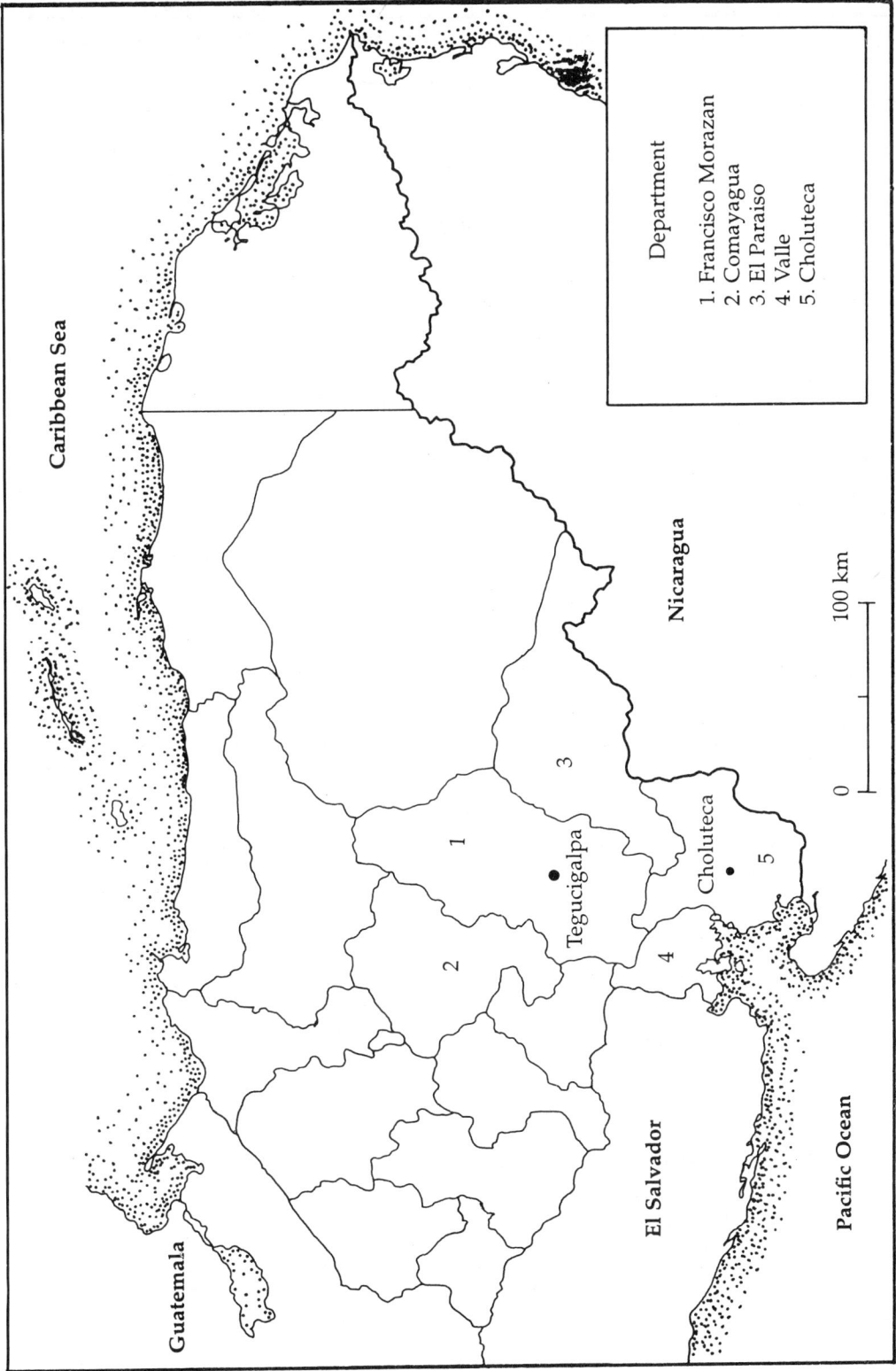

Figure 1. The Honduras.

Table 1. Diseases and pathogens identified on *Sorghum bicolor* in central and southern Honduras, 1983 and 1984 cropping seasons.

Disease	Pathogen	Region[1]
Acremonium wilt	*Acremonium strictum*	C,Y,CM,EP,FM
Gray leaf spot	*Cercospora sorghi*	C,Y,CM,EP,FM
Ladder spot	*Cercospora fusimaculans*	C,CM,FM
Anthracnose	*Colletotrichum graminicola*	C,Y,EP
Grain mold	*Curvularia lunata* and *Fusarium moniliforme*	C,Y,CM,EP,FM
Leaf blight	*Exserohilum turcicum*	C,Y,CM,EP,FM
Head blight, Pokkah-boeng, and stem rot	*Fusarium moniliforme*	C,EP
Zonate leaf spot	*Gloeocercospora sorghi*	C,CM,EP
Charcoal rot	*Macrophomina phaseolina*	C
Sorghum downy mildew	*Peronosclerospora sorghi*	CM, EP
Bacterial Stripe	*Pseudomonas andropogoni*	C
Rust	*Puccinia purpurea*	C,Y,CM,EP,FM
Oval leaf spot	*Ramulispora sorghicola*	C,Y,CM
Sheath blight	*Sclerotium rolfsii*	C
Covered smut	*Sporisorium sorghi*	C
Loose smut	*Sphacelotheca cruenta*	C,Y,EP
MCDV	Maize chlorotic dwarf virus	EP
MDMY	Maize dwarf mosaic virus	C,Y,CM,EP,FM
SCMY	Sugarcane mosaic virus	C,Y,CM,EP,FM

1. C = Choluteca, Y = Yalle, CM = Comayagua, EP = El Paraiso, and FM = Francisco Morazan, geographic departments of the Honduras.

each of these pathogens includes maize, sorghum, and sugarcane. All of these crops are commonly grown throughout the country.

In Texas, Natural et al. (1982), using paired-plant comparisons, found that AW caused a 50% yield reduction on a hybrid sorghum (ATx 623 × 77 CS 1). Frederiksen (1984) reported naturally-occurring AW incidence on a number of experimental sorghum lines at Choluteca in 1982, ranging from 0 to 100%. Traditional landraces show symptoms of this disease, such as vascular discoloration and death of sheath and leaf tissue, but they do not generally appear as severe as on certain susceptible Texas cultivars. Alexander et al. (1984, 1985) reported yield loss caused by MDM on susceptible cultivars in Texas. Both AW and MDM are diseases which lend themselves well to paired-plant comparisons, due to their systemic nature.

Materials and methods

Sorghum plants naturally affected by AW were identified and tagged early in the growing season (Oct 1983). Healthy plants adjacent to diseased ones were also tagged forming 100 pairs each in cultivar BTx 623 and in landrace, or 'maicillo,' plants.

In the same fashion, maicillo plants with MDM were paired with healthy plants and tagged in Oct 1983, 30 days after sowing. These experiments took place at La Lujosa Experiment Station in Choluteca.

The fungus *Acremonium strictum* W. Gams was isolated from wilted plants. All diseased and healthy plants were growing, and the identity of MDMV was verified by serological methods in various seed-production or yield trials.

Tagged plants were harvested at the end of the growing season. Those tagged as diseased with AW were cut and inspected for vascular discoloration.

Plant height, panicle length, panicle mass, grain mass, and 100-seed mass were measured. Panicle exertion appeared to be affected by AW, and it was also recorded. Seed number was calculated from grain mass and 100-seed mass. Panicle exertion was measured from base of the flag leaf to base of the panicle.

Results

Paired-plant comparisons revealed that diseased plants of BTx 623 suffered a 5% reduction in plant height, had a 3% shorter panicle length, and a 52% shorter panicle exertion. They had 24% fewer seeds, a 16% reduction in 100-seed mass, and 36% lower yield (Table 2). Healthy plants were superior in each of these parameters.

Table 2. Difference (%) between healthy and acremonium wilt infected sorghum plants for six parameters.

Parameter	Maicillo criollo	BTx 623
Plant height	3	5**
Grain yield	33**	36**
Grain head⁻¹	26**	24**
Mass (100 grains)⁻¹	9	16**
Panicle length	11*	3*
Panicle exertion	9	52**

*, ** Significant at 5% and 1%, in paired t-tests.

Comparisons between acremonium-wilt and healthy maicillo plants showed real differences in only three of the six parameters. Panicle length was reduced by 11% and seed number by 26%, resulting in a yield reduction of 33%. MDM affected maicillos, reducing plant height by 9%, panicle length by 13%, seed number by 43%, 100-seed mass by 25%, and yield by 52% (Table 3).

Foliar Diseases on Landrace Cultivars

The sites selected in southern Honduras provided differing prevalences of sorghum diseases in farmers' fields. At the first, there was a severe gray leaf spot (*Cercospora sorghi*) epidemic. The second site was selected because only rust (*Puccinia purpurea*) was observed, although it was at a relatively low severity. A third site provided an oval leaf spot (*Ramulispora sorghicola*) epidemic.

Materials and methods

Data collection was carried out between 12 and 17 Dec 1985, at various locations in Choluteca, the Honduras. Plants were found at stages 11.2–11.3 of Feekes' scale (Zadoks and Schein 1979). At site 1, in El Rebase, many plants had 100% gray leaf spot severity. At site 2, in Lajero Blanco, the severity or rust was not as high as that of gray leaf spot in site 1. At site 3, in La Joyada, oval leaf spot predominated and many plants had 100% leaf damage.

The parameters measured in the field were: plant height, number of green leaves, leaf length, leaf width, panicle length, and disease

Table 3. Means and differences for six parameters observed on 34 pairs of MDMV-diseased and healthy plants of landrace sorghum cultivars, or "maicillos criollos."

Treatment	Plant height	Panicle exertion	Panicle length	Grain mass	100-seed mass	Seed number
	--------------------(cm) --------------------			------------- (g) ------------		(Ha)
Healthy	318	7.2	25	23	2.1	1050
Diseased	290	10.4	22	11	1.6	602
Difference	28**	–3.2	3*	12**	0.5**	448*
Difference (%)	–9%	31%	–13%	–52%	–25%	–43%

*, ** Significant at 5% and 1% in paired t-tests.

severities. Standard area diagrams were prepared and used to aid in estimating disease severities. Leaf length and width were taken from the central leaves.

An average of 45 plants selected to cover all disease-severity levels at the site were selected, and the panicles tagged and harvested. They were later weighed and threshed individually. Humidity readings were then taken and seed mass per panicle determined.

These data were used to develop empirical mathematical models to explain variations in yield at each site and to estimate yield losses due to these diseases. Data were also combined to produce a general mathematical model based on principal components of plant height, panicle length, and disease severities.

Results

Yield estimates for minimum and maximum disease-severity levels at mean plant height and panicle length reflect the effect of the change in disease severity, independent of possible effects of disease severity on panicle length and plant height (Table 4). Only gray leaf spot showed a real difference in yield levels due to disease effects alone; a 7.2% difference between the yield estimates for plants with minimum and with maximum disease levels, at mean plant height and panicle length values, was recorded. Yield estimates for rust, oval leaf spot, and smut had

no real differences between low and high severity levels, since confidence limits overlapped.

When yield differences were estimated allowing plant height and panicle length to vary along with disease severities, there was a 14.6% difference in estimated yields between plants with less than 20% gray leaf spot and plants with 100%. There was a 3.6% difference between estimated yields for plants with less than 5% rust and with 25% rust. A difference of 5.5% was calculated between plants with less than 20% oval leaf spot severity and plants with 100% (Table 4). Observed versus estimated yields for the general model had a correlation coefficient of 0.792.

Foliar Disease Severities and Yields in Fungicide-treated and Nontreated Plots

During the 1983 cropping season, zonate leaf spot was the predominant disease at the La Lujosa Experiment Station in Choluteca. This disease is caused by the fungus (*Gloeocercospora sorghi*). In an experiment to study the effect of this disease on sorghum yields; comparisons were made between plots protected with fungicide applications and nonprotected plots. A similar study was done at the same location during the 1984 cropping season. In this season, the predominant foliar disease was gray leaf spot caused by the fungus *Cercospora sorghi*.

Table 4. Sorghum yield loss estimates in farmers' fields at three sites in southern Honduras, 1985.

Disease severity	Plant height and panicle length		Plant height and panicle length	
	Yield estimate [g (plant)-1]	Yield difference (%)	Yield estimate [g (plant)-1]	Yield difference (%)
CERC[1] 20%	116.3		121.4	
CERC 100%	107.8	7.2	103.1	14.6
Rust 5%	113.2		114.2	
Rust 25%	109.8	–	110.0	3.6
OVAL[2] 10%	107.5		110.0	
OVAL 100%	106.7		104.0	5.5

1. CERC = Gray leaf spot.
2. OVAL = Oval leaf spot.

Materials and methods

The 1983 experiment was sown on 27 Aug. The design was a split plot with six replications. Main plot treatments were four sorghum cultivars (Dorado, Tortillero, Liberal, and Coludo). Subplot treatments were fungicide-protected and nonprotected. Main plots consisted of five 10-m rows, with 0.8 m between rows; subplots had five 5-m rows. Blocks were separated by spreader rows. Rows were arranged perpendicular to prevailing wind direction.

Fungicide was applied with a backpack sprayer at 15-day intervals, starting from 22 Sep. Rates were 1 g benomyl plus 4 g mancozeb plus 0.3 mL surfactant L^{-1} of solution, applied to runoff.

Disease-severity ratings were carried out at monthly intervals for the upper and lower leaves. Area under the disease progress curve (AUDPC) was calculated from upper and lower leaves, and for the total plant by calculating a mean (Wall 1986).

In 1984, the experimental design was changed to make use of more limited space; this was a completely randomized design, with two replications and three treatments, consisting of no fungicide protection, and with fungicide applications at 15- and 30-day intervals.

Results

In the trial zonate leaf spot caused a reduction in yield of nearly 14%. Nonprotected plots averaged 3.37 t ha^{-1} yield, while those protected from zonate leaf spot with fungicide averaged 3.92 t ha^{-1} (Fig. 2). Disease severity in the latter was considerably reduced. AUDPC for fungicide-protected plots had a mean value of 310.34, while the value for nonprotected plots was 993.03. The improved cultivars Dorado and Tortillero yielded higher than the landraces Coludo and Liberal, although the improved cultivars had higher disease severities (Fig. 3).

In 1984, controlling gray leaf spot resulted in a significant yield increase of the cv Tortillero. Without fungicide treatment, this cultivar suffered a 21% loss of leaf area from gray leaf spot, and yielded 4.5 t ha^{-1} (Table 5). Treated plants produced 5.6 t ha^{-1}.

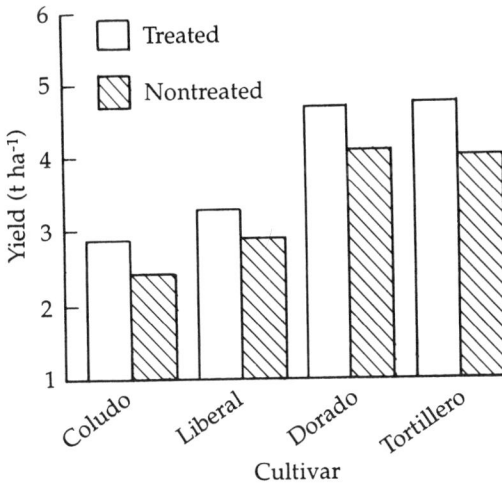

Figure 2. Yield levels of two landraces and two improved sorghum cultivars grown during the 1983 cropping season in La Lujosa, Choluteca, the Honduras, and affected by zonate leaf spot disease.

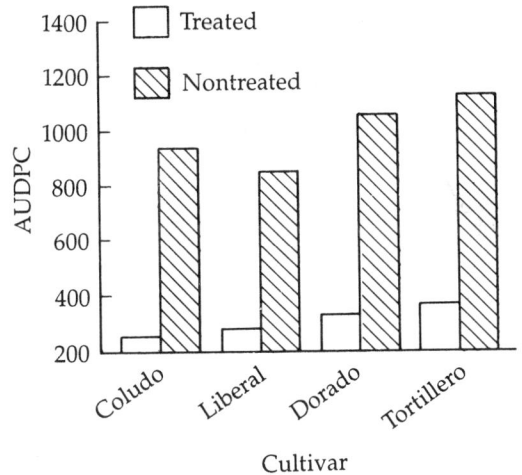

Figure 3. Zonate leaf spot severity levels, as reflected by the Area under disease progress curve (AUDPC), on two landraces and two improved sorghum cultivars treated and nontreated with fungicidal sprays during the 1983 cropping season in Choluteca, the Honduras.

Table 5. Yield and gray leaf spot severity levels for cv Tortillero, grown during the 1984 cropping season at La Lujosa Experiment Station, Choluteca, the Honduras.

Treatment	Yield (t ha⁻¹)	Severity (%)
Fungicide every 2 weeks	5.6250a[1]	1.0b[2]
Fungicide every 4 weeks	5.5650a	7.5b
Nonprotected control	4.5250b	21.0a

1. Means with different letter are different at .05 level (Duncan's Multiple Range Test).
2. Means with different letter are different at .01 level (Duncan's Multiple Range Test).

Effect of Downy Mildew on Grain Yield

Peronosclerospora sorghi (Weston and Uppal) C.G. Shaw causes the disease sorghum downy mildew (SDM) on *Sorghum bicolor*. Plants systemically infected by SDM are usually barren. The disease is generally recognized as one of the most important sorghum diseases (Frederiksen et al. 1973). Sorghum downy mildew was first reported in the Honduras in 1974, and is well established in Comayagua.

Materials and methods

Three near-isogenic sorghum populations were created with F_3 progeny from a cross between the resistant line SC 414-12 and the susceptible inbred line Tx 412 (Craig et al. 1988). The F_3 families were tested for reaction to disease, and were combined into a resistant, an intermediate, and a susceptible population. In the latter, the families showed more than 90% SDM incidence, the intermediate family showed 25%; and the resistant population showed less than 10%.

To test for yield differences between populations, all three were sown in a complete randomized-block design with four replications at Choluteca, where SDM was believed absent. Plots were four 5-m rows, with 0.8 m between

rows and 0.15 m between plants. Plots were sown on 3 Sep and harvested on 11 Dec, 1983.

An identical test was sown 19 Sep in Comayagua, where SDM is endemic. Fertilizer application and cultivation were carried out in the same manner in both locations. The two central rows were harvested in each plot. The Comayagua experiment was harvested on Dec 27.

The number of plants, panicles, plants with SDM 21 days after emergence, plants with SDM at harvest, panicle mass, and grain mass were recorded for each plot.

The SDM (%) data from Comayagua were subjected to an arcsine transformation, then to analysis of variance. Grain weight (yield) data were also subjected to analysis of variance. There were significant differences in SDM (%) and in yield between populations, so regression analysis were performed, using the nontransformed SDM (%) data as the independent variable and yield as the dependent variable.

Results

In Comayagua SDM significantly reduced yields. These losses were directly proportional with the incidence of disease (Fig 4). In Choluteca, where there was no SDM, there were no yield differences between the near-isogenic sorghum populations.

The resistant population in Comayagua, with a mean of 2.8% SDM, yielded 1.5 t ha⁻¹. The

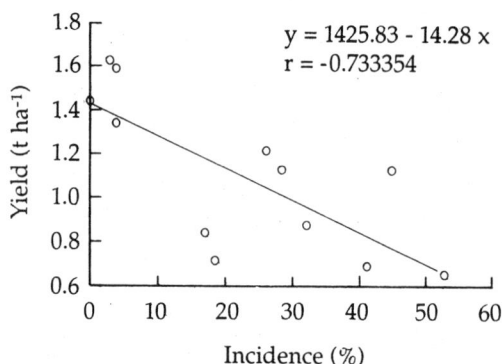

Figure 4. Regression of yield on sorghum downy mildew incidence on near-isogenic sorghum populations sown at Comayagua, the Honduras, in 1983.

$$y = 1425.83 - 14.28 x$$
$$r = -0.733354$$

intermediate population had 22.5% SDM incidence and yielded 1 t ha⁻¹. The susceptible population had 42.9% SDM incidence, yielding only 0.8 t ha⁻¹, 44.3% less than the resistant population.

Evaluation of Sorghum Cultivars for Resistance

For the farmer, sowing disease-resistant material is the most cost-efficient way of reducing losses due to diseases (Mundt and Browning 1985). Most of the sorghum in the Honduras is produced by resource-poor farmers, so it is desirable to have sources of disease resistance in the Honduras identified, so that these traits can be incorporated into future sorghum varieties and hybrids.

International disease nurseries are an excellent source of disease-resistant cultivars. These nurseries are generally put together with the most outstanding sources of resistance for a variety of diseases. Many such disease nurseries are available, and several were sown at two locations in 1983 and 1984 in the Honduras. They were evaluated for disease resistance to naturally occurring diseases during these growing seasons. Other nurseries, sown by researchers working with the Honduran sorghum project, were also evaluated to identify possible sources of disease resistance which later may be used in the sorghum improvement program there.

Cultivars that show disease resistance in a number of locations are likely to be more stable than those having specific types of resistance, which may be resistant at one site but susceptible at another (Wiese 1980). Different specific resistances can also be used in combination (Mundt and Browning 1985). The basis of any gene-deployment scheme is to have known resistance of different types. A lesson to be learned from the past is avoiding the use of only one source of specific resistance. Selection of cultivars to be used as sources of resistance should be based on several trials through time and space, preferably tested in the areas of anticipated sowing.

Results and discussion

Resistance was found for AW, SDM, and gray, ladder, oval, and zonate leaf spot diseases.

Among the landraces, resistance was identified for SDM, gray leaf spot, ladder spot, and oval leaf spot. Eighteen sorghum entries, including improved lines and landraces, were found to possess resistance to two or more diseases (Table 6).

As more testing is done, information available regarding the stability of these resistant sources will become more reliable. If possible, diverse sources of resistance must be used to safeguard against rapid "breakdown" to a given disease.

The stable source of resistance should be put into a uniform nursery along with experimental materials. This permits the constant monitoring of the elite resistant sources, at the same time facilitating direct comparison of experimental materials with elite lines. Cultivars, particularly those also well adapted to climatic conditions in the Honduras, with resistance to more than one disease should prove useful in saving valuable breeding time.

Table 6. Sorghum entries selected for resistance to several diseases from various nurseries in the Honduras 1983–85.

Cultivar/designation/ entry	Disease resistance[1]
SC 326-6	AC, GR, LD, ZN (AN)
82 CS 447	AC, GR, LD, ZN
80 B 2892	GR, LD, ZN
R 3338	GR, LD, ZN
82 EON 112	AC, SDM, ZN
San Miguel 1	GR, LD, OV
SC 748-5	GR, ZN (AN)
77 CS 1	GR, LD
79 HW 207	AC, ZN
Tx 435	AC, SDM
Tx 2794	LD, SDM
81 B 6078	LD, SDM
Pom Pom	GR, LD
Las Lajas	GR, LD
VG 146	GR, SDM
82 BH 5718	SDM, ZN
81 EON 69	AC, ZN
Brandes	ZN (AN)

1. AC = Acremonium wilt, GR = Gray leaf spot, LD = Ladder spot, (AN) = Reported resistant to anthracnose elsewhere, OV = Oval leaf spot, SDM = Sorghum downy mildew, ZN = Zonate leaf spot.

Cultivar CS 3541 is not resistant to SDM in Comayagua. This cultivar is the genetic base for most, if not all, improved varieties and hybrids produced to date by the Honduran sorghum project. This is risky, because crop-threatening SDM epidemics could occur due to the lack of resistance in the improved cultivars. To prevent this situation, it is important that more genetic diversity be used in the formation of future releases in the country.

Disease Severities Under Different Cropping Systems

Many cropping systems are used by sorghum producers in the Honduras. It is not uncommon to find several species intercropped in small fields, because this permits more efficient use of land, labor, and other resources. Land resources are scarce for subsistence farmers in the Honduras.

Microclimate is thought to be an important factor in host-parasite interactions, and in epidemic development (Wiese 1980). The cropping systems practiced by subsistence farmers in producing sorghum in the Honduras may have important microclimatic and various other differences associated with them. These could have viable effects on plant health in the course of one cropping season, particularly in the case of polycyclic diseases.

Several cropping systems involving sorghum are found in other American tropical areas (Rao 1985; SRRNN 1984). The most widely practiced sorghum cropping system in Central America is that of sorghum-maize intercropping. This is the most popular cropping system in both El Salvador and the Honduras, accounting for more than 90% of the sorghum production in these two countries. Other systems include sorghum sown alone or intercropped with maize and beans, beans, cowpeas, 'ayote' (*Cucurbita* sp), sesame, and cassava.

Sorghum-maize intercropping is popular as a risk-reducing system where there is danger of droughts. From a standpoint of human nutrition, sorghum-bean is a desirable farming system, as noted by De Walt and De Walt (1984). McCulloch and Futrell (1983), in a socio-economic and nutritional study in southern Honduras, concluded that the most nutritionally efficient cropping systems were those with the highest yields of sorghum and beans. In both of these studies, the term 'beans' is used to mean any of several *Phaseolus* and *Vigna* species. Beans are an important part of the diet in the Honduras and in southern Honduras; so is sorghum. However, national production has been insufficient to meet demand for either commodity, and importation has become necessary (SRRNN 1984). Seeing this need, in addition to sorghum-maize and sole sorghum, the sorghum-bean cropping system was also studied with respect to disease development.

Results and discussion

Sorghum-bean intercropping showed no increase in incidence of sorghum disease. Landraces yielded more than improved cultivars in this system. This was not the case in sole sorghum plots in other studies (Fig. 3). The advantage of the improved cultivars is manifested at higher plant densities. Landrace cultivars or maicillos are better adapted at lower plant densities because they tiller profusely; when intercropping, plant density for any of the component species is less, so it is not surprising that landraces adapt better to such systems. Plant height may also contribute, as the taller maicillos suffer no competition for light from the beans.

Oval leaf spot was more severe in sorghum alone than in the sorghum-maize system. This was not the case for gray leaf spot. One difference may lie in the dispersal mechanisms. Gray leaf spot is more widespread than oval leaf spot although, where it occurs, the latter can be more severe. This may be due to a more pronounced influence of inoculum originating outside of the field in question for gray leaf spot, and a larger dependence of inoculum originating within the field for oval spot. Sorghum-maize intercropping imposes greater distances between sorghum plants and fewer sorghum plants per unit area.

Improved cultivars appeared to have fewer foliar diseases, but not lower severities, than maicillos when sown in postrera (second sowing season) under subsistance farmers' conditions, i.e., with little or no fertilization. Yield levels of the farmers' sowings, however, were quite low; panicle size was very small. With sufficient fertilizer, improved cvs sown at higher densities out-yield landraces, but they suffer higher disease severities. Two questions arise: what happens to improved cultivars if sown in the early season, so as to obtain two crops in a single year? The problem then is grain molds, rather than foliar diseases; improved cultivars, almost totally insensitive to photoperiod, are sown in May or June and flower in July or August, normally the period of heaviest rainfall. The second question: what happens after several years of sowing improved cultivars in a field? Less biomass, but more inoculum is returned to the soil.

Will this combination of factors result in higher disease severities? This needs study. It is clear that the improved cultivars now in use are subject to greater disease severities than are the maicillos. If the former are to replace the latter, their levels of disease resistance need to be improved to prevent serious yield losses. Research is also needed to explore cultural practices that may reduce disease problems.

Conclusions and Recommendations

In the course of this study, 21 diseases were identified on sorghum growing in different areas of the Honduras during surveys from 1983 to 1985. Disease importance varied from region to region. Cultivars intended mainly for use in one of these regions need to be resistant only to diseases important in that region. Those intended for use in all of the sorghum-producing regions should possess adequate levels of resistance to the five diseases found to be important at all locations in the Honduras: MDM, gray leaf spot, rust, SDM, and AW. Although SDM was not found in Choluteca, it has spread to new areas since the studies began. SDM reduces yields drastically, and occurs on maize as well as sorghum.

Survey data were used in deciding upon the priority diseases for yield-loss studies. These were carried out for: AW, MDM, gray leaf spot, oval leaf spot, rust, covered kernel smut, SDM, and zonate leaf spot. All, except for covered smut, were found to reduce yields on sorghum.

Experiments on AW and MDM permitted an evaluation of the effect of these diseases on various yield components. Acremonium-infected maicillo plants yielded 33% less than healthy ones. The more susceptible BTx 623 had a 36% reduction in yield due to AW. MDM was found

to reduce yields by 52% on maicillos. It is still necessary to study how different incidence levels of each of these two diseases may affect yield levels in a sorghum field.

Results from the SDM studies confirmed the devastating nature of this disease. The highest incidence, 42.9%, reduced yield by 43.3%. Results of this study also show that disease resistance is not necessarily obtained at a cost in yield potential, as generally believed; yields in the resistant and susceptible populations did not differ at Choluteca, where no SDM occurred.

Yield loss studies with foliar diseases showed that yields in farmers' fields are reduced by as much as 15%.

Gray leaf spot severity was negatively correlated with plant height and panicle length in farmers' fields; the interaction of these three variables produced a 14.6% reduction in yield. The partial effect of gray leaf spot severity was a 7% yield reduction. This suggests that the disease affected yield by reducing plant height and panicle length, and that it had a separate effect, in addition to these, as well.

In the survey conducted during 1985, the overall severity mean for gray leaf spot was 17%. This degree of disease severity on cultivar Tortillero reduces yields by 13.3%. The overall mean for SDM incidence (%) in the same survey produces a 11% reduction in yield. In areas where both diseases occur, this implies that yield losses may well be more than 20%, although precise yield-loss estimates in such a case would require a simultaneous study of the two diseases.

Zonate leaf spot was an important disease in Choluteca. Not only were its severity and prevalence among the highest for that region, but it was also shown to reduce yields by almost 14% on improved and on landrace cultivars.

There is no doubt that sorghum production requires effective disease-control strategies. Several questions arose in the course of these studies, which if addressed, could lead to the development of practical means of control. The roles of plant height, as well as those of related traits, in disease development need to be clarified. Plant height and disease severity were correlated in several of the studies. Selection of the right combination of morphological traits in a cultivar could help keep disease levels low.

Plant density and its effect on disease development likewise needs further study. Since it is likely that different cultivars require different population densities for optimum yields, disease and yield levels could be simultaneously studied with density traits on all cultivars.

There is a high degree of heterogeneity for a number of traits in maicillos. This gives maicillos the stability that allows production of some grain, even under quite unfavorable conditions. With respect to disease reaction, heterogeneity in a population can slow down epidemic development (Robinson 1975). As changes are made in maicillos through breeding programs, sufficient heterogeneity should be maintained. The more susceptible genotypes can be eliminated, replacing them with new sources of resistance. The resulting plant population would then have the original sources of resistance (if any), plus the new ones, and fewer or no extremely susceptible types.

The intercropping of sorghum with maize seems to reduce the severity of certain foliar diseases; however, MDM had higher incidence under maize-sorghum intercropping than under sorghum monoculture (Table 7). Cultivars of either species intended for use in intercropping should have a high degree of resistance to MDM. Sources of inoculum, vectors, and the disease cycle should be studied to develop effective control strategies.

Sorghum-bean intercropping has been studied and recommended, based on socioeconomic and human nutrition considerations (De Walt and De Walt 1984; McCulloch and Futrell 1983). There are no known phytopathological reasons why sorghum-bean intercropping should not be encouraged. With nutritional, social, and crop-protection recommendations in agreement, it should prove an exciting system to promote. Central Americans have practiced intercropping for many years (Moreno 1985), so adoption should not be a problem.

Improved cultivars had fewer disease problems than maicillos when sown in postrera, the last season. In other experiments in this study, maicillos had lower disease severities than improved cultivars (Fig. 3). Improved cultivars showed no AW or leaf spot incidence; landrace cultivars did. In the case of AW, it is likely that among the improved cvs sown, AW resistance is the rule rather than the exception. Apparently, landrace cultivars have lower levels of resistance to this disease; though it is clear that they can be improved and AW incidence could be reduced through genetic resistance, much like in the improved cultivars.

In the experimental plots mentioned above, plant densities were higher than in farmers' fields surveyed, and fertilizer application was almost certainly higher as well. Those farmers who adopt the improved varieties in place of the landraces they now use will have to sow at

Table 7. Mean severity of oval and gray leaf spot diseases and incidence (%) of MDMV on sole sorghum and on sorghum/maize intercrop at Sta. Irene and Sn. Francisco, the Honduras, 1984, and mean severity of zonate leaf spot at Sta. Irene, the Honduras, 1984.

| Location | Sorghum/ variety | Disease | Mean severity | | | |
			Sorghum/ maize intercrop (%)	Standard deviation	Sole sorghum (%)	Standard deviation
Santa Irene, Choluteca	Peloton	Oval	15.6	6.18	29.7	7.12
		Gray	5.0	3.88	5.0	2.48
		MDMV[1]	7.6	7.61	1.7	0.82
		Zonate	2.0	2.28	5.5	6.44
Sn. Francisco 1984	Gigante	Oval	10.0	7.22	15.1	7.77
		Gray	6.8	4.06	7.1	1.88
		MDMV[1]	25.0	8.86	5.0	7.07

1. Incidence (%).

314

higher densities and apply fertilizer. Otherwise, yields of the improved varieties will not be higher than the maicillos they are to replace. Under such conditions, as in the experimental plots, improved cultivars will have more disease problems with gray leaf spot and zonate leaf spot than do the landrace cultivars. If a farmer sows susceptible materials year after year in the same field, it may eventually lead to increased initial inoculum levels, creating potentially serious problems.

It was not possible to test the reaction of improved cultivars to oval leaf spot, so it is not known if they possess adequate levels of resistance, or if the absence of this disease in the improved cultivars sampled is due to other reasons. However, 2 of 16 maicillos tested were resistant to oval leaf spot (Table 8). Until resistant materials are identified, these can be used in areas where oval leaf spot is likely to be detrimental to production.

Table 8. Sorghum landrace cultivars resistant to oval leaf spot at Santa Irene, Choluteca, the Honduras, 1984.

Cultivar	Source	Mean severity (%)
San Miguel 1	CENTA	3.04
MC 3	RRNN	5.46

If it were true that the improved varieties now being used do possess resistance to both AW and oval leaf spot, they could eventually become reinforced with gray and zonate leaf spot resistance; and step by step, yield losses caused by several diseases can be prevented and transformed into higher yields.

Sources of resistance were found for AW, gray leaf spot, SDM, oval leaf spot, and zonate leaf spot. The variety Sureno (VG 146), already released, was found resistant to SDM and to gray leaf spot. The maicillos San Miguel 1, Las Lajas, and Pom pom were resistant to both species of *Cercospora*. In addition, San Miguel 1 was resistant to oval leaf spot. These materials can be recommended for use where these diseases may be a problem. In the future these and other resistance sources may serve as donors for breeding programs designed to incorporate this resistance into superior cultivars.

Currently, it is feasible to grow maicillos, improved varieties, or hybrids already released that possess disease resistance as a means of reducing losses. In SDM areas, for instance, the hybrid Catracho and the improved varieties Sureno and Dorado can be recommended; the maicillos MC 97, MC 177, MC 122, and MC 136 could also be recommended for farmers who prefer to sow the maicillos, particularly in sorghum-maize intercropping.

The sorghum accessions that were resistant to SDM in Comayagua should be evaluated for their reaction to downy mildew in El Paraiso. Two isolates of *Peronosclerospora sorghi*, from Comayagua and from El Paraiso, were shown to have differing virulence (Pawar et al. 1985). More recently, different pathotypes have been observed to occur in Comayagua (Fernandez and Meckenstock 1987). Therefore, it is important that SDM resistance-screening in the Honduras be carried out in both Comayagua and El Paraiso. When sources of resistance at both locations are identified, they can be used to incorporate resistance into future releases—improved varieties and hybrids or landraces—by backcrossing and retesting at both locations. One can also use sources of resistance identified elsewhere, such as any of the 10 cultivars found resistant to isolates from different parts of the world (Pawar et al. 1985). However, those sorghum materials already adapted to climatic conditions in the Honduras would be the most desirable sources of resistance.

Ideally, different sources of resistance should be used. Relying on resistance from the same genetic background for all varieties and hybrids must be avoided. Several approaches to the production of reliable SDM-resistant cultivars are possible. Various approaches of resistance could be combined into one cultivar. A different resistance source could be incorporated into each of several cultivars grown in the same region. If heterogeneity is acceptable in a cultivar, as it is the case of maicillos, then various sources of disease resistance can be introduced without having to combine these in the same plants.

The fact that two resistance sources are of different origin is no guarantee that they are truly different. Mechanisms of resistance constitute a better criterion for selecting truly different sources of resistance. There are differences, for example, in plant growth stage at which SDM resistance is expressed (Pawar et al. 1985).

There are other measures available for control of SDM; seed treatment with metalaxyl is effective in protecting seedlings from infection, and in preventing seedborne distribution of the pathogen. Processed seed should be treated with metalaxyl because seed treatment is easily done in seed processing plants. It would be more expensive to treat seed at the farm level, because the cost of the chemical and the cost of application would probably be much higher on a small-unit basis.

Reducing disease severities can also be achieved through means other than resistance. Just as it was possible to create high levels of disease incidence with spreader rows between experimental plots in some of the studies presented here, it should be possible to reduce disease levels through cultural means. More studies are needed in this regard. Studies should include cropping arrangements differing in time as well as in space. This study has already shown that various diseases are in fact reducing sorghum yields. The next step in research is to find ways of reducing the losses.

References

Alexander, J.D., Toler, R.W., and Giorda, L.M. 1985. Correlation of yield reductions with severities of disease symptoms in sorghum infected with sugarcane mosaic or maize dwarf mosaic virus. Pages 109–112 in Proceedings, Grain Sorghum Research and Utilization Conference, 17–20 Feb 1985, Lubbock, TX, USA.

Alexander, J.D., Toler, R.W., and Miller, F.R. 1984. Effects of maize dwarf mosaic virus strain B infection on growth and yield of susceptible sorghum. Sorghum Newsletter 27:124–126.

Chambers, R., and Ghildyial, B.P. 1985. Agricultural research for resource-poor farmers: the farmer-first-and-last model. Brighton, East Sussex, UK: DP 203, University of Sussex, Institute of Development Studies. 29 pp.

Craig, J., Odvody, G.N., Wall, G.C., and Meckenstock, D.H. 1988. Sorghum downy mildew loss assessment with near-isogenic sorghum populations. Phytopathology 78.

De Walt, B.R. 1982. Farming systems research in southern Honduras. University of Kentucky INTSORMIL Report. Lexington, KY, USA: University of Kentucky. 83 pp.

De Walt, B.R., and De Walt, K.M. 1984. Sistemas de cultivo en Pespire, sur de Honduras: un enfoque de agroecosistemas. Instituto Hondureno de Antropología e Historia. 88 pp.

FAO (Food and Agricultural Organization). 1985. FAO Production Yearbook, Vol. 39. Rome, Italy: FAO.

Fernandez, L.D., and Meckenstock, D.H. 1987. Virulencia de Peronosclerospora sorghi en Honduras. En Memoria de la XXXIII reunión anual del PCCMA (Programa Cooperativo Centroamericano para el Mejoramiento de Cultivos Alimenticios), 30 Mar to 4 Apr, 1987. Guatemala, Guatemala: ICTA.

Frederiksen, R.A. 1984. Acremonium wilt. Pages 49–51 in Sorghum root and stalk rots, a critical review: proceedings of the Consultative Group Discussion of Research Needs and Strategies for Control of Sorghum Root and Stalk Rot Diseases, 27 Nov-2 Dec 1983, Bellagio, Italy. Patancheru, Andhra Pradesh 502 324, India: International Crops Research Institute for the Semi-Arid Tropics.

Frederiksen, R.A. (ed.). 1986. Compendium of Sorghum Diseases. St. Paul, MN, USA: American Phytopathological Society. 82 pp.

Frederiksen, R.A. 1973. Sorghum downy mildew, a disease of maize and sorghum. Texas Agricultural Experiment Station, Research Monograph 2. College Station, TX, USA: Texas A&M University. 32 pp.

Guiragossian, V. 1983. Reporte sobre el estado de CLAIS y su futuro. Pages 15–19 in Memoria Anual de la II Reunión Anual Comisión Latinoamericana de Investigadores en Sorgo (CLAIS), Tegucigalpa, the Honduras, 13–18 Nov 1983.

McCulloch, E.R., and Futrell, M. 1983. Socioeconomic and nutrition research concerning sorghum production and use in southern Honduras. Department of Sociology and Anthropology and Department of Home Economics

Summary Report for 1981–1983. Mississippi State, MI, USA: Mississippi State University. 45 pp.

Moreno, R.A. 1985. Plant pathology in the small farm context. Annual Review of Phytopathology 23:491–512.

Mundt, C.C., and Browning, J.A. 1985. Genetic diversity and cereal rust management. Pages 527–560 in The Cereal Rust, Vol. II. New York, NY, USA: Academic Press.

Natural, M.P., Frederiksen, R.A., and Rosenow, D.T. 1982. Acremonium wilt of sorghum. Plant Disease 66:863–865.

Navarro, L. 1979. Restricciones socio-economicas reflejadas en sistemas de cultivo practicado por pequenons agricultores. Turrialba, Costa Rica: CATIE.

Nolasco, R. 1983. El estado y potencial de sorgo en Honduras. Pages 7–11 in Memoria de la II Reunión, Anual de la Comisión Latinoamericana de Investigadores en Sorgo (CLAIS). Tegucigalpa, Honduras, 13–18 Nov 1983.

Pawar, M.N., Frederiksen, R.A., Mughogho, L.K., and Bonde, M.R. 1985. Survey of virulence in Peronosclerospora sorghi isolates form India, Ethiopia, Nigeria, Texas (USA.), Honduras, Brazil, and Argentina. Phytopathology 75:1374. (Abstract).

Rao, M.R. 1985. Perspectivas del sorgho y milo en los sistemas de cultivo en el noroeste de Brasil. Pages 87–106 in Sistemas de Producción de sorgo en America Latinao (INTSORMIL). CIMMYT, Mexico.

Robinson, R.A. 1975. Plant Pathosystems. Berlin, Germany: Springer-Verlag. 184 pp.

SRRNN (Secretaria de Recursos Naturales) 1984. Memoria Tecnica Anual 1983, Programa Nacional de Sorgo y Frijol. Tegucigalpa, the Honduras.

Wall, G.C. 1986. A study of sorghum diseases in Honduras, their importance under different cropping systems and strategies for their control. Ph.D. dissertation, Texas A&M University, College Station, TX, USA. 108 pp.

Wiese, M.V. 1980. Comprehensive and systematic assessment of crop yield determinants. Pages 262–269 in proceedings, E.C. Stalkman Commemorative Symposium. Miscellaneous Publication 7-1980. Minneapolis—St. Paul, MN, USA: University of Minnesota Agricultural Experiment Station.

Zadoks, J.C., and Schein, R.D. 1979. Epidemiology and Plant Diseases Management. New York, NY, USA: Oxford University Press. 427 pp.

Using Germplasm from the World Collection in Breeding for Disease Resistance

D.T. Rosenow[1]

Abstract

Breeding for disease resistance with emphasis on use of germplasm in the World Collection is discussed. It deals primarily with sorghum, but includes references to pearl millet. Suggestions regarding the collection, maintenance, storage, and distribution of seed of the Collection, as well as database accumulation, storage, and dissemination are presented. The Sorghum Conversion Program is discussed along with usefulness as a germplasm enhancement technique. For effective use, germplasm sources possessing high levels of stable, heritable resistance must be identified and appropriate screening techniques must be available. Breeding procedures are discussed and inheritance patterns of major diseases given. The use of diverse germplasm, using large field screening nurseries, simultaneous selection for disease resistance, good agronomic traits, other desirable plant and grain traits, and the use of quick, visual rating are encouraged. The need for close cooperation between breeder and pathologist is emphasized.

Introduction

Host-plant resistance is undoubtedly the most important single element in the control of diseases of sorghum and millet. Optimal benefit requires cooperative research by pathologists and breeders. It often seems a panacea to say the solution to the problem is resistant varieties. However, achieving the goal of developing a disease-resistant variety that possesses the other traits needed to make it a useful and acceptable variety is not easy.

Four requirements essential to a successful disease resistance breeding program are: (1) ability to screen and identify sources of resistance; (2) sources of resistance must have sufficiently high levels of resistance; (3) the resistance must be heritable so that it can be transferred; and (4) the resistance must be sufficiently stable across environments and with time.

The genetic base of the sorghum species is broad and diverse. Approximately 30 000 accessions reside in the World Collection. The pearl millet collection has approximately 19 000 accessions. Effective use of the World Germplasm Collection in breeeding for disease resistance requires consideration of three major components: (1) the Germplasm Collection itself; (2) disease resistance techniques including screening, identification of sources of resistance and determining stability of resistance, and documentation; and (3) the actual breeding and incorporation of disease resistance into useful cultivars or hybrids.

Germplasm Collection

These elements include collection, maintenance, storage, description, and classification of the material, as well as database storage, retrieval, and distribution. Although ICRISAT has done admirable work with the collection, acquisition of useful information on the lines, and charac-

---

1. Professor, Texas A&M University Agricultural Research and Extension Center, Lubbock, Tx 74901, USA.

Rosenow, D.T. 1992. Using germplasm from the world collection in breeding for disease resistance. Pages 319–324 *in* Sorghum and millets diseases: a second world review. (de Milliano, W.A.J., Frederiksen, R.A., and Bengston, G.D., eds). Patancheru, A.P. 502 324, India: International Crops Research Institute for the Semi-Arid Tropics.

terization of the lines for important traits, easy access to seed is still a problem. Quarantine restrictions on seed entry into many countries is of course mandatory, but such restrictions certainly complicate easy and timely transfer of germplasm. Many collections made by various individuals or groups are not sent to ICRISAT or even to the germplasm repository within their own country. Thus parts and bits of various collections exist in locations throughout the world.

United States Department of Agriculture's National Germplasm System is making a concerted effort to enhance the use of germplasm, both in short and long term, within the country. The system includes the National Seed Storage Laboratory, Fort Collins, Colorado; the Regional Plant Introduction Station at Griffin, Georgia; and the GRIN system. GRIN, a database system to catalog all sorghum introductions in the USA, contains descriptive information and the known ratings of important traits for each accession. The information is available to anyone.

Closer cooperation between the U.S. National Germplasm System and the ICRISAT Germplasm Unit is encouraged. Additional collections are needed from certain areas of the world. Also needed are consolidation, cataloging, classification, and gathering of data on useful traits on material in the many collections now maintained in various locations.

Assembling seed and information into central repositories for long-term storage, maintenance, and source seed is an urgent need. Many landrace varieties from tropical areas are tall and photoperiod sensitive, and thus too late for effective use in temperate regions. The cooperative Texas A&M, Texas Agricultural Experiment Station, and USDA/ARS Sorghum Conversion Program converts tall, photoperiod sensitive, exotic lines to shorter, early (nonphotoperiod sensitive) lines for use by all sorghum improvement programs, public or private, worldwide. The program utilizes short winter days in Puerto Rico for making crosses and backcrosses, with selection for short, early plants during long summer days in Texas. After four backcrosses, most converted lines are sufficiently similar to the original exotics and are increased and released as converted versions of the original lines, with a "C" added to the original IS number. To date, 423 fully converted lines have been released. A total of 1385 lines have been entered into the Conversion Program. New items entered in the past few years include selected Zera-zeras from Ethiopia, selected Guineaense lines from western Africa, lines from sandy, very low rainfall zones of Sudan and Niger, and selected individual lines with specific traits identified by sorghum researchers throughout the world. Sorghum workers from around the world are encouraged to suggest sorghums with elite trait(s) that should be converted.

The conversion project is very effective in providing new, diverse germplasm with useful traits in useful form to sorghum workers. In USA, most of the best currently used and widely adapted sources of resistance to head smut, downy mildew, anthracnose, grain mold/weathering, charcoal rot, midge, lodging, and stay-green trait (postflowering drought tolerance) are materials from the Sorghum Conversion Program.

A conversion program to change the height and maturity of pearl millet, using procedures similar to those effective with sorghum, has been initiated by the USDA/ARS in Georgia, Glen Burton, and Wayne Hanna, USDA/ARS, Tifton, Georgia (Personal communication). Long-term pollen storage, rather than short winter days, is used to achieve crossing of photosensitive lines.

Disease Resistance

The second major component is disease resistance itself, determined by screening to identify sources of resistance, and documentation of findings. Screening has identified sources of high levels of resistance for most sorghum diseases. Many are listed in Sorghum Diseases, a World Review (Williams et al. 1980), and in papers by Frederiksen and Rosenow (1980) and are screened during or immediately following conversion for disease resistance and other traits. Rosenow and Clark (1987) list newly released converted lines resistant to head smut, downy mildew, and anthracnose.

Although good use has been made of identified sources of resistance, improvement is needed. New sources of resistance are always needed as well as information regarding the differences, if any, in genetic makeup. Uniformity and reliability of rating schemes is an urgent need. A standardized procedure for prompt documentation of the improved material, and for accurate evaluation of resistance levels and of differing pathotypes or races are required. The biotype/race situation now experienced with pathogens of downy mildew, anthracnose, and many other diseases create many problems with identification. Knowledge of sources and/or types of resistance with the highest probability of remaining stable is immensely important, but very difficult to obtain.

Breeding Procedures

The third major component of effective use of the World Germplasm Collection is the actual incorporation of resistance into improved lines and/or hybrids that are agronomically elite and possess the other traits necessary for the adoption and use by producers.

To breeders, disease and resistance to disease are simply additional traits with which a breeder must struggle. In most cases, more than 20 genetically controlled important traits will be involved, and the number can be as high as 50.

Disease resistance should be a basic component of a plant breeding program. Success requires close cooperation between breeders and pathologists. It must be more than just a breeder sending his breeding lines to the pathologist for screening and the pathologist rating the lines and writing up a report. A common nursery— planned, established, and managed cooperatively in every way—is essential.

Knowledge of the inheritance of resistance, or at least a working knowledge of the ease of transfer, is essential. Inheritance of resistance to most important sorghum diseases is summarized in Table 1. Selection of specific breeding and screening techniques requires consideration of the nature of the host-parasite interaction and apparent vulnerability to each disease. These are summarized by Rosenow and Frederiksen (1982).

Several papers (Rosenow 1980; Frederiksen and Rosenow 1980; Rosenow and Clark 1987; Rosenow and Frederiksen 1982) discuss the diseases, list sources of resistance, and present discussions on utilization of exotic germplasm with sorghum. Frederiksen and Rosenow (1980) detail inoculation, inoculum preparation procedures, and screening techniques. Many kinds of breeding procedures, such as pedigree line breeding, backcross breeding, and population

Table 1. Summary of inheritance of resistance of some diseases of sorghum.

Disease	Inheritance pattern
Grain mold	Intermediate/ dominant
Downy mildew	Dominant
Anthracnose	Dominant
Charcoal rot	Recessive/ intermediate
Maize dwarf mosaic	Dominant/ recessive
Head smut	Dominant/ intermediate
Fusarium head blight	Dominant/ intermediate
Fusarium stalk rot	Recessive/ intermediate
Rust	Dominant
Gray leaf spot	Recessive
Ladder spot	Recessive
Leaf blight	Dominant
Sooty stripe	Recessive/ intermediate
Bacterial stripe	Recessive
Bacterial streak	Recessive/ intermediate
Acremonium wilt	Recessive/ intermediate
Zonate leaf spot	Recessive/ intermediate
Long smut	?
Sugary disease	?
Covered kernel smut	Dominant
Loose kernel smut	Dominant

breeding have been successfully used to breed for disease resistance. The most overall desirability of the line in which resistance is found, and the type of material, line, hybrid, or population into which the resistance is to be incorporated.

When dealing with a disease resistance source that is good agronomically and the resistance is easily transferable, pedigree breeding involving a single direct cross to an existing adapted elite cultivar is an ideal method. In many cases, however, the disease resistant source will have poor agronomic traits. In such cases, a modified procedure may be required. One example would be crossing disease resistance sources "A" with elite line "B". As soon as progeny "C" is stabilized for the desired disease resistance and hopefully some improved agronomic traits, it is immediately crossed to elite line "D" which possesses several desirable traits missing in "C". Strong selection pressure should be applied in early generations of resulting progeny "E" for both agronomic performance and disease resistance and for other desirable plant and grain traits. This provides an opportunity to select away from low yield traits associated with the disease-resistant source. The opportunity to develop a cultivar which combines desirable traits in unique combinations is now available.

Population breeding can be a very effective tool for enhancing disease resistance at the population level, especially if individual plants of S_1 progeny can be effectively screened. However, it is difficult to simultaneously maintain or concentrate a large number of desirable traits in a population. Extracting individual lines for direct use becomes real problem due to the lack of control over the many undesirable traits usually found in a population, as well as the seemingly dominance of lower yield. Many of the cultivated sorghums possess weedy traits—good for survival, but not for high yield. Population breeding can be a very effective tool to develop diverse germplasm with high levels of resistance to certain diseases. Likewise, it is a useful approach for maintaining a large number of diverse sources of resistance in a segregating situation, from which individual disease-resistant selections may be extracted. In pearl millet, population improvement is very useful since most present varieties are actually random mating populations.

A number of sorghum groups such as Guineaense, Dochna-Subglabresent derivatives, and Nervosum (Kaoliang) tend to be low yielding with the low yield behaving much as a domi-nant trait. Breeding with such lines requires using other breeding techniques such as additional backcrossing and intergression.

One example of an indirect method of selecting for disease resistance involves charcoal rot and head blight. Resistance or tolerance to post-flowering drought stress correlates very well with resistance to charcoal rot, since it is a disease that requires predisposition by drought stress during grain development. In the Texas A&M drought program, we select genotypes which do not prematurely senesce and thus tolerate this late-season stress, possessing what we call the "stay-green" trait. At the same time we exert heavy selection pressure for stalk and peduncle (weak neck) lodging resistance. Resulting lines developed in this way show a very high degree of resistance to charcoal rot and head blight. Discussion of this may be found in Rosenow (1977, 1984) and Rosenow et al. (1983).

When breeding parental lines for use in hybrids, special techniques are required, but the same principles apply. If disease resistance or other traits are dominant, it is somewhat easier to combine them by using hybrids. However, recessive traits are very difficult when dealing with hybrids as both parental lines must possess the trait.

Efficient and effective screening of a large number of genotypes is critical in a successful disease-resistance breeding program. Simultaneous selection for disease and agronomic traits is the ideal. I strongly encourage field screening whenever possible, using natural infection if possible, but inoculation if needed. Large field screening nurseries, combining disease screening and breeding at one or more location where several important diseases naturally occur would be an ideal facility.

In the humid south Texas, we sow large sorghum nurseries where many internationally important diseases are endemic. We observe excellent development of grain mold, downy mildew, head smut, and often head blight, anthracnose, charcoal rot, zonate leaf spot, gray leaf spot and MDMV in nearly every growing season. In other areas we sow field screening nurseries for charcoal rot and MDMV. The best breeding lines are then identified from nurseries and tested extensively in Texas and other areas in the All Disease and Insect Nursery (ADIN) and in various head smut, downy mildew, anthracnose, charcoal rot, and lodging tests. The most elite lines are placed eventually in the International Disease and Insect Nursery (IDIN) and distributed worldwide to anyone desiring seed.

ICRISAT's large and effective multilocational uniform disease screening program contributes greatly to the overall disease resistance breeding effort. ICRISAT sorghum and millet pathologists and breeders combine efforts in the Institute's cereals improvement programs.

I primarily deal here with sorghum, but the same principles hold for pearl millet, keeping in mind that pearl millet cross-pollinates. A good reference on disease-resistant sources and breeding for disease resistance has been prepared by Andrews et al. (1985). Other excellent references on diseases include Thakur and Chahal (1987), Singh et al. (1987) and Kumar and Rao (1987). These three papers appear in the Proceedings of the International Pearl Millet Workshop (ICRISAT 1987).

Conclusion

A productive disease-resistance program will:

1. Utilize large amounts of genetically diverse material.
2. Sow, at a few prime locations, large field screening nurseries of disease material.
3. Utilize sources with high levels of resistance that are easily (if possible) transferable, stable, and of good agronomic types.
4. Know good and bad traits of all its germplasm and select parents carefully to complement each other.
5. Select simultaneously, if possible, for disease resistance and other desirable traits.
6. Recombine the best sources of resistance, even among early generation sources.
7. Reflect full cooperation between all scientists involved. Breeders must be able to rate for disease resistance and   pathologists must work in breeding material with the breeder in the field.

Acknowledgment. This report was supported in part by a grant from the United States Agency for International Development.

References

Andrews, D.J. King, S.B., Witcombe, J.R. Sing, S.D., Rail, K.N., Thakur, R.R., Talukdar, B.S., Chavan, S.B., and Singh, P. 1985. Breeding for disease resistance and yield in pearl millet. Field Crops Research 11:241–258.

Frederiksen, R.A., and Rosenow, D.T. 1980. Breeding for disease resistance in sorghum. Pages 137–176 in Proceedings of the International Conference in Host Plant Resistance, 22 Jul to 4 Aug 1979, Texas A&M University, College Station, TX, USA: Texas Agricultural Experiment Station MP-1452. 605 pp.

ICRISAT (International Crops Research Institute for the Semi-Arid Tropics). 1987. Proceedings of the International Pearl Millet Workshop, 7–11 Apr 1986. ICRISAT Center, India, Patancheru, A.P. 502 324, India: ICRISAT.

Kumar, K. Anand, and Rao, S. Appa. 1987. Diversity and utilization of pearl millet. Pages 69–82 in Proceedings of the International Pearl Millet Workshop, 7–11 Apr 1986. ICRISAT Center, India, Patancheru, A.P. 502 324, India: International Crops Research Institute for the Semi-Arid Tropics.

Rosenow, D.T. 1977. Breeding for lodging resistance in sorghum. Pages 171–185 in Report of the 32nd Annual Corn and Sorghum Research Conference, 6–7 Dec 1977, Chicago, Illincis, Washington D.C., USA: American Seed Trade Association.

Rosenow, D.T. 1980. Stalk rot resistance breeding in Texas. Pages 306–314 in Sorghum diseases: a world review: proceedings of the International Workshop on Sorghum Diseases, 11–15 Dec 1978, ICRISAT, Hyderabad, India. Patancheru, A.P. 502 324, India: International Crops Research Institute for the Semi-Arid Tropics.

Rosenow, D.T. 1984. Breeding for resistance to root and stalk rot in Texas. Pages 209–218 in Sorghum root and stalk rots, a critical review: proceedings of the Consultative Group Discussion on research needs and strategies for control of sorghum root and stalk rot diseases, 27 Nov to 2 Dec 1983, Bellagio, Italy. Patancheru, A.P.

502 324, India: International Crops Research Institute for the Semi-Arid Tropics.

Rosenow, D.T., and Frederiksen, R.A. 1982. Breeding for disease resistance in sorghum. Pages 447–455 *in* Sorghum in the Eighties: Proceedings of the International Symposium on Sorghum, 2–7 Nov 1981, Patancheru, A.P. 502 324, India: International Crops Research Institute for the Semi-Arid Tropics.

Rosenow, D.T., and Clark, L.E. 1987. Utilization of exotic germplasm in breeding for yield stability. Biennial Grain Sorghum Resistance and Utilization Conference. 15:39–46.

Rosenow, D.T., Quisenberry, J.D., Wendt, C.W., and Clark, L.E. 1983. Drought tolerant sorghum and cotton germplasm. Agricultural Water Management 7:207–222.

Singh, S.D., Ball, S., and Thakur, D.P. 1987. Problems and strategies in the control of downy mildew. Pages 161–172 *in* Proceedings of the International Pearl Millet Workshop, 7–11 Apr 1987, ICRISAT Center, India. Patancheru, A.P. 502 324, India: International Crops Research Institute for the Semi-Arid Tropics.

Thakur, R.P., and Chahal, S.S. 1987. Problems and strategies in the control of ergot and Smut in pearl millet. Pages 173–182 *in* Proceedings of the International Pearl Millet Workshop, 7–11 Apr 1986, Patancheru, A.P. 502 324, India: International Crops Research Institute for the Semi-Arid Tropics.

Williams, R.J., Frederiksen, R.A., Mughogho, L.K., and Bengtson, G.D. (eds). 1980. Sorghum diseases, a world review: proceedings of the International Workshop on Sorghum Diseases, 11–15 Dec 1978, ICRISAT, Hyderabad, India. Patancheru, A.P. 502 324, India: International Crops Research Institute for the Semi-Arid Tropics. 478 pp.

Using the World Germplasm Collection in Breeding for Disease Resistance at ICRISAT

S.Z. Mukuru[1]

Abstract

Disease-control strategies at ICRISAT emphasize the development of breeding materials resistant to the important cereals diseases. This effort is interdisciplinary, with breeders and pathologists cooperating. The process involves developing efficient, reliable field and laboratory screening tecniques; identifying sources of stable and durable resistance in the sorghum germplasm collection and other breeding materials; testing stability of resistance with various races of the pathogens in multiple environments; and then incorporating the stable resistance into high-yielding agronomically desirable genotypes.

Introduction

A wide range of abiotic and biotic stress factors limit sorghum production in the semi-arid tropics (SAT). Of these, the most important and widespread are drought, insect pests, and diseases. For resource-poor farmers in the semi-arid tropics the most practical control method is growing resistant cultivars, if available. This is a simple and an effective means of improving stability of yield.

In India, the most important diseases of sorghum are grain molds, stalk rots, downy mildew, and several leaf diseases (see Anahosur, this publication). Disease-control strategies at ICRISAT emphasize the development of breeding material resistant to these diseases. At ICRISAT this effort is interdisciplinary and involves breeders and pathologists. The procedure follows:

1. Develop efficient, fast, and reliable field and laboratory screening techniques.
2. Use these techniques to identify sources of stable and durable resistance in the sorghum germplasm collection and breeding material.
3. Test stability of resistance in different environments and with races of the pathogens.
4. Incorporate identified stable resistance into high-yielding agronomically desirable genotypes.

At ICRISAT Center we have a genetically diverse collection, numbering in the hundreds, of germplasm of indigenous cultivars and weedy and wild relatives of sorghum. We have developed effective techniques to screen the entire collection to identify those resistant to various diseases.

Resistance Screening Techniques and Sources of Resistance

The key to the identification of sources of stable resistance is the development of reliable, efficient, and epidemiologically sound screening techniques which discriminate between resistant and susceptible genotypes (Mughogho 1981). ICRISAT screening techniques have been effective for grain molds, downy mildew, stalk rots, rusts, and anthracnose (ICRISAT 1984).

---

1. Principal Plant Breeder, Cereals Program, ICRISAT Center, Patancheru, Andhra Pradesh 502 324, India.

Mukuru, S.Z. 1992. Using the world germplasm collection in breeding for disease resistance at ICRISAT. Pages 325–328 *in* Sorghum and millets diseases: a second world review. (de Milliano, W.A.J., Frederiksen, R.A., and Bengston, G.D., eds). Patancheru, A.P. 502 324, India: International Crops Research Institute for the Semi-Arid Tropics. (CP 736).

Breeding for resistance

The main objective of the sorghum breeding program at ICRISAT is to develop high-yielding breeding material, acceptable in food quality and resistant to insect pests and diseases. The program at ICRISAT includes projects on diseases, *Striga*, insect pests, multifactor resistant populations, sorghum conversion, and development of hybrids and hybrid parents. At present, our single project on disease resistance focuses on grain molds. Sources of resistance to downy mildew, stalk rots, and leaf diseases are used in all the breeding projects, of course, and are selected simultaneously with other specific traits.

Heritabilities for resistance to downy mildew and leaf diseases are fairly high, therefore frequency of these resistance genes in the breeding material generated should be high. Rosenow and Frederiksen (1982) summarized the inheritance of resistance to several diseases. Mold resistance is dominant in some F_1 hybrids, and over-dominant or intermediate in some others. Resistance to downy mildew, anthracnose, and rusts is dominant. That of charcoal rot is recessive to intermediate.

Grain Molds

Grain mold (GM) has become a major and widespread disease of sorghum in the SAT, especially where flowering and grain development and maturity coincide with the rainy warm weather. Grain molds significantly reduce yields and grain quality. Growing of high-yielding, early-maturing, mold-resistant cultivars is the only practical control method suited to farmers in the SAT.

At ICRISAT Center, the sorghum germplasm collection was screened between 1975 and 1978 for sources of GM resistance; lines consistently resistant were selected for breeding. Selections from crosses involving these low-level sources of mold resistance yielded improved breeding lines with moderate grain yield potential but with low levels of resistance. The mold resistance of these lines is not sufficient for effective mold control under moderate to high mold-disease pressure. Several low susceptible lines were intermated to generate variability and concentrate the scattered resistance genes, but this did

not provide significant improvement in GM-resistance in the white-grained sorghums.

In the early 1960s, high-level GM resistance was identified in colored-grain sorghums. These maintain their resistance for 2 to 3 weeks following physiological maturity (Bandyopadhyay et al. 1987). We involved these colored-grain lines in crosses and intensively screened the segregating progenies for white-grained segregates with high levels of mold resistance. Five white-grained selections with levels of mold resistance as high as their colored-grained parental lines were identified, specifically ICS × 62 K 140 B 3-1, ICS × 62 K 140 B 2-1, ICS × 119 K 19 W 1-6-1, ICS × 119 K 64 W 1-2-1, and ICS × 119 K 19 W 1-4-1.

These lines, however, are tall and low-yielding. So we are now using them as sources of high mold resistance in developing improved breeding material with good yield potential and desirable agronomic traits.

Mold resistance in the white-grained selections appears to be associated with grain hardness, (Mukuru; Waniska et al., this publication) while mold resistance in colored-grain is associated with tannin and flavan-4-ol content or grain hardness, or a combination of these factors. High flavan-4-ol content has not been identified in white-grained sorghum lines.

We routinely screen for resistance to grain molds in all breeding lines generated in the many breeding projects at ICRISAT Center; all have been found to be mold-susceptible. (Mukuru; Waniska et al; Forbes et al., this publication).

Other Diseases

The other sorghum diseases of importance in India are charcoal rot, downy mildew, anthracnose, and rust. The sorghum germplasm collection has been screened, and resistance to these diseases identified (ICRISAT 1984).

Charcoal rots caused by a common soilborne fungus, *Macrophomina phaseolina* is a destructive disease of sorghums in the SAT (see Pande on Stalk Rots, this publication) Plant lodging and reduction of yields and quality of the grain result from stalk rot infections. Charcoal rot-resistant germplasm lines have been identified and are used in various breeding projects. All breeding material generated is grown during the post-rainy season; and those nonsenescing genotypes that do not lodge are selected. Postrainy season-improved breeding lines showing low susceptibility to charcoal rots include ICSV 576, ICSV 588, ICSV 603, ICSV 606, ICSV 612, ICSV 616, ICSV 628, ICSV 630, ICSV 645, ICSV 646, and ICSV 802.

Adequate screening for sorghum downy mildew (SDM) resistance at ICRISAT Center is impossible, because environmental conditions for its infection and spread are not ideal. However, at Dharwad, India, SDM is endemic. At Dharwad, ICRISAT sorghum pathologists have succeeded in developing an effective large-scale field-screenign technique, using conidial showers from infector rows (Mughogho 1981). A number of lines have been identified as resistant to SDM. One of the lines, QL 3, and its sister lines are immune to SDM. QL 3 has been used extensively in crosses, and selections from these crosses have yielded breeding lines similar to QL 3 in SDM resistance, are agronomically superior, tan in plant color, and resistant to anthracnose. Parents of hybrids are likewise screened for SDM resistance, and several have been found to possess moderate to high levels. The ICSA lines 31, 37, 53, 60, 62, and M 36257 show good levels of SDM resistance.

Several leaf diseases occur in India, but rarely cause economic loss (Rao et al. 1980), although anthracnose and rust are harmful to forage quality in some years. ICRISAT sorghum pathologists have developed an effective large-scale field-screening technique for anthracnose at Pantnagar, in northern India, and for rust at Dharwad in eastern India (Mughugho 1981). The environment for expression of these diseases is better in these areas. The screening techniques for anthracnose and rust were used to screen all the breeding material generated in various breeding projects. A large number of lines have been found to have high levels of resistance to anthracnose and rust. Some breeding lines, including the ICSV numbers 1, 108, and 120 and the M numbers 35610, 36170, 36172, 36190, 36248, 36257, 36266, and 60328 show resistance to more than one disease. One female parent, ICSA 11, has high combining ability for yield and shows high levels of resistance to anthracnose.

Multiple Resistance

If resistance is all that mattered, resistance breeding would often be easy; the difficulty comes in combining sufficient resistance with other characters required to make a satisfactory variety (Simmonds 1979). At ICRISAT, we feel that we have made progress in developing agronomically elite breeding materials possessing sufficient resistace levels of diseases, Striga, and insect pests. These should be intermated to develop improved population from which cultivars with multiple resistance can be extracted. We have initiated a long-term population-improvement program aimed at compositing three broadbased multifactor resistant (MFR) populations, incorporating improved sources of disease and insect resistance and desirable agronomic characteristics. Compositing of these populations will be completed in 1988, then improvement by recurrent selection will begin.

Three MFR populations are CSP 1 BR/MFR, ICSP 2 BR/MFR, and ICSP 3 BR/MFR. ICSP 1 BR/MFR will be improved by recurrent selection for resistance to grain molds, stem borer, shoot fly, and midge; ICSP 2 BR/MFR for resistance to grain mods, Striga, and stand establishment. ICSP 3 BR/MFR will be improved for postrainy season adaptation.

References

Bandyopadhyay, R., Mughogho, L.K., and Prasad Rao, K.E. 1988. Sources of resistance to sorghum grain molds. Plant Disease 72: 504–508.

ICRISAT (International Crops Research Institute for the Semi-Arid Tropics). 1984. Sorghum Pathology Progress Report No. SP/G/1/84. Sources of resistance to sorghum diseases in India. Patancheru, Andhra Pradesh 502 324, India: ICRISAT.

Mughogho, L.K. 1981. Strategies for sorghum disease control. Pages 273–282 in Sorghum in the Eighties: Proceedings of the International Symposium on Sorghum, 2–7 Nov 1981. Patancheru, Andhra Pradesh 502 324, India: International Crops Research Institute for the Semi-Arid Tropics.

Rao, N.G.P., Vidyabhushanam, R.V., Rana, B. S., Jaya Mohan Rao, V., and Vasudeva Rao, M.J. 1980. Breeding sorghums for disease resistance in India. Pages 430–443 in Sorghum diseases, a world review: proceedings of the First International Workshop on Sorghum Diseases, 11–15 Dec 1978, ICRISAT, Hyderabad, India. Patancheru, Andhra Pradesh 502 324, India: International Crops Research Institute for the Semi-Arid Tropics.

Rosenow, D.T., and **Frederiksen, R.A.** 1982. Breeding for disease resistance in sorghum. Pages 447–455 *in* Sorghum in the eighties: proceedings of the International Symposium on Sorghum, 2–7 Nov 1981, ICRISAT Center, India. Vol 1. Patancheru, Andhra Pradesh 502 324, India: International Crops Research Institute for the Semi-Arid Tropics.

Simmonds, N.W. 1979. Principles of Crop Improvement. London; New York, NY, USA: Longman. 399 pp.

Part 6

Building for the Future

Successful Transfer of ICRISAT Downy Mildew Resistance Screening Technology: an Example of Transfer of Technology

S. Pande[1] and S.D. Singh[1]

Abstract

Technology transfer is an ICRISAT mandate. Successful transfer of sorghum downy mildew disease resistance technique is presented as an example of technology transfer. Components of the technique were demonstrated and successfully transferred to national and regional programs in SAT countries.

Introduction

"Technology-Transfer" can be defined as an extension activity in which results of scientific investigation in any area of production and protection (especially in crop improvement) are applied to practical use in other, usually distant, places.

An intended technique may be transferred in two ways: by training an individual from the "targeted" area at a place where the technique is successful and well established, so that scientist can establish and use the technology upon returning to his or her area of operation; and by demonstrating various components of the technology to individuals at their own facilities, using the resources available. The second method has been the most successful as receiving individuals appear to gain confidence by modifying, as needed, the locally available facilities to establish a particular technique.

At ICRISAT, large-scale field-screening techniques for evaluating sorghum resistance to downy mildew (*Peronosclerospora sorghi*), grain molds (fungal complex), and ergot (*Sphacelia sorghi*), were developed. Similarly, pearl millet resistance to artificially inoculated downy mildew (*Sclerospora graminicola*), ergot (*Claviceps fu-*

siformis), and smut (*Tolyposporium penicillariae*) diseases in field nurseries was identified.

Although these field-screening techniques were routinely explained and demonstrated to visitors from Africa, the Americas, and Asia enrolled in certain training programs at ICRISAT Center, they have been adopted at only a few locations outside ICRISAT. The main reason for this in many places is lack of basic facilities and experienced hands.

The successful establishment of the ICRISAT sorghum downy mildew field-screening technique in Zimbabwe and Zambia during the 1986/87 and 1987/88 rainy seasons, respectively, is discussed here.

Screening for SDM Resistance

A large-scale screening technique for sorghum downy mildew (SDM) resistance was standardized by scientists at ICRISAT Center, substation Dharwad, Karnataka state, India, during the 1982 rainy season. Temperatures and humidities at Dharwad are favorable for sorghum DM development and spread.

The infector-row field-screening technique for sorghum downy mildew is based on wind-

---

1. Plant Pathologists, Cereals Program, ICRISAT Center, Patancheru, Andhra Pradesh 502 324, India.

Pande, S., and **Singh, S.D.** 1992. Successful transfer of ICRISAT downy mildew resistance screening technology: an example of transfer of technology. Pages 331–334 *in* Sorghum and millets diseases: a second world review. (de Milliano, W.A.J., Frederiksen, R.A., and Bengston, G.D., eds). Patancheru, A.P. 502 324, India: International Crops Research Institute for the Semi-Arid Tropics. (CP 737).

borne conidia (asexual spores) of *Peronosclerospora sorghi*, the causal organism of SDM. Conidial showers, blown by wind from infector rows onto test materials, provide the inoculum.

Two especially important aspects of the successful employment of this technique are establishment of the disease in infector rows, and favorable temperatures and humidities for abundant conidial production by plants in the infector rows. Screening should be conducted, therefore, at a location known to be favorable for the disease, as is the Dharwad installation in India.

Source and off-season maintenance of inoculum

Most of the downy mildew diseases have evolved in close association with a number of noncultivated grasses. The DMs have become a concern only when extensive sowings of susceptible genotypes are established in areas where a DM is introduced and/or endemic on a native grass species. In general, grass species that are hosts for DMs are native to the southeastern Asia, and generally of the tribes Andropogoneae and Maydee. In North America, johnsongrass (*Sorghum halepense*) and shattercane (a feral *S. bicolor*) can become infected naturally in the field by *P. sorghi*. Since oospores and conidia are produced in both species, these spp often serve as sources of primary inoculum. Perennial wild sorghums are common in borders of cultivated fields and in drainage and irrigation ditches. The sources of primary conidial inoculum used for establishment of screening techniques at three locations were:

Dharwad, India. During May 1981, two volunteer plants of a sorghum cv with systemic disease were located. We used these two plants as conidial sources, and increased the inoculum by use of the sandwich inoculation procedure. The conidia inoculum in systemically infected plants was carried over in the off season. Just before the main growing season (rainy season), this inoculum was increased for use in establishing infector rows.

Matopos, Zimbabwe. The primary inoculum was collected from systemically infected ratoons of cultivated sorghums at Kadoma. Infection (plants with systemic disease) on some entries was as high as 60%. These ratoons surely were infected by conidia blown from other infected plants, that had been infected by either oospores or conidia from weedy and wild sorghums. This inoculum was used to infect 20 sorghum lines, using the sandwitch inoculation procedure and some 20 plants with systemic infection were obtained (Pande 1987); these in turn were used to establish infector row screening during the 1986/87 season (Singh 1987) using modified sandwich inoculation. Natural temperatures were found to be congenial for sporulation and infection.

Golden Valley, Zambia. During 1986/87, a severe outbreak of sorghum downy mildew was observed at Golden Valley and some of the entries showed as high as 80% systemic infection (B.N. Verma and W.A.J. de Milliano, personal communication). During the last week of November, few systemically infected weedy and wild sorghums (*S. halepense*) were observed near the irrigation canal. After the first rains, we collected thousands of systemically infected *S. halepense* and shattercane plants from many locations around Golden Valley. Inoculum from these plants was used to establish the infector rows.

Establishing infector rows

The most important and vital component of the ICRISAT SDM screening technique is the infector rows. To establish these, we sow germinated and downy mildew-infected seeds of a highly susceptible cultivar (DMS 652, IS 643, Marupantse, or Sugardrip). On these cvs, the pathogen produces abundant conidia. Pregerminated seeds were inoculated following the sandwich inoculation technique.

In sandwich inoculation, germinated seeds are incubated by placing them on the adaxial surface of a piece of systemically infected leaf, and covering them with another piece of systemically infected leaf, so that the germinated seeds are sandwiched. The procedure is carried out in petri dishes lined with moist filter paper; dishes are incubated in darkness at 18–20°C for 12 to 16 h. By this time, the seedlings are covered with a mycelial mat of the fungus, indicating that the fungus has sporulated and conidia might have caused the systemic infection in the growing plumule. However, one cannot be sure until the emergence of systemically infected plants in the infector rows.

If incubators and petri dishes are not available, one can modify this seedling inoculation technique (Singh 1987). In this method, germinated seeds are spread on a polyethylene sheet (approximately 1 to 2 m long × 0.75 m wide). Systemically infected potted plants are laid on the seedlings so that the lower leaf surfaces are in contact with the germinating seed. The infected plants are covered with moist gunny bags and finally with polyethylene sheets, to maintain a saturated relative humidity.

These chambers are left overnight under natural conditions for the infection process. The advantage of this technique is that infected plants can be used, repeatedly if "rested" about 12 h between use, and 2 to 3 kg of germinated seeds can be inoculated at one time. However, the effectiveness of this method of inoculation depends upon the availability of systemically infected leaves. If all leaves of the conidia-donor plants are completely systemically infected, screening is more successful because the possibility of uneven infection of the infector rows is reduced.

Recently at Golden Valley, the senior author used seed beds as a substitute for petri dishes, replacing the moist filter papers with moist newspaper. Systemically infected leaves of wild sorghums and 3 to 5 kg of germinated seeds were layered as in the sandwich technique. However, he finally covered the seed bed with polyethylene sheets to maintain the humidity. This inoculation procedure produced 90 to 100% systemically infected plants in the infector rows.

Natural environment can replace incubators if seasonal night temperatures are favorable at the location where SDM infector row field screening is to be established. Incubators may be available when petri dishes are not; then one can use two dinner plates, or two serving trays of equal size. We have successfully used these in this season at Golden Valley.

Seedlings inoculated by either of the above methods are planted in every six or tenth row, depending upon wind velocity at the location. At Golden Valley we have sowed each sixth row as an infector row. However, at Matopos (Singh 1987; De Milliano 1987) each seventh row was sown as an infector row. Once the infector-row

plants are established and show heavy DM sporulation (about 20 to 25 days from sowing), the test material should be seeded.

Sowings of test materials

Sow five or nine rows (depending on distance between the infector rows) of test material. The center row of the trial is sown to the same DM susceptible variety as the test rows; it serves as an indicator of disease pressure and as a susceptible control for the test rows between it and the infector rows. The arrangement of infector rows, test rows, and indicator rows may be adjusted to accommodate local situations—such as the availability of land and the number of entries to be screened.

Evaluating resistance

The control row and the test materials are evaluated for DM at the seedling and flowering stages; sorghum lines with no more than 5% of the plants showing systemic disease are regarded as resistant.

With this procedure, 4500 sorghum breeding and germplasm lines were evaluated at Matopos; of these, 1008 were found to be resistant. In this season, senior author has sown more than 5000 sorghum and maize lines, including material from ICRISAT Center, SADCC/ICRISAT, and Zambia's national sorghum- and maize-improvement programs at Golden Valley; of these 928 sorghum and four maize lines were found to be resistant.

Additional Adoptions

Components of this technique have been adopted by several national programs, such as the All-India Coordinated Sorghum Improvement Program at the University of Agricultural Sciences, Dharwad, Karnataka state, India (Anahosur, personal communication). The sandwich technique was successfully used to establish infector rows for SDM screening in Rwanda (Bandyopadhyay 1987).

Several other techniques, such as pearl millet ergot and smut inoculations were conducted at Matopos (Thakur 1986). Singh (1985) successfully established the pearl millet downy mildew screening technique in Mali, and Bandyopadhyay (1987) demonstrated the different phases of sorghum ergot screening technique at Rwanda.

Conclusion

In all, good progress has been made by ICRISAT scientists in demonstrating the successful transfer of resistance-screening techniques in places where these diseases are important constraints in sorghum and millet improvement.

References

Bandyopadhyay, R. 1987. Report of visit to Rwanda. 2 Sep–2 May 1987. Patancheru, Andhra Pradesh 502 324, India: International Crops Research Institute for the Semi-Arid Tropics. 26 pp. (Limited distribution.)

de Milliano, W.A.J. 1987. Zambia Report, SADCC/ICRISAT SMIP. Bulawayo, Zimbabwe: SADCC/ICRISAT. 16 pp. (Limited distribution.)

Pande, S. 1987. Report of visit to Zambia and Zimbabwe, 3–18 Oct 1986. Patancheru, Andhra Pradesh 502 324, India: International Crops Research Institute for the Semi-Arid Tropics. 32 pp. (Limited distribution.)

Singh, S.D. 1985. Report on trip to Mali, 12 Jul–12 Aug 1985. Patancheru, Andhra Pradesh 502 324, India: International Crops Research Institute for the Semi-Arid Tropics. 15 pp. (Limited distribution.)

Singh, S.D. 1987. Report on trip to Zimbabwe, 26 Oct 1986–30 Apr 1987. A technology transfer mission of ICRISAT Center. Patancheru, Andhra Pradesh 502 324, India: International Crops Research Institute for the Semi-Arid Tropics. 15 pp. (Limited distribution.)

Thakur, R.P. 1986. Report on a visit to Zambia and Zimbabwe, 26 Feb–9 Mar 1986. Patancheru, Andhra Pradesh 502 324, India: International Crops Research Institute for the Semi-Arid Tropics. 12 pp. (Limited distribution.)

Recommendations

Seventeen recommendation committees met and addressed issues of concern regarding disease problems and disease management tactics.

1. Smut diseases
2. Sorghum ergot
3. Pearl millet ergot
4. Bacterial diseases
5. Virus diseases
6. Nematology
7. Seed rots and seedling diseases
8. Grain mold
9. Sorghum downy mildew
10. Leaf diseases
11. Anthracnose
12. Sorghum stalk rot
13. *Striga*
14. Disease management in sorghum and pearl millet
15. Downy mildew in pearl millet
16. Foliar diseases of pearl millet
17. Utilization of disease resistance for sorghum and millets

Smut Diseases

Covered and loose kernel smuts

These two diseases are considered low priority areas. They are controlled by seed dressing. It is realized, of course, that loose smut infection can occur in the ratoon crop, but the impacts of loose and covered smuts are too low to warrant extensive programs for their control.

National scientists in areas suffering from these two diseases are encouraged to conduct seed-dressing demonstration trials to show the benefits of this practice.

Head smut

This is a potentially serious disease that will require an integrated approach. Development of resistant seed parents has been underway for the past decade. It is recommended that research workers:

1. Breed elite cultivars for affected regions.
2. Monitor pathotype variation.
3. Exploit cultural, biological, and chemical means of control.

Long smut (Sorghum and millet)

1. Evaluate effectiveness of natural spread, using an inoculated spreader variety, in screening for resistance in areas where the disease is endemic.
2. Continue with current screening techniques, and the search for new sources of resistance.
3. The taxonomic position of smut species occurring on millet needs clarification and verification; screening programs for resistance are needed.

Sorghum Ergot

Sorghum ergot was recognized as increasing in importance in Africa and India, and also in Thailand, because of the increased use of male steriles in hybrid production. An acute lack of knowledge (dependent on fundamental and applied research coordinated in and between the regions concerned) presently restricts advances in control procedures.

1. Study the extent of significant variability in the pathogen occurring from the eastern tip of Africa to its southern region, and the relationship of the fungus occurring in Nigeria, Burkina Faso, and Senegal. Significant factors include the rate of infection of ovaries; efficiency of differentiation of sclerotial tissues. alkaloid pattern (probably more as an indicator of genetic variability than as a toxicity concern); longevity of asexual conidia on plant debris; and the role of secondary conidiation in epidemiology (by providing a potential windborne inoculum).
2. Evaluate, in different countries, the significance of differences in pathogens of the same species.
3. Identify and evaluate significance of collateral hosts, particularly in India and Africa.

4. Using male sterile and fertile cultivars, determine the precise period of susceptibility of fertilized gynoecia of pathogens from Africa, India, and Thailand in several environments.
5. Survey farmers' practices, particularly with respect to knowledge of ergot contamination (sclerotia and/or sexual conidia) of seed and to postharvest land management.
6. Standardize resistance-screening techniques and evaluation of resistant genotypes, notably including Ethiopian sorghum. Resistance in this genotype was claimed in the 1978 Workshop at Hyderabad.
7. Determine role of sclerotia in epidemiology of ergot disease.
8. Assess losses to ergot in sorghum.

Pearl Millet Ergot

1. Determine the importance, to farmers, of pearl millet ergot with respect to grain yield losses and to toxicity of the sclerotia to humans, livestock, and poultry.
2. Fully test (as alternatives to resource-consuming ergot-resistant breeding programs for pearl millet) other possibilities, including use of existing ergot-resistant populations as finished varieties, pollen management (pollen-donor variety intersown with a conventional hybrid) at the farmers' level, and hybrids not based on sterile cytoplasm (e.g., topcross hybrids).
3. Study (where resources are available) the biology/epidemiology of pearl millet ergot and resistance mechanisms, including postpollination stigmatic constriction or similar aging phenomena.
4. Investigate doubled haploids as a means of fixing short protogeny (ergot resistance) in pearl millet lines.
5. Encourage institutions funded for basic research to study the mechanism of postpollination stigmatic constriction, which excludes pathogenic parasitization of the ovary via the stigma.

Bacterial Diseases

1. Document incidence and importance of bacterial diseases by publishing in appropriate journals, newsletters, and data bases.

2. Determine, where possible, sources of resistance from hybrids, cultivars, or landraces.
3. Determine effect of plant-pathogenic bacteria on stalk rot.
4. Develop techniques such as semiselective media or serological procedures for accurate identification of bacterial causal agents.
5. Determine role of seed transmission in dissemination of bacterial diseases.
6. Identify portals of entry of plant pathogenic bacteria into plants through study of hydathodes, stomates, and other plant structures. Study effects of wounds.
7. Publish and distribute illustrated handbooks on bacterial diseases to national scientists working with sorghum and pearl millet in the field.

Virus Diseases

1. Continue survey and identification programs for sorghum viruses and virus diseases.
2. Study crop losses to virus diseases.
3. Study epidemiology host/vector/virus interactions.
4. Standardize inoculation techniques and virus-severity rating scales.
5. Develop short- and long-term strategies for managing sorghum virus diseases.
6. Study genetics of resistance: its mechanisms and the role of inheritance.
7. Develop an index of 'resistance' as a criterion for selection.
8. Standardize and publish techniques for research workers in sorghum and millet virology:
   a. ELISA
   b. electron microscopy
   c. inoculation
   d. rating scales
   e. screening
9. Collaboration and training
   a. Provide training opportunities for plant virologists in developing nations.
   b. International cooperation in serology and electron microscopy for identification of sorghum and millet virus diseases.
   c. Expand training in molecular aspects of virus identification.
   d. Organize and conduct an international workshop on sorghum and millet viruses and virus diseases.

Nematology

1. Data on the role of nematodes in sorghum production in the developing nations are extremely limited. A major reason is the small number of scientists, in most of the developing nations, who can adequately assess nematode problems. National scientists with postgraduate training in plant nematology are urgently needed.
2. Identification and documentation of the pathogenicity and relative aggressiveness of nematodes associated with sorghum production is incomplete.
3. Distribution of pathogenic nematode species among sorghum-producing nations of the world is poorly understood. Frequency distributions and population densities are needed to assess the worldwide impact, relative to other pathogens and pests, of nematodes on sorghum.
4. Diversity in nematode population dynamics attributable to farming practices (traditional vs. recently adopted practices) is unknown or poorly understood. The influence of cropping sequences, mixed cropping vs. intercropping, cultural practices, and practices to improve moisture-holding capacity, modify soil structure, or influence biological activity in soils on nematode population densities requires investigation in many nations.
5. Literature on plant-parasitic nematodes and the diseases they cause is poorly distributed; in some cases it does not exist. Development, publication, and distribution of field guides covering the symptomatology of nematode diseases is needed.
6. Identification of resistance/tolerance sources in sorghum germplasm deserves high priority. The nematode tolerance/resistance characteristics of sorghum germplasm have not been determined. Resistance/tolerance in a sorghum genotype will be species specific. New techniques to screen germplasm effectively and efficiently may be required.

Seed Rots and Seedlings Diseases

1. There are sufficient instances of seedling diseases being a contributing factor in poor stand establishment in sorghum and occasionally in millet, but occurrence is sporadic—indicating a strong environmental interaction.
2. With this in mind, research is needed to:
   a. Determine the relative contribution of biotic and abiotic components in stand-establishment problems, specifically (i) host-related genetic factors (e.g., seedling vigor) and seed quality, (ii) role and identity of pathogens and saprophytes and their interactions, and (iii) environmental factors and frequency of deleterious or conducive environments and their interactions with host and pathogen development.
   b. Establish economic importance of (i) seed dressings, (ii) high-quality seeds, (iii) seedbed preparation, and (iv) development of resistance to indicated pathogens and/or tolerance to negative soil environments.

Grain Mold (GM)

1. Fungi have been implicated in grain deterioration, but how, when, and in which tissues these fungi cause damage is poorly understood.

   Thus there is need for experiments designed to:
   a. Determine the location of infection and colonization of specific fungi at different stages of host maturity.
   b. Determine the response of specific host-plant tissues to microbial attack and elucidate the respective roles of maternal and true seed tissues in microbial deterioration of grain. This approach should include, but not be limited to, the use of histochemical and histoserological techniques.
2. Procedures used in different parts of the world to evaluate grain deterioration should be compared and, if possible, standardized. Accurate assessment of host maturity at the time of evaluation is critical.
3. Select a set of cultivars to be grown, using a standard procedure, in different parts of the world for evaluation of GM damage. Current techniques for inducing microbial seed deterioration (e.g., *Fusarium moniliforme* inoculation at anthesis, sprinkling, etc.) should be used in these nurseries.
4. Continue studies of the heritability and number of genes involved in specific mechanisms

of resistance to grain deterioration (e.g., early infection, endosperm texture, phenol, etc.).

5. Design experiments on biochemical mechanisms of host-plant resistance. Utilize histological study techniques, including measurement of microbial presence and damage. Use chemical (ergosterol, ELISA, etc.) as well as histological techniques.

6. Continue searching for sources of resistance. Screen landraces, local varieties, and photoperiod-sensitive hybrids.

7. Study the epidemiology of microbial degradation of seeds.

8. Compare routinely the GM situation on sorghum with microbial deterioration of other grains of other crop species.

Sorghum Downy Mildew (SDM)

1. Variability of virulence in *Peronosclerospora sorghi* is a major concern. We recommend continuation of the current ICRISAT multilocational survey for virulence in *P. sorghi*, with expansion to include additional entries of resistant materials from participating countries.

2. Initiation and continuation of research to identify the type and amount of variability among the genetic factors in sorghum that confer resistance to SDM, as well as studies to determine the genetic relationships among resistant genotypes.

3. Host resistance to *P. sorghi* conferred by oligogenic or polygenic is preferable to monogenic resistance. We recommend that genetic combinations be found (or developed, if necessary) and utilized in sorghum-improvement programs.

4. Plant characters active in enabling sorghum to escape DM, but not directly involved in the host-pathogen interaction, may provide a means of controlling SDM without placing selection pressure for virulence on the pathogen population. We recommend research to identify these plant characters and determine their potential in SDM-control programs.

5. The inability to induce consistent germination of *P. sorghi* oospores restricts the use of this spore form in SDM-resistance screening, and hinders research on the genetics of virulence in *P. sorghi*. We recommend the development of effective oospore-germination techniques for use with *P. sorghi*.

6. Metalaxyl provides effective control of SDM, but the possibility that the pathogen will develop metalaxyl-resistant strains exists. We recommend the development of an alternative chemical control.

7. We recommend continued formal and informal cooperation and communication between organizations and individuals engaged in SDM research. Much of the considerable success to date is attributable to a spirit of working together.

Leaf Diseases

1. Effects of diseases on sorghum plants differ at different growth stages. This knowledge for each pathogen would be very valuable.

2. Development of a standard international rating scale, based on leaf area infected, and sufficiently flexible and sensitive for use in detection of resistance and in disease progress studies is vital. Growth stages at which plants are to be rated must be standardized.

3. We recommend that studies be initiated on techniques for estimating the effect of prevalent diseases on grain and fodder weights in given localities.

4. In view of the importance in evaluating sorghums for resistance, development of effective, practical, and reliable inoculation techniques for this purpose is essential.

5. Continuation and expansion of international cooperation in multilocational evaluation and characterization of germplasm response to foliar pathogens, along with exchange of elite germplasm, is essential. We recommend that entries should include international and local susceptible controls.

6. Efforts to identify, collect, and preserve landraces and other sorghum germplasm should be intensified, as should be the search for resistant sources.

7. Development of foliar disease resistance, i.e.,:

   a. effective across variable environments;
   b. durable through reduced selection pressure for physiological variants within a pathogen population; and
   c. active against several foliar pathogens, especially during vulnerable growth stages of the host.

8. Use of field and controlled environment screening and other research techniques to obtain information on:
   a. host/pathogen interaction,
   b. germplasm response to foliar pathogens, and
   c. pathogen variability.
9. Research to understand the biological mechanism influencing survival and function of pathogen propagules.
10. Determination of influences of naturally occurring and introduced epiphytic and other leaf microflora on host, pathogen, and host/pathogen interactions be studied.

Anthracnose

1. Investigate host/parasite/environment interactions.
2. Evaluate known sources of resistance worldwide and screen nontested world collection items at known 'hot spots.'
3. Monitor pathogen variability, using standard differential varieties and regionally important cultivars
4. Determine relationships of grain and foliar anthracnose to stalk rot.
5. Determine mechanisms of anthracnose resistance.
6. Evaluate chemicals (e.g., fungicides, plant-growth regulators, etc.), cultural practices, and biological controls with the aim of developing crop-management strategies for control.
7. Develop more effective and relevant screening techniques for identifying resistant materials.
8. Determine role of seedborne inoculum.
9. Identify and utilize improved and exotic sources of resistance by classical or innovative (biotechnological, genetic engineering, etc.) methods.
10. Determine the inheritance and heritability of resistance.
11. Elucidate the etiology and epidemiology of the disease.
12. Evaluate collected isolates from graminaceous hosts at containment facilities or in a temperate climate with less sorghum hectarage to determine the variability of the pathogen and host range.

13. Investigate taxonomy of *Colletotrichum graminicola* on different graminaceous hosts.
14. Establish an 'Anthracnose on the Gramineae' working group as part of the International Society of Plant Pathologists to promote cooperation in research activity with anthracnose on graminaceous hosts.

Sorghum Stalk Rot

Because of the need for reliable methodology to screen genotypes resistant to stalk rot and a multilocational cooperative program for the identification of sources of resistance to this major sorghum disease, the Committee recommends:

1. Methodology for stalk-rot screening:
   a. Use of nonsenescence and resistance to lodging as indicators of stalk rot resistance.
   b. Use uniform criteria to evaluate genotypes. Use a 1-to-5 scale for both parameters, with a plant free of observable symptoms receiving a rating of 1, and a severely diseased or dead plant receiving a rating of 5. The lodging scale should be based on percentage of lodged plants.
   c. Make stalk rot reaction comparisons only within entries of similar maturities.
   d. Take readings any time between physiological maturity and harvest if differential reactions permit comparisons among genotypes.
   e. Record flowering date and plant height for each entry.
   f. Because of the relationship between susceptibility to stalk rot and high yield, we recommend that yield or a yield rating be recorded, particularly in resistant entries. This may reveal genotypes that combine high yield and stalk rot resistance.
   g. In a stalk rot-resistance evaluation sowings, we recommend entries be seeded in 4-m rows. In advanced lines, use replications and multiple sowing dates and locations as extensively as appropriate. We also recommend including local susceptible cultivars, sown strategically through the screening nursery. For early generation evaluations where replications are not necessary, we suggest sowing at least one susceptible control for each five entries.
2. Multilocational Cooperative Program

a. We recommend that ICRISAT assemble and coordinate an International Stalk Rot Nursery (ISSRN).
b. Entries should number between 30 and 35, including lines or hybrids considered resistant to stalk rot.
c. Photoperiod-sensitive and nonsensitive reactions, when known, should be assigned to each particular entry.
d. Experimental design and methodology of evaluation should be consistent with the principles stated above.

Striga

A. Statement of the problem

1. The increasing severity of the problem with *Striga* spp and the lack of interest or involvement by crop-protection scientists and related researchers threatens losses in sorghum, millet, and maize production, bringing in turn economic and sociological disaster and famine in much of the semi-arid tropics, especially in developing nations. The unique growth and survival traits of the *Striga* spp and the present widespread tendency toward cereal monocropping indicate that the only hope for control of this extremely important weed species lies in integrated programs. A variety of complementary control strategies and techniques are available.

B. We recommend:

1. Identification of research efforts on *Striga* and its control, and expansion of laboratory and nursery breeding and disease programs. An example is screening techniques and their adaptation to local conditions.
2. Intensification of training programs for research and extension workers in *Striga*-endemic areas.
3. Expansion of farmer-education programs that show how simple strategies may be used to reduce losses to *Striga*. Sorghum varieties and hybrids highly susceptible to *Striga* should not be released.

4. Continued cooperation and collaboration of all groups concerned—NARSs, IARCs, and national governments—and intensification of efforts to enlist breeders and pathologists in *Striga*-control programs. Continued collaboration across national and discipline borders is imperative.
5. Assumption, by ICRISAT, of responsibility for basic and strategic *Striga* research programs.
6. Formation of a '*Striga* Research and Control Network' featuring an intercontinental action program strong in regional components and national commitments to research, training, and extension for control of *Striga* spp.

Disease Management in Sorghum and Pearl Millet

1. Determine importance and prevalence of diseases in the working area.
2. Determine that portion of the disease situation that may be amenable to change.
3. Determine type of control most likely to be effective—
   a. Use resistant varieties where available.
   b. Use chemical control where practical and environmentally sound.
   c. Control through cultural practices and cropping procedures.
   d. Integrate use of appropriate host/plant resistance with complementary cultural and cropping practices to (i) minimize losses to economically important diseases, (ii) Decrease potential for pathogen variability, and (iii) avoid practices that discriminate pathogens.
4. Quarantine! Keep out pathogens in the working area.
5. Disseminate pertinent information on disease control.

Pearl Millet Downy Mildew

1. Determine distribution and severity of pearl millet DM, especially in southern Africa.
2. Conduct studies that increase understanding of the biology, especially oospore biology, of *S. graminicola*, and the precise environmental factors that favor development of diseases and disease epidemics.

3. Determine the influence of host, pathogen, and environment on symptom expression, including recovery resistance.
4. Study the genetics of host resistance and pathogen virulence, and the dynamics of virulence in pathogen populations.
5. Quantify the relationship between DM incidence and crop yield loss.
6. Strengthen a multilocational network of reliable resistance screening, especially in Africa.
7. Develop reliable and efficient screening techniques that are easily integrated with breeding programs.
8. Increased interaction of breeders and pathologists in identifying resistance and its use in breeding.
9. Investigate control measures that complement genetic resistance.

Foliar Diseases of Pearl Millet

1. Develop reliable rust-resistance screening procedures; classify taxonomic identities of reported rust pathogens.
2. Develop techniques for accurately identifying viral, fungal, and bacterial causal agents on plants in the field. Publish and distribute bulletins, handbooks, or compendia listing foliar diseases and showing photographs of diseased plants in accurate color.
3. Document epidemics of foliar diseases involving loss of millet yields.
4. Ascertain the prevalence of false mildew of pearl millet in southern and eastern Africa.
5. Determine the cause of yellow blotch of millet, maize, and sorghums.
6. Determine the incidence and severity of bacterial leaf stripe and bacterial leaf streak of millet in western Africa. The potential role of seed in dissemination of the bacterial causal agents of these diseases needs clarification.

Utilization of Disease Resistance for Sorghum and Millets

1. Collaboration of breeders and pathologists in developing screening techniques and identifying sources of resistance (especially stable types), particularly for developing disease-resistant cultivars.

2. Determine the genetic basis for resistance. Identify sources of resistance continuously so that the number of different genes for resistance may be increased.
3. Where there is danger of breakdown, avoid or use vertical resistance only with extreme caution in breeding programs.
4. Use a level of disease resistance consistent with achieving adequate control; using a level higher than necessary is resource-consuming and may exert unnecessary selection pressure on the pathogen.

Priority Diseases

Committees of all delegates attempted to rank the importance of sorghum and millet diseases by geographic area: (i) Africa, (ii) Asia, and (iii) Europe and the the Americas.

Group 1:

Priority diseases in sorghum and infrastructure for Research in African Countries

Chairperson: L.K. Mughogho
Rapporteurs: M.D. Thomas
            Y. Kebede

After several minutes of discussion on appropriate criteria that would identify the priority diseases, seven were selected and ranked in the order of importance. Seventeen diseases were rated, with 1 designating low priority; 2, intermediate priority; and 3, high priority. A fourth (non-numerical) category, indicated by the letter A, signified a lack of knowledge for a particular criterium.

The criteria were assigned numerical ranking; the least value signifying lesser importance and the highest value signifying greater importance. The 17 diseases were as follows:
1. Grain mold
2. Downy mildew
3. Charcoal rot
4. Leaf blight
5. Rust
6. Gray leaf spot
7. Anthracnose
8. Sooty stripe
9. Zonate leaf spot

10. Bacterial stripe
11. Ergot
12. Head smut
13. Long smut
14. Covered smut
15. Loose smut
16. Nematodes
17. *Striga* spp

The criteria and their numerical ranking from the least to the most important follow:

Criteria	Rating
Health hazard	1
Farming system	2
Method of spread/overseason-ing	3
Pathogen variability	4
Lack of control methods	5
Prevalence/distribution	6
Effect of yield/quality	7

Each member of the discussion group was asked to develop a table for his or her country, with the 17 diseases on a vertical column and the seven criteria on the top of the table, and match each of the 17 diseases with each of the seven criteria, using the 1-to-3 scale.

These tables were analyzed by multiplying each disease rating by the numerical rating (multiplication factor) of the corresponding criterium to which it was matched. In that way a set of values indicating the importance of a disease in a given region was obtained.

It was thought that this procedure was a more objective approach to determination of the research priorities of the 17 diseases listed.

The results obtained by grouping individual tables into regions for the six important diseases are given below. The diseases are listed below in order of priority based on the average numerical value obtained.

A. Western Africa:
1. *Striga* spp
2. Grain mold
3. Long smut
4. Anthracnose
5. Gray leaf spot
6. Sooty stripe

B. Eastern Africa:
1. *Striga* spp
2. Anthracnose
3. Covered smut
4. Leaf blight
5. Loose smut
6. Grain mold

C. Southern Africa:
1. Grain mold
2. Leaf blight
3. Anthracnose
4. Covered smut
5. Ergot
6. Downy mildew

Group 2:

Priority Diseases in Sorghum and Infrastructure for Research in the Americas and Europe

Chairperson: R. R. Duncan
Rapporteurs: L. E. Claflin
G. Forbes

Attendees: Craig
Frederiksen
McGee
Narro
Odvody
Rosenow
Toler

Foliar diseases:
1. Anthracnose
2. Gray leaf spot (USA), ladder spot (USA, Brazil), leaf blight (USA, Mexico), bacterial stripe (Argentina), bacterial streak (Mexico and Kansas, and Texas in USA), bacterial spot (high altitudes, Canada)
3. Rust, sooty stripe, target spot, rough spot, sheath blights (pink, banded)

Root and stalk rots:
1. *Fusarium* spp. (*F. graminearum* more important than *F. moniliforme*), charcoal rot
2. Anthracnose red rot
3. Bacterial fungus (*Pythium*)
4. Acremonium wilt
5. Periconia

Panicle diseases:
1. Anthracnose (more damaging); head smuts (more difficult)
2. Head blight

Virus diseases:
1. MDMV-A (more prevalent)
   MDMV-B (less prevalent)
2. Sorghum yellow banding

Comment: Maize chlorotic dwarf virus and brome mosaic virus (BMV) have no effect on the plant. BMV is a potential threat should it reach epidemic proportions (from the *Compendium*).

Systemic diseases:
Downy mildew, crazy top

Mycoplasma diseases:
Yellow sorghum stunt

Overall priority ratings:
1. Anthracnose
2. Grain mold complex
3. Root and stalk rots
4. Head smut
5. Viruses
6. Downy mildew
7. Head blight
8. Foliar diseases

Group 3:

Priority diseases in sorghum and infrastructure for research in Asia and the Far East

Chairperson: J. S. Kanwar
Rapporteurs: K. M. Anahosur
               N. G. Tangonan

Attendees: T. Boon-Long
               S.C. Dalmacio
               T.B. Garud
               T. Kimigafukuru
               P.G. Mantle
               S. Pande
               R.V. Vidyabhushanan

I. Priority diseases in grain sorghum

A. India:
   1. Postrainy (*Rabi*)
      a. Stalk rots
      b. Viruses (Somo/Mpomo)
      c. Rust

   2. Rainy (*Kharif*)
      a. Grain molds
      b. Ergot
      c. Stalk rots
      d. Leaf spots
      e. Downy mildew

   3. Overall
      a. Grain molds
      b. Stalk rots
      c. Ergot
      d. Downy mildew

B. Thailand
   1. Grain molds
   2. Ergot
   3. Leaf diseases (gray leaf spot and anthracnose)

C. Philippines
   1. Foliar diseases (tar spot, gray, target, banded leaf and sheath blight)
   2. Grain molds

II. Priority diseases in fodder sorghum

A. India
   1. Leaf spots
   2. Downy mildew
   3. Ergot

B. Japan
   1. Banded leaf and sheath blight
   2. Leaf spots (target or zonate)

III. Priority infrastructure

A. Collaboration between scientists within regions and between regions
B. Training
C. Funding for publication in widely circulated and commonly abstracted journals

Appendix

Discussion

Guiragossian: Is there a price difference between white and red seed sorghums?

House: In the 1985–86 season, the price of red- and white-grained sorghum was U.S. $180 t⁻¹, equal to that of maize. Because of the surplus the price of red-grained sorghum was dropped to U.S. $100 t⁻¹, while the price of white-grained sorghum and maize remained at U.S. $180 t⁻¹. These are purchase prices by the Grain Marketing Board; the selling price is U.S. $220 t⁻¹ for red- and white-grained sorghum and for maize.

Balasubramanian: Is the bacterial pathogen (*Pseudomonas andropogonis*), for which South Africa has imposed a quarantine, seedborne?

de Milliano: The pathogens of bacterial leaf streak (*Xanthomonas campestris* pv *holcicola*) and bacterial leaf stripe (*Pseudomonas andropogonis*) have both been recorded as being seedborne. (See also Claflin et al. in this Publication.)

Vidyabhushanam: It was mentioned that seed rots and seedling blight may reduce stands by as much as 50%. What measures have been evolved to overcome this problem?

de Milliano: This information was in relation to downy mildew sick plot conditions, and does not necessarily apply to farmers' fields.

Mushonga: The people in Zimbabwe prefer the white-grained sorghums, if they can purchase them. The reason why red sorghums are grown by commercial farmers is that they need them for brewing opaque beer.

Qhobela: What specific virus diseases are prevalent in the region?

de Milliano: I do not know. We believe that we have a strain of maize dwarf mosaic virus.

Vidyabhushanam: Dr Hulluka, your generalized statement that diseases are more serious on high-yielding cultivars than on local landraces needs substantiation with specific examples!

Hulluka: We have experienced this situation in many instances. Improved lines are more prone to attack by anthracnose, leaf blight, and ergot in Ethiopia, specially when they are used in large-scale farming.

House: Not all landraces have resistance and not all introductions are susceptible. While it is true that some introductions have greater susceptibility, it is also true that some of the most useful sources of resistance are from introductions.

Guiragossian: Smuts (covered and loose) seem to be a problem in eastern Africa. We all know there are chemicals that can be used to overcome smuts, yet farmers do not use them. Therefore, I recommend that we identify genotypes with genetic resistance as an alternative method for kernel smut control.

Hulluka: I have no information regarding breeding for smut resistance. Since these diseases can be effectively controlled by seed-dressing chemicals, screening for resistance to these diseases may not be justified.

King: Covered smut and loose smut are very effectively and economically controlled by seed treatment. I would hope that this discussion would not favor breeding for resistance to these diseases; breeding for resistance is a far greater undertaking for a resource-poor country than extending purchase of seed-treatment fungicides to farmers.

Esele: Covered and loose smut are the most important smuts in the eastern African region. Since sorghum is a subsistence-level crop, seed treatment is not economical. Utilization of host resistance is the best control, because farmers keep their own seed for sowing.

Teri: (With regard to Dr King's comment.) On the usefulness of seed treatment. This is alright where farmers buy seed. What about situations where farmers grow their own seeds? There we may need to encourage farmers to buy seeds or develop technologies that can enable farmers to treat their own seeds.

Ahmed Sheikh: In Somalia, at the Bonka Research Station, we found that Fernasan D® is very effective against covered kernel smut. We plan to give out or sell 10-g sachets to farmers. We find that the disease is very low in the research area and in farmers' fields where they use Fernasan D®.

Balasubramanian and Duncan: Could you categorize the different isolates of *Colletotrichum graminicola* into specific groups based on their reaction to differential sorghum lines?

Hulluka: There is not much information on categorizing pathogens of *C. graminicola* in eastern Africa. Based on studies conducted in Ethiopia, we have identified three pathotypes differing in their reaction on a set of differential varieties obtained from Texas.

Anahosur: Is *Macrophomina phaseolina* the only pathogen associated with charcoal rot?

Hulluka: In Ethiopia, *Macrophomina phaseolina* causes charcoal rot.

Obilana: Specifically, damage from *Striga* is one pest problem that is accentuated by changes in genotypes and farming systems. In Sudan and India, the introduction of improved genotypes has been followed· by an increase in *Striga* incidence and damage.

Njuguna: Are there landraces that you may have noticed in farmers' fields that are resistant to long smut?

Thomas: There may be some landraces that are resistant to long smut, as the incidence is low in the farmers' fields.

Theuri: In your talk on diseases in western Africa, you did not mention virus or virus-like diseases on sorghum. Does this indicate that these diseases are not present, or that sorghums are not grown in the highland areas where virus diseases occur?

Thomas: I have not observed virus or virus-like diseases on sorghum in western Africa, although a virus disease has been reported from Nigeria.

Vidyabhushanam: Many of the improved lines and breeding lines being developed in India are rated as resistant to foliar disease; some of these should be screened in your region to identify sources of resistance to these diseases.

Thomas: Certainly the lines from India can be screened in western Africa for leaf diseases. In fact, we have noticed that some of these lines show susceptible reaction to gray leaf spot, leaf anthracnose, and sooty stripe in western Africa.

Vidyabhushanam: Framida is mentioned as *Striga*-resistant in the literature.

Thomas: The situation with *Striga* and Framida is not clear. Framida is tolerant to *Striga*, but our evidence is incomplete. That is, yield potential remains the same in spite of high levels of *Striga*. In some locations, for example in northern Togo, Framida supports a high population of *Striga* but Framida is low compared with E 35-1 which is clearly susceptible.

Mughogho: So far you have given us an overview of sorghum diseases as seen by ICRISAT scientists in the region. It would be useful to know what research is conducted by national programs, say, in Cameroon, Nigeria, and Burkina Faso where we know that national sorghum programs exist.

Thomas: Except for perhaps Nigeria, I believe most national programs emphasize screening available materials for resistance to the prevalent diseases in their location.

Dangi: Breeding work on grain-mold resistance and foliar diseases is in progress in the national sorghum-improvement program of Cameroon. Lines like CS 95, S 35, CS 54, and CS 61 were identified as resistant to long smut, *Striga*, and leaf diseases in Cameroon. These cultivars are now in extension programs for on-farm testing and adoption.

Adipala: What *Striga* species did you have in western Africa? Did you use any other cover crop apart from groundnuts?

Thomas: 1. *Striga hermonthica*; and 2. Only groundnut was tested.

Obilana: Primarily you are right in saying that intercropping reduces *Striga* in sorghum and groundnuts. Secondly, what is observed is not the result of cooler soil temperature, but the effect of groundnut as a trap crop, causing suicidal germination of the *Striga* seed, and thus reducing the number of *Striga* plants emerging. Please cooperate closely with the *Striga* workers (ICRISAT and national scientists) active in the western and southern African regions and the ICRISAT Center.

Guiragossian: Some diseases—such as gray leaf spot, rust, or leaf blight that appear after flowering—might not affect yield significantly, yet are important because they reduce acceptability by farmers. Therefore one should conduct a survey with farmers to find out whether such diseases affect farmers' decisions to accept or reject a variety or hybrid.

Thomas: This is true. We do intend to continue screening for gray leaf spot in hot-spot locations. Our final materials must have acceptable levels of resistance to all leaf diseases. However, a disease like gray leaf spot, which we suspect causes little damage, will have the lesser emphasis in our program.

Balasubramanian:

1. Have you developed an inoculation technique to screen sorghum lines against *Rhizoctonia solani*?
2. Does the cultivation of rice affect the incidence of *Rhizoctonia* blight of sorghum?

Kimigafukuro:

1. We inoculate with 10 g of *R. solani* cultivated on barley grain at the base of each sorghum seedling and then cover with vinyl film to keep the plant moist.
2. The increase of *Rhizoctonia* disease of sorghum has a major effect on the cultivation of rice, because the causal organisms of banded leaf and sheath blight on rice and on sorghum both belong to anastomosis group 1 of *R. solani*.

Anahosur: Elaborate on the inoculation technique and maintenance of inoculum of *Phyllachora sacchari* in sorghum.

Dalmacio: *P. sacchari* is maintained by screening sorghum continuously. Tar spot development is promoted through creation of a humid environment. Infected sorghum leaves serve as inoculum and inoculation is done either by spreading infected leaves along the rows or by attaching infected leaf sections to leaves of test material using a paper clip.

Adipala: Have you carried out cross-inoculation trials with the *Rhizoctonia solani* isolates from rice and sorghum?

Dalmacio: No, but we believe that *R. solani* isolates infecting rice also infect sorghum.

Theuri: What form of rotation is used on sorghum with respect to *Rhizoctonia*? Do the *Rhizoctonia* isolates from rice attack soybeans?

Dalmacio: Sorghum is used in rotation with corn and in some cases with rice. Where used as such, sorghum sowing coincides with dry environment and *Rhizoctonia* does not develop to a level that it damages sorghum.

Balasubramanian:

1. Have you developed an inoculation technique to screen sorghum lines against *Rhizoctonia solani*?
2. Does the cultivation of rice have any influence on the incidence of *Rhizoctonia* blight of sorghum?

Dalmacio:

1. Yes, we have developed the leaf-sheath inoculation technique; it consists of growing the fungus on sterilized sorghum grains and placing 5 to 10 infested sorghum grains at the axil of the leaf blade two internodes above the ground.
2. No, because when sorghum is rotated with rice, sorghum sowing coincides with the dry season, an unfavorable time for development of *Rhizoctonia solani*.

House: Why has the production of sorghum declined so much in the last few years?

Dalmacio: Mainly due to lack of market. Seed millers are somewhat reluctant to utilize sorghum as a feed ingredient.

Guiragossian: Since we have several new diseases that are not included in the Sorghum and Pearl Millet Disease Handbook, maybe it is time to update the handbook?

Is tar spot on sorghum different from the tar spot on maize?

Dalmacio: I have not seen the *Phyllachora* isolate from sorghum attacking maize in the Philippines.

Giorda:

1. What are the parameters that you use to determine the relative importance of a disease? (You are considering sugarcane mosaic a minor disease, when it causes an estimated yield loss of about 50%?)
2. What strains of SCMV do you have in your area?.

Dalmacio:

1. Parameters used were incidence and severity of disease under natural conditions. The 50% yield loss from sugarcane mosaic virus was based on artificial inoculation, not on natural infection.
2. We do not know the strains of SCMV occurring in sorghum in the Philippines.

Mughogho: It might be useful if you would inform us about your extensive research program conducted at various centers in India.

Anahosur: In India the different centers, situated in different agroclimatic zones, have been working on diseases of regional importance primarily and major diseases in general. These are detailed in Tables 2 and 3 of my paper.

Vidyabhushanam: In view of the increased incidence of virus diseases, some strategy needs to be developed to contain them.

Anahosur: Some work is being carried out at Parbhani by Dr Garud. This program will be strengthened in the future.

Theuri: You have shown excellent control of downy mildew in India by the use of fungicides. Has somebody looked at the economic returns realized from sorghum harvests treated with these fungicides? Would you recommend the use of these fungicides to farmers, especially because many of the farms are small-scale?

Anahosur: We primarily look for resistant genotypes, and farmers prefer these. Fungicides are used to treat susceptible lines used in experiments, to maintain experimental material, and for seed production.

Dalmacio: Is the other ingredient combined with Ridomil® also effective against the downy mildew fungus? So as to minimize the occurrence of Ridomil®-resistant strain?

Anahosur: The other ingredient with metalaxyl (mancozeb or ziram) may help minimize the development of a metalaxyl-resistant strain as well as reduce damage caused by other pathogens.

Laing:

1. You have explained that you are doing research with metalaxyl against sorghum downy mildew; are you worried about metalaxyl-resistant races?
2. Have you tested etridiazole, propamocarb, or hymexazol in this role?

Anahosur:

1. We are worried about metalaxyl-resistant races but we are using Ridomil ZM®, a metalaxyl/mancozeb mixture, and this should slow down the development of resistant races.
2. No.

Laing: Is there any evidence of the monogenic resistance to sorghum rust being differential/vertical resistance and therefore vulnerable to new virulent pathogenic races?

Anahosur: Yes, the resistance is monogenic or digenic, but we are looking for polygenic resistance. It also depends on the genetic background into which it is bred.

Frederiksen: Have you seen ladder spot of sorghum in Brazil?

Casela: Yes, I have.

Garud: Is sugarcane acting as a source of inoculum for (SCMV) in sorghum? Which was the important strain of SCMV occurring on sorghum?

Wall: Serology tests were performed only on infected sorghum plants. MDMV strains A and B predominated.

Frederiksen: Do the moderately resistant sorghums in your conidial-inoculation trials for reaction to sorghum downy mildew have similar levels of resistance in the field?

Teyssandier: We have not demonstrated the relationship between laboratory and field infection. We believe that genotypes that consistently show low sporulation rates under artificial inoculation will probably have less inoculum pressure in the field, preventing or delaying an epidemic.

Mughogho: In your evaluation of sorghum lines for resistance to the new virulent pathotype of sorghum downy mildew, you showed QL 3 to be resistant and susceptible. What does this mean?

Teyssandier: That means that occasionally, even using QL 3 India, we obtained about 4 plants out of 25 inoculated with SDM. It could indicate that the line is not homozygous for this character. We need to analyze the line for genetic purity, as suggested by Dr Craig.

Thomas: It would perhaps be useful if you give a brief description of the screening techniques you use for grain mold and sprouting.

Teyssandier: We use a field-lab technique where sorghum entries are inoculated at anthesis with a *Fusarium moniliforme* suspension. Panicles are harvested at maturity and 400 seeds per entry are germinated in a blotter for 1 week. We classified the entries as tolerant or susceptible, using a 1 to 5 scale according to the percentage of molded grains. A score of 1 means no fungal colonization, and a score of 5 means more than 75% of the grains have fungal colonies.

To screen for tolerance to sprouting, we expose mature panicles from different genotypes to stable environmental conditions with very high relative humidity and temperature between 25 and 28°C. The speed of symptoms development is used as a selection criterion. Entries sprouting in 3–4 days are considered susceptible, the ones that don't sprout in at least 10 days are considered tolerant.

Qhobela: What was the bacterial pathogen that you inoculated into sorghum stalks?

Teyssandier: We have not identified the agent.

Njuguna: You have mentioned that *Acremonium strictum* is an important pathogen in Central America. Does this organism infect maize and, does it not cause an important problem in maize? Are there sorghum cultivars resistant to *Acremonium strictum*?

Wall: (in response to the query regarding resistant sorghum cvs) Yes!

Mughogho: (commenting on cross-inoculation) I have not come across any publication on cross-inoculation studies of *Acremonium strictum* infection on sorghum and maize. However, taxonomically it appears that the pathogen of maize is similar to that of sorghum. There is this obvious need for cross-inoculation studies.

Laing: The breeding approach adopted in Mexico appears to be perfect for the development of vertical resistance and therefore "boom and bust" cycles. I question the utility of this in view of the pathogens faced and the needs of subsistence farmers for stable resistance. I suggest a more valuable approach is to seek horizontal resistance (as was done in the FAO Programs) such as those successfully conducted by Martin Beek and Walter de Milliano on wheat in Brazil and Zambia.

Laing: Please confirm that the Mexican breeding program for downy mildew resistance is to seek a resistance source and then backcross that source of resistance into agronomically desirable cultivars.

Narro: The breeding strategy adopted is to cross Texas resistant materials into locally adapted material, and then test the backcrossed material in a Mexican "hot spot".

Vidyabhushanam: What are the disease-resistant factors identified in A2 cytoplasm mentioned in your presentation?

Narro: We have just started using A2 cytoplasm as well as A1 cytoplasm for hybrid production. We have no information, at this stage, about disease resistance using A2 cytoplasm.

Rosenow: With regard to the question by Vidyabhushanam on what specific disease resistance does Narro obtain from A2 cytoplasm in Mexico, and how is A2 related to A1.

To my knowledge, there are no specific resistances to diseases that are directly related to the A2 cytoplasm, but are merely related to the germplasm which we have been able to sterilize in A2. The A2 is considerably different from A1, and much different in fertility restoration from A1.

Laing: How much of a problem are mycotoxins in sorghum?

Frederiksen: Not enough work has been done, but we do know that the sorghum grains are colonized by many fungi which compete with the mycotoxin-producing fungi. The situation with regards to *Phoma* sp in Nigeria is not clear. Teyssandier points out that T2 toxins are produced by *Fusarium* spp; these toxins can be dangerous in feeds for swine and chicks.

McGee: Following up on Frederiksen's response as to why aflatoxin is not a problem in sorghum:

Aflatoxin is a severe problem in maize, peanuts, and cotton. All can be extensively invaded by field fungi. It seems to me that the answer to this question is more complex, because physiological and ecological factors are involved.

Zummo: Part of the reason that aflatoxin is not a problem in sorghum is because aflatoxin is a by-product of the substrate that the fungus grows on. Sorghum is apparently not as good a substrate for aflatoxin production as maize, cotton seed, and groundnuts.

Frederiksen: Mycotoxins have not generally been a serious problem in sorghum. We do not understand why.

Theuri: With reference to the use of tolerant varieties for control of MDM, is it not risky to use these varieties since there is a buildup of virus in the tolerant variety? Why are resistant varieties not used?

Frederiksen: Tolerance permits the survival of the host and pathogen without producing a screening tool for the selection of another virus strain that may damage the host.

Mbaye: You have shown one table in which you put (-) behind foliar diseases in France. Does it mean that there is no foliar disease in France?

Forbes: No. The (-) means that I can present no evidence for foliar diseases, either as a current or potential problem. I am suggesting that foliar diseases pose no potential threat because environmental conditions are not conducive to foliar diseases. This situation is similar to that found in the northern sorghum-growing regions in USA.

Mughogho: Maize is affected by downy mildew in Somalia. Do you know if downy mildew also occurs on sorghum, or if the downy mildew on maize also infects sorghum?

Hassan: Downy mildew occurs on sorghum, but we do not know if it is the same as the one infecting maize in the irrigated areas of Somalia.

Anahosur: You mentioned head smut control with a fungicide. What method of treatment and fungicide have you used?

Hassan: I was referring to covered kernel smut, and the chemical is Fernasan D®.

Anahosur: Are there any fungicides that can be used to control head smut on sorghum?

Njuguna: There is experimental evidence to suggest that Bayleton®, manufactured by Bayer, is effective when used as a seed dressing for head smut of maize.

McGee: Carboxin® controls head smut in maize.

Theuri (to King): You illustrated the concept of "Recovery Resistance" with downy mildew, a disease which requires moisture for its development. Do you exclude the possibility that reduced moisture may have terminated the development of this disease? What would've happened if the same plants were grown in a greenhouse under controlled conditions?

King: The recovery phenomenon is not dependent on low moisture for expression. It occurs in humid greenhouses and humid field conditions.

Vidyabhushanam: What is the role of alternative hosts in initiating an ergot infection?

King: We do not know the importance of alternative hosts in the epidemiology of pearl millet ergot. Ergot can develop to a high degree in the absence of infected alternative hosts. Presumably the inoculum is ascospores from germinating sclerotia in the soil.

Laing: In your data for several differential interactions between cultivars and sites of testing; do you see this as a vertical resistance (VR) result, arising from a cultivar × pathotype interaction; or a cultivar × environment interaction?

de Milliano: I only suggest that there may be VR; however, a cultivar × environment interaction may also be involved.

Laing: Are you actively seeking to expose vertical resistance?

de Milliano: Yes, we would like to screen at one location for the entire region. With VR, this is not possible when different pathotypes are present in the region.

Mbaye: Are there morphological differences in stigmas of ergot-resistant and ergot-susceptible cultivars? Are there differences in speed of germination of conidia and pollen grain?

Mantle: There may well be variations in the size of the exerted stigmas, and small stigmas are likely to favor escape from ergot. The constriction response to pollination seems to be a constriction feature in pearl millet. Ergot conidia germinate much more slowly than do pollen grains. If both are applied at the same time, pollen grains achieve fertilization before the ergot conidia germinate.

Vidyabhushanam: Would you explain the effects of the resistance mechanism through stylar constriction after fertilization or the ability of the fungal hyphae to cause disease.

Mantle: In pearl millet, invasion of the ovary by a pollen tube or a pathogen hyphae via the stylodia is presented by stigmata constriction in response to initial pollination. At the first sign of constriction, fertilization is probably complete. If fungal hyphae were able to enter the ovary at this stage or probably for a few days thereafter there is no reason to believe, by analogy with other ergot fungus/host examples, that disease could not be fully and effectively established. This constriction of stigmas in pearl millet is an exceptionally effective barrier to infection when initiated by pollination and provides a unique mechanism. You have probably confused fertilization as a genetic event, and one which does not alter the structure of the ovule or ovary as far as a potential pathogen is concerned. The differentiation and structural changes which proceed subsequently, over several days, provide physical barriers through which a pathogen invading via the stigmas would be unable to penetrate.

McGee: Do you have electromicrographs to show the infection tubes are blocked by the constriction in the style?

Mantle: Yes, the tissue in the collapsed region has its special integrity and the transmission tracts cease to be prominent. The stigmas distal to the constriction desiccate quickly on account of a failure to maintain turgor through the collapsed stylodal region.

Njuguna: From your talk, it is evident that ergot millet does not constitute a serious health hazard in humans; is this true also about ergot in rye?

Mantle: No, ergot of rye (*Claviceps purpurea*) often contains alkaloids which have potent pharmacological properties additional to those expressed by the millet ergot fungal alkaloids. Some of the *C. purpurea* alkaloids have an established place in modern medicine and are used extensively in obstetrics and in the prophylaxis and treatment of migraine. Excessive dose in the human is certainly hazardous.

Guiragossian: Farmers and children in Rwanda consume honey dew (ergot) at early stages and they claim that people who consume a lot of honey dew get a bad smell in their mouth originating from their stomach. Would you know what causes the bad smell?

Mantle: Millet ergot honey dew has a distinctive aroma which is certainly attractive to flies and other insects in Africa. These volatiles may be more readily released in the stomach. They are typical *C. fusiformis* alkaloids, but probably not enough in quantity to cause gastric disturbances or other malaise.

Laing: Do the alkaloids in ergot of sorghum and millet act as human contraceptives?

Mantle: There is evidence that the alkaloids of the ergot of millet are active with respect to human conception or pregnancy. From first principles, it may be deducted that they would not be particularly active. It is possible, that ergot of sorghum in Africa is no significant hazard to human health nor do the alkaloids have useful biological properties.

Adipala: Does *Xanthomonas campestris* on pearl millet attack other millets (like finger millet) in India? Do you have intercropping of pearl millet and finger millet, and if so, what is the effect on the incidence of *Xanthomonas campestris*?

Balasubramanian: We do not have *X. campestris* on pearl millet in India, nor do we intercrop pearl millet and finger millet.

Qhobela: The *Xanthomonas campestris* pathovar that infects pearl millet, will also infect proso millet but it is not a pathogen on finger millet or Japanese millet.

Balasubramanian (to Claflin): In the photograph of gels you showed, are the bands isozymes or proteins?

Claflin: The slide was a protein profile of *Pseudomonas avenae, P. andropogonis,* and the pearl millet isolate (*P. avenae*) determined by a SDS-page gel.

Mughogho: During your survey in Nigeria, were you able to identify the bacterium that causes the yellow leaf blotch of sorghum first reported by N. Zummo?

Claflin: Two predominant bacterial-colony types were reported by Zummo as the cause of yellow-leaf blotch. We identified the yellow colony type as *Erwinia herbicola*, a common saprophyte on plant tissue. The white colony type was identified as a pseudomonad, as reported by Zummo, but our culture was lost while attempting to identify the species.

King: On the survey you mentioned in Nigeria, in which almost all pearl millet plants had bacterial symptoms, would you be able to assess the possible damage to yields?

Claflin: At the time of the survey, the pearl millet plants were at the flowering stage of development and it was unknown if the bacterial lesions continued to progress. To say that this bacterial stripe affected yields would be only conjecture on my part.

Teyssandier: Could you tell me if it is possible to distinguish symptoms produced by *Pseudomonas andropogonis* from those produced by *Pseudomonas alboprecipitans* on sorghum?

Claflin: I am unable to detect any difference in symptoms produced by *P. andropogonis* and *P. avenae* (Syn. *P. alboprecipitans*). They are easily distinguished by physiological, biological, and serological tests.

de Milliano: How important are bacertia on sorghum in Lesotho?

Claflin: A high incidence of bacterial streak was observed in Lesotho, particularly in forage sorghums. It is unknown if bacterial diseases are involved in sorghum yield reductions in Lesotho. Only *Xanthomonas campestris* pv *holcicola* was isolated from sorghum tissue.

Balasubramanian: Will the airgun inoculation technique damage cells before the virus enters? Is it not an unnatural method to introduce the virus into sorghum plants in the field?

Toler: The airgun device, if used at the right pressure, will not kill cells on the leaf. The carborundum particles tear holes in the cell wall;

these are very similar to wounds caused by insects.

Balasubramanian: The green islands you have shown in your photograph, of maize dwarf mosaic, are they similar to the green island effects as a result of rust infection in sorghum?

Toler: The cells that are green in the mottled area do not contain virus particles. This is a cellular phenomenon not associated with host-plant resistance.

de Milliano: Do you know of symptomless carriers of maize dwarf mosaic virus?

Toler: No.

Njuguna: In terms of symptoms, is there difference in the expression of virus, rather than viroids, in plant tissue?

Toler: No! The symptoms are not necessarily different; the difference is the absence of a protein coat for a viroid.

Teyssandier: I would like to know if there is a vector involved in transmission of yellow-banding virus.

Toler: The vector at present is unknown.

Teyssandier: I would like to know if it is possible to establish the relative importance of a sorghum-virus disease using a comparison of incidence and distinction on yield loss.

Toler: Yes, losses and distribution can be used to evaluate virus diseases.

Vidyabhushanam: What is the role of seed transmission with specific reference to MDMV in sorghum?

Toler: Maize dwarf mosaic virus is not seedborne in sorghum, but is seedborne in maize.

Vidyabhushanam: Can you indicate the resistant genotypes of sorghum to MDM?

Toler: Tx 090, BTx 623, QL 3 India Sel., BTx 51.

Homan: Insects have a tendency to feed down the row, so your half-row techniqe for inoculation could develop a delayed transmission of the virus in the nontreated part of the row.

Toler: True, but with the half-row method, where half of the row is the control, an infected control plant shows that natural infection is taking palce.

Giorda: Have you considered the possibility of a mixed infection of SCMV-fg and *Acremonium* spp? Isolation from the plant infected with the virus is necessary to detect the presence of the fungus.

Garud: The repeated isolations from roots of such plant yielded no *Acremonium* and thus *Acremonium* infection is overruled. (Editor's note: *Acremonium strictum* can produce foliar systems and wilting even though not infecting roots.) The transmission with *R. maydis* and *L. saccharin* yielded virus transmission.

Guiragossian: Has work been done on the genetic inheritance of resistance to MDMV?

Giorda: Yes! Teakle et al. in Australia; Giorda et al. in Texas.

Wall: Is there a characteristic difference between the red stripe reaction you showed us and acremonium wilt symptoms (for field identification)?

Garud: Yes! In red stripe, mosaic will later develop characteristic interveinal symptoms not associated with acremonium wilt.

Mughogho: I visited field experiments of Dr Garud in January 1985 and did not see symptoms of Acremonium wilt in the trials at that time.

Laing (to Garud): Is the sorghum resistance you spoke of in relation to sugarcane mosaic virus alone, or to aphid resistance as well? Or to the mosaic virus and other things?

Garud: It is to virus only.

Cole: How does soil type affect the incidence of seedling diseases?

Odvody: Heavier (clay) soils retain deposited water for longer periods than do sandy soils and provide a more conducive environment for attack by pathogens.

Mushonga: Our experience in southern Africa is that more seedling problems are noticed in sandy soils than in heavier types, Does this differ with your experience?

Odvody: The stand-establishment problem is a complex syndrome, involving many interacting factors. There could be several factors, other than bacterial or fungal pathogens, causing stand loss in sandy soils. One could be nematodes.

Anahosur: In India, seed rot and seedling blight due to *Fusarium* spp and *Curvularia* spp, *Rhizoctonia* spp, *Aspergillus* spp, etc., have been found. Seed treatment with thiram has helped to reduce this malady. Only when seeds are not treated, this problem is noticed.

Odvody. I agree, except where seed has been significantly damaged by these fungi (excluding *Rhizoctonia*). The fungicide may control further fungal attack on these seeds, but the probable low vigor in seedlings could have increased vulnerability to other soilborne pathogens.

King: Some field and pot experiments using nematacides at ICRISAT's Sahelian Centre in Niger increased seedling establishment in groundnut and pearl millet. It was thought that this might to some degree have been due to reduction in populations of parasitic nematodes.

Mughogho: There is some information on seedling diseases of sorghum in other countries, such as India and Nigeria, in addition to the information provided by Odvody on the work done in Texas.

Odvody: My report does include work done in areas other than Texas. Many reports were not included because they simply corroborated existing information or they were discounted because they failed to adequately establish associations with seedling disease in the field. A few reports could have been missed if they appeared in journals that are hard to access.

Mughogho: You have talked only about soil-borne seedling diseases. What is the role of seedborne pathogens in seed rots and seedling blights/diseases?

Odvody: I have talked about seedborne pathogens. The grain mold fungi (*Fusarium moniliforme* and *Curvularia lunata*) are active as pathogens in seed. They cause direct damage, as when that seed is sown, can cause seed rots, blighting, and seedling disease. *Colletotrichum graminicola* can be transmitted by seed, but its association with seedling disease appears to be limited and it is not reported as causing seed rots. I did not mention it in my presentation, but Watanabe's report of seedborne *Glaeocercospora sorghi* causing subsequent seedling disease is the only well-documented case for this pathogen, although its seedborne nature is often reported. Many papers report seedborne associations of common sorghum pathogens and other fungi, but their subsequent roles in seed rots and seedling diseases are not documented. Sometimes their role in seedling disease is evaluated in sterile soil or through inoculation, but that does not prove an association with seed rots and seedling diseases in the field where soils have competitive microflora. I do not believe that listing these types or reports serves a constructive purpose as we seek to delineate the active seedborne and soilborne biotic components related to seed rots and seedling diseases.

Laing: Are there *Rhizoctonia solani* groups causing seedling establishment problems in sorghum?

Odvody: *R. solani* has been isolated from sorghum seedlings, but it is not a problem. The root-attacking *R. solani* isolates are of a different anastomosis group than those attacking foliage.

Wall: In Central America, stand establishment problems are mainly related to ants and birds, and to some extent grain molds, but mostly to the first two.

Starr: With regard to effects of nematicides on stand establishment, it should be noted that many nematicides have some fungicidal activity. This fungicidal activity is particularly evident at the higher rates of chemical applications used in experimental studies.

Starr: Although not normally classified under the category of seedling diseases, plant-parasitic nematodes can affect seedling vigor and stand establishment.

Odvody: Yes, they can be very important, as the additional stress factor may make seedlings more vulnerable to seedling diseases.

Frederiksen: Have you found literature on the biotic effects on stand establishment in the warmer tropical soils?

Odvody: No, not on stand establishment, but there are reports on fungal associations with roots.

Vidyabhushanam: Is there data to indicate that one leaf disease has a dominant effect over other(s)?

Odvody: I am not aware of an active mechanism, other than possibly competition for infection sites, since multiple diseases can occur on a single leaf. Currently, I have a graduate student evaluating the potential of inducing resistance in susceptible sorghum by initial inoculation with avirulent or low-virulence isolates of *C. graminicola*. After allowing time for a host biochemical response, the same and distal leaves are challenge-inoculated with a virulent isolate of *C. graminicola* to determine if there is increased resistance to the pathogen.

Axtell: Sorghum is used in the midsouthern United States in rotation with soybeans, as a control of soybean cyst nematode. Is it known if sorghum has a negative effect on soybean cyst nematode populations, or is it merely the absence of soybeans in the sorghum year that helps nematode control?

Starr: I am unaware of specifics with regard to direct affects of sorghum on *Heterodera glycines*. Sorghum is a nonhost for *H. glycines*, and therefore nematode populations would be expected to decline under sorghum cultivation.

Sithole: Has the effect of nematodes on sorghum yield been quantified? If not, are there plans to do so?

Starr: There has been little effort to quantify ef-

fects of nematodes on yield. This is a high-priority item; we only lack people who want to work on nematode problems in sorghum.

Sithole: Are nematodes really a problem in sorghum production?

Starr: We lack sufficient data to support general statement, but the data available suggest that nematodes do affect yield. We need data on frequency of fields infested with populations of pathogenic nematode species.

Sithole: How feasible is the idea of breeding for resistance where nematodes are the problem? Or, is breeding for tolerance a possibility here?

Starr: Tolerance (where there is high nematode reproduction without host damage) is possible and should be part of management systems if identified.

Resistance can be difficult to identify, especially for the less-specific ectoparasitic species. But based on progress made in identifying resistance to nematodes in other crop species, and in sorghum to other pathogens, I believe that development of cvs resistant to important nematodes should be attempted.

Page: In Zimbabwe, our survey showed that it is the farmers with the least land who have the most serious pest/disease problems, including nematodes. These farmers live in the most marginal areas of the country, where the soil is sandy, exhausted, and lacking in organic matter, and rainfall is low. For these farmers even the possibility of controlling nematode pests by rotation is not open, as they have insufficient land and are forced to continuously crop with cereals. The only solution for these farmers is land reform.

Thomas: Was population density of nematodes in Zimbabwe estimated, and if so, what kind of numbers are we talking about?

Starr: In Botswana, sorghum in fields with populations of *Tylenchorhyncus* spp of about 600 nematodes in 500 cm^3 soil and *Longidorus pisi* of about 100 nematodes in 500 cm^3 soil exhibited poor growth.

Page: In Zimbabwe, populations of *Pratylenchus*

zeae of about 1000 nematodes in 1000 cm³ soil were associated with poor sorghum growth.

Soil populations of plant-parasitic nematodes are considered to be at a level that is causing damage when they are more numerous than 1000 in 1 L of soil. Known nematode pests found in the roots, e.g., soil, *Pratylenchus zeae*, are always assumed to be causing some damage.

Adipala: Dr Page, on the Zimbabwe side, were nematodes actually isolated from the finger millet and was the pathogenicity of these isolates tested?

Page: Pot experiments have shown that *Meloidogyne javanica* can reproduce and cause stunting in finger millet.

Mushonga: Has there been an effort by nematologists to associate plant type and species of nematodes?

Page: We did not record whether the red or the white sorghum types were most affected by nematodes during our survey in Zimbabwe. We have not been able to test the pathogenicity of these nematodes that we found, during the survey on sorghum, due to staff shortages. We have been concentrating our research on maize, as it is the preferred crop in Zimbabwe.

Kebede: When we talk about *Striga* spread, is it a physical spread, or susceptible varieties being grown, or optimum condition being present for *Striga* germination? Resistance in any given variety may have to be qualified in view of the existence of biotype, ecotype, and strain.

Obilana: *Striga* spread can be both physical or cultural. Physical as to movement by soil erosion water and cultural by growing susceptible/tolerant cultivars that allow *Striga* to form seeds for next season. We are always specific when we say a cultivar of sorghum/millet is resistant, either to species or at a specific location or environment.

Dube: What conditions are necessary for the solarization control strategy to be fully effective against *Striga*?

Obilana: High temperature, high humidity, and a polyethylene sheet. The first two conditions

are easily attained in irrigated fields or in southern Sudan or in the Guinea Savanna of western Africa where rainfall reaches 1200 mm per annum.

Dube: You talk of exposing fields in the solarization strategy—just what time period are you implying — 2 weeks, 4 weeks, or 6 months?

Obilana: As I had indicated, the solarization method of *Striga* control is new and is being investigated. The exact period for polyethylene sheet cover is under investigation.

Thomas: In your discussion on survival mechanisms, you mentioned the ability of *Striga* to form biotypes. In your opinion, how strong is the evidence for the existence of biotypes? Do biotypes of *Striga* explain why at some locations Framida has very low levels of *Striga* and can be classified as tolerant or resistant, while in other locations Framida is heavily attacked by *Striga*?

Obilana: Evidence for existence of biotypes is strong; it abounds in the literature. It explains the differential reactions of sorghum cultivars, like Framida, in different locations-with different biotypes of *Striga*.

Vidyabhushanam: The crop loss estimated to *Striga* in India of 20% is not realistic. Since *Striga* incidence never follows a definite pattern, this kind of comment cannot be made. Please correct your paper.

Obilana: These estimates are from the recent literature on crop loss estimates for *S. asiatica* in India [Vasudeva Rao, Chidley, and House (1987)].

El Hilu Omer: Do antitranspirants have any advantage over contact herbicides? Do they need to be used ferquently for effective control?

Obilana: Yes. Antitranspirants are less toxic than contact herbicides, and they are less expensive.

Guiragossian: You referred to manure application as no input. Please explain!

Obilana: In this paper for your proposed control strategies, by input I mean activity or supply that has signigicant financial implications. Manure (cattle, sheep, goat) would not be considered as input under this situation if the animals referred to are housed on the farm and their manure is applied to the land by the farmer.

Anahosur: You mentioned about integrated control of *Striga* which means more than one control measures are brought together to control *Striga*, but you have not given prime importance to genotype. Genotype is the important component of integrated control.

Obilana: Genotypes (resistant ones) are implicated in variety substitution or crop rotation as used in our control-strategy proposals.

House: We are interested in getting sweet sorghum for evaluation as a fuel source. How can this be done?

Duncan: The Southern Regional Plant Introduction Station at Griffin (Experiment), Georgia, USA, 30223-1797, has the complete sweet sorghum collection. Contact Gill Lovell, sorghum curator, or R.R. Duncan if you desire this collection. Include all information plus the import permit required by your country.

Saadan: *Striga asiatica* is mainly found along the coastal areas as well as the central region of the country. *S. hermonthica* is found around Lake Victoria.

Guiragossian: You mentioned that local varieties are resistant to ergot. Could you confirm that?

Saadan: We have not confirmed that local varieties are resistant to ergot, however, the diseases are more prevalent on male-sterile lines.

King: Did I understand you correctly, that grain mold occurs on pearl millet in Tanzania?

Saadan: Yes, grain mold occurs in Tanzania in the central region when there is rainfall after the grain maturation. Rain is not common after grain maturation, but there are times when these conditions occur.

Andrews: Was honey dew also seen on genetic male steriles (GMS) in Tanzania as well as on cytoplasmic male steriles (CMS)?

Saadan: I have not noticed honey dew in random mating populations. Dr Mukuru has noted that it didn't matter whether it was GMS or CMS—both were susceptible to ergot.

Diourte: What are the strategies for controlling charcoal rot in Tanzania?

Saadan: We use resistant varieties. These resistant varieties come from the ICRISAT Programs.

Claflin: What is the causal agent of bacterial blight of finger millet in Uganda?

Adipala: *Xanthomonas coracanae* Desai et al. This pathogen is very similar to some menbers of the *X. campestris* group.

Adipala: Is the *Pyricularia* infection more severe than the *Helminthosporium* infection in finger millet? What are the causes of the seedling blight? Could it be *Pyrenophora*, *Pyricularia*, or *Sclerotium rolfsii*?

Kuwite: We have not determined which is the most severe of the two diseases. In this instance it seems that *Helminthosporium* is more likely to be seedborne than *Pyricularia* spp. The seedling blight was apparently caused by *Helminthosporium* spp and *Fusarium* spp.

Thomas: Would you rate oval leaf spot in terms of leaf area infected compared with sooty stripe and leaf blight? Also, you mentioned only *Fusarium moniliforme* when you talked about grain mold. Have you noticed other fungi-infected grains?

Chintu: Oval leaf spot, which can cause up to 25% of the tissue damaged is less important than sooty stripe and leaf blight with 10% and 15% tissue damage, respectively. Grain mold caused by *Curvularia lunata* has also been severe, but not as common as that caused by *Fusarium moniliforme*.

Thomas: I was not referring to actual data or figures on the leaf area infected by oval leaf spot. I wanted you to rank oval leaf spot in relation to say, sooty stripe and leaf blight. You mentioned that both sooty stripe and leaf blight are important, because they affect large areas of the leaf.

Chintu: Oval leaf spot, compared with sooty stripe and leaf blight, is less important because it occurs after the crop is mature.

King: Is the covered smut you mentioned caused by *Tolyposporium penicillariae*? Did you say that you can control it by a seed treatment? This would be very unusual.

Muza: Johnson, D.T. (1968) seems to indicate so. I refer to his "Bulrush millet (*Pennisetum typhoides*)—An important traditional grain." Bulletin 2509, Zimbabwe Agricultural Journal, Pages 19–23.

Thomas: In one of your tables you showed some lines that were resistant at Kamboinse but susceptible at Fada. Kamboinse and Fada are only 260 km apart. How do you explain this?

Paco: It is probably due to the difference in isolates in the two locations. We have also some lines which are resistant at Fada and susceptible at Kamboinse.

Andrews: In your last table, how much yield loss does 43% incidence mean?

Paco: Yield loss was not measured in this trial-but in another trial on farmers' fields, metalaxyl treatment increased yields by 20%.

Mtisi: Would you please explain what was used in treatment 5?

Paco: Metalaxyl is the active ingredient of Apron 35®.

Adipala: Have you examined *Xanthomonas campestris* isolates from finger millet? We have issued isolates from Uganda deposited with CMI which we think are members of *X. campestris* group.

Qhobela: No, we have not.

King: What is the significance of sowing the tall photoperiod sensitivity African variety every 25 rows in your anthracnose nursery?

Duncan: To reduce wind movement and hold morning dews on the plants for a long period. In effect, it creates a field-based incubation chamber where natural build-up and spread of *Colletotrichum graminicola* occurs, and it helps reduce replication × replication variation among the trials.

de Milliano: Do you think it is really better to use four pathotypes of *C. graminicola* instead of one? Possibly, you may be increasing the genetic variability by increasing the number of pathotypes.

Duncan: Yes, it is a possibility. I could be creating more pathotypes by the screening techniques that I use. But I have learned to readjust my breeding techniques to the point that I have a better probability now of having resistance in this breeding material if a new pathotype emerges. I need the four differentials (susceptibles) in my nursery to monitor development of the disease on a year by year basis for evaluation purposes.

Mohamed Ali: You mentioned that johnsongrass can serve as a reservoir host for the sorghum isolates of *C. graminicola*. In our cross-inoculation studies, different sorghum isolates with varying levels of virulence did not infect johnsongrass.

Duncan: All our isolates originate from johnsongrass. That is the rule for us instead of the exception. Perhaps the one strain of johnsongrass that you used was resistant. Try two or three different strains from around the area. Johnsongrass is quite variable itself.

Frederiksen: Dr. Duncan, we have noted your populations of *C. graminicola* vary from year to year.

Duncan: I have quite a unique pathogen population situation which varies according to environmental (moisture, heat) conditions. The research that Dr. Cardwell has conducted will hopefully shed some light on the dynamics of this pathogen population.

Mughogho: I would like to make a correction in the presentation. Anthracnose was not first described in Texas in 1911; but in Togo in 1902 and was named *Colletotrichum sublineola.*

Mohamed Ali: I would appreciate if Dr Mughogho will give me the name of the reference for this correction.

Mughogho: The information can be found in Sutton, B.C. 1980. Coelomycetes. Commonwealth Mycological Institute, Kew, UK.

Mukuru: You mentioned that the antidote increases downy mildew infection in a DM-susceptible cultivar. Have you used the antidote on a resistant cultivar? If so, what is the reaction?

Craig: The herbicide antidote Concep II® does not increase downy mildew severity in a resistant cultivar.

El Hilu Omer: What is the name of the herbicide associated with increased disease incidence and how does it bring about increased incidence? Is it through increased susceptibility of the host or enhanced spore germination?

Craig: It is not a herbicide, but a herbicide antidote named Concep II®. Apparently the increased disease incidence is caused by inhibition of maturation of host tissue. The chemical has no effect on spore germination of the pathogen.

de Milliano: Does Concep II® increase severity of disease both for oospores and conidia?

Craig: Yes!

Duncan: Have you had an opportunity to work with herbicide-resistant plants in relation to downy mildew incidence in order to circumvent the herbicide-antidote problem? Are you going to research this area in the future?

Craig: No, I have not worked with herbicide-resistant plants, and do not plan to do this in the future.

Vidyabhushanam: How do you explain the occurrence of susceptible F_2 plants in a cross between two resistant parents, i.e., SC 414-14 × QL 3? Downy mildew resistance has been reported as dominant in both the parents.

Craig: The susceptible phenotypes in the F_2 generation occur because the loci of the resistant factors in QL 3 differ from the locus of the resistant factor in SC 414-12.

Balasubramanian: The types of resistance to downy mildew—is one to oospore and the other to conidia?

Craig: Not to my knowledge. I have seen no case in which cultivars resistant to conidia were susceptible to oospores. Because of our inability to get consistent infection using oospore inoculum in the greenhouse, I have not been able to test the reverse possibility.

Andrews: You said that a nondominant complex of minor genes would offer the best type of genetic resistance. How feasible do you think this would be to breed for? It might be possible in varieties— but wouldn't it be very difficult for hybrids, where the resistance has to be incorporated in both parents?

Craig: Generally it would be difficult. It would involve going back and generating a new germplasm pool with diversity of resistance and then select, avoiding major gene resistance.

Laing: Have you tried propamocarb (Previcar N®) from Scheuring on sorghum downy mildew?

Craig: Dr Odvody has told us that propamocarb did not work in his downy mildew trials when applied as a drench.

Nowell: A comment on horizontal resistance (HR). We have found that gains in HR are equally quick in composites and in pedigree breeding, or gene-stacking as it is sometimes called. Possibly the most important factor is the genetic diversity of the material that is initially used to create the source material.

Craig: Yes, this is true. Our difficulty with minor genes for controlling downy mildew is in recognizing these minor factors. I have not devised a method for identifying these factors.

Putter: In reply to a previous question, you explained that you do not have a selection criterion that will reveal what you call polygenic, minor gene resistance. You also explained that the selection criterion that you have chosen is a yes/no type test: the pathogen either does or does not sporulate on day 6 after infection. The single specific time you choose to evaluate the latent period is but one (arbitrary) point observation of a process (the latent period that varies in time). Thus, if you do want to "tease out" variation, as you put it earlier, it can readily be achieved by making more observations on other days.

It is axiomatic that the latent period varies its quantitative expression in time. Thus, without changing the target of your selection process — evaluation of the latent period — you could obtain the variability that you seek.

Please note that I am not necessarily recommending a selection process based on the latent period; rather, I am commenting about arbitrary, categorical yes/no tests using your current commitment to evaluating the latent period as the starting point.

Balasubramanian: Do you find any varieties in populations of *M. phaseolina* in sorghum with reference to the history of previous crops, such as maize, sorghum or castor?

Pande: We have not studied the variability in *M. phaseolina* isolates.

Laing: In your list of research objectives, you mentioned an examination for pathogenic specialization. *Macrophomina* attacks more than 400 species, so it is safe to assume that there is specialization and it is therefore only of academic interest to pursue this issue. It also suggests that it will probably be unproductive to pursue vertical resistance against this disease.

Pande: Recent reports suggest that *M. phaseolina* has pathogenic variation. Soybean isolates have preference to soybeans, but not for other hosts— such as sorghum and maize. In the light of this information, it is important to study the variability of the organism (*M. phaseolina*).

Odvody: Recently, Pearson in USA was able to demonstrate that a *Macrophomina phaseolina* population in soil may contain strains differing in their sensitivity to chlorate. Growth of *M. phaseolina* isolates from soybeans was inhibited in a basal medium containing chlorate, but isolates from maize were not. In some soils with a predominant population of chlorate-insensitive isolates the soybeans were preferentially infected with chlorate-sensitive isolates. I provided Pearson with sorghum and maize isolates of *M. phaseolina* from Texas which he determined to be chlorate-insensitive. These results do not indicate that cross infectivity is impossible but the host preferences indicate some virulence difference on these hosts.

Pande: Thanks for the comment and information. We would like to follow the methodology particularly with respect to the sensitivity of *M. phaseolina* isolates to chlorate in cross-inoculation tests with isolates of *M. phaseolina* from other hosts.

Teyssandier: Do you believe that drought allows *Macrophomina* sp to colonize host tissue because it has a competitive advantage over other root/crown inhabiting fungi?

Pande: *Fusarium* spp are always associated with sorghum root, crown, and nodal tissue throughout the life-span of sorghum plant; their frequency increases after drought stress, but the end product is the symptom produced by *M. phaseolina*. I believe that *Fusarium* spp may predispose sorghums to attack by *M. phaseolina*.

Butler: The relationship between root exudates and the organisms which produce stalk rot was not clear. The reason I ask is that under laboratory conditions we find that the production of root exudate acting as a germination stimulant for *Striga* is greatly increased by drought stress. There could be a relationship between the effect of drought stress on exudate production and the incidence of this disease.

Pande: I was referring to the nearly nonexistent information on the mechanism of infection by *M. phaseolina* in sorghum. However, there are reports that root exudate triggers the germination of sclerotia of *M. phaseolina* and, thereby, these germinated sclerotia colonize roots of sorghum.

Balasubramanian: Is the production of sclerotia in sorghum ergot controlled by geontype or by environment or by genotype-environmental interaction?

Mughogho: The literature contains no information on this, and I know of no research program in this important area.

Mantle: Commenting partly in response to the previous question on variability in the expression of ergot sclerotium formation on sorghum, it is assumed that the review aspect of Dr Mughogho's paper has been a synthesis of the sparse literature concerning both India and Africa. If so, from the experience of our comparative studies in London on ergot isolates form India and Africa, and taking into account Dr Boon-Long's conference report concerning sorghum pathology in Thailand, it seems probable that there are some significant geographic differences in the pathogen. In Thailand, sclerotial differentiation is absent or slight, while in Africa sclerotia are frequently produced and are of a size roughly comparable to that of seed. In contrast, the Indian spur-like parasitic bodies produced during the dry season consist mainly of a distal portion which is actually sphacelial tissue. Only the lesser proximal portion is a true sclerotium. The shorter parasitic body produced during the rainly season may simply be the sclerotial component, the sphacelial tissue having been washed away by rain.

Vidyabhushanam: Is there any evidence to indicate that the seed obtained from ergot-infested (with honeydew only) heads carrying the honeydew initiate disease in the next generation?

Mughogho: There is no information in the literature on this. However, I would not expect seedborne honeydew, inoculum, which is known to survive for up to 7 months, to be important because once the seed is sown, the inoculum would not survive in the soil. In addition since infection occurs through florets and not through roots, seedborne inoculum would not initiate the disease.

Andrews: What is the relative importance of ergot compared with other disease? Is it a major problem on perfectly fertile cultivars or is it an induced problem with male steriles or imperfectly fertile hybrids?

Mughogho: Ergot is important in both cases. In Rwanda, I have seen epidemics of ergot in per-

fectly fertile varieties in farmers' fields where landrace cultivars were grown. We may consider it more important in hybrid seed production because of the loss in seed for the next crop. It is this aspect of the disease that has become prominent in India and Thailand.

Laing: What is the origin of the quantitative, nonspecific resistance against head smut? You mentioned new resistant cultivars with high-level, race-specific resistance being deployed in the USA. This is classical vertical resistance (VR). Do you think this VR can be utilized without creation of "boom and bust" cycles? Do you think this VR can be used usefully in subsistence agriculture?

Frederiksen: The source of the quantitative resistance is BTS 399 (Wheatland). Tx 399 is an old stiff-stalked inbred used for seed production. No, and I argue against deployment of VR in areas where smut is endemic.

King: I believe it is extremely important to determine yield loss associated with long smut. A panicle can look as rough, as if it is heavily infected with long smut, even though only a low proportion of the florets are producing sori compared to those producing grains. I believe long smut is a disease in which actual yield loss can easily be over estimated.

Guiragossian: Have there been studies on the inheritance of resistance to long smut of sorghum?

Frederiksen: Some studies are being proposed in Niger, Mali, and Sudan.

Hulluka: I observe long-smut occurrence when we have drier seasons. Would that be due to differential growth rates of the host and pathogen?

Frederiksen: The disease occurs in very dry sorghum-growing regions of Africa and Asia, and frequently it is more severe in the very dry years.

Mushonga: What is the effect of grain mold on animal feed?

Forbes: Some studies show little or no effect on

cattle, but other studies suggest that feed contaminated with toxins produced by *Fusarium* spp are deleterious to poultry and swine.

Thomas: Could you clarify the distinction between grain mold and weathering?

Forbes: First let me say that this distinction is not made by all grain mold scientists. Grain mold occurs before physiological maturity and involves very few fungi acting as parasites. Weathering occurs after physiological maturity, and includes many field fungi acting in conjunction with climatic factors.

Adipala: In terms of yield loss, how do you compare loss due to grain mold with weathering? Does it appear that the genes for resistance to grain mold are the same as those that work against weathering?

Forbes: First you have to establish two relationships. One is the relationship between grain mold severity and loss and the other is the relationship between weathering severity and loss. To do this you have to isolate the effects of grain mold from those of weathering. There are several theoretical ways of doing this, but it is difficult to do in the filed. Once these relationships are established you can try to estimate yield loss posed on grain mold and weathering severity. This is very hypothetical and I would guess that for crop-loss assessment both weathering and grain mold would be considered together. In regard to your second question, it appears that grain-mold resistance is generally related to weathering resistance. In some cultivars, however, the two types of resistance seem to be independent.

Rosenow: Regarding the data presented on flavan-4-ols analysis, were white resistant cultivars evaluated or are all white sorghums very low in these compounds? If all are low, this analytical procedure would not be useful among white types.

Waniska: Both GM-susceptible and GM-resistant cultivars evaluated to date have low levels of flavan-4-ols. This compound is high in grain-mold resistant red pericarp cultivars. Resistant cultivars with a pigmented testa can have either high flavan-4-ol levels, high tannin levels, or both.

Guiragossian: In my experience, there are no yellow-endosperm sorghums resistant to grain mold in Mexico. All yellow endosperms tested for grain mold have been, without exception, susceptible.

Mukuru: This may be true. However we need to screen for mold resistance in yellow endosperm types with hard and vitreous grains.

Mushonga: Did you try to cross the brown-grain type with a yellow-endosperm type to see if there was resistance in the progeny?

Mukuru: No, we did not.

Vidyabhushanam: Have the hard-grain types reported to have mold resistance been evaluated for food quality?

Mukuru: There are reports which indicate that vitreous grains have low digestibility. We need to evaluate the digestibility and food quality of mold-resistant sorghums. Vitreous grains are preferred for regate in eastern Africa.

Mbaye: Are there differences in resistance to grain mold from different types of sorghum, i.e., between guineaense and caudatum types?

Mukuru: Resistance to grain mold has been identified in various types. Resistance to grain mold for colored grain types without the testa layer has been identified in guineaense types.

Obilana: In terms of utilization of germplasm and disease resistance, how much have you implicated the different sorghum races like the guineas (hard grain) and farafara - caudatum (medium/soft grain) which are both white (and some red without pericarp in the guineaense). in mold resistance?

Mukuru: The mold-resistant sources identified are diverse in geographic origin, as well as taxonomic races. The colored-grain mold-resistant sources without the testa layer, however, are predominantly the guineaense, or kafir, types.

Obilana: Following Dr Mukuru's answer to my first question, I recommended that these long-season tall and photosensitive sorghums (guineaense and caudatum farafara) be evaluated for

GM-resistance in the savannas of southern Guinea, western Africa, possibly by ICRISAT.

Mukuru: I agree that screening for a mold-resistant photoperiod-sensitive germplasm collection should be taken up at locations where this can be done adequately and reliably.

Mughogho: The seedborne diseases of sorghum listed by Noble and Richardson contain references to publications which have not been reviewed or meet the criteria for seedborne diseases. There is need to ensure that only those reports of scientifically proven associations are listed as seedborne.

McGee: I agree it will be an enormous task to validate the literature on seedborne diseases of sorghum.

Claflin: Bacterial streak and bacterial stripe are reported to be seedborne by Noble and Richardson, yet their source of information is based on conjecture. Therefore, how and why are they included on the list?

McGee: The publication indicates that an association between a microorganism and seed has been reported.

Duncan: What is the potential of viruses being introduced by sorghum seed? This problem recently emerged when peanut mottle virus was introduced on groundnuts from China, even though proper health testing and quarantine regulations were observed.

McGee: There is always a hazard, for any crop, of an exotic virus on other microorganism being introduced by seed. Even with vigorous genetics directed against well-defined pathogens, there may be loopholes.

Claflin: Can scientists be intimidated when conducting research on seedborne diseases if the potential results would be harmful to the industry?

McGee: This is possible.

Lating: In your talk on screening techniques, you suggest that selection for resistance should use Robinson's single-pathotype-susceptible genotypes for starting. The conventional procedure is to use resistant material to start with and expose it to as many pathotypes as are available. Please explain the limitations of the conventional approach and the Robinson approach a little.

Putter: Robinson is the original author of the one-pathotype technique. An explanation of the reasons why it serves its purpose is provided by Vanderplank. The technique was also later endorsed by Parlevliet in a letter to the editor of Phytopathology, 1983, 73:379.

Craig: How does one handle this avoidance of numerical values in your scoring when dealing with diseases in which severity and incidence are the same, i.e., systemic phase of sorghum downy mildew?

Putter: Incidence/severity transformations are accommodated by the multiple-infection transformation developed by Gregory. Incidence percentages can be used directly, but they assume that all plants have the same severity of disease and that they were equally (i.e., stochastically equal) exposed to infection. Systemic diseases that involve vectors create problems due to the need to differentiate between pathogen and vector resistance, respectively.

Downy mildew does not involve a vector, you are dealing almost exclusively with systemic infections developing from soilborne oospores. If, as you suggest, from your experience, severity and incidence are the same, then this is equivalent to saying that all plants with systemic symptoms were infected by the same number of oospores or that severity is equal and representative of the number of separate oospore infection events. Obviously there is room for clarifying these points.

Would it not be logical to assume that one could prepare soil with known—or at least standard—concentrations of oospores. Batches of seedlings to be screened can then be exposed to this "sick soil" and the incidence of successful infections can be measured directly. Adequate replication and the inclusion of a reference variety should (in the normal statistical way) provide a means of interpreting the results in spite of variability and other factors that contribute to experimental error. The incidence counts will be the data for analysis of variance.

If it is argued that oospore infection is too variable to use in this technique as the basis of a screening procedure, then I must point out that this difficulty will also apply to your current selection procedure: the outcome of my suggestion and your outcome are both predicated on the same process and set of infection events.

Survivors of my suggested selection/screening procedure should not be recorded as escapes. With respect to inoculum, my suggestion meets the *ceteris paribus* requirement and the differences—if there are any—from the analysis of variance (assuming that the rules of biometry are followed) can be assumed to measure differences in the resistance response to infection.

Balasubramanian: In the light of Webber-Fechner law, the classification of disease-rating scale in the field will geometrically take care of many errors. Why should we not continue the testing scale which exists today in many laboratories (25% of plants) to quantify disease?

Putter: The principle of scaling which I demonstrated comes from Mathew and Jinks. It shows the influence of varous methods of transforming the data points or scale values after these have been collected. The Webber-Fechner law and its logarithmic implications, as these were explored by Horsfall and Barratt, influence the way in which we arrive at the values that will subsequently be submitted to scale-transformation processes of the kind described by Mathew and Jinks and which I have emphasized in my presentation.

Osmanzai: In your presentation of yield regressed on SDM, the dots looked dispersed. Have you checked the significance of r^2?

Wall: It was significant at the 0.01 level, and it was included in the statistics presented. Other models, such as quadratic, polynomial, and exponential were tried, but were not satisfactory.

Osmanzai: Do you consider the inclusion of climate and sowing data as an important parameter in the model for assessing yield loss because of SDM?

Wall: I dealt with critical point models, and furthermore, with models that allowed practical application. The climatic influence is there, only it is not separated. Climatic factors are important, and they can be included in models.

de Milliano: How do you think your crop-loss assessments apply to other regions in the world?

Wall: The particular models we developed are applicable, strictly speaking, to the specific combination of factors that existed during that season and in that location. To simplify a model, either in time or season, or in regions or locations, one would have to take data for various seasons or at various locations, and include them when developing the model.

Zummo: Why is it that there is not much importance attached to *Striga* research?

Chivinge: There is too little documentation on the importance of *Striga* and its effects on crop yield and the lack of manpower to specialize on *Striga* in developing countries.

Guiragossian: National programs in eastern Africa have realized that *Striga* is an important problem in sorghum production. For that reason the national programs of Ethiopia and Sudan are taking the problem seriously and have hired experts to develop strategies to overcome the problem. The national program scientists should request that ICRISAT take *Striga* seriously.

Hulluka: I suspect that *Striga* has received less attention than other pathogens because it is not a problem in the developed countries.

Mbwaga: When we say a plant is resistant to *Striga* does it mean the plant stops producing *Striga*-germinating stimulants?

Kebede: It could be a low stimulant-producing line or the cell walls of the root may be difficult for *Striga* to penetrate.

Craig: Did the F_1 and F_2 plants of a given R x S cross come from the same seed lots for the 2 yearly tests conducted?

Beninati: No. The F_2 populations all come from the same parental plants, but not necessarily from the same F_1 plant.

King: Was the millet host only *Setaria* sp, and not pearl millet? At ICRISAT Center, India, we almost always find some intermittent streaking (broken lines) associated with symptoms, especially at an early stage. Do you usually find these symproms with SYB in Texas?

Giorda: The millet host was *Setaria italica* cv "German strain R". Yes, I usually find streakings as a few broken lines, which coalesce forming the typical yellow banding. In addition, a severe stunting has been observed in plants infected following mechanical inoculation.

Mughogho: At ICRISAT Center, it is suspected that the virus is transmitted by the leafhopper *Peregrinus maidis*. Have you tried transmission tests with this insect?

Giorda: I will not discard the possibility of leafhopper transmission.

Claflin: Since the sorghum yellow banding virus is a spherical virus, is it vectored by beetles?

Giorda: The fact that SYBV is a spherical virus does not imply that it has to be vectored by beetles. For example, the cucunovirus group is isometric and transmitted by aphids in a non-persistent manner. In the luteovirus group, spherical viruses belonging to this group are aphid-transmitted in a persistent manner.

Laing: Tolremicity is a characteristic which could easily be missed by the application of too high a level of inoculum—a level that could swamp this resistance mechanism.

Putter: If the characteristic is a qualitative character, such as a glume being tightly shut, then the inoculum load will not affect it. However, with a quantitative characheristic the inoculum dose levels will be critical to the selection of tolremicity.

Anahosur: Is it possible to look for tolerance in systemic diseases such as SDM?

Frederiksen: If you accept the concept of the ability of the host to produce in spite of disease and if you consider the host as a population of individuals, those with a low frequency of systemically infected plants could be considered as more tolerant than those with a higher frequency of disease.

Giorda: You have mentioned resistance and tolerance in virus; these terms are "defined" based on symptoms and yield. You have not mentioned the virus content as another parameter to cosider in this definition. The spread of the virus depends in part of the amount of virus present in the host, since transmission of virus depends on vectors and this in turn affects resistance. There are different host-virus-vector interactions in the sorghum *syste*, thus the effect of virus content on resistance will differ. What do we mean by field resistance? Have you considered escape and nonpreference as elements of fieldresistance?

Frederiksen: Certainly!

Balasubramanian: Preselective mutants in a pathogen population increase with the deployment of matching cvs and decrease with the withdrawal of the cultivar. Can we utilize deployment of genotypes in sorghum and millet disease?

Frederksen: Absolutely!

Anahosur: When we use high inoculum levels to select for resistance, we may loose minor gene resistance and we need to look for both major and minor resistance.

Qhobela: How do you feel about the deployment of "single vertical resistance" genes that give phenotypes that could be classified as susceptible in certain environments, thus allowing the pathogen to "grow"? Then you are not selecting for more-virulent races of the pathogen.

Frederiksen: Probably about the same as I feel about the deployment of any other "single vertical resistance"gene. Be careful, watch out, and have a backup system.

King: In India we have a situation where pearl millet hybrids are breaking down to downy mildew after 4 of 5 years of intensive cultivation by farmers. The genetics of resistance is not well understood, but reports generally indicate that resistance is dominant, but usually involves some minor genes as well. The genetics of virulence has not been studied.

Craig: On the basis of our experience in Texas, it seems futile to expect satisfactory control with the monogenic resistance factors we have used. We should be examining the possibility of combinations of "minor" genes, or better still, escape mechanisms (for example, faster maturation of seedling tissues or possible morphological characters that prevent or reduce the probability of invasion by the pathogen).

Putter: You point out that both horizontal resistance (HR) and partial vertical resistance (VR) exert an effect upon the rate of disease progress. How can one select for HR and avoid particular quantitative VR when "dilatory resistance" could be either?

Frederiksen: We are attempting to evaluate the effects of various sources of host resistance on changes on pathogen populations. Those host genotypes that do not show directional selection pressure are preferred.

de Milliano: In our selection for resistance to downy mildew, conidia resulted in more disease than oospores. Do you think new pathotypes are more likely to develop from conidia or from oospores? Should we infect mainly with conidia, or should we also keep on adding oospores?

Craig: In our experience in Texas, resistance to conidia provides resistance to oospores; I do not believe shift in pathogen virulence is necessarily related to sexual recombinations in the oospore stage, but rather represents shifts in frequencies of virulent genotypes in a heterogenous population in response to the host genotype present. If so, conidial inoculum should give the same range of virulence as oospores.

King: I just want to include the concept of recovery resistance to downy mildew in pearl millet in this discussion. It is a phenomenon that may be useful for desirable resistance, but we don't know much about it. We are investigating durable resistance and should know more about its potential value in another year.

Laing: Downy mildew and anthracnose of sorghum show great genetic plasticity. How do you suggest we should tackle these diseases?

Frederiksen: With anthracnose we are looking at

levels of dilatory resistance and studying the effect of these resistance traits on changes in the pathogen population. Sources of dilatory resistance with minimal effects on the pathogen population will be preferred.

Kuwite: Breeding for horizontal resistance or tolerance is an excellent approach when breeding at the national level, but with international exchange of genetic material it could introduce some pathogens that are seedborne. Exclusion-control techniques, like meristem-tip culture, are inadequate for handling thousands of lines sent to national programs from international institutions. There is therefore a chance of introducing new pathogens in areas where they do not exist.

Frederiksen: Fortunately, there are few seedborne pathogens of sorghum and these, naturally, are widely distributed. Seed exchanges should not present unusual hazards to plant protection under normal circumstances.

Claflin: Seemingly, information concerning anatomical differences and/or gene products when breeding for resistance is limited. Perhaps other programs, such as Dr Rosenow's at Texas where he has developed genotypes with anatomical traits that greatly reduce logging, would be beneficial. Consulting biochemists and plant anatomists for input into the breeding effort might pay considerable dividends.

Balasubramanian: Physical barriers with sclerified cells in *Striga*-resistant sorghums have been demonstrated since 1977. Why can't we exploit this character, because most of the physical resistance traits are easily introduced?

Frederiksen: If the resistance works and can be recognized among segregating individuals in a population, it will be used.

Toler: We believe that the resistance to viruses includes five mechanisms: tolerance, field resistance, immunity, resistance to infections, and multiple resistance.

Guiragossian: In our long smut research, we inoculate plants with abundant spores without controlling concentration of the inoculum. Do you think we should conduct an experiment using differing inoculum concentrations in order to detect differences among genotypes?

Frederiksen: Seems like a very useful approach to me. You may learn much about levels of resistance and infection thresholds.

Laing: Many USA postgraduate programs have an extensive involvement with high technology: genetic engineering, crop-loss assessments, ELISA. Does the INTSORMIL program take into account that many researchers from developing countries will not have access to such technologies? How is the training of scientists from the developing world tailored to be appropriate to the available technologies of their home countries?

Rosenow/Frederiksen: That is a good question; while extremely important, we are not in a position to discuss these issues at this forum. The issue is addressed by each INTSORMIL graduate advisor based on the needs and resources of each collaborator and each graduate student.

Andrews: Do breeders include *Striga* resistance in their breeding program?

de Milliano: Not necessarily. However, certain agronomically acceptable sorghums with resistance have been released, and I believe they will be used both in production fields and in breeding programs.

Andrews: It is desirable to use pathology concepts and practices in dealing with *Striga*. For instance in an area where a particular disease was prevalent, all material would be tested for undue susceptibility to that disease prior to release (if there had been no breeding programs specifically for that disease). This principle is not the same as breeding for resistance. Even with unreliable *Striga* incidence, the highly susceptible are often obvious and do not escape.

Gupta: There are no reliable techniques for screening pearl millet for resistance to *Striga*. *Striga* epidemics in western Africa are sporadic, consequently we often have failures and occasionally the levels of *Striga* do not permit us to differentiate among host reaction classes.

King: One of the reasons that cultivars are re-leased without *Striga* resistance is that there are no reliable screening techniques. Also, I believe that *Striga* control depends on more than deployment of resistance genes. Cultural practices are involved as well.

House: At ICRISAT Center, cultivars and hybrids have been developed without an effort to incorporate *Striga* resistance. However, a separate program designed to select for resistance was successful. Improvement in yield was selected in resistant material. Resistant lines are used by farmers in endemic areas. Outside endemic areas, these lines would not be competitive.

Homans: We must use caution and not be concerned with only one disease or groups of diseses. We must also be concerned with insects, birds, nematodes, rodents, and even human pilfering. Our challenge is to guide our world with infinite wisdom so that our goal of producing more food for the expanding populations that of the world may be achieved. A holistic approach must be used in all of our research.

Obilana: Now that pathologists have been encouraged to do more research on *Striga* in collaboration with the breeders and agronomists, we need to complete the inheritance patterns in *Striga*. I will help to provide some of the information.

We should also add the conversion program being done at ICRISAT on the guineaense and roxburghii sorghums from Nigeria which, when successful, will make available good sources of hard-grain traits and mold resistance to the sorghum researchers of the world.

House: Over the past 25 years there has been a substantial improvement in sorghum. The collection and conversion programs have each been valuable. It would be useful for this body to include a recommendation supporting the collection and evaluation and the conversion programs alike.

Anahosur: We have tested several germplasm lines of sorghum and have found disease resistant genotypes, but they are often quite susceptible to insect pests. Therefore it is necessary to consider both in the crop-improvement process. What should be the strategy for this?

Vidyabhushanam: In the case of diseases, control is based entirely on resistance/tolerance that is fortunately advantageous for (attempting control of) most of the diseases. However, with insects we are operating at a moderate level of tolerance and insect control is mainly based on avoidance and cultivation practices. Nevertheless in the breeding programs for crop improvement all aspects—like yield, pest and disease resistance, damage tolerance, and grain quality—are taken into account.

Obilana: Since many disease-resistant cultivars are susceptible to insects, it is pertinent to suggest that we utilize population-improvement techniques, which I see as the most plausible approach to incorporation of resistances to both insects and diseases.

The sorghum-population breeding programs at SADCC/ICRISAT Sorghum and Millets Improvement Progarm and at ICRISAT Center are commendable but they need to be upgraded in research time towards providing adaptable breeding stock for national breeding programs.

Photos by the author unless
otherwise credited.

ROTTWEILER

The Publisher would like to thank Jennifer Grosiak and Annette
Krochmaluk for participating in the training photos for this book.

Distributed in the UNITED STATES to the Pet Trade by T.F.H. Publications,
Inc., One T.F.H. Plaza, Neptune City, NJ 07753; on the Internet at
www.tfh.com; in CANADA Rolf C. Hagen Inc., 3225 Sartelon St. Laurent-
Montreal Quebec H4R 1E8; Pet Trade by H & L Pet Supplies Inc., 27 Kingston
Crescent, Kitchener, Ontario N2B 2T6; in ENGLAND by T.F.H. Publications,
PO Box 15, Waterlooville PO7 6BQ; in AUSTRALIA AND THE SOUTH PACIFIC
by T.F.H. (Australia), Pty. Ltd., Box 149, Brookvale 2100 N.S.W., Australia; in
NEW ZEALAND by Brooklands Aquarium Ltd. 5 McGiven Drive, New Ply-
mouth, RD1 New Zealand; in SOUTH AFRICA, Rolf C. Hagen S.A. (PTY.) LTD.
P.O. Box 201199, Durban North 4016, South Africa; in Japan by T.F.H.
Publications, Japan—Jiro Tsuda, 10-12-3 Ohjidai, Sakura, Chiba 285, Japan.
Published by T.F.H. Publications, Inc.
MANUFACTURED IN THE UNITED STATES OF AMERICA
BY T.F.H. PUBLICATIONS, INC.

contents

INTRODUCTION

I have been teaching dog obedience classes for over 20 years and see, on the average, just over a thousand dogs and owners per year. Through the years, I have seen a number of different breeds rise and fall in popularity. For a few years, German Shepherd Dogs were the rage, Poodles came and went, then Cocker Spaniels. Today, Rottweilers are the breed of choice for many people. Rottweilers, unfortunately, are also becoming very trendy.

As an obedience instructor, I always have misgivings when people buy the "popular" breed and Rotties are no exception. Now, don't get me wrong, Rotties are nice dogs! However, they are not the right dog for everyone. Prospective dog owners need to understand the breed they are considering—

Orginally a powerful cattle drover and protection dog, the Rottweiler has recently grown in popularity as a family pet.

rottweiler

A Rottweiler will need plenty of socialization and training to fulfill his potential as a loving companion.

whether it is a Rottie, Lab, Golden, Cocker, or Doberman Pinscher.

Prospective owners also need to know that there are other options than simply buying a puppy. Puppies are a lot of work and take a long time to grow up. Adopting an older puppy, a young adult, or even a mature dog might be a better choice for many people.

In this book, I have tried to present Rottweiler owners (and prospective Rottie owners) with choices about choosing a dog and good effective information as to how to make that dog a well-behaved member of the family. Rotties can be fun-loving happy dogs, as well as serious protective working dogs. When in the right home and paired with a knowledgeable owner who is willing to do the necessary training, a Rottweiler can be a friend, a partner, and a great companion.

SELECTING
the Right Dog for You

WHAT ARE ROTTWEILERS?

As with many very ancient breeds, the Rottweiler's origins are shrouded in mystery. Several different books state that the Rottweiler accompanied the Romans as they conquered Europe. While drovers taking herds of cattle to market once used powerful dogs, Rottweilers, to help move and protect the cattle. In addition, when the cattle were sold, the drovers would fasten a purse around the dog's neck so that bandits could not steal the proceeds of the sale.

Photo by Robert Smith.

A hardy dog of formidable size, the Rottweiler is known for his loyal and protective nature.

much of the breed's history has been lost in the passage of time, there is a town called Rottweil that lies at the crossroads of old trade routes used by the Romans. The

Today, Rottweilers are popular dogs for many purposes. As a pet, they are extremely loyal and protective. The breed has also excelled in obedience competition, carting, therapy dog work, and as an able partner to police officers.

Temperament

"The Rottweiler is basically a calm, confident, and courageous dog, with a self-assured aloofness that does not lend itself to immediate and indiscriminate friendships," reads the American Kennel Club's breed standard for the Rottweiler. It continues by saying that the Rottweiler has

Despite his imposing appearance, the Rottweiler should have an even temperament and an adaptable and amiable nature.

With a dog as large as the Rottweiler, proper training is essential to ensure a well-mannered and trustworthy pet.

an inherent desire to protect home and family and is an intelligent dog of extreme hardiness and adaptability with a strong desire to work, making him especially suited as a companion, guardian, and general all-purpose dog.

Very people-oriented, Rottweilers need to spend time with their family and need time to play, exercise, learn obedience skills and social rules, and even just quiet time. This desire to be with people affects all aspects of his life. A Rottie is not a good dog to be left alone in the backyard all day. When bored and lonely, the Rottie may become destructive, digging up the yard, ripping up your garden, or tearing up the garden furniture. In a desire to be with people, some Rotties become incredible escape artists; jumping over or digging under fences to get out of the yard.

Rottweilers can excel in many activities, particularly in police, military, and protection work.

Photo by Isabelle Francais.

However, this love for family has positive aspects, too. Rotties can be incredible working dogs, sometimes serious to a fault. Rottweilers are the breed of choice for many law enforcement and military dog schools and, in many schools, have replaced the German Shepherd Dog, a breed that held the position of supreme working dog for many years. Rotties today are serving as drug detection dogs, bomb detection dogs, arson detection dogs, and much much more.

Don't let this ability to work fool you, though. Although they are serious working dogs, Rotties can still find the fun in anything and will. Puppies will play with anything; chasing butterflies, balls, toys, and even airplanes high in the sky. Even grown-up mature adults retain a sense of fun and will turn into a clown when given any excuse to do so. If you decide to add a Rottweiler to your life, you better like to laugh!

Physical Appearance

Rottweilers are large dogs, usually between 22 and 27 inches tall. The general appearance should be of a